Ergebnisse der Mathematik und ihrer Grenzgebiete

Band 72

I. I. Gihman · A. V. Skorohod

Stochastic Differential Equations

Springer-Verlag Berlin Heidelberg GmbH
1972

I. I. Gihman, A. V. Skorohod
Academy of Sciences of Ukrainian SSR, Mathematical Institute, Kiev

Title of the Russian Original Edition: Stokhasticheskie differentsialnie uravneniya
Edited by Yu. A. Mitropolski
Publisher: Naukova Dumka, Kiev/USSR 1968

AMS Subject Classifications (1970):
Primary 60H10
Secondary 34F05, 60H05, 60J25, 60J60, 62N15, 93E15

ISBN 978-3-642-88266-1 ISBN 978-3-642-88264-7 (eBook)
DOI 10.1007/978-3-642-88264-7

Library of Congress Catalog Card Number 72-86885.

Translator's Preface

Stochastic differential equations whose solutions are diffusion (or other random) processes have been the subject of lively mathematical research since the pioneering work of Gihman, Itô and others in the early fifties. As it gradually became clear that a great number of real phenomena in control theory, physics, biology, economics and other areas could be modelled by differential equations with stochastic perturbation terms, this research became somewhat feverish, with the results that a) the number of theroretical papers alone now numbers several hundred and b) workers interested in the field (especially from an applied viewpoint) have had no opportunity to consult a systematic account.

This monograph, written by two of the world's authorities on probability theory and stochastic processes, fills this hiatus by offering the first extensive account of the calculus of random differential equations defined in terms of the Wiener process. In addition to systematically abstracting most of the salient results obtained thus far in the theory, it includes much new material on asymptotic and stability properties along with a potentially important generalization to equations defined with the aid of the so-called random Poisson measure whose solutions possess jump discontinuities.

Although this monograph treats one of the most modern branches of applied mathematics, it can be read with profit by anyone with a knowledge of elementary differential equations armed with a solid course in stochastic processes from the measure-theoretic point of view.

Aside from the correction of a number of typographical errors (many of which were detected by the authors) and the compilation of an index, this translation differs in no essential way from the Russian original. In particular, I have resisted, as did the authors, the temptation to add to the bibliography, which contains only references to new mathematical results. Readers interested in further literature on applications should consult the relevant journals on control theory and applied probability.

I take this opportunity to thank Professors Gihman and Skorohod for their careful reading of the manuscript which resulted in many suggestions for improvement, and Miss Stanislava Flégrová for assistance with several syntactical subtleties of the Russian language.

Heidelberg, Summer 1972

Kenneth Wickwire

Table of Contents

Introduction 1

Part I. One-dimensional Stochastic Differential Equations of First Order

Chapter 1. Stochastic Integrals and Differentials 7
§ 1. The Wiener Process 7
§ 2. The Stochastic Integral 11
§ 3. Properties of Stochastic Integrals as Function of the Upper Limit 16
§ 4. Stochastic Integrals with Random Limits. 27

Chapter 2. The Solutions of Stochastic Differential Equations. . . 33
§ 5. Stochastic Differential Equations of First Order. 33
§ 6. Existence and Uniqueness of the Solutions 39
§ 7. Stochastic Equations which Depend on a Parameter . . . 50
§ 8. Dependence of the Solutions of Stochastic Differential Equations on the Initial Data 59

Chapter 3. Solutions of Stochastic Differential Equations and Markov Diffusion Processes 63
§ 9. Markov Processes. Diffusion Processes 63
§ 10. Diffusion Processes as Solutions of Stochastic Equations . 67
§ 11. Kolmogorov's Equation 73
§ 12. Measures in Function Space Induced by Diffusion Processes 80
§ 13. Formulas for Transition Density Functions. 91
§ 14. Kolmogorov's Equation for the Transition Probability Density 99
§ 15. Time-homogeneous Solutions of Stochastic Differential Equations 105

Chapter 4. Asymptotic Behavior of the Solutions of Stochastic Equations 114
§ 16. Bounded and Unbounded Solutions of Stochastic Equations 114

§ 17. Theorems on the Asymptotic Behavior of Solutions . . . 124
§ 18. Ergodic Theorems. 134
§ 19. Stability of Solutions 145
§ 20. Some Other Limit Theorems 151

Chapter 5. Stochastic Differential Equations on a Finite Spatial Interval 159
§ 21. Boundary Conditions at the Ends of the Interval 159
§ 22. Processes with Absorption at the Boundary. 165
§ 23. Instantaneous Reflection at the Boundary 178
§ 24. Delayed Reflection at the Boundary. 193
§ 25. Processes with Jump Reflection at the Boundary 205

Part II. Systems of Stochastic Differential Equations

Chapter 1. Vector Stochastic Differential Equations 215
§ 1. Stochastic Line Integrals 216
§ 2. Stochastic Line Integrals as Function of the Upper Limit. 232
§ 3. Stochastic Differential Equations 237

Chapter 2. Stochastic Differential Equations without After-effect . 245
§ 4. Preliminary Remarks 245
§ 5. Some Special Types of Stochastic Integrals 247
§ 6. The Generalized Itô Formula for Stochastic Differentials 263
§ 7. Stochastic Differential Equations without After-effect . . 273
§ 8. Stochastic Differential Equations Depending on a Parameter. Differentiability w.r.t. the Initial Data 275
§ 9. Solutions of Stochastic Differential Equations as Markov Processes 288
§ 10. The Distribution of Functionals of the Solutions of Stochastic Differential Equations 300
§ 11. Some Problems Connected with Homogeneous Stochastic Differential Equations 304

Chapter 3. Asymptotic Behavior of the Solutions of Stochastic Differential Equations 310
§ 12. Stability of Solutions 310
§ 13. Boundedness of the Solutions of Stochastic Differential Equations 330
§ 14. Limit Theorems for Stochastic Differential Equations . . 333

Bibliography 348

Index 351

Introduction

The study of differential equations with random terms has received increasing attention in mechanics, automatic control, radio engineering and many areas of theoretical physics. The number of papers devoted to various aspects of specific equations met in these fields is very large (in our bibliography we have included only those papers which contain significantly new mathematical results).

In our treatment we distinguish between two cases. In the first, the random terms appearing in the differential equations are sufficiently smooth and most of the questions concerning the properties of their solutions can be answered using the classical methods of the theory of ordinary differential equations. An exception is provided by the problem of finding the finite-dimensional distributions (or other probabilistic properties) of the solution. Up till now, this problem has been solved only in the (rather simple) case of linear equations; for the nonlinear case there are no general methods for the solution of such problems.

In the second case we consider differential equations containing generalized random processes of the "white noise" type. Such equations can be viewed as the result of a limiting process from equations which describe systems subject to rapidly changing disturbances. (Examples of such situations are the shot effect in radio systems, the chaotic thermal movement of molecules leading to Brownian motion, etc.) The classical methods are not applicable to such situations and it was necessary to introduce the special theory of stochastic differential equations. In contrast to the first case, effective methods of determining the finite-dimensional distributions of the solutions exist here. These methods are based on the fact that the solutions turn out to be Markov processes. Thus, the theory of stochastic differential equations has proved to be closely connected with one of the most important parts of the contemporary theory of random processes — Markov processes.

The term "stochastic differential equation" was introduced by S. N. Bernstein [4, 5] when referring to a certain difference scheme obtained in a limit passage from a Markov chain. Bernstein was not actually interested in the limiting process, but rather in the existence of a limiting distribution for the sequence of Markov chains. Although the methods developed by Bernstein have found applications in the theory of stochastic differential equations (according to their contemporary definition), to consider him the founder of this theory would be exaggerating.

An important role in the birth of the theory was played by the paper of Bogolubov and Krylov [7]. They first considered an equation for the limiting behavior of a dynamical system under the influence of a random force which converges to a process with independent values. They showed that a Markov process is obtained in the limit and derived the Fokker-Planck equation for its transition probability. However, their limiting procedure was insufficiently justified. I. I. Gihman presented a rigorous justification in [14, 15]; this was the beginning of the systematic study of differential equations with random terms. In [21, 22], Gihman presented the general notion of a stochastic differential equation. He proved existence and uniqueness theorems for such equations, demonstrated the differentiability of their solutions with respect the initial data and introduced Kolmogorov's equation for their transition probabilities.

At the same time, the Japanese mathematician K. Itô, considering continuous Markov processes, independently constructed a similar theory based on the notion of a stochastic integral. He later generalized his equations to also include discontinuous one-dimensional Markov processes.

Since these developments, many authors have investigated such equations: additional existence and uniqueness conditions have been given, stochastic equations for processes with boundaries constructed, dependence of solutions on a parameter investigated, asymptotic expansions developed, Bogolubov's method of averaging justified for such equations, equations for discontinuous multi-dimensional Markov processes derived, stability properties studied and ergodic theorems proved. Summaries of the theory of stochastic differential equations appearing in previous books on the theory of random processes [16, 17, 27, 53] cannot begin to reflect its contemporary state.

The authors have undertaken to fill this gap with the present text[1]. It consists of two parts. The first presents the theory of one-dimensional stochastic equations whose solutions are Markov diffusion processes. A treatment of such equations is justified by two circumstances: In the first place, the basic concepts of the theory are easier to see in this

[1] After the manuscript of this volume was completed, the authors learned of the existence of a book by De Dju Gen, [14] which is also devoted to the theory of stochastic differential equations.

simpler case; in the second, we can obtain more and completer results for these equations.

The second part contains a more general definition of a stochastic differential equation based on the idea of a line integral along a random curve. Such a definition allows the inclusion of all the various types of stochastic equations and the answering of questions of existence, uniqueness, dependence on a parameter, and so forth. Here, equations in several dimensions are also treated along with those defined in terms of discontinuous processes.

Part I. One-dimensional Stochastic Differential Equations of First Order

Chapter 1. Stochastic Integrals and Differentials

§ 1. The Wiener Process

A process $w(t)$ will be called a *Wiener* process if it is defined for $t \geqq 0$ and is a homogeneous Gaussian process with independent increments for which $w(0)=0$, $\mathbb{M}\, w(t)=0$ and $\operatorname{Var} w(t)=t$. From this definition it follows that the distribution of $w(t+h)-w(t)$ coincides with the distribution of $w(h)$ and is normal with mean zero and variance h, i.e.,

$$\mathbb{P}\{a<w(h)<b\}=\frac{1}{\sqrt{2\pi h}}\int_a^b e^{-\frac{u^2}{2h}}\,du,$$

and the characteristic function of the variable $w(h)$ is given by the formula

$$\mathbb{M}\, e^{izw(h)}=e^{-\frac{z^2h}{2}}.$$

It is obvious that a necessary and sufficient condition for a process $w(t)$ to be Wiener is the following: for arbitrary n, $0=t_0<t_1<\cdots<t_n$ and $z_0, z_1, \ldots, z_n$

$$\mathbb{M}\exp\left\{i\sum_{k=1}^{n} z_k[w(t_k)-w(t_{k-1})]+iz_0 w(t_0)\right\}=\exp\left\{-\tfrac{1}{2}\sum_{k=1}^{n} z_k^2(t_k-t_{k-1})\right\}. \quad (1)$$

This formula will be used to check whether or not a given process is a Wiener process. In what follows we will consider only separable Wiener processes which are continuous with probability one. We present without proof the following

Lemma 1. *For each $T>a$ and $x>0$ the following formula holds:*

$$\mathbb{P}\{\sup_{0\leqq t\leqq T} w(t)>x\}=2\,\mathbb{P}\{w(t)>x\}=\frac{2}{\sqrt{2\pi}}\int_x^{\infty} e^{-\frac{u^2}{2T}}\,du.$$

Corollary 1. *For each $\varepsilon>0$ and $m>0$ there exists a constant $L_m(\varepsilon)$ such that*

$$\mathbb{P}\{\sup_{0\leqq s\leqq t}|w(s)|>\varepsilon\}\leqq L_m(\varepsilon)\, t^m.$$

Indeed, it follows from the symmetry of the process $w(t)$ $(-w(t)$ has the some distribution as $w(t))$, that

$$\mathbb{P}\{\sup_{0\leqq s\leqq t}|w(s)|>\varepsilon\}\leqq\mathbb{P}\{\sup_{0\leqq s\leqq t}w(s)>\varepsilon\}+\mathbb{P}\{\sup_{0\leqq s\leqq t}(-w(s))>\varepsilon\}$$
$$=\frac{4}{\sqrt{2\pi t}}\int_{\varepsilon}^{\infty}e^{-\frac{u^2}{2t}}du\leqq\frac{4}{\sqrt{2\pi t}}\int_{\varepsilon}^{\infty}\frac{u^{2m}}{\varepsilon^{2m}}e^{-\frac{u^2}{2t}}du$$
$$\leqq\frac{4t^m}{\sqrt{2\pi}\,\varepsilon^{2m}}\int_0^{\infty}u^{2m}e^{-\frac{u^2}{2}}du,$$

thus

$$\mathbb{P}\{\sup_{0\leqq s\leqq t}|w(s)|>\varepsilon\}\leqq\frac{2^{m+1}\Gamma\left(\frac{2m+1}{2}\right)}{\sqrt{\pi}\,\varepsilon^{2m}}t^m,$$

and the claim is proved.

The main result of this section is the following theorem, proved by J.L. Doob.

Theorem 1. *Let the process $\xi(t)$ be defined and continuous with probability one for $t\geqq 0$, $\xi(0)=0$ and for all $t\geqq 0$ let the σ-algebras $\mathfrak{F}_t$ be defined with $\mathfrak{F}_{t_1}\subset\mathfrak{F}_{t_2}$ for $t_1<t_2$. If the following three assumptions hold:*

1) *For all $t\geqq 0$ the variable $\xi(t)$ is measurable with respect to $\mathfrak{F}_t$;*

2) $\mathbb{M}([\xi(t+h)-\xi(t)]/\mathfrak{F}_t)=0$ *w.p. 1 (with probability one) for all $t\geqq 0$ and $h>0$;*

3) $\mathbb{M}([\xi(t+h)-\xi(t)]^2/\mathfrak{F}_t)=h$ *w.p. 1 for all $t\geqq 0$, $h>0$, then $\xi(t)$ is a Wiener process.*

Proof. We calculate the conditional characteristic function of the variable $\xi(t+h)-\xi(t)$ with respect to $\mathfrak{F}_t$

$$\mathbb{M}[e^{iz(\xi(t+h)-\xi(t))}/\mathfrak{F}_t].$$

Put

$$w_{k+1}^n=\xi\left(t+\frac{k+1}{n}h\right)-\xi\left(t+\frac{k}{n}h\right).$$

Then

$$\mathbb{M}(\exp\{iz[\xi(t+h)-\xi(t)]\}/\mathfrak{F}_t)$$
$$=\mathbb{M}\left(\exp\left\{iz\sum_{k=1}^{n}w_k^n\right\}/\mathfrak{F}_t\right)$$
$$=e^{-\frac{hz^2}{2}}+\sum_{r=0}^{n-1}\mathbb{M}\left[\left(e^{iz\sum_{k=1}^{r+1}w_k^n}-e^{iz\sum_{k=1}^{r}w_k^n-\frac{hz^2}{2n}}\right)/\mathfrak{F}_t\right]e^{-\frac{n-r}{2n}hz^2}.$$

Next we evaluate the difference

$$\mathbb{M}\left[\exp\left\{iz\sum_{k=1}^{r} w_k^n\right\}\left(e^{izw_{r+1}^n}-e^{-\frac{hz^2}{2n}}\right)/\mathfrak{F}_t\right]$$

$$=\mathbb{M}\left[\exp\left\{iz\sum_{k=1}^{r} w_k^n\right\}\mathbb{M}\left(e^{izw_{r+1}^n}-e^{-\frac{hz^2}{2n}}/\mathfrak{F}_{t+\frac{r}{n}h}\right)/\mathfrak{F}_t\right]$$

$$=\mathbb{M}\left[\exp\left\{iz\sum_{k=1}^{r} w_k^n\right\}\mathbb{M}\left(e^{izw_{r+1}^n}-1-izw_{r+1}^n+\frac{z^2}{2}(w_{r+1}^n)^2/\mathfrak{F}_{t+\frac{r}{n}h}\right)/\mathfrak{F}_t\right]$$

$$+\mathbb{M}\left[\exp\left\{iz\sum_{k=1}^{r} w_k^n\right\}\left(1-\frac{z^2}{2n}h-e^{-\frac{hz^2}{2n}}\right)/\mathfrak{F}_t\right].$$

Taking into account the relation $e^{-\alpha}-1+\alpha=O(\alpha^2)$, we obtain

$$\left|\mathbb{M}(\exp\{iz[\xi(t+h)-\xi(t)]\}/\mathfrak{F}_t)-e^{-\frac{hz^2}{2}}\right|$$

$$\leqq O\left(\frac{h^2}{n}\right)+\sum_{r=0}^{n-1}\mathbb{M}\left\{\left|\mathbb{M}\left(e^{izw_{r+1}^n}-1-izw_{r+1}^n+\frac{z^2}{2}(w_{r+1}^{(n)})^2/\mathfrak{F}_{t+\frac{r}{n}h}\right)\right|/\mathfrak{F}_t\right\}.$$

We will now show that $\mathbb{M}[(\xi(t+h)-\xi(t))^4/\mathfrak{F}_t]\leqq 4h^2$. Since for $\delta>0$

$$\sum_{k=1}^{n}(w_k^n)^{2+\delta}\leqq\max_k(w_k^n)^\delta\sum_{k=1}^{n}(w_k^n)^2,$$

the variables $\sum_{k=1}^{n}(w_k^n)^2$ are almost surely bounded because $\mathbb{M}\sum_{k=1}^{n}(w_k^n)^2=h$ and $\max_k(w_k^n)^\delta\to 0$ w.p. 1 due to the continuity of $\xi(t)$.

Thus,

$$[\xi(t+h)-\xi(t)]^4=\lim\left[\left(\sum_{k=1}^{n} w_k^n\right)^4+3\sum_{k=1}^{n}(w_k^n)^4-4\sum_{k=1}^{n}(w_k^n)^3\sum_{k=1}^{n} w_k^n\right]$$

in the sense of convergence in probability. This means

$$\mathbb{M}\{[\xi(t+h)-\xi(t)]^4/\mathfrak{F}_t\}$$

$$\leqq\mathbb{M}\left[\lim_{n\to\infty}\left\{\left(\sum_{k=1}^{n} w_k^n\right)^4+3\sum_{k=1}^{n}(w_k^n)^4-4\sum_{k=1}^{n}(w_k^n)^3\sum_{k=1}^{n} w_k^n\right.\right.$$

$$\left.\left.+2\sum_{k<j}(w_k^n)^2(w_j^n)^2\right\}/\mathfrak{F}_t\right]$$

$$\leqq\lim_{n\to\infty}\mathbb{M}\left\{4\left(\sum_{k<j} w_k^n w_j^n\right)^2+2\sum_{k=1}^{n}(w_k^n)^2\left(\sum_{j\neq k} w_j^n\right)^2/\mathfrak{F}_t\right\}$$

$$=8\lim_{n\to\infty}\mathbb{M}\left(\sum_{k<j}(w_k^n)^2(w_j^n)^2/\mathfrak{F}_t\right)$$

$$=8\lim\frac{n(n-1)}{2}\,\frac{h^2}{n^2}=4h^2$$

because

$$\mathbb{M}[w_k^n w_j^n w_l^n w_r^n/\mathfrak{F}_t]=0 \quad \text{for } k\leqq j\leqq l<r$$

and

$$\mathbb{M}[w_k^n w_j^n [w_l^n]^2/\mathfrak{F}_t]=\frac{h}{n}\mathbb{M}(w_k^n w_j^n/\mathfrak{F}_t)=0 \quad \text{for } k<j<l.$$

Consequently, we find

$$\begin{aligned}&\mathbb{M}(|\xi(t+h)-\xi(t)|^3/\mathfrak{F}_t)\\&\quad\leqq\sqrt{\mathbb{M}[(\xi(t+h)-\xi(t))^2/\mathfrak{F}_t]\,\mathbb{M}(|\xi(t+h)-\xi(t)|^4/\mathfrak{F}_t)}\leqq 2h^{\frac{3}{2}}.\end{aligned}$$

This means

$$\mathbb{M}\left(|w_k^n|^3/\mathfrak{F}_{t+\frac{k-1}{n}h}\right)=O\left(\frac{h}{n}\right)^{\frac{3}{2}}.$$

Since

$$\left|e^{izx}-1-izx+\frac{z^2x^2}{2}\right|\leqq\frac{|zx|^2}{6},$$

$$\sum_{r=0}^{n-1}\mathbb{M}\left\{\left|e^{izw_{r+1}^n}-1-izw_{r+1}^n+\frac{z^2}{2}(w_{r+1}^n)^2\right|/\mathfrak{F}_t\right\}\leqq\frac{|z|^3}{6}\,n\,O\left(\frac{h}{n}\right)^{\frac{3}{2}}=O\left(\frac{1}{\sqrt{n}}\right).$$

Thus we have shown that

$$\mathbb{M}[\exp\{iz(\xi(t+h)-\xi(t))\}/\mathfrak{F}_t]=e^{-\frac{z^2h^2}{2}}.$$

Hence, for $0=t_0<t_1<\cdots<t_n$ we have

$$\begin{aligned}&\mathbb{M}\exp\left\{i\sum_{k=1}^{n}z_k(\xi(t_k)-\xi(t_{k-1}))\right\}\\&\quad=\mathbb{M}\exp\left\{i\sum_{k=1}^{n-1}z_k[\xi(t_k)-\xi(t_{k-1})]\right\}\mathbb{M}(\exp\{iz_n[\xi(t_n)-\xi(t_{n-1})]\}/\mathfrak{F}_{t_{n-1}})\\&\quad=\mathbb{M}\exp\left\{i\sum_{k=1}^{n-1}z_k[\xi(t_k)-\xi(t_{k-1})]\right\}e^{\frac{-z_n^2(t_n-t_{n-1})}{2}},\end{aligned}$$

i.e., formula (1) holds for $\xi(t)$. □

Remark 1. It follows from Theorem 1 that if η is $\mathfrak{F}_0$-measurable, then

$$\mathbb{M}\exp\left\{iz_0\eta+i\sum_{k=1}^{n}z_k(\xi(t_k)-\xi(t_{k-1}))\right\}=\mathbb{M}\,e^{iz_0\eta}\exp\left\{-\tfrac{1}{2}\sum_{k=1}^{n}(t_k-t_{k-1})\,z_k^2\right\},$$

therefore the process $\xi(t)$ doesn't depend on the σ-algebra $\mathfrak{F}_0$.

§ 2. Stochastic Integrals

In this section we will define the stochastic integral

$$\int_0^T f(t)\, dw(t) \tag{1}$$

with respect to the Wiener process.

First we determine the class of random functions (processes) for which the integral (1) will be defined.

For all $t \in [0, T]$ let the σ-algebras of events $\mathfrak{F}_t$ be defined so as to possess the properties: a) for $t_1 < t_2$, $\mathfrak{F}_{t_1} \subset \mathfrak{F}_{t_2}$; b) $w(t)$ is $\mathfrak{F}_t$-measurable; c) the process $w_t(s) = w(t+s) - w(t)$ (t fixed, s is the argument) does not depend on the σ-algebra $\mathfrak{F}_t$. If a probability space and process $w(t)$ are given, then as the σ-algebra $\mathfrak{F}_t$ we can take the collection of all events which do not depend on the process $w_t(s)$. The symbol $H_2[0, T]$ will designate the space of random functions $f(t)$ defined for $t \in [0, T]$ and $\mathfrak{F}_t$-measurable for each t, and for which

$$\int_0^T f^2(t)\, dt$$

is finite w.p. 1. We will show that for all $f(t)$ in $H_2[0, T]$ we can define the integral (1) in such a way that it possesses the following properties.

I. If f_1 and $f_2 \in H_2[0, T]$ and α_1, α_2 are random variables for which $\alpha_1 f_1(t) + \alpha_2 f_2(t) \in H_2[0, T]$, then

$$\int_0^T [\alpha_1 f_1(s) + \alpha_2 f_2(s)]\, dw(s) = \alpha_1 \int_0^T f_1(s)\, dw(s) + \alpha_2 \int_0^T f_2(s)\, dw(s).$$

II. If $\chi_{[\alpha, \beta)}$ is the characteristic function of the interval $[\alpha, \beta)$ contained in $[0, T]$, then

$$\int_0^T \chi_{[\alpha, \beta)}(t)\, dw(t) = w(\beta) - w(\alpha).$$

III. If $f \in H_2[0, T]$ and $\int_0^T \mathbf{M} f^2(t)\, dt < \infty$, then

$$\mathbf{M} \int_0^T f(t)\, dw(t) = 0, \qquad \mathbf{M}\left(\int_0^T f(t)\, dw(t)\right)^2 = \int_0^T \mathbf{M} f^2(t)\, dt.$$

IV. For all $f \in H_2[0, T]$, $C > 0$ and $N > 0$

$$\mathbf{P}\left\{\left|\int_0^T f(t)\, dw(t)\right| > C\right\} \leq \mathbf{P}\left\{\int_0^T f^2(t)\, dt > N\right\} + \frac{N}{C^2}.$$

Properties I and II are those of the usual Stieltjes integral. From them it follows that, for a step function from $H_2[0,T]$, i.e., a function $f(t)$ for which there exists a partition of $[0,T]$ $(0=t_0<t_1<\cdots<t_m=T)$ such that $f(t)=f(t_k)$ with $t_k\leqq t<t_{k+1}$, the integral (1) must have the form

$$\int_0^T f(t)\,dw(t)=\sum_{k=0}^{m-1} f(t_k)[w(t_{k+1})-w(t_k)].$$

Let us verify Property III for step functions. We note that $f(t_k)$ and $w(t_{k+1})-w(t_k)$ are independent since $f(t_k)$ is $\mathfrak{F}_{t_k}$-measurable. Hence,

$$\mathbb{M}f(t_k)[w(t_{k+1})-w(t_k)]=\mathbb{M}f(t_k)\mathbb{M}[w(t_{k+1})-w(t_k)]=0,$$

when $\mathbb{M}|f(t_k)|<\infty$, and then $\mathbb{M}\int_0^T f(t)\,dw(t)=0$. From $\int_0^T \mathbb{M}f^2(t)\,dt<\infty$ it follows that $\mathbb{M}f^2(t_k)<\infty$ since

$$\int_0^T \mathbb{M}f^2(t)\,dt=\sum_{k=0}^{m-1}\mathbb{M}f^2(t_k)(t_{k+1}-t_k).$$

Moreover,

$$\mathbb{M}f^2(t_k)\big[w[t_{k+1})-w(t_k)\big]^2=\mathbb{M}f^2(t_k)(t_{k+1}-t_k) \tag{2}$$

since $f^2(t_k)$ and $[w(t_{k+1})-w(t_k)]^2$ are independent. Thus, for $k<j$

$$\mathbb{M}\,|f(t_k)(w(t_{k+1})-w(t_k))f(t_j)|\leqq\sqrt{\mathbb{M}f^2(t_k)[t_{k+1}-t_k]}\sqrt{\mathbb{M}f^2(t_j)}<\infty.$$

Thus, using the independence for $k<j$ of the variables

$$f(t_k)[w(t_{k+1})-w(t_k)]\,f(t_j)$$

(this variable is $\mathfrak{F}_{t_j}$-measurable) and $w(t_{j+1})-w(t_j)$ and also the existence of their expectations, we can write

$$\begin{aligned}&\mathbb{M}f(t_k)(w(t_{k+1})-w(t_k))f(t_j)(w(t_{j+1})-w(t_j))\\&\quad=\mathbb{M}f(t_k)(w(t_{k+1})-w(t_k))f(t_j)\mathbb{M}(w(t_{j+1})-w(t_j))=0.\end{aligned}$$

From (2) and (3) we have

$$\begin{aligned}\mathbb{M}\left(\int_0^T f(t)\,dw(t)\right)^2&=\mathbb{M}\left(\sum_{k=0}^{m-1} f(t_k)(w(t_{k+1})-w(t_k))\right)^2\\&=\mathbb{M}\sum_{k=0}^{m-1} f^2(t_k)(w(t_{k+1})-w(t_k))^2\\&\quad+2\mathbb{M}\sum_{k<j} f(t_k)(w(t_{k+1})-w(t_k))f(t_j)(w(t_{j+1})-w(t_j))\\&=\int_0^T \mathbb{M}f^2(t)\,dt.\end{aligned}$$

We now show that, for step functions, Property IV follows from Property III. We associate with each step function $f(t)$ another step function $\varphi_N(t)$ as follows: if $f(t)$ is constant on $[t_k, t_{k+1})$, $k=0,\dots,m-1$, then on this interval we set $\varphi_N(t)=f(t)$ if $\sum_{j=0}^{k} f^2(t_j)(t_{j+1}-t_j)\leqq N$ and set $\varphi_N(t)=0$ if $\sum_{j=0}^{k} f^2(t_j)(t_{j+1}-t_j)>N$. The function $\varphi_N(t)$ belongs to $H_2[0,T]$ if $f(t)\in H_2[0,T]$. It is easy to see that

$$\int_0^T \varphi_N^2(t)\,dt=\sum_{j=0}^{\nu} f^2(t_j)(t_{j+1}-t_j),$$

where ν is the largest number for which $\sum_{j=0}^{\nu} f^2(t_j)(t_{j+1}-t_j)\leqq N$. Hence

$$\mathbb{M}\int_0^T \varphi_N^2(t)\,dt=\int_0^T \mathbb{M}\,\varphi_N^2(t)\,dt\leqq N.$$

Moreover, $f(t)-\varphi_N(t)=0$ for all $t\in[0,T]$ if $\int_0^T f^2(t)\,dt\leqq N$. Thus

$$\begin{aligned}
\mathbb{P}\left\{\left|\int_0^T f(t)\,dw(t)\right|>C\right\} &\leqq \mathbb{P}\left\{\left|\int_0^T \varphi_N(t)\,dw(t)\right|>C\right\}\\
&\quad+\mathbb{P}\left\{\int_0^T f(t)\,dw(t)\neq\int_0^T \varphi_N(t)\,dw(t)\right\}\\
&\leqq\frac{\mathbb{M}\left[\int_0^T \varphi_N(t)\,dw(t)\right]^2}{C^2}+\mathbb{P}\left\{\int_0^T f^2(t)\,dt>N\right\}\\
&\leqq\frac{N}{C^2}+\mathbb{P}\left\{\int_0^T f^2(t)\,dt>N\right\}.
\end{aligned}$$

Property IV is thus established for step functions.

Now let $f(t)\in H_2[0,T]$ be such that there exists a sequence of step functions $g_n(t)\in H_2[0,T]$ for which

$$\lim_{n\to\infty}\mathbb{M}\int_0^T [f(t)-g_n(t)]^2\,dt=0.$$

Then also

$$\lim_{\substack{n\to\infty\\ m\to\infty}}\mathbb{M}\int_0^T [g_n(t)-g_m(t)]^2\,dt=0,$$

and so

$$\lim_{\substack{n\to\infty\\ m\to\infty}} \mathbb{M}\left[\int_0^T g_n(t)\,dw(t) - \int_0^T g_m(t)\,dw(t)\right]^2 = \lim_{\substack{n\to\infty\\ m\to\infty}} \mathbb{M}\int_0^T [g_n(t)-g_m(t)]^2\,dt = 0.$$

Thus the sequence of random variables $\int_0^T g_n(t)\,dw(t)$ is fundamental (or Cauchy) in the sense of convergence in mean square. Hence it converges to some limit, which we will denote by $\int_0^T f(t)\,dw(t)$. It is easy to see that the values of this limit for two different sequences $g_n(t)$ and $\tilde{g}_n(t)$ coincide w.p. 1 so that the definition of the stochastic integral is independent of the choice of the sequence $g_n(t)$. In the case of convergence of random variables in mean square, convergence of their first two moments also holds. Thus

$$\mathbb{M}\int^T f(t)\,dw(t) = \lim_{n\to\infty} \mathbb{M}\int_0^T g_n(t)\,dw(t) = 0,$$

$$\mathbb{M}\left(\int_0^T f(t)\,dw(t)\right)^2 = \lim_{n\to\infty} \mathbb{M}\left(\int_0^T g_n(t)\,dw(t)\right)^2$$

$$= \lim_{n\to\infty} \int_0^T \mathbb{M}\,g_n^2(t)\,dt = \int_0^T \mathbb{M}f^2(t)\,dt.$$

Thus III is valid for the class of functions $f(t)$ under consideration. Property IV follows from III just as for step functions if we set

$$\varphi_N(t) = f(t)\,\chi_N\left(\int_0^t f^2(s)\,ds\right),$$

where $\chi_N(x)=1$ for $x \leqq N$ and $\chi_N(x)=0$ for $x>N$.

Now assume $f(t)\in H_2[0,T]$. Then we can construct a sequence of step functions $g_n(t)$ for which $\int_0^T [g_n(t)-f(t)]^2\,dt \to 0$ in probability. Then for arbitrary $\varepsilon>0$ we have

$$\lim_{\substack{n\to\infty\\ m\to\infty}} \mathbb{P}\left\{\int_0^T [g_n(t)-g_m(t)]^2\,dt > \varepsilon\right\} = 0.$$

Taking arbitrary $\varepsilon>0$ and $\rho>0$ and using Property IV (whose validity for step processes has been verified), we can write the inequality

$$\mathbb{P}\left\{\left|\int_0^T g_n(t)\,dw(t) - \int_0^T g_m(t)\,dw(t)\right| > \varepsilon\right\} \leqq \rho + \mathbb{P}\left\{\int_0^T [g_n(t)-g_m(t)]^2\,dt > \varepsilon^2\rho\right\},$$

from which it follows that the sequence of random variables $\int_0^T g_n(t)\,dw(t)$ is fundamental in the sense of convergence in probability. Thus it converges (in probability) to some limit which is independent of the choice of $g_n(t)$. This limit is equal by definition to the integral

$$\int_0^T f(t)\,dw(t).$$

Properties I, II and IV for the general case follow from their validity for step functions and the possibility of proceeding to the limiting case.

Remark 1. If $f(t)\in H_2[0,T]$ and $\chi_{[\alpha,\beta)}(t)$ is the characteristic function of the interval $[\alpha,\beta)\in[0,T]$, then the symbol $\int_\alpha^\beta f(t)\,dw(t)$ will denote $\int_0^T \chi_{[\alpha,\beta)}(t)\,f(t)\,dw(t)$. Let ξ be an arbitrary $\mathfrak{F}_\alpha$-measurable variable and $\mathbb{M}\int_0^T f^2(t)\,dt<\infty$. Then

$$\xi\,\chi_{[\alpha,\beta)}(t)\,f(t)\in H_2[0,T] \quad\text{and}\quad \int_0^T \mathbb{M}\,\xi^2\,\chi^2_{[\alpha,\beta)}(t)\,f^2(t)\,dt<\infty.$$

Hence from Property III,

$$\mathbb{M}\,\xi\int_\alpha^\beta f(t)\,dw(t)=0,$$

$$\mathbb{M}\,\xi^2\left(\int_\alpha^\beta f(t)\,dw(t)\right)^2=\mathbb{M}\,\xi^2\int_\alpha^\beta f^2(t)\,dt=\mathbb{M}\,\xi^2\int_\alpha^\beta \mathbb{M}(f^2(t)/\mathfrak{F}_\alpha)\,dt.$$

From these relations we obtain the following sharper form of Property III:

III*. If $f(t)\in H_2[0,T]$ and $\int_0^T \mathbb{M}f^2(t)\,dt<\infty$, then for $[\alpha,\beta)\in[0,T]$

$$\mathbb{M}\left(\int_\alpha^\beta f(t)\,dw(t)/\mathfrak{F}_\alpha\right)=0,$$

$$\mathbb{M}\left(\left[\int_\alpha^\beta f(t)\,dw(t)\right]^2/\mathfrak{F}_\alpha\right)=\int_\alpha^\beta (\mathbb{M}(f^2(t)/\mathfrak{F}_\alpha))\,dt.$$

Remark 2. If $f(t)$ is a continuous process in $H_2[0,T]$ w.p.1, then

$$\int_0^T f(t)\,dw(t)=\lim_{\lambda\to 0}\sum_{k=0}^{n-1} f(t_k)\,[w(t_{k+1})-w(t_k)],$$

where $0=t_0<t_1<\dots<t_n=T$, $\lambda=\max_k(t_{k+1}-t_k)$.

§ 3. Properties of Stochastic Integrals as Function of the Upper Limit

Let $f(t) \in H_2[0, T]$. We investigate properties of the integral

$$I(t) = \int_0^t f(s)\, dw(s) \tag{1}$$

considered as function of the upper limit. For each t the quantity $I(t)$ is only defined w.p. 1, i.e., the process $I(t)$ is defined up to stochastic equivalence. Since every process is stochastically equivalent to a separable process, we will assume in what follows that $I(t)$, for different t, has been "coordinated" in such a way that it is a separable process.

Now we consider some auxiliary lemmas.

Lemma 1. *Let the random variables* $\xi_1, \xi_2, \ldots, \xi_n$ *be such that* $\mathbb{M}|\xi_k| < \infty$ *and for arbitrary k assume*

$$\mathbb{M}(\xi_n - \xi_k / \xi_1, \ldots, \xi_k) = 0.$$

Set $\zeta = \sup\{0, \xi_1, \ldots, \xi_n\}$ *and* $\xi_n^+ = \frac{1}{2}(\xi_n + |\xi_n|)$. *Then for* $\alpha > 1$

$$\mathbb{M}\,\zeta^\alpha \leqq \left(\frac{\alpha}{\alpha-1}\right)^\alpha \mathbb{M}(\xi_n^+)^\alpha.$$

Proof. Let $a > 0$ and $\chi_k(a) = 1$ if $\xi_1 < a, \ldots, \xi_{k-1} < a, \xi_k \geqq a$; $\chi_k(a) = 0$ otherwise. Since $\sum_{k=1}^n \chi_k(a) = 1$ for $a \leqq \zeta$, $\sum_{k=1}^n \chi_k(a) = 0$ for $\zeta > a$, we find for $\alpha > 1$ that

$$\zeta^\alpha = \alpha \int_0^\infty a^{\alpha-1} \sum_{k=1}^n \chi_k(a)\, da.$$

From the definition of $\chi_k(a)$ it follows that $a\chi_k(a) \leqq \xi_k \chi_k(a)$. Thus,

$$a \sum_{k=1}^n \chi_k(a) \leqq \sum_{k=1}^n \xi_k \chi_k(a) \quad \text{and} \quad a^{\alpha-1} \sum_{k=1}^n \chi_k(a) \leqq \sum_{k=1}^n \xi_k a^{\alpha-2} \chi_k(a).$$

Since $\chi_k(a)$ is a measurable function of the variables $\xi_1, \ldots, \xi_k$, we have

$$\mathbb{M}(\xi_n - \xi_k)\chi_k(a) = \mathbb{M}\,\chi_k(a)\,\mathbb{M}(\xi_n - \xi_k / \xi_1, \ldots, \xi_k) = 0,$$

and

$$\mathbb{M}\, a^{\alpha-1} \sum_{k=1}^n \chi_k(a) \leqq \mathbb{M} \sum_{k=1}^n a^{\alpha-2} \chi_k(a)\, \xi_n \leqq \mathbb{M} \sum_{k=1}^n a^{\alpha-2} \chi_k(a)\, \xi_n^+.$$

Integrating the last relation with respect to a from zero to infinity, we find

$$\mathbb{M}\frac{1}{\alpha}\zeta^\alpha \leqq \mathbb{M}\frac{1}{\alpha-1}\zeta^{\alpha-1}\xi_n^+.$$

Finally, using Hölder's inequality,

$$\mathbb{M}\zeta^{\alpha} \leqq \frac{\alpha}{\alpha-1} \mathbb{M}\zeta^{\alpha-1}\xi_n^+ \leqq \frac{\alpha}{\alpha-1} [\mathbb{M}\zeta^{\alpha}]^{\frac{\alpha-1}{\alpha}} [\mathbb{M}(\xi_n^+)^{\alpha}]^{\frac{1}{\alpha}}$$

which proves the lemma.

Remark 1. If we note that $\mathbb{P}\{\zeta > a\} = \sum_{k=1}^{n} \mathbb{M}\chi_k(a)$, then from the relation

$$a^r \sum_{k=1}^{n} \chi_k(a) \leqq \sum_{k=1}^{n} (\xi_k^+)^r \chi_k(a) \qquad (r=1,2)$$

we can obtain the inequalities

$$\mathbb{P}\{\zeta > a\} \leqq \frac{\mathbb{M}\xi_n^+}{a}; \qquad \mathbb{P}\{\zeta > a\} \leqq \frac{\mathbb{M}(\xi_n^+)^2}{a^2}.$$

Corollary 1. *Applying the lemma to the variables* $-\xi_1, -\xi_2, \ldots, -\xi_n$, *we find that*

$$\mathbb{M}\zeta_-^{\alpha} \leqq \left(\frac{\alpha}{\alpha-1}\right) \mathbb{M}(\xi_n^-)^{\alpha},$$

where $\zeta_- = \max\{0, -\xi_1, \ldots, -\xi_n\}$ *and* $\xi_n^- = \dfrac{|\xi_n| - \xi_n}{2}$. *Then, taking into account the identities* $\max_k |\xi_k| = \max(\zeta, \zeta_-)$ *and* $|\xi_n|^{\alpha} = (\xi_n^+)^{\alpha} + (\xi_n^-)^{\alpha}$, *we obtain*

$$\mathbb{M}(\max_k |\xi_k|)^{\alpha} \leqq \left(\frac{\alpha}{\alpha-1}\right)^{\alpha} \mathbb{M}|\xi_n|^{\alpha}.$$

Lemma 2. *Let* $f(t)$ *be a step function from* $H_2[0,T]$ *and assume* $\int_0^T \mathbb{M} f^2(t)\,dt < \infty$. *Then*

$$\mathbb{P}\left\{\sup_{0 \leqq t \leqq T} \left| \int_0^t f(s)\,dw(s) \right| > a \right\} \leqq \frac{1}{a^2} \int_0^T \mathbb{M} f^2(s)\,ds, \tag{2}$$

$$\mathbb{M} \sup_{0 \leqq t \leqq T} \left| \int_0^t f(s)\,dw(s) \right|^2 \leqq 4\mathbb{M}\left[\int_0^T f(t)\,dw(t)\right]^2 = 4\int_0^T \mathbb{M} f^2(t)\,dt. \tag{3}$$

Proof. Let a sequence of points t_{nk}, $k = 0, 1, \ldots, n$ for each $n > n_0$ generate a partition of the interval $[0, T]$, $0 = t_{n0} < t_{n1} < \cdots < t_{nn} = T$ in such a way that $f(t)$ is constant on $[t_{nk}, t_{nk+1})$. Furthermore, we assume that if Λ_n denotes the set $\{t_{n0}, \ldots, t_{nn}\}$, then $\Lambda_n \subset \Lambda_{n+1}$ and $\bigcup_n \Lambda_n$ is the separability set for the process

$$I(t)=\int_0^t f(s)\,dw(s).$$

Then the number

$$\zeta=\sup_{0\leq t\leq T}\left|\int_0^t f(s)\,dw(s)\right|$$

is the limit of the sequence

$$\zeta_n=\sup_k\left|\int_0^{t_{nk}} f(t)\,dw(t)\right|.$$

Since the variables $\int_0^{t_{nk}} f(t)\,dw(t)$ for $k\leq j$ are $\mathfrak{F}_{t_{nj}}$-measurable,

$$\mathbb{M}\left(\int_0^{t_{nn}} f(t)\,dw(t)-\int_0^{t_{nj}} f(t)\,dw(t)/\int_0^{t_{n1}} f(t)\,dw(t),\ldots,\int_0^{t_{nj}} f(t)\,dw(t)\right)=0$$

and for $\int_0^{t_{nk}} f(t)\,dw(t)$ the assumptions of Lemma 1 hold. On the basis of Remark 1 and Corollary 1 we can write

$$\mathbb{P}\{\zeta_n>a\}\leq\frac{1}{a^2}\int_0^T \mathbb{M}f^2(t)\,dt,$$

$$\mathbb{M}\,\zeta_n^2\leq 4\int_0^T \mathbb{M}f^2(t)\,dt.$$

Proceeding to the limit as $n\to\infty$, we have the lemma.

Let $f(t)\in H_2[0,T]$ be such that $\int_0^T \mathbb{M}f^2(t)\,dt<\infty$. We choose a sequence of step functions $f_n(t)$ such that

$$\lim_{n\to\infty}\int_0^T \mathbb{M}(f(t)-f_n(t))^2\,dt=0.$$

Then for arbitrary $t\in[0,T]$

$$\int_0^t f(s)\,dw(s)=\lim_n\int_0^t f_n(s)\,dw(s)$$

in the sense of convergence in probability. We now choose a subsequence n_k for which

$$\int_0^T \mathbb{M}[f(t)-f_{n_k}(t)]^2\,dt\leq\frac{1}{2^k}.$$

Then

$$\int_0^T \mathbb{M}[f_{n_k}(t)-f_{n_{k+1}}(t)]^2\,dt$$

$$\leqq 2\int_0^T \mathbb{M}[f_{n_k}(t)-f(t)]^2\,dt+2\int_0^T \mathbb{M}[f_{n_{k+1}}(t)-f(t)]^2\,dt \leqq \frac{3}{2^k}.$$

The function $f_{n_k}(t)-f_{n_{k+1}}(t)$ is a step function, thus

$$\mathbb{P}\left\{\sup_{0\leqq t\leqq T}\left|\int_0^t f_{n_k}(s)\,dw(s)-\int_0^t f_{n_{k+1}}(s)\,dw(s)\right|>\frac{1}{k^2}\right\}$$

$$\leqq k^4\int_0^T \mathbb{M}[f_{n_k}(t)-f_{n_{k+1}}]^2\,dt \leqq \frac{3k^4}{2^k}.$$

Since the series $\sum_{k=1}^{\infty}\frac{3k^4}{2^k}$ converges, on the basis of the Borel-Cantelli lemma there exists w.p.1 a (in general random) number k beginning with which

$$\sup_{0\leqq t\leqq T}\left|\int_0^t f_{n_k}(s)\,dw(s)-\int_0^t f_{n_{k+1}}(s)\,dw(s)\right|\leqq\frac{1}{k^2}.$$

Thus the series

$$\int_0^t f_{n_1}(s)\,dw(s)+\sum_{k=1}^{\infty}\int_0^t (f_{n_{k+1}}(s)-f_{n_k}(s))\,dw(s)$$

converges uniformly w.p. 1. Consequently, the sum of this series, as sum of continuous functions, will also be continuous. We have also convinced ourselves that when $\int_0^t \mathbb{M}f^2(t)\,dt<\infty$, then among those processes which are stochastically equivalent to $\int_0^t f(t)\,dw(t)$, there exist w.p. 1 continuous ones. Continuous processes are w.p. 1 separable with respect to an arbitrary set of values of their argument. It is easy to see that when two processes coincide (at least one of which is one-sided continuous) on their common separability set, then they are identical w.p. 1. Thus each process which is stochastically equivalent to $I(t)$ will be continuous w.p. 1.

Considering the uniform convergence of

$$\int_0^t f_{n_k}(s)\,dw(s) \quad\text{to}\quad \int_0^t f(s)\,dw(s)$$

and Lemma 2, we can confirm that (2) and (3) are valid if $\int_0^t \mathbb{M}f^2(t)\,dt<\infty$.

Theorem 1. *Assume* $f(t)\in H_2[0,T]$ *and* $\int_0^t \mathbb{M} f^2(t)\,dt<\infty$. *Then the separable process* $I(t)=\int_0^t f(s)\,dw(s)$ *is continuous w.p. 1 and satisfies for* $a>0$ *the inequalities*

$$\mathbb{P}\left\{\sup_{0\leqq t\leqq T}\left|\int_0^t f(s)\,dw(s)\right|>a\right\}\leqq\frac{1}{a^2}\int_0^T \mathbb{M} f^2(t)\,dt,$$

$$\mathbb{M}\sup_{0\leqq t\leqq T}\left|\int_0^t f(s)\,dw(s)\right|^2\leqq 4\int_0^T \mathbb{M} f^2(t)\,dt.$$

Let $f(t)$ be some function $\in H_2[0,T]$. Set $\chi_N(z)=0$ for $z>N$, $\chi_N(z)=1$ for $z\leqq N$. Then the function

$$f_N(t)=f(t)\,\chi_N\left(\int_0^t f^2(s)\,ds\right)$$

belongs to $H_2[0,T]$ and $\int_0^T f_N^2(t)\,dt\leqq N$, so that $\int_0^T \mathbb{M} f_N^2(t)\,dt<\infty$. Consequently, the process $\int_0^t f_N(s)\,dw(s)$ is continuous w.p. 1. But this process coincides with the process $\int_0^t f(s)\,dw(s)$ on the set $\left\{\omega\colon \int_0^T f^2(t)\,dt\leqq N\right\}$. Since $\mathbb{P}\left\{\int_0^T f^2(t)\,dt<\infty\right\}=1$, almost all ω belong to one of the sets $\left\{\omega\colon \int_0^T f^2(t)\,dt\leqq N\right\}$. Taking

$$\int_0^t f(s)\,dw(s)=\int_0^t f_N(s)\,dw(s),$$

if $\omega\in\left\{\omega\colon \int_0^T f^2(t)\,dt<N\right\}$, we obtain a process stochastically equivalent to $I(t)$ and continuous w.p. 1. Summarizing the above considerations we conclude that the continuous process obtained is almost surely identical to the separable process $I(t)$.

Theorem 2. *If* $f\in H_2[0,T]$, *then the separable process* $I(t)=\int_0^t f(s)\,dw(s)$ *is continuous w.p. 1; moreover*

$$\mathbb{P}\left\{\sup_{0\leqq t\leqq T}\left|\int_0^t f(s)\,dw(s)\right|>C\right\}\leqq\mathbb{P}\left\{\int_0^T f^2(t)\,dt>N\right\}+\frac{N}{C^2}. \tag{4}$$

We only need to show (4) for the proof. It follows from Theorem 1, the separability of the processes $\int_0^t f(t)\,dw(t)$, $\int_0^t f_N(t)\,dw(t)$ and the coincidence of these processes on the set $\left\{\omega\colon \int_0^T f^2(t)\,dt\leqq N\right\}$.

We now introduce the notion of a stochastic differential. Let the process $\zeta(t)$ satisfy for all $0 \leqq t_1 < t_2 \leqq T$ the relation

$$\zeta(t_2) - \zeta(t_1) = \int_{t_1}^{t_2} a(t)\,dt + \int_{t_1}^{t_2} b(t)\,dw(t),$$

where $\sqrt{|a(t)|} \in H_2[0, T]$ and $b(t) \in H_2[0, T]$ (the first integral is an ordinary one). We will then say that the process $\zeta(t)$ has a stochastic differential on $[0, T]$:

$$d\zeta(t) = a(t)\,dt + b(t)\,dw(t).$$

This operation of differentiation is linear. The formulas for the differentiation of the product and composition of functions for this differential differ, however, from the usual ones. In order to introduce them, we establish

Lemma 3. *Let* $t_1 < t_2$,

$$t_1 = t_{n0} < t_{n1} < \cdots \leqq t_{nn} = t_2, \qquad \lim_{n \to \infty} \max_k (t_{nk+1} - t_{nk}) = 0.$$

Then $\sum_{k=0}^{n-1} [w(t_{nk+1}) - w(t_{nk})]^2$ *converges in probability to* $t_2 - t_1$.

Proof. Set

$$\zeta_n = \sum_{k=0}^{n-1} [w(t_{nk+1}) - w(t_{nk})]^2.$$

Then

$$\mathbb{M}\zeta_n = t_2 - t_1.$$

From the independence of the variables $[w(t_{nk+1}) - w(t_{nk})]^2$ it follows that

$$\operatorname{Var} \zeta_n = \sum_{k=0}^{n-1} \operatorname{Var} [w(t_{nk+1}) - w(t_{nk})]^2$$

but

$$\operatorname{Var} [w(t_{nk+1}) - w(t_{nk})]^2 \leqq \mathbb{M} [w(t_{nk+1}) - w(t_{nk})]^4 = 3(t_{nk+1} - t_{nk})^2,$$

thus

$$\operatorname{Var} \zeta_n \leqq 3 \sum_{k=0}^{n-1} (t_{nk+1} - t_{nk})^2 \leqq 3 \max_k (t_{nk+1} - t_{nk})(t_2 - t_1) \to 0.$$

From Chebychev's inequality follows

$$\mathbb{P}\{|\zeta_n - \mathbb{M}\zeta_n| > \varepsilon\} \leqq \frac{\operatorname{Var} \zeta_n}{\varepsilon^2}.$$

The lemma is proved.

Corollary 2. *For $t_1 < t_2$ we have*

$$\int_{t_1}^{t_2} w(t)\,dw(t) = \tfrac{1}{2}[w(t_2)]^2 - \tfrac{1}{2}[w(t_1)]^2 - \tfrac{1}{2}(t_2 - t_1).$$

Indeed, if $t_1 = t_{n0} < t_{n1} < \cdots t_{nn} = t_2$,

$$\begin{aligned}\int_{t_1}^{t_2} w(t)\,dw(t) &= \lim_{n\to\infty} \sum w(t_{nk})\,[w(t_{nk+1}) - w(t_{nk})]\\ &= \lim_{n\to\infty} \tfrac{1}{2} \sum_{k=1}^{n-1} \{(w(t_{nk+1}))^2 - (w(t_{nk}))^2 - [w(t_{nk+1}) - w(t_{nk})]^2\}\\ &= \tfrac{1}{2}[w(t_2)]^2 - \tfrac{1}{2}[w(t_1)]^2 - \tfrac{1}{2} \lim_{n\to\infty} \sum_{k=0}^{n-1} [w(t_{nk+1}) - w(t_{nk})]^2.\end{aligned}$$

Theorem 3. *Let*

$$d\zeta_1(t) = a_1(t)\,dt + b_1(t)\,dw(t),$$

$$d\zeta_2(t) = a_2(t)\,dt + b_2(t)\,dw(t),$$

then

$$d(\zeta_1(t)\,\zeta_2(t)) = \zeta_1(t)\,d\zeta_2(t) + \zeta_2(t)\,d\zeta_1(t) + b_1(t)\,b_2(t)\,dt.$$

Proof. Assume first that the functions $a_1(t) = a_1$, $b_1(t) = b_1$, $a_2(t) = a_2$, $b_2(t) = b_2$ are constant. To prove the theorem in this case it is sufficient to show that

$$d(t\,w(t)) = w(t)\,dt + t\,dw(t), \tag{5}$$

$$d[w(t)]^2 = 2w(t)\,dw(t) + dt. \tag{6}$$

For the proof of (5) we take $t_1 < t_2$ and a sequence partitioning the interval $t_1 = t_{n0} < \cdots < t_{nn} = t_2$, for which $\lim_{n\to\infty} \max_k (t_{nk+1} - t_{nk}) = 0$, then

$$\int_{t_1}^{t_2} w(t)\,dt = \lim_{n\to\infty} \sum_{k=0}^{n-1} w(t_{nk+1})\,[t_{nk+1} - t_{nk}],$$

$$\int_{t_1}^{t_2} t\,dw(t) = \lim_{n\to\infty} \sum_{k=0}^{n-1} t_{nk}\,[w(t_{nk+1}) - w(t_{nk})]$$

(the second limit exists in the sense of convergence in probability). Therefore,

$$\begin{aligned}\int_{t_1}^{t_2} w(t)\,dt + \int_{t_1}^{t_2} t\,dw(t) &= \lim_{n\to\infty} \sum_{k=0}^{n-1} [t_{nk+1}\,w(t_{nk+1}) - t_{nk}\,w(t_{nk})]\\ &= t_2\,w(t_2) - t_1\,w(t_1).\end{aligned}$$

This equality is equivalent to (5) by the definition of the differential. Equation (6) follows from Corollary 2. From (5) and (6) the proof of the theorem is finished for constant $a_1(t)$, $b_1(t)$, $a_2(t)$ and $b_2(t)$, whence immediately also for step functions. The general case is easily carried through by a limit passage from step functions.

Lemma 4. *For all* $m \geqq 2$

$$d(w(t))^m = m(w(t))^{m-1}\,dw(t) + \frac{m(m-1)}{2}(w(t))^{m-2}\,dt. \tag{7}$$

Proof. The formula follows for $m=2$ from Theorem 3. Assume it holds for some k. We will show that it then holds for $m=k+1$. Using Theorem 3 we can write

$$\begin{aligned} d(w(t))^{k+1} &= (w(t))^k\,dw(t) + w(t)\,d(w(t))^k + k(w(t))^{k-1}\,dt \\ &= (w(t))^k\,dw(t) + w(t)\left[k(w(t))^{k-1}\,dw(t) + \frac{k(k-1)}{2}(w(t))^{k-2}\,dt\right] \\ &\quad + k(w(t))^{k-1}\,dt \\ &= (k+1)(w(t))^k\,dw(t) + \frac{k(k+1)}{2}[w(t)]^{k-1}\,dt, \end{aligned}$$

whence the validity of (7) for $m \geqq 2$.

Corollary 3. *If* $P(x)$ *is a polynomial in* x, *then*

$$dP(w(t)) = P'(w(t))\,dw(t) + \tfrac{1}{2}P''(w(t))\,dt, \tag{8}$$

where $P'(x)$ *and* $P''(x)$ *denote the first and second derivatives of* $P(x)$.

Indeed, the set of polynomials for which (8) is valid is linear and contains all powers according to Lemma 4.

Corollary 4. *Let* $f(x)$ *be twice continuously differentiable. Then*

$$df(w(t)) = f'(w(t))\,dw(t) + \tfrac{1}{2}f''(w(t))\,dt. \tag{9}$$

To prove (9) we construct a sequence of polynomials $q_n(x)$ which converges uniformly on finite intervals to $f''(x)$. Set

$$Q_n(x) = f(0) + f'(0)\,x + \int_0^x (x-y)\,q_n(y)\,dy.$$

It is easy to see that $Q_n(x)$ and $Q_n'(x)$ converge uniformly on finite intervals to $f(x)$, resp. $f'(x)$. Moreover, $Q_n''(x) = q_n(x)$. Proceeding to the limit for $n \to \infty$ in

$$Q_n(w(t_2)) - Q_n(w(t_1)) = \int_{t_1}^{t_2} Q_n'(w(t))\,dw(t) + \tfrac{1}{2}\int_{t_1}^{t_2} Q_n''(w(t))\,dt,$$

we conclude that $\lim\limits_{n\to\infty} \int_{t_1}^{t_2} [Q_n'(w(t)) - f'(w(t))]^2\,dt = 0$ and as with formula (4), we obtain (9).

Lemma 5. *Let the function $\Phi(t,x)$, defined for $t\in[0,T]$, $x\in(-\infty,\infty)$ be continuously differentiable with respect to t and twice continuously differentiable w.r.t. x. Then*

$$d\Phi(t,w(t))=\left[\Phi'_t(t,w(t))+\tfrac{1}{2}\Phi''_{xx}(t,w(t))\right]dt+\Phi'_x(t,w(t))\,dw(t). \qquad (10)$$

Proof. Assume that $\Phi(t,x)=g(t)\,\varphi(x)$, where $g(t)$ is continuously differentiable and $\varphi(x)$ is twice continuously differentiable. From Theorem 3:

$$\begin{aligned}d\Phi(t,w(t))&=\varphi(w(t))\,g'(t)\,dt+g(t)\,d\varphi(w(t))\\&=\left[\varphi(w(t))\,g'(t)+\tfrac{1}{2}g(t)\,\varphi''(w(t))\right]dt+g(t)\,\varphi'(w(t))\,dw(t).\end{aligned}$$

Therefore, (10) holds in this case. As in Corollary 4 we can show that the set of functions $\Phi(t,x)$ for which (10) is valid is linear and closed w.r.t. uniform convergence of $\Phi(t,x)_n$, $\Phi'_t(t,x)_n$, $\Phi'_x(t,x)_n$ and $\Phi''_{xx}(t,x)_n$ to $\Phi(t,x)$, $\Phi'_t(t,x)$, $\Phi'_x(t,x)$ and $\Phi''_{xx}(t,x)$ on finite x-intervals. It remains to note that, for any function $\Phi(t,x)$ satisfying the conditions of the lemma, we can define a sequence of functions $\Phi(t,x)_n$ of the form $\sum_{k=1}^{n} g_k(t)\,\varphi_k(x)$, also satisfying the conditions of the lemma, such that for arbitrary $N>0$

$$\begin{aligned}\lim_{n\to\infty}\sup_{|x|\le N}\sup_{0\le t\le T}\big\{&|\Phi(t,x)-\Phi(t,x)_n|+|\Phi'_t(t,x)-\Phi'_t(t,x)_n|\\&+|\Phi'_x(t,x)-\Phi'_x(t,x)_n|+|\Phi''_{xx}(t,x)-\Phi''_{xx}(t,x)_n|\big\}=0.\end{aligned}$$

The lemma is proved.

Remark 2. Lemma 5 remains valid if the function $\Phi(t,x)$ also depends on ω, assuming $\Phi(t,w,(t))$ is $\mathfrak{F}_t$-measurable and for each ω, the conditions of Lemma 5 are satisfied.

Theorem 4. *Assume the process $\xi(t)$ has the stochastic differential*

$$d\xi(t)=a(t)\,dt+b(t)\,dw(t),$$

and that the function $f(t,x)$ is continuous and has continuous derivatives $f_t(t,x)$, $f'_x(t,x)$, $f''_{xx}(t,x)$. Then the process $f(t,\xi(t))$ also possesses a stochastic differential and

$$\begin{aligned}df(t,\xi(t))=&\left[f'_t(t,\xi(t))+\tfrac{1}{2}f''_{xx}(t,\xi(t))\,b^2(t)+f'_x(t,\xi(t))\,a(t)\right]dt\\&+f'_x(t,\xi(t))\,b(t)\,dw(t).\end{aligned} \qquad (11)$$

Formula (11) is called Itô's formula and in what follows we will mention it by this name.

Proof. We treat first the case of constant a and b. Then $\xi(t)=\xi(0)+a\,t+b\,w(t)$, and consequently, $f(t,\xi(t))=f(t,\xi_0+a\,t+b\,w(t))$, i.e., $f(t,\xi(t))=\Phi(t,w(t))$, where $\Phi(t,x)=f(t,\xi_0+a\,t+b\,x)$. Here

$$\Phi_t'(t,x)=f_t'(t,\xi_0+a\,t+b\,x)+f_x'(t,\xi_0+a\,t+b\,x)\,a;$$

$$\Phi_x'(t,x)=f_x'(t,\xi_0+a\,t+b\,x)\,b;$$

$$\Phi_{xx}''(t,x)=f_{xx}''(t,\xi_0+a\,t+b\,x)\,b^2.$$

Applying Lemma 5 and Remark 2, we convince ourselves of the correctness of (11) for constant $a(t)=a$ and $b(t)=b$. From this follows its validity for step functions $a(t)$ and $b(t)$. The proof is completed by passage to the limit from step functions.

Remark 3. Let $f(t,x_1,x_2,\ldots,x_r)$ have continuous derivatives f_t', f_{x_i}'', $f_{x_i x_j}''$ $(i,j=1,2,\ldots,r)$ and $\xi_1(t),\ldots,\xi_r(t)$ be random processes with stochastic differentials

$$d\xi_i(t)=a_i(t)\,dt+b_i(t)\,dw(t).$$

Then

$$\begin{aligned} df(t,\xi_1(t),\ldots,\xi_r(t)) = \Big[&\frac{\partial f}{\partial t}(t,\xi_1(t),\ldots,\xi_r(t)) \\ &+\sum_{i=1}^{r} f_{x_i}'(t,\xi_1(t),\ldots,\xi_r(t))\,a_i(t) \\ &+\tfrac{1}{2}\sum_{i,j=1}^{r} f_{x_i x_j}''(t,\xi_1(t),\ldots,\xi_r(t))\,b_i(t)\,b_j(t)\Big]\,dt \\ &+\sum_{i=1}^{r} f_{x_i}'(t,\xi_1(t),\ldots,\xi_r(t))\,b_i(t)\,dw(t). \end{aligned} \tag{12}$$

If a_i and b_i are constants, then for the derivation of (12) it is sufficient to apply Lemma 5 to the function

$$\Phi(t,x)=f(t,\xi_1(0)+a_1\,t+b_1\,x,\ldots,\xi_r(0)+a_r\,t+b_r\,x).$$

The extension of (12) to the general case is carried out as in Theorem 4. Formula (12) is a generalization of (11) and is also called Itô's formula. We remark that Theorem 3 is a consequence of (12).

We will use the derived differentiation rules to calculate moments and estimates for moments of stochastic integrals.

Theorem 5. *Let $b_1(t)$ and $b_2(t)$ belong to $H_2[0,T]$ and*

$$\int_0^T \mathbb{M}\,b_1^2(t)\,dt<\infty, \qquad \int_0^T \mathbb{M}\,b_2^2(t)\,dt<\infty.$$

Then

$$\mathbb{M}\int_0^T b_1(t)\,dw(t)\int_0^T b_2(t)\,dw(t)=\int_0^T \mathbb{M}\,b_1(t)\,b_2(t)\,dt.$$

Proof. Property III for stochastic integrals implies

$$\begin{aligned}\mathbb{M}\int_0^T b_1(t)\,dw(t)\int_0^T b_2(t)\,dw(t)&=\tfrac{1}{2}\mathbb{M}\left[\int_0^T (b_1(t)+b_2(t))\,dw(t)\right]^2\\&\quad-\tfrac{1}{2}\mathbb{M}\left(\int_0^T b_1(t)\,dw(t)\right)^2-\tfrac{1}{2}\mathbb{M}\left(\int_0^T b_2(t)\,dw(t)\right)^2\\&=\tfrac{1}{2}\int_0^T \mathbb{M}(b_1(t)+b_2(t))^2\,dt\\&\quad-\tfrac{1}{2}\int_0^T \mathbb{M}\,b_1^2(t)\,dt-\tfrac{1}{2}\int_0^T \mathbb{M}\,b_2^2(t)\,dt\\&=\int_0^T \mathbb{M}\,b_1(t)\,b_2(t)\,dt.\end{aligned}$$

Theorem 6. *Assume* $f(t)\in H_2[0,T]$ *and* $\int_0^T \mathbb{M}f^{2m}(t)\,dt<\infty$ *(m is a natural number). Then*

$$\mathbb{M}\left[\int_0^T f(t)\,dw(t)\right]^{2m}\leqq[m(2m-1)]^{m-1}\,T^{m-1}\int_0^T \mathbb{M}f^{2m}(t)\,dt.$$

Proof. Applying Itô's formula to the function $\varphi(x)=x^{2m}$ (m is a non-negative whole number) and $\xi(t)=\int_0^t f(s)\,dw(s)$ we obtain

$$\begin{aligned}\left[\int_0^t f(s)\,dw(s)\right]^{2m}&=2m\int_0^t\left(\int_0^s f(u)\,dw(u)\right)^{2m-1} f(s)\,dw(s)\\&\quad+\frac{2m(2m-1)}{2}\int_0^t\left[\int_0^s f(u)\,dw(u)\right]^{2m-2} f^2(s)\,ds.\end{aligned}$$

We assume initially that $f(s)$ is a step function bounded by a constant not depending on chance. Then it's easy to see that

$$\int_0^T \mathbb{M}\left(\int_0^s f(u)\,dw(u)\right)^{4m-2} f^2(s)\,ds<\infty,$$

thus

$$\mathbb{M}\left[\int_0^t f(s)\,dw(s)\right]^{2m}=\frac{2m(2m-1)}{2}\int_0^t \mathbb{M}\left[\int_0^s f(u)\,dw(u)\right]^{2m-2} f^2(s)\,ds. \tag{13}$$

Applying Hölder's inequality, we have

$$\mathbb{M}\left[\int_0^T f(s)\,dw(s)\right]^{2m} \leqq \frac{2m(2m-1)}{2}\left\{\int_0^T \mathbb{M}\left[\int_0^t f(u)\,dw(u)\right]^{2m} dt\right\}^{\frac{2m-2}{2m}} \times\left\{\int_0^T \mathbb{M} f^{2m}(s)\,ds\right\}^{\frac{2}{2m}}.$$

From (13) it follows that $\mathbb{M}\left[\int_0^t f(s)\,dw(s)\right]^{2m}$ is increasing in t. Hence,

$$\mathbb{M}\left[\int_0^T f(t)\,dw(t)\right]^{2m} \leqq \frac{2m(2m-1)}{2}\left\{\int_0^T \mathbb{M}\left[\int_0^T f(s)\,dw(s)\right]^{2m}\right\}^{1-\frac{1}{m}} \times\left\{\int_0^T \mathbb{M} f^{2m}(t)\,dt\right\}^{\frac{1}{m}}.$$

Taking m-th powers and simplifying both sides of the inequality to $\left\{\mathbb{M}\left[\int_0^T f(s)\,dw(s)\right]^{2m}\right\}^{m-1}$, we have proved the theorem for the case where $f(t)$ is a step function bounded above by a nonrandom constant. The general case can be obtained by passage to the limit from functions of the type already considered.

§ 4. Stochastic Integrals with Random Limits

If $f(t)\in H_2[0,T]$ and ζ is some random variable for which $\mathbb{P}\{0\leqq\zeta\leqq T\}=1$, then by the integral $\int_0^\zeta f(t)\,dw(t)$ we will understand the variable $I(\zeta)$, where $I(t)=\int_0^t f(s)\,dw(s)$ is w.p. 1 a continuous process. We note that in the case where $I(t)$ is right continuous w.p. 1, $I(\zeta)$ is a random variable measurable w.r.t. the initial space as the limit of the variables $I(\zeta_n)$, where

$$\zeta_n=\frac{k+1}{n} \quad \text{for} \quad \frac{k}{n}\leqq\zeta<\frac{k+1}{n}.$$

It is easy to see that the stochastic integral on a random interval possesses Properties I and II of § 2 but need not satisfy III. For example, assuming $\zeta=0$ if $\sup\limits_{0\leqq t\leqq T} w(t)<1$ and $\zeta=\inf\{t\colon w(t)\geqq 1\}$ if $\sup\limits_{0\leqq t\leqq T} w(t)\geqq 1$ and taking $f(t)=1$, we have

$$\mathbb{M}\int_0^\zeta f(t)\,dw(t)=\mathbb{P}\left\{\sup_{0\leqq t\leqq T} w(t)\geqq 1\right\}\neq 0.$$

We can, however, define a class of random variables for which Property III (in the usual formation) also holds. This class of variables is formed by the *Markov times* w.r.t. the collection of σ-algebras $\mathfrak{F}_t$ (in the sequel we will call them simply Markov times).

Definition. A random variable ζ, assuming nonnegative values, is called a *Markov time* (w.r.t. the σ-algebras $\mathfrak{F}_t$), if for arbitrary $t \in [0, T]$ the event $\{\zeta > t\}$ belongs to the σ-algebra $\mathfrak{F}_t$.

In other words, in order to determine whether or not the time ζ has occurred up to t it is sufficient to know the events in $\mathfrak{F}_t$, i.e., the event $\{\zeta > t\}$ is not influenced by values of $w(s)$ for $s \geqq t$. We note that the variable ζ in the example considered above is not a Markov time since the occurrence of the event $\{\zeta > 0\}$ is determined by the behavior of $w(t)$ for $0 \leqq t \leqq T$.

Typical examples of Markov times are first passage times:

$\zeta_1 = \inf\{t: w(t) = a\}$ – the first time the process $w(t)$ attains the value a;

$\zeta_2 = \inf\left\{t;\ \int_0^t f(u)\,dw(u) = a\right\}$ – the first time the process $\int_0^t f(u)\,dw(u)$ reaches the value a.

If we consider the processes on the interval $[0, T]$ and the value a is not attained, then we say that $\zeta = T$ (if $T = +\infty$, then ζ can also assume the value $+\infty$).

Let ζ_1 and ζ_2 be random variables for which $\mathbb{P}\{0 \leqq \zeta_1 \leqq \zeta_2 \leqq T\} = 1$. Then we have

$$\int_{\zeta_1}^{\zeta_2} f(t)\,dw(t) = \int_0^{\zeta_2} f(t)\,dw(t) - \int_0^{\zeta_1} f(t)\,dw(t).$$

(Integrals with random upper limits have already been defined.)

We will show that when the limits of a stochastic integral are Markov times we can modify the formulas of Property III in such a way that they retain their validity for this case.

Theorem 1. *Assume that* $f(t) \in H_2[0, T]$ *and* $\int_0^T \mathbb{M} f^2(t)\,dt < \infty$, *and that* ζ_1 *and* ζ_2 *are Markov times for which* $\mathbb{P}\{0 \leqq \zeta_1 \leqq \zeta_2 \leqq T\} = 1$. *Then*

$$\mathbb{M} \int_{\zeta_1}^{\zeta_2} f(t)\,dw(t) = 0,$$

$$\mathbb{M}\left[\int_{\zeta_1}^{\zeta_2} f(t)\,dw(t)\right]^2 = \mathbb{M} \int_{\zeta_1}^{\zeta_2} f^2(t)\,dt.$$

Proof. Introduce the random function $\chi_i(t) = 1$ for $t = \zeta_i$ and $= 0$ for $t \geqq \zeta_i$. It follows from the definition of a Markov time that $\chi_i(t)$ is

$\mathfrak{F}_t$-measurable. Moreover,

$$\begin{aligned}\int_{\zeta_1}^{\zeta_2} f(t)\,dw(t) &= \int_0^{\zeta_2} f(t)\,dw(t) - \int_0^{\zeta_1} f(t)\,dw(t)\\ &= \int_0^T \chi_2(t)\,f(t)\,dw(t) - \int_0^T \chi_1(t)\,f(t)\,dw(t)\\ &= \int_0^T f(t)\,[\chi_2(t)-\chi_1(t)]\,dw(t).\end{aligned}$$

Obviously, $\chi_i(t)\,f(t) \in H_2[0, T]$ and

$$\int_0^T \mathbb{M}\,\chi_i^2(t)\,f^2(t)\,dt \leqq \int_0^T \mathbb{M} f^2(t)\,dt < \infty.$$

Consequently,

$$\mathbb{M}\int_{\zeta_1}^{\zeta_2} f(t)\,dw(t) = \mathbb{M}\int_0^T f(t)\,[\chi_2(t)-\chi_1(t)]\,dw(t) = 0,$$

$$\begin{aligned}\mathbb{M}\left\{\int_{\zeta_1}^{\zeta_2} f(t)\,dw(t)\right\}^2 &= \mathbb{M}\left\{\int_0^T f(t)\,[\chi_2(t)-\chi_1(t)]\,dw(t)\right\}^2\\ &= \int_0^T \mathbb{M} f^2(t)\,[\chi_2(t)-\chi_1(t)]^2\,dt\\ &= \mathbb{M}\int_0^T f^2(t)\,[\chi_2(t)-\chi_1(t)]\,dt = \mathbb{M}\int_{\zeta_1}^{\zeta_2} f^2(t)\,dt\end{aligned}$$

(that $[\chi_2(t)-\chi_1(t)]^2 = \chi_2(t)-\chi_1(t)$ follows from the fact that $\chi_2(t)-\chi_1(t)$ can assume only the values 0 and 1: $\chi_2(t)-\chi_1(t)=1$ for $\zeta_1 \leqq t < \zeta_2$ and $\chi_2(t)-\chi_1(t)=0$ otherwise). The theorem is proved.

Remark 1. If the process $f(t)$ and the σ-algebra $\mathfrak{F}_t$ are defined for all $t \geqq 0$, if for all $T>0$, $f(t) \in H_2[0, T]$ and if ζ is Markov time for which $\mathbb{M}\int_0^{\zeta} f^2(t)\,dt < \infty$, then

$$\mathbb{M}\int_0^{\zeta} f(t)\,dw(t) = 0, \qquad \mathbb{M}\left\{\int_0^{\zeta} f(t)\,dw(t)\right\}^2 = \mathbb{M}\int_0^{\zeta} f^2(t)\,dt.$$

To prove this consider the function $f_N(t) = f(t)$ for $t \leqq N$ and $|f(t)| \leqq N$ and equal to zero otherwise. From Theorem 1

$$\mathbb{M}\int_0^{\zeta} f_N(t)\,dw(t) = 0,$$

$$\mathbb{M}\left\{\int_0^{\zeta} f_N(t)\,dw(t)\right\}^2 = \mathbb{M}\int_0^{\zeta} f_N^2(t)\,dt.$$

Proceeding to the limit for $N \to \infty$ we obtain the desired result.

Remark 2. Denote by $\mathfrak{F}_{\zeta_1}$, where ζ_1 is some Markov time, the minimal σ-algebra containing all events of the form $\mathfrak{A}_t \cap \{\zeta_1 > t\}$, with $\mathfrak{A}_t \in \mathfrak{F}_t$. If η is an arbitrary, bounded $\mathfrak{F}_{\zeta_1}$-measurable variable and ζ_1 and ζ_2 are defined as in Theorem 1, then for each $t \in [0, T]$ the variable $\eta[\chi_2(t) - \chi_1(t)]$ is $\mathfrak{F}_t$-measurable. We assume first that η is equal to one if the event $\{\zeta_1 > s\} \cap \mathfrak{A}_s$ occurs and equal to zero, if it doesn't. Then $\eta = \chi_{\mathfrak{A}_s} \chi_1(s)$, where $\chi_{\mathfrak{A}_s}$ is the indicator of the event $\mathfrak{A}_s$. Then

$$\eta[\chi_2(t) - \chi_1(t)] = \chi_{\mathfrak{A}_s} \chi_1(s)[\chi_2(t) - \chi_1(t)].$$

If $s \leqq t$, then all factors are $\mathfrak{F}_t$-measurable; if $s > t$, then the product is equal to zero, i.e., it is constant since

$$0 \leqq \chi_1(s)[\chi_2(t) - \chi_1(t)] \leqq \chi_1(t)[\chi_2(t) - \chi_1(t)] \leqq \chi_1(t)[1 - \chi_2(t)] = 0$$

and for $s > t$ is $\mathfrak{F}_t$-measurable. Since an arbitrary event from $\mathfrak{F}_{\zeta_1}$ can be obtained from the events $\{\zeta_1 > s\} \cap \mathfrak{A}_s$ and their complements with the help of the operations of union and intersection, for all $\mathfrak{A} \in \mathfrak{F}_{\zeta_1}$ the variable $\chi_{\mathfrak{A}}[\chi_2(t) - \chi_1(t)]$, where $\chi_{\mathfrak{A}}$ is the indicator of $\mathfrak{A}$, will be $\mathfrak{F}_t$-measurable. Since an arbitrary bounded variable η can be represented as the uniform limit of variables $\sum c_k \chi_{\mathfrak{A}_k}$, where $\chi_{\mathfrak{A}_k}$ is the indicator of events in $\mathfrak{F}_{\zeta_1}$, we have that $\eta[\chi_2(t) - \chi_1(t)]$ is also $\mathfrak{F}_t$-measurable. Making use of this fact as we did in Theorem 1, we obtain

$$\mathbb{M}\, \eta \int_{\zeta_1}^{\zeta_2} f(t)\, dw(t) = 0,$$

$$\mathbb{M}\left[\eta \int_{\zeta_1}^{\zeta_2} f(t)\, dw(t)\right]^2 = \mathbb{M}\, \eta^2 \int_{\zeta_1}^{\zeta_2} f^2(t)\, dt.$$

Hence, under the conditions of Theorem 1, we have w.p. 1

$$\mathbb{M}\left[\int_{\zeta_1}^{\zeta_2} f(t)\, dw(t) / \mathfrak{F}_{\zeta_1}\right] = 0,$$

$$\mathbb{M}\left[\left(\int_{\zeta_1}^{\zeta_2} f(t)\, dw(t)\right)^2 / \mathfrak{F}_{\zeta_1}\right] = \mathbb{M}\left[\int_{\zeta_1}^{\zeta_2} f^2(t)\, dt / \mathfrak{F}_{\zeta_1}\right].$$

We use Remark 2 for the proof of an important property of the Wiener process.

Theorem 2. *If τ is a Markov time, then the process $\tilde{w}(t) = w(t+\tau) - w(\tau)$ is also a Wiener process and does not depend on $\mathfrak{F}_\tau$.*

This property is called the *strong Markov property*.

Proof. It's clear that the time $\tau_h = \tau + h$ is a Markov time. Moreover, from Remark 2 with $h_1 < h_2$ there follows w.p. 1

$$\mathbb{M}\left[\int_{\tau+h_1}^{\tau+h_2} dw(t)/\mathfrak{F}_{\tau_{h_1}}\right] = 0,$$

$$\mathbb{M}\left[\left(\int_{\tau+h_1}^{\tau+h_2} dw(t)\right)^2 /\mathfrak{F}_{\tau_{h_1}}\right] = \mathbb{M}\left[\int_{\tau+h_1}^{\tau+h_2} dt/\mathfrak{F}_{\tau_{h_1}}\right] = h_2 - h_1.$$

But $\int_{\tau+h_1}^{\tau+h_2} dw(t) = w(\tau+h_1) - w(\tau+h_2) = \tilde{w}(h_2) - \tilde{w}(h_1)$. Whence the process $\tilde{w}(t)$ and the collection of σ-algebras $\mathfrak{F}_{\tau_t} = \tilde{\mathfrak{F}}_t$ satisfy all the assumptions of Theorem 1 and Remark 1 in § 1. Thus $\tilde{w}(t)$ is a Wiener process independent of the σ-algebra $\tilde{\mathfrak{F}}_0 = \mathfrak{F}_\tau$. The theorem is proved.

It turns out that stochastic integrals can be considered as values of a Wiener process at random times.

Theorem 3. *Let the process $f(t)$ be defined for $t \geqq 0$ and for each $T > 0$ let $f(t) \in H_2[0, T]$. We assume that w.p. 1 $\int_0^\infty f^2(t)\,dt = +\infty$ and set $\tau_t = \inf\left\{s \colon \int_0^s f^2(u)\,du \geqq t\right\}$. Then the process $\zeta_t = \int_0^{\tau_t} f(s)\,dw(s)$ is a Wiener process.*

Proof. Clearly, τ_t is a Markov time for each $t \geqq 0$ and $\tau_{t_1} < \tau_{t_2}$ for $t_1 < t_2$. Since

$$\int_0^{\tau_t} f^2(s)\,ds = t, \qquad \text{we have } \mathbb{M}\int_0^{\tau_t} f^2(s)\,ds < \infty,$$

hence from Remark 2 for $t_1 < t_2$

$$\mathbb{M}\left[\int_{\tau_{t_1}}^{\tau_{t_2}} f(s)\,dw(s)/\mathfrak{F}_{\tau_{t_1}}\right] = 0,$$

$$\mathbb{M}\left(\left[\int_{\tau_{t_1}}^{\tau_{t_2}} f(s)\,dw(s)\right]^2 /\mathfrak{F}_{\tau_{t_1}}\right) = \mathbb{M}\left(\int_{\tau_{t_1}}^{\tau_{t_2}} f^2(s)\,ds/\mathfrak{F}_{\tau_{t_1}}\right) = t_2 - t_1.$$

We shall show that ζ_t is a continuous process w.p. 1. It is continuous at all points of continuity of the process τ_t. Since τ_t is w.p. 1 monotone, its only discontinuities are at most jumps. Thus, the only discontinuities of ζ_t are also at most jumps; moreover, they can occur only at the discontinuity points of the process τ_t. Assuming that $\tau_{t-0} < \tau_{t+0}$, then we have in spite of this

$$\int_{\tau_{t-0}}^{\tau_{t+0}} f^2(s)\,ds = 0 \quad \text{and} \quad \int_{\tau_{t-0}}^{\tau_{t+0}} f(s)\,dw(s) = 0,$$

i.e., $\zeta_{t-0}=\zeta_{t+0}$. The continuity of ζ_t is thus established. Hence, ζ_t and the σ-algebras $\mathfrak{F}_{\tau_t}$ satisfy all the assumptions of Theorem 1 in § 1. This implies that ζ_t is a Wiener process.

Remark 3. If $f(t)$ is only defined for $t\in[0,T]$, then for the application of Theorem 3 we can extend it by setting $f(t)=1$ for $t\geqq T$.

Corollary 1. *If $f(t)\in H_2[0,T]$, then*

$$\mathbb{P}\left\{\int_0^T f^2(s)\,ds\leqq h\right\}\geqq 2\mathbb{P}\{w(h)>a\}-\mathbb{P}\left\{\sup_{0\leqq t\leqq T}\int_0^T f(s)\,dw(s)>a\right\}. \quad (1)$$

Let $\int_0^T f^2(s)\,ds=\xi$. Then in the notation of Theorem 3

$$\sup_{0\leqq t\leqq T}\int_0^t f(s)\,dw(s)=\sup_{0\leqq t\leqq \xi}\zeta_t.$$

Assume τ_a is the first time ζ_t attains a. Then

$$\mathbb{P}\{\xi>h\}=\mathbb{P}\{\xi>h,\tau_a\leqq h\}+\mathbb{P}\{\xi>h,\tau_a\geqq h\}\leqq\mathbb{P}\{\xi>\tau_a\}+\mathbb{P}\{\tau_a>h\}.$$

Thus,

$$\mathbb{P}\{\xi\leqq h\}\geqq\mathbb{P}\{\tau_a\leqq h\}-\mathbb{P}\{\xi>\tau_a\}.$$

But

$$\mathbb{P}\{\xi>\tau_a\}=\mathbb{P}\{\sup_{0\leqq t\leqq\xi}\zeta_t>a\},$$

$$\mathbb{P}\{\tau_a\leqq h\}=\mathbb{P}\{\sup_{0\leqq t\leqq h}\zeta_t>a\}=2\mathbb{P}\{\zeta(h)>a\}$$

because of Lemma 1 in § 1. Since ζ_t is a Wiener process, $\mathbb{P}\{\zeta_h>a\}=\mathbb{P}\{w(h)>a\}$. To complete the proof of formula (1) we need only note that

$$\mathbb{P}\{\xi\leqq h\}=\mathbb{P}\left\{\int_0^T f^2(s)\,ds\leqq h\right\}.$$

Chapter 2. The Solution of Stochastic Differential Equations

§ 5. Stochastic Differential Equations of First Order

We consider the stochastic differential equation

$$d\eta(t)=a(t,\eta(t))\,dt+\sigma(t,\eta(t))\,dw(t). \tag{1}$$

The functions $a(t,x)$ and $\sigma(t,x)$, defined and measurable for $t\in[0,T]$ and $x\in(-\infty,\infty)$ are called coefficients of Eq. (1); $\eta(t)$ is the solution of this equation and $w(t)$ is a Wiener process. The random process $\eta(t)$ will be said to be a solution of (1) on $[0,T]$ if it satisfies the following conditions:

1) denoting by $\mathfrak{F}_t$, $0\leqq t\leqq T$ the minimal σ-algebra with respect to which the variables $\eta(s)$ for $s\leqq t$ and $w(s)$ for $s\leqq t$ are measurable, the process $w_t(s)=w(t+s)-w(t)$ does not depend on $\mathfrak{F}_t$;

2) denoting by $H_2[0,T]$ the space of measurable random functions $\varphi(t)$ which, for each $t\in[0,T]$ are $\mathfrak{F}_t$-measurable and for which the integral $\int_0^T \varphi^2(t)\,dt$ is w.p. 1 finite, $|a(t,\eta(t))|^{\frac{1}{2}}$ and $\sigma(t,\eta(t))$ belong to $H_2[0,T]$;

3) the process $\eta(t)$ has on $[0,T]$ the stochastic differential $d\eta(t)=\bar{a}(t)\,dt+\bar{\sigma}(t)\,dw(t)$, also, for all $t\in[0,T]$ we have w.p. 1 $\bar{a}(t)=a(t,\eta(t))$, $\bar{\sigma}(t)=\sigma(t,\eta(t))$. If $\eta(t)$ is a solution of (1), then it follows from 2) that $\eta(0)$ is independent of $w(t)-w(0)$ for $t>0$. Using the definition of a stochastic integral, we can express (1) in the following integral form:

$$\eta(t)=\eta(0)+\int_0^t a(s,\eta(s))\,ds+\int_0^t \sigma(s,\eta(s))\,dw(s). \tag{2}$$

Eqs. (1) and (2) are equivalent and are to be solved for given $\eta(0)$, with $\eta(0)$ chosen independent of $w(t)$. If $\sigma(t,x)=0$, Eq. (1) can be considered as an ordinary differential equation and solved for each ω.

Using Itô's formula, we consider transformations of (1). Let $\eta(t)$ be a solution of (1) and let $f(t,x)$ be a monotone (in x), continuous function, defined for $t\in[0,T]$, $x\in(-\infty,\infty)$, for which the derivatives

$f_t'(t,x)$, $f_x'(t,x)$, and $f_{xx}''(t,x)$ exist and are continuous. For each $t\in[0,T]$ there exists a function $g(t,x)$ inverse to $f(t,x)$, i.e., a function for which $f(t,g(t,x))=x$, $g(t,f(t,x))=x$. Set $\zeta(t)=f(t,\eta(t))$. Then $\eta(t)=g(t,\zeta(t))$ and

$$d\zeta(t)=\left[f_t'(t,\eta(t))+f_x'(t,\eta(t))\,a(t,\eta(t))+\tfrac{1}{2}f_{xx}''(t,\eta(t))\,\sigma^2(t,\eta(t))\right]dt$$
$$+f_x'(t,\eta(t))\,\sigma(t,\eta(t))\,dw(t).$$

Hence the process $\zeta(t)$ will satisfy

$$d\zeta(t)=\bar{a}(t,\zeta(t))\,dt+\bar{\sigma}(t,\zeta(t))\,dw(t),$$

where

$$\begin{aligned}\bar{a}(t,x)=&f_t'(t,g(t,x))+f_x'(t,g(t,x))\,a(t,g(t,x))\\&+\tfrac{1}{2}f_{xx}''(t,g(t,x))\,\sigma^2(t,g(t,x)),\end{aligned}\tag{3}$$

$$\bar{\sigma}(t,x)=f_x'(t,g(t,x))\,\sigma(t,g(t,x)).\tag{4}$$

Such a substitution of an undetermined function allows the transformation of the differential equation (by a change of its coefficients) to a more convenient form. We mention two such substitutions: a) that for which $\bar{\sigma}(t,x)=1$, i.e.,

$$f_x'(t,x)\,\sigma(t,x)=1,\qquad f_x'(t,x)=\frac{1}{\sigma(t,x)},$$

$$f(t,x)=\int_0^x\frac{1}{\sigma(t,y)}\,dy.\tag{5}$$

Such a substitution is always possible if $\sigma(t,x)>0$ and $\sigma_x'(t,x)$ is continuous; b) the substitution for which $\bar{a}(t,x)=0$, in this case the function $f(t,x)$ necessarily satisfies the equation

$$\frac{\partial f(t,x)}{\partial t}+a(t,x)\frac{\partial f(t,x)}{\partial x}+\frac{1}{2}\sigma^2(t,x)\frac{\partial^2 f(t,x)}{\partial x^2}=0.\tag{6}$$

If the coefficients $a(t,x)$ and $\sigma(t,x)$ are independent of t, one can find an explicit expression for $f(t,x)$. This function is then also independent of t and (6) assumes the form

$$a(x)\,f'(x)+\tfrac{1}{2}\sigma^2(x)\,f''(x)=0,$$

so that

$$f(x)=C_1+C_2\int_0^x\exp\left\{-\int_0^z\frac{2a(u)}{\sigma^2(u)}\,du\right\}dz.$$

We consider some very simple examples of stochastic differential equations.

Example 1. An equation whose coefficients depend only on t,

$$d\eta(t)=a(t)\,dt+\sigma(t)\,dw(t).\tag{7}$$

The solution of (7) has the form

$$\eta(t)=\eta(0)+\int_0^t a(s)\,ds+\int_0^t \sigma(s)\,dw(s).$$

This will be a process with independent Gaussian increments since $\int_{t'}^{t''} \sigma(s)\,dw(s)$ is normally distributed. We have

$$\mathrm{M}[\eta(t)-\eta(0)]=\int_0^t a(s)\,ds, \qquad \operatorname{Var}[\eta(t)-\eta(0)]=\int_0^t \sigma^2(s)\,ds.$$

Example 2. An equation which reduces to the form (7) after substitutuion of an undetermined function.

We consider an equation of the general form (1) and find conditions which allow it to be reduced to (7) by a substitution of an undetermined function $\zeta(t)=f(t,\eta(t))$. We assume that the coefficients of the resulting equation are sufficiently smooth, so that all derivatives which occur in the following calculations exist. In order that the process $\zeta(t)$ satisfy the equation $d\zeta(t)=\bar{a}(t)\,dt+\bar{\sigma}(t)\,dw(t)$ it is necessary (because of (3) and (4)) that

$$f_t'(t,x)+a(t,x)\,f_x'(t,x)+\tfrac{1}{2}f_{xx}''(t,x)\,\sigma^2(t,x)=\bar{a}(t),$$

$$f_x'(t,x)\,\sigma(t,x)=\bar{\sigma}(t).$$

(Since the right-hand sides of these equations do not depend on x, we can replace $g(t,x)$ by x.) From the second equation we find

$$f_x'(t,x)=\frac{\bar{\sigma}(t)}{\sigma(t,x)}.$$

Differentiating the first w.r.t. x:

$$f_{tx}''(t,x)+\frac{\partial}{\partial x}\left[a(t,x)\,f_x'(t,x)+\tfrac{1}{2}\sigma^2(t,x)\,f_{xx}''(t,x)\right]=0.$$

Substituting in this the values $f_x'(t,x)$,

$$f_{tx}''(t,x)=\frac{\bar{\sigma}_t'(t)\,\sigma(t,x)-\bar{\sigma}(t)\,\sigma_t'(t,x)}{\sigma^2(t,x)}$$

and

$$f_{xx}''=-\frac{\bar{\sigma}(t)\,\sigma_x'(t,x)}{\sigma^2(t,x)},$$

we obtain

$$\frac{\bar{\sigma}_t'(t)}{\sigma(t,x)}-\bar{\sigma}(t)\left[\frac{\sigma_x'(t,x)}{\sigma(t,x)}-\frac{\partial}{\partial x}\left(\frac{a(t,x)}{\sigma(t,x)}\right)+\frac{1}{2}\sigma_{xx}''(t,x)\right]=0,$$

whence

$$\frac{\bar{\sigma}_t'(t)}{\bar{\sigma}(t)}=\sigma(t,x)\left[\frac{\sigma_t'(t,x)}{\sigma^2(t,x)}-\frac{\partial}{\partial x}\left(\frac{a(t,x)}{\sigma(t,x)}\right)+\frac{1}{2}\sigma_{xx}''(t,x)\right]. \tag{8}$$

Differentiating both sides w.r.t. x we obtain the condition for the reducibility of (1) to (7)

$$\frac{\partial}{\partial x}\left\{\sigma(t,x)\left[\frac{\sigma_t'(t,x)}{\sigma^2(t,x)}-\frac{\partial}{\partial x}\left(\frac{a(t,x)}{\sigma(t,x)}\right)+\frac{1}{2}\sigma_{xx}''(t,x)\right]\right\}=0. \tag{9}$$

If (9) is satisfied, then the right side of (8) depends only on t and from (8) we can determine $\bar{\sigma}(t)$ and $f(t,x)$ from $f_x'(t,x)=\dfrac{\bar{\sigma}(t)}{\sigma(t,x)}$. Moreover, (8) is equivalent to

$$\frac{\partial}{\partial x}\left[f_t'(t,x)+a(t,x)f_x'(t,x)+\tfrac{1}{2}\sigma^2(t,x)f_{xx}''(t,x)\right]=0,$$

therefore the expression in square brackets is independent of t and it can be taken as $\bar{a}(t)$.

We now mention separately those cases for which the coefficients of the equation are independent of t, i.e., $a(t,x)=a(x)$ and $\sigma(t,x)=\sigma(x)$. Here (8) assumes the form

$$\frac{\bar{\sigma}'(t)}{\bar{\sigma}(t)}=\sigma(x)\left[\frac{1}{2}\sigma''(x)-\frac{d}{dx}\left(\frac{a(x)}{\sigma(x)}\right)\right].$$

Since the left side of this equation depends only on t, and the right only on x, we have

$$\frac{\bar{\sigma}'(t)}{\bar{\sigma}(t)}=\sigma(x)\left[\frac{1}{2}\sigma''(x)-\frac{d}{dx}\left(\frac{a(x)}{\sigma(x)}\right)\right]=c, \tag{10}$$

where c is some constant. Consequently, $\sigma(t)=e^{ct}$ and $f(t,x)$ can be taken as $f(t,x)=e^{ct}\int\limits_0^x\dfrac{dy}{\sigma(y)}$. Hence

$$a(t)=e^{ct}\left[c\int\limits_0^x\frac{dy}{\sigma(y)}+\frac{a(x)}{\sigma(x)}-\frac{1}{2}\sigma'(x)\right],$$

whereby the expression in brackets is constant since its derivative, by (10), is zero.

Example 3. A linear equation.

It is natural to call linear equations for which the functions $a(t,x)$ and $\sigma(t,x)$ depend linearly on x, i.e., equations of the form

$$d\eta(t)=[\alpha(t)+\beta(t)\eta(t)]\,dt+[\gamma(t)+\delta(t)\eta(t)]\,dw(t). \tag{11}$$

In the case where $\alpha(t)=0$ and $\gamma(t)=0$, the linear equation will be called homogeneous. We consider first the solution of the homogeneous equation

$$d\eta(t)=\beta(t)\,\eta(t)\,dt+\delta(t)\,\eta(t)\,dw(t). \tag{12}$$

We assume that $\eta(0)>0$. Then, because of continuity, $\eta(t)>0$ on a certain interval. Set $\zeta(t)=\log\eta(t)$. Using Itô's formula we obtain

$$d\zeta(t)=\frac{1}{\eta(t)}\,\beta(t)\,\eta(t)\,dt-\frac{1}{2}\cdot\frac{1}{\eta^2(t)}\,\delta^2(t)\,\eta^2(t)\,dt+\frac{1}{\eta(t)}\,\delta(t)\,\eta(t)\,dw(t).$$

Thus

$$\begin{aligned}
d\zeta(t)&=[\beta(t)-\tfrac{1}{2}\delta^2(t)]\,dt+\delta(t)\,dw(t),\\
\zeta(t)&=\zeta_0+\int_0^t[\beta(s)-\tfrac{1}{2}\delta^2(s)]\,ds+\int_0^t\delta(s)\,dw(s),\\
\eta(t)&=\eta(0)\exp\left\{\int_0^t[\beta(s)-\tfrac{1}{2}\delta^2(s)]\,ds+\int_0^t\delta(s)\,dw(s)\right\}.
\end{aligned} \tag{13}$$

It is easy to see that formula (13) represents the solution of (12) so long as this solution doesn't vanish. And since the right side (13) does not vanish for $t>0$, (13) represents an arbitrary solution of (12) when $\eta(0)>0$. Applying an analogous argument to $\eta(t)$ with $\eta(0)<0$, we see that (13) holds if $\eta(0)\neq 0$. Proceeding to the solution of the general Eq. (11), we introduce a new unknown function $\zeta(t)$, related to $\eta(t)$ by

$$\zeta(t)=\eta_0(t)\,\eta(t),$$

where

$$\eta_0(t)=\exp\left[-\int_0^t[\beta(s)-\tfrac{1}{2}\delta^2(s)]\,ds-\int_0^t\delta(s)\,dw(s)\right].$$

A simple calculation shows that

$$d\eta_0(t)=\eta_0(t)\left[(-\beta(s)+\delta^2(s))\,ds-\delta(s)\,dw(s)\right].$$

Hence,

$$\begin{aligned}
d\zeta(t)&=\eta(t)\,d\eta_0(t)+\eta_0(t)\,d\eta(t)-[\gamma(t)+\delta(t)\,\eta(t)]\,\eta_0(t)\,\delta(t)\,dt\\
&=\eta(t)\,\eta_0(t)\,[\delta^2(s)-\beta(s)]\,ds-\eta(t)\,\eta_0(t)\,\delta(s)\,dw(s)\\
&\quad+\eta_0(t)\,[\alpha(t)+\beta(t)\,\eta(t)]\,dt+\eta_0(t)\,[\gamma(t)+\delta(t)\,\eta(t)]\,dw(t)\\
&\quad-\delta(t)\,[\gamma(t)+\delta(t)\,\eta(t)]\,\eta_0(t)\,dt\\
&=\eta_0(t)\,[\alpha(t)-\gamma(t)\,\delta(t)]\,dt+\eta_0(t)\,\gamma(t)\,dw(t)
\end{aligned}$$

and

$$\zeta(t)=\zeta(0)+\int_0^t\eta_0(s)\,[\alpha(s)-\gamma(s)\,\delta(s)]\,ds+\int_0^t\gamma(s)\,\eta_0(s)\,dw(s).$$

Substituting $\eta_0(t)$ in the relation for $\zeta(t)$ and expressing $\eta(t)$ in terms $\zeta(t)$, we obtain finally

$$\eta(t)=\exp\left\{\int_0^t [\beta(s)-\tfrac{1}{2}\delta^2(s)]\,ds+\int_0^t \delta(s)\,dw(s)\right\}$$
$$\times\left[\eta(0)+\int_0^t \exp\left\{-\int_0^s [\beta(u)-\tfrac{1}{2}\delta^2(u)]\,du-\int_0^s \delta(u)\,dw(u)\right\}\right.$$
$$\times[\alpha(s)-\gamma(s)\,\delta(s)]\,ds$$
$$\left.+\int_0^t \exp\left\{-\int_0^s [\beta(u)-\tfrac{1}{2}\delta^2(u)]\,du-\int_0^s \delta(u)\,dw(u)\right\}\gamma(s)\,dw(s)\right].$$

Example 4. Equations reducible to the linear case.

We consider now the case in which the coefficients in (1) and (11) are independent of t. If the equation is to reduce to the linear case with the help of the substitution $\zeta(t)=f(\eta(t))$ and $g(x)$ is the inverse function to $f(x)$, then it is necessary that there exist constants α, β, γ and δ for which

$$a(g(x))\,f'(g(x))+\tfrac{1}{2}\sigma^2(g(x))\,f''(g(x))=\alpha+\beta x,$$
$$\sigma(g(x))\,f'(g(x))=\gamma+\delta x.$$

Substituting $x=f(z)$ into these equations we find

$$a(x)\,f'(x)+\tfrac{1}{2}\sigma^2(x)\,f''(x)=\alpha+\beta f(x),\qquad \sigma(x)\,f'(x)=\gamma+\delta f(x). \tag{14}$$

Assume that $\delta\neq 0$. Then taking $B(x)=\int_0^x \frac{1}{\sigma(z)}\,dz$, we have

$$f(x)=C\,e^{\delta B(x)}-\frac{\gamma}{\delta},$$

where C is some constant. Substituting this expression into the first equation at (14), we obtain

$$\left\{\frac{a(x)}{\sigma(x)}\delta+\frac{1}{2}\sigma^2(x)\left[\frac{\delta^2}{\sigma^2(x)}-\frac{\delta\,\sigma'(x)}{\sigma^2(x)}\right]\right\}C\,e^{\delta B(x)}=\beta\,C\,e^{\delta B(x)}-\frac{\beta\gamma}{\delta}+\alpha,$$

whence

$$\left[\frac{a(x)}{\sigma(x)}\delta+\frac{1}{2}\delta^2-\beta-\frac{\delta}{2}\sigma'(x)\right]e^{\delta B(x)}=\frac{\alpha\delta-\gamma\beta}{\delta C}.$$

Set $Z(x)=\dfrac{a(x)}{\sigma(x)}-\dfrac{1}{2}\sigma'(x)$. Eliminating by differentiation the unknown constants we find the following condition for the reduction of an equation to the linear case:

$$\frac{d}{dx}\left[\frac{\frac{d}{dx}(\sigma(x)\,Z'(x))}{Z'(x)}\right]=0. \tag{15}$$

It is easy to verify that when (15) is satisfied, the substitution

$$f(x)=C\exp\{\delta B(x)\}, \qquad \text{where } \delta=-\frac{\frac{d}{dx}(\sigma(x)\,Z'(x))}{Z'(x)}$$

for some C reduces the initial equation to linear form. If $\delta=0$, then in the same notation

$$f(x)=\gamma B(x)+C, \qquad \left[\frac{a(x)}{\sigma(x)}-\frac{1}{2}\sigma'(x)\right]\gamma=\beta\gamma B(x)+C_1,$$

thus the condition for reducibility to linearity will be $\frac{d}{dx}[\sigma(x)\,Z'(x)]=0$.
It is obvious that the latter equality implies (15), thus the general condition for reducibility of an equation with coefficients independent of t is given by (15).

We remark that in all of the cases considered, the form of the solution of the stochastic equation was found under the single assumption that such a solution exists. From the equivalence of all the transformations it is easy to see that the resulting expressions are actually solutions of the corresponding equations. Thus, in the cases considered, the existence and uniqueness of the solutions of the stochastic differential equations with given initial conditions is proved.

§ 6. Existence and Uniqueness of the Solutions of Stochastic Differential Equations

We consider the equation

$$d\eta(t)=a(t,\eta(t))\,dt+\sigma(t,\eta(t))\,dw(t), \tag{1}$$

whose coefficients $a(t,x)$ and $\sigma(t,x)$ are defined and measurable for $t\in[0,T]$ and $x\in(-\infty,\infty)$ and prove the existence and uniqueness of the solution to (1) for given initial condition. In what follows it will be clear that questions of existence and uniqueness play an important role in the construction of apparatus for determining the probabilistic characteristics of the relevant solutions (such an apparatus will be set up in Chapter 3). In the case of ordinary differential equations, which result from (11) when $\sigma(t,x)=0$, one proves the simplest existence and uniqueness theorem for right hand side with coefficient $a(t,x)$ satisfying a Lipschitz condition in x and bounded w.r.t. t for some x. We show that for the existence and uniqueness of the solution of a stochastic differential equation it is sufficient that both coefficients of the equation possess the properties refered to.

Theorem 1. *If the following assumptions are satisfied:*

1) *The functions* $a(t,x)$ *and* $\sigma(t,x)$ *are defined for* $t\in[0,T]$ *and* $x\in(-\infty,\infty)$ *and measurable w.r.t. all their arguments;*

2) *There exists a constant K such that for* $t\in[0,T]$ *and* $x, y\in(-\infty,\infty)$

$$|a(t,x)-a(t,y)|+|\sigma(t,x)-\sigma(t,y)|\leqq K|x-y|,$$
$$|a(t,x)|^2+|\sigma(t,x)|^2\leqq K^2(1+|x|^2);$$

3) $\eta(0)$ *does not depend on* $w(t)$ *and* $\mathbb{M}\,\eta(0)^2<\infty$. *Then there exists a solution of* (1) *satisfying the conditions:*

A) $\eta(t)$ *is continuous w.p. 1 and* $\eta(t)=\eta(0)$ *for* $t=0$;

B) $\sup\limits_{0\leqq t\leqq T}\mathbb{M}\,\eta(t)^2<\infty$.

If $\eta_1(t)$ *and* $\eta_2(t)$ *are two solutions of* (1) *satisfying A) and B), then*

$$\mathbb{P}\{\sup_{0\leqq t\leqq T}|\eta_1(t)-\eta_2(t)|=0\}=1.$$

Proof. We first establish the uniqueness. Let

$$\eta_i(t)=\eta(0)+\int_0^t a(s,\eta_i(s))\,ds+\int_0^t \sigma(s,\eta_i(s))\,dw(s),\qquad i=1,2,$$

then

$$\begin{aligned}\mathbb{M}[\eta_1(t)-\eta_2(t)]^2&=\mathbb{M}\Big[\int_0^t(a(s,\eta_1(s))-a(s,\eta_2(s))\,ds\\&\qquad+\int_0^t(\sigma(s,\eta_1(s))-\sigma(s,\eta_2(s)))\,dw(s)\Big]^2\\&\leqq 2\mathbb{M}\Big[\int_0^t(a(s,\eta_1(s))-a(s,\eta_2(s)))\,ds\Big]^2\\&\qquad+2\mathbb{M}\Big[\int_0^t(\sigma(s,\eta_1(s))-\sigma(s,\eta_2(s)))\,dw(s)\Big]^2\\&\leqq 2\mathbb{M}\,t\int_0^t[a(s,\eta_1(s))-a(s,\eta_2(s))]^2\,ds\\&\qquad+2\mathbb{M}\int_0^t[\sigma(s,\eta_1(s))-\sigma(s,\eta_2(s))]^2\,ds\end{aligned}$$

$$\leqq 2tK^2 \int_0^t \mathbb{M}\,|\eta_1(s)-\eta_2(s)|^2\,ds$$

$$+2K^2 \int_0^t \mathbb{M}\,|\eta_1(s)-\eta_2(s)|^2\,ds$$

$$\leqq L \int_0^t \mathbb{M}\,|\eta_1(s)-\eta_2(s)|^2\,ds,$$

where

$$L=2(T+1)K^2.$$

Lemma 1. *Let $\varphi(t)$ and $\alpha(t)$ be measurable bounded functions and for some $L>0$ assume*

$$\varphi(t)\leqq\alpha(t)+L\int_0^t \varphi(s)\,ds.$$

Then

$$\varphi(t)\leqq\alpha(t)+L\int_0^t e^{L(t-s)}\alpha(s)\,ds. \tag{2}$$

Proof. We denote the right side of (2) by $\psi(t)$. Then $\psi(t)$ satisfies

$$\psi(t)=\alpha(t)+L\int_0^t \psi(s)\,ds, \qquad \psi(0)=\alpha(0),$$

so that setting $\Delta(t)=\psi(t)-\varphi(t)$ we can write

$$\Delta(t)\geqq L\int_0^t \Delta(s)\,ds\geqq L^2\int_0^t\int_0^s \Delta(u)\,du$$

$$=L^2\int_0^t (t-u)\,\Delta(u)\,du\geqq L^3\int_0^t (t-u)\int_0^u \Delta(s)\,ds$$

$$=L^3\int_0^t \frac{(t-s)^2}{2}\Delta(s)\,ds\geqq\cdots\geqq\frac{L^n}{(n-1)!}\int_0^t (t-s)^{n-1}\Delta(s)\,ds.$$

Since the right side of the last inequality tends to zero for $n\to\infty$, $\Delta(t)\geqq 0$. □

We return to the proof of Theorem 1 and establish that

$$\mathbb{M}\,|\eta_1(t)-\eta_2(t)|^2\leqq L\int_0^t \mathbb{M}\,|\eta_1(s)-\eta_2(s)|^2\,ds.$$

Applying Lemma 1, we find that $\mathbb{M}\,|\eta_1(t)-\eta_2(t)|=0$. Consequently, for each $t\in[0,T]$ $\mathbb{P}\{\eta_1(t)=\eta_2(t)\}=1$. From this it follows that for each countable subset $N\subset[0,T]$

$$\mathbb{P}\{\sup_{t\in N}|\eta_1(t)-\eta_2(t)=0\}=1.$$

If N is everywhere dense in $[0, T]$, then from the continuity of $\eta_1(t)$ and $\eta_2(t)$ there results

$$\mathbb{P}\{\sup_{0 \leqq t \leqq T} |\eta_1(t)-\eta_2(t)|=0\}=\mathbb{P}\{\sup_{t \in N} |\eta_1(t)-\eta_2(t)|=0\}=1.$$

The uniqueness of the solution is proved.

We now demonstrate the existence of a solution to (1) satisfying A) and B). Set $\eta_0(t)=\eta(0)$ and

$$\eta_n(t)=\eta(0)+\int_0^t a(s, \eta_{n-1}(s))\, ds+\int_0^t \sigma(s, \eta_{n-1}(s))\, dw(s) \tag{3}$$

with the help of estimates analogous to those used in the uniqueness proof one shows

$$\begin{aligned}\mathbb{M}[\eta_{n+1}(t)-\eta_n(t)]^2 &= \mathbb{M}\Big[\int_0^t (a(s, \eta_n(s))-a(s, \eta_{n-1}(s)))\, ds \\ &\qquad + \int_0^t (\sigma(s, \eta_n(s))-\sigma(s, \eta_{n-1}(s)))\, dw(s)\Big]^2 \\ &\leqq L \int_0^t \mathbb{M}|\eta_n(s)-\eta_{n-1}(s)|^2\, ds\end{aligned}$$

(L has the same significance as before). Iterating this inequality we find

$$\mathbb{M}|\eta_{n+1}(t)-\eta_n(t)|^2 \leqq L \int_0^t \frac{(t-s)^{n-1}}{(n-1)!} \mathbb{M}|\eta_1(s)-\eta_0(s)|^2\, ds,$$

but

$$\begin{aligned}\mathbb{M}|\eta_1(s)-\eta_0(s)|^2 &= \mathbb{M}\Big[\int_0^t a(s, \eta(0))\, ds\Big]^2 + \mathbb{M}\int_0^t \sigma^2(s, \eta(0))\, ds \\ &\leqq LTK^2(1+\mathbb{M}\eta^2(0)).\end{aligned}$$

Hence there exists a constant C such that

$$\mathbb{M}|\eta_{n+1}(t)-\eta_n(t)|^2 \leqq C \frac{(LT)^n}{n!}.$$

We note that

$$\begin{aligned}\sup_{0 \leqq t \leqq T} |\eta_{n+1}(t)-\eta_n(t)| &\leqq \int_0^T |a(s, \eta_n(s))-a(s, \eta_{n-1}(s))|\, ds \\ &\qquad + \sup_{0 \leqq t \leqq T} \left|\int_0^t (\sigma(s, \eta_{n+1}(s))-\sigma(s, \eta_n(s)))\, dw(s)\right|.\end{aligned}$$

Consequently, using Cauchy's inequality, Theorem 1 in §3 and the Lipschitz condition we can write

$$\mathbb{M} \sup_{0 \leq t \leq T} |\eta_{n+1}(t) - \eta_n(t)|^2$$
$$\leqq 2T \int_0^T \mathbb{M} K^2 |\eta_n(s) - \eta_{n-1}(s)|^2 \, ds + 8K^2 \int_0^T \mathbb{M} |\eta_n(t) - \eta_{n-1}(s)|^2 \, ds$$
$$\leqq \frac{C_1 L^{n-1} T^{n-1}}{(n-1)!},$$

where $C_1 = K^2(2T+8)T$. From the convergence of

$$\sum_{n=1}^{\infty} \mathbb{P}\left\{ \sup_{0 \leq t \leq T} |\eta_{n+1}(t) - \eta_n(t)| > \frac{1}{n^2} \right\} \leqq \sum_{n=1}^{\infty} \frac{C_1 L^{n-1} T^{n-1}}{(n-1)!} n^4$$

follows the uniform convergence w.p. 1 of

$$\eta(0) + \sum_{n=0}^{\infty} [\eta_{n+1}(t) - \eta_n(t)].$$

The sum of this series is w.p. 1 the uniform limit of $\eta_n(t)$ which means that $\eta_n(t)$ converges to some random process $\eta(t)$. Proceeding to the limit for $n \to \infty$ in (3) we see that $\eta(t)$ is a solution of (1), whereby $\eta(t)$ is measurable w.r.t. the σ-algebra $\mathfrak{F}_t$, the minimum σ-algebra with respect to which $\eta(0)$ and $w(s)$ are measurable for $s \leqq t$ (with respect to this σ-algebra all $\eta_n(t)$ are measurable and therefore so also is the limit of $\eta_n(t)$). The continuity of $\eta(t)$ is a result of the fact that it is equal w.p. 1 to the uniform limit of continuous processes. Moreover,

$$\mathbb{M}\eta_n^2(t) \leqq 3\left\{ \mathbb{M}\eta^2(0) + \mathbb{M}\left[\int_0^t a(s, \eta_{n-1}(s))\, ds\right]^2 + \mathbb{M}\left[\int_0^t \sigma(s, \eta_{n-1}(s))\, dw(s)\right]^2 \right\}$$
$$\leqq 3\mathbb{M}\eta^2(0) + 3L \int_0^t \mathbb{M}\eta_{n-1}^2(s)\, ds.$$

Applying this estimate successively to $\eta_{n-1}(s)$, $\eta_{n-2}(s)$ etc., we obtain

$$\mathbb{M}\eta_n^2(t) \leqq 3\mathbb{M}\eta^2(0) + 3\mathbb{M}\eta^2(0)\, 3Lt + (3L)^2 \int_0^t (t-s)\mathbb{M}\eta_{n-2}^2(s)\, ds$$
$$\leqq 3\mathbb{M}\eta^2(0) + 3Lt\, 3\mathbb{M}\eta^2(0) + 3\mathbb{M}\eta^2(0) \frac{(3Lt)^2}{2} + \cdots$$
$$\leqq 3\mathbb{M}\eta^2(0)\, e^{3Lt}.$$

Proceeding to the limit as $n \to \infty$, we find

$$\sup_{0 \leq t \leq T} \mathbb{M}\eta^2(t) \leqq 3\mathbb{M}\eta^2(0)\, e^{3LT}. \quad \square$$

In order to weaken somewhat the conditions assuring the existence of a unique solution of (1) we prove a theorem on the local dependence of the solution on the coefficients. This theorem will be important in the study of several properties of solutions of stochastic equations.

Theorem 2. *Assume that the coefficients* $a_1(t,x)$, $\sigma_1(t,x)$, $a_2(t,x)$ *and* $\sigma_2(t,x)$ *of the equations*

$$d\eta_i(t)=a_i(t,\eta_i(t))\,dt+\sigma_i(t,\eta_i(t))\,dw(t), \qquad i=1,2 \tag{4}$$

satisfy the conditions of Theorem 1 and that for some $N>0$ *with* $|x|\leqq N$ $a_1(t,x)=a_2(t,x)$ *and* $\sigma_1(t,x)=\sigma_2(t,x)$. *If* $\eta_1(t)$ *and* $\eta_2(t)$ *are solutions of* (4) *with the same initial condition* $\eta_1(0)=\eta_2(0)=\eta_0$, $\mathbb{M}\eta_0^2<\infty$, *and* τ_i *is the largest t for which* $\sup_{0\leqq s\leqq t}|\eta_i(t)|\leqq N$, *then* $\mathbb{P}\{\tau_1=\tau_2\}=1$ *and*

$$\mathbb{P}\{\sup_{0\leqq s\leqq \tau_1}|\eta_1(s)-\eta_2(s)|=0\}=1\,.$$

Proof. We define $\varphi(t)$ as follows:

$$\varphi_1(t)=1 \quad \text{if } \sup_{0\leqq s\leqq t}|\eta_1(s)|\leqq N,$$

$$\varphi_1(t)=0 \quad \text{if } \sup_{0\leqq s\leqq t}|\eta_1(s)|> N$$

(that is, $\varphi_1(t)$ is the indicator function of the interval $[0,\tau_1]$). We can now write

$$\begin{aligned}\varphi_1(t)[\eta_1(t)-\eta_2(t)]&=\varphi_1(t)\int_0^t[a_1(s,\eta_1(s))-a_2(s,\eta_2(s))]\,ds\\&\quad+\varphi_1(t)\int_0^t[\sigma_1(s,\eta_1(s))-\sigma_2(s,\eta_2(s))]\,dw(s)\\&=\varphi_1(t)\int_0^t[a_1(s,\eta_1(s))-a_2(s,\eta_1(s))]\,ds\\&\quad+\varphi_1(t)\int_0^t[\sigma_1(s,\eta_1(s))-\sigma_2(s,\eta_1(s))]\,dw(s)\\&\quad+\varphi_1(t)\int_0^t[a_2(s,\eta_1(s))-a_2(s,\eta_2(s))]\,ds\\&\quad+\varphi_1(t)\int_0^t[\sigma_2(s,\eta_1(s))-\sigma_2(s,\eta_2(s))]\,dw(s).\end{aligned}$$

We note that from $\varphi_1(t)=1$ follows $a_1(s,\eta_1(s))=a_2(s,\eta_1(s))$ and $\sigma_1(s,\eta_1(s))=\sigma_2(s,\eta_1(s))$ for $s\leqq t$. Thus

$$\begin{aligned}\varphi_1(t)[\eta_1(t)-\eta_2(t)]^2&\leqq 2\varphi_1(t)\left\{\int_0^t[a_2(s,\eta_1(s))-a_2(s,\eta_2(s))]\,ds\right\}^2\\&\quad+2\varphi_1(t)\left[\int_0^t[\sigma_2(s,\eta_1(s))-\sigma_2(s,\eta_2(s))]\,dw(s)\right]^2.\end{aligned}$$

Taking into account that $\varphi_1(t)=1$ implies $\varphi_1(s)=1$ for $s \leqq t$ we can write

$$\varphi_1(t)[\eta_1(t)-\eta_2(t)]^2 \leqq 2\left\{\int_0^t \varphi_1(s)[a_2(s,\eta_1(s))-a_2(s,\eta_2(s))]\,ds\right\}^2 + 2\left[\int_0^t \varphi_1(s)[\sigma_2(s,\eta_1(s))-\sigma_2(s,\eta_2(s))]\,dw(s)\right]^2.$$

Taking the expectation and then using the Lipschitz condition and Cauchy's inequality, we can show that there exists a constant L for which

$$\mathbb{M}\varphi_1(t)[\eta_1(t)-\eta_2(t)]^2 \leqq L\int_0^t \mathbb{M}\varphi_1(s)[\eta_1(s)-\eta_2(s)]^2\,ds.$$

From Lemma 1 we have that $\mathbb{M}\varphi_1(t)[\eta_1(t)-\eta_2(t)]^2=0$. Considering the continuity of $\eta_1(t)$ and $\eta_2(t)$ we can establish

$$\mathbb{P}\{\sup_{0\leqq t\leqq T}\varphi_1(t)[\eta_1(t)-\eta_2(t)]^2=0\}=1.$$

It follows that on the interval $[0,\tau_1]$ the processes $\eta_1(t)$ and $\eta_2(t)$ coincide with probability one. Hence $\mathbb{P}\{\tau_2\geqq\tau_1\}=1$. Interchanging in the proof of the theorem the indices "1" and "2", we can show similarly that $\mathbb{P}\{\tau_1\geqq\tau_2\}=1$. The theorem is proved.

Remark 1. It is easy to see that Theorem 2 can be generalized to the case where the coefficients of the two equations coincide in some arbitrary region. Then up to the time of their exit from this region, solutions of the two equations coincide, provided they were equal at the initial moment of time.

Theorem 3. *Let the coefficients of* (1) *be defined and measurable for* $t\in[0,T]$, $x\in(-\infty,\infty)$ *and satisfy the conditions*

1) *for some* K

$$|a(t,x)|^2+|\sigma(t,x)|^2\leqq K^2(1+x^2);$$

2) *for each* N *there exists an* L_N *for which*

$$|a(t,x)-a(t,y)|+|\sigma(t,x)-\sigma(t,y)|\leqq L_N|x-y|$$

with $|x|\leqq N$, $|y|\leqq N$. *Then* (1) *has a solution satisfying the initial condition* $\eta(t)|_{t=0}=\eta(0)$, *this solution is unique in the sense of Theorem 1.*

Proof. We first show the existence of such a solution. Set $\eta_N(0)=\eta(0)$ for $|\eta(0)|\leqq N$; $\eta_N(0)=N\operatorname{sign}\eta(0)$ for $|\eta(0)|>N$; $a_N(t,x)=a(t,x)$ for $|x|\leqq N$; $a_N(t,x)=a(t,N\operatorname{sign}x)$ for $|x|>N$; $\sigma_N(t,x)=\sigma(t,x)$ for $|x|\leqq N$ and $\sigma_N(t,x)=\sigma(t,N\operatorname{sign}x)$ for $|x|>N$.

Denote by $\eta_N(t)$ the solution of

$$d\eta_N(t)=a_N(t,\eta_N(t))\,dt+\sigma_N(t,\eta_N(t))\,dw(t) \tag{5}$$

with initial condition $\eta_N(t)|_{t=0}=\eta_N(0)$. For (5) with its initial condition all the assumptions of Theorem 1 are satisfied. Hence, there exists a solution $\eta_N(t)$ of (5) satisfying A) and B) of that theorem. Let τ_N be the largest value of t for which $\sup_{0\leqq s\leqq t}|\eta_N(s)|\leqq N$. Let $N'>N$. Since $a_N(t,x)=a_{N'}(t,x)$ and $\sigma_N(t,x)=\sigma_{N'}(t,x)$ on $[-N,N]$, from Theorem 2 $\eta_N(t)=\eta_{N'}(t)$ w.p. 1 on $[0,\tau_N]$. Hence for $N'>N$

$$\mathbb{P}\{\sup_{0\leqq t\leqq T}|\eta_N(t)-\eta_{N'}(t)|>0\}\leqq\mathbb{P}\{\tau_N>T\}=\mathbb{P}\{\sup_{0\leqq t\leqq T}|\eta_N(t)|>N\}.$$

If we can show that $\mathbb{P}\{\sup_{0\leqq t\leqq T}|\eta_N(t)|>N\}\to 0$ for $N\to\infty$, then it will clearly follow that $\eta_N(t)$ converges uniformly w.p. 1 to some limit $\eta(t)$ as $N\to\infty$. Going to the limit in

$$\eta_N(t)=\eta_N(0)+\int_0^t a_N(s,\eta_N(s))\,ds+\int_0^t\sigma_N(s,\eta_N(s))\,dw(s)$$

we convince ourselves that $\eta(t)$ is equal w.p. 1 to the continuous solution to (1) satisfying the initial condition $\eta(t)|_{t=0}=\eta(0)$. Hence, to finish the proof of the existence of a solution it remains to show that

$$\lim_{N\to\infty}\mathbb{P}\{\sup_{0\leqq t\leqq T}|\eta_N(t)|>N\}=0. \tag{6}$$

Set $\psi(x)=\dfrac{1}{1+x^2}$. Then

$$\psi(\eta(0))\sup_{0\leqq t\leqq T}|\eta_N(t)|^2\leqq 3\psi(\eta(0))\,\eta_N^2(0)+3\psi(\eta(0))\,T\int_0^T a_N^2(s,\eta_N(s))\,ds$$
$$+3\psi(\eta(0))\sup_{0\leqq t\leqq T}\left[\int_0^t\sigma_N(s,\eta_N(s))\,dw(s)\right]^2.$$

Using first the conditional expectation for fixed $\eta(0)$ and then the unconditional expectation we obtain

$$\mathbb{M}\psi(\eta(0))\sup_{0\leqq t\leqq T}|\eta_N(t)|^2\leqq 3+3TK^2\int_0^T\mathbb{M}\psi(\eta(0))[1+|\eta_N(t)|^2]\,dt$$
$$+12K^2\int_0^T\mathbb{M}\psi(\eta(0))[1+|\eta_N(t)|^2]\,dt,$$

but

$$\mathbb{M}\psi(\eta(0))|\eta_N(t)|^2\leqq 3+3tK^2\int_0^t\mathbb{M}\psi(\eta(0))[1+|\eta_N(s)|^2]\,ds$$
$$+3K^2\int_0^t\mathbb{M}\psi(\eta(0))[1+|\eta_N(s)|^2]\,ds.$$

Applying Lemma 1 we can show that for some C depending only on K and T (and not on N), $\mathbb{M}\psi(\eta(0))|\eta_N(t)|^2 \leqq C$. Thus for some C_1, independent of N the following inequality holds

$$\mathbb{M}\psi(\eta(0)) \sup_{0\leqq t\leqq T} |\eta_N(t)|^2 \leqq C_1.$$

We can, moreover, write

$$\begin{aligned}\mathbb{P}\{\sup_{0\leqq t\leqq T} |\eta_N(t)| > N\} &= \mathbb{P}\{\psi(\eta(0)) \sup_{0\leqq t\leqq T} |\eta_N(t)|^2 > N^2 \psi(\eta(0))\}\\ &\leqq \mathbb{P}\{\psi(\eta(0)) \sup_{0\leqq t\leqq T} |\eta_N(t)|^2 > \delta N^2\} + \mathbb{P}\{\psi(\eta(0)) \leqq \delta\}\\ &\leqq \frac{C_1}{\delta N^2} + \mathbb{P}\{\psi(\eta(0)) \leqq \delta\},\end{aligned}$$

consequently,

$$\overline{\lim_{N\to\infty}} \mathbb{P}\{\sup_{0\leqq t\leqq T} |\eta_N(t)| > N\} \leqq \mathbb{P}\{\psi(\eta(0)) \leqq \delta\}.$$

Since δ is an arbitrary positive number and $\mathbb{P}\{\psi(\eta(0))=0\}=0$, (6) results from the preceding relation. This completes the proof of the existence of a solution to (1).

We now prove the uniqueness. Let $\eta_1(t)$ and $\eta_2(t)$ be two solutions of (1) which are w.p. 1 continuous and which satisfy the unitial condition $\eta_1(0)=\eta_2(0)=\eta(0)$. Denoting by $\varphi(t)$ the variable equal to one if $\sup_{s\leqq t}|\eta_1(s)| \leqq N$, $\sup_{s\leqq t}|\eta_2(s)| \leqq N$, and equal to zero otherwise, we can write using Condition 2

$$\begin{aligned}\mathbb{M}(\eta_1(t)-\eta_2(t))^2 \varphi(t) &\leqq 2\mathbb{M}\varphi(t)\left\{\int_0^t [a(s,\eta_1(s)) - a(s,\eta_2(s))]\, ds\right\}^2\\ &\quad + 2\mathbb{M}\varphi(t)\left[\int_0^t [\sigma(s,\eta_1(s)) - \sigma(s,\eta_2(s))]\, dw(s)\right]^2\\ &\leqq 2t\mathbb{M}\int_0^t \varphi(s)[a(s,\eta_1(s)) - a(s,\eta_2(s))]^2\, ds\\ &\quad + 2\mathbb{M}\int_0^t \varphi(s)[\sigma(s,\eta_1(s)) - \sigma(s,\eta_2(s))]^2\, ds\\ &\leqq (2T+2)L_N^2 \int_0^t \mathbb{M}\varphi(s)[\eta_1(s)-\eta_2(s)]^2\, ds.\end{aligned}$$

Then from Lemma 1 $\mathbb{M}(\eta_1(t)-\eta_2(t))^2\varphi(t)=0$, i.e.,

$$\mathbb{P}\{\eta_1(t) \neq \eta_2(t)\} \leqq \mathbb{P}\{\sup_{0\leqq s\leqq T} |\eta_1(s)| > N\} + \mathbb{P}\{\sup_{0\leqq s\leqq T} |\eta_2(s)| > N\}.$$

From the continuity of $\eta_i(t)$ follows its boundedness. Hence the probabilities on the right side of this inequality tend to zero as $N\to\infty$, i.e., for all $t\in[0,T]$ $\mathbb{P}\{\eta_1(t)=\eta_2(t)\}=1$ from which the uniqueness follows as in the proof of Theorem 1. □

Remark 2. For the uniqueness proof we only used Condition 2.

Remark 3. In connection with the existence of a solution, Condition 1 can be replaced by the following: for some K

$$x\,a(t,x)+\sigma^2(t,x)\leqq K^2(1+x^2).$$

As in the proof of Theorem 3 it is sufficient to verify that for some C $\mathbb{M}\psi(\eta(0))\,|\eta_N(t)|^2\leqq C$ (C independent of N). From Itô's formula

$$d\eta_N^2(t)=2a_N(t,\eta_N(t))\,\eta_N(t)\,dt+\sigma_N^2(t,\eta_N(t))\,dt+2\sigma_N(t,\eta_N(t))\,\eta_N(t)\,dw(t).$$

From this we obtain

$$\begin{aligned}&\mathbb{M}\eta_N^2(t)\,\psi(\eta(0))-\mathbb{M}\eta_N^2(0)\,\psi(\eta(0))\\&\quad=\mathbb{M}\psi(\eta(0))\int_0^t\left[2a_N(s,\eta_N(s))\,\eta_N(s)+\sigma_N^2(s,\eta_N(s))\right]ds\\&\quad\leqq 2K^2\left[t\mathbb{M}\psi(\eta(0))+\int_0^t\mathbb{M}\psi(\eta(0))\,\eta_N^2(s)\,ds\right].\end{aligned}$$

Using Lemma 1 we find

$$\mathbb{M}\psi(\eta(0))\,\eta_N^2(t)\leqq 2K^2\,\mathbb{M}\psi(\eta(0))\left[t+\int_0^t e^{2K^2(t-s)}s\,ds\right],$$

which proves the claim.

Remark 4. Theorem 1, Theorem 3, and Theorem 3 along with Remark 3 give more general conditions for the existence and uniqueness of the solutions of stochastic differential equations with given initial conditions. These theorems will frequently be referred to in the sequel as *the existence and uniqueness theorems.*

We now establish estimates for the moments of solutions of (11).

Theorem 4. *Assume the coefficients of* (1) *satisfy the conditions of Theorem 3 (Condition 1 can be replaced by that of Remark 3) and that* $\mathbb{M}[\eta(0)]^{2m}<\infty$ *(*$m>0$ *an integer). Then we can find a* C*, depending only on* m*,* K *and* T*, for which*

$$\mathbb{M}[\eta(t)]^{2m}\leqq\mathbb{M}\left[1+(\eta(0))^{2m}\right]e^{Ct}. \tag{7}$$

If only the conditions of Theorem 3 are satisfied, then there exists a constant $\bar{K}$ *depending on* m*,* K *and* T *for which*

$$\mathbb{M}[\eta(t)-\eta(0)]^{2m}\leqq\bar{K}\left[\mathbb{M}(\eta(0))^{2m}+1\right]e^{Ct}\,t^m. \tag{8}$$

Proof. We will employ the notation of Theorem 3. From Itô's formula

$$d[\eta_N(t)]^{2m}=[\eta_N(0)]^{2m}+\int_0^t [2m(\eta_N(s))^{2m+1} a_N(s,\eta_N(s)) \\ +m(2m-1)(\eta_N(s))^{2m-2}\sigma_N^2(s,\eta_N(s))]\,ds \\ +\int_0^t 2m(\eta_N(s))^{2m-1}\sigma_N(s,\eta_N(s))\,dw(s). \tag{9}$$

Since

$$\eta_N(t)=\eta_N(0)+\int_0^t a_N(s,\eta_N(s))\,ds+\int_0^t \sigma_N(s,\eta_N(s))\,dw(s)$$

and $\mathbb{M}[\eta_N(0)]^{2m}<\infty$ and $a_N(t,x)$ and $\sigma_N(t,x)$ are bounded, we have $\mathbb{M}[\eta_N(t)]^{2m}<\infty$.

Taking mathematical expectations in (9) we have

$$\mathbb{M}[\eta_N(t)]^{2m}=\mathbb{M}[\eta_N(0)]^{2m}+\int_0^t \mathbb{M}[2m\,a_N(s,\eta_N(s))\,\eta_N(s) \\ +m(2m-1)\,\sigma_N^2(s,\eta_N(s))]\,(\eta_N(s))^{2m-2}\,ds \\ \leqq \mathbb{M}[\eta(0)]^{2m}+(2m+1)\,mK^2\int_0^t \mathbb{M}(1+\eta_N^2(s))\,[\eta_N(s)]^{2m-2}\,ds.$$

Using the inequality $[\eta_N(s)]^{2m-2}\leqq 1+[\eta_N(s)]^{2m}$ we can show

$$\mathbb{M}[\eta_N(t)]^{2m}\leqq \mathbb{M}[\eta(0)]^{2m}+(2m+1)\,mK^2\int_0^t \{1+2\mathbb{M}[\eta_N(s)]^{2m}\}\,ds.$$

From Lemma 1

$$\mathbb{M}[\eta_N(t)]^{2m}\leqq \mathbb{M}[\eta(0)]^{2m}+m(2m+1)\,K^2\,t \\ +2m(2m+1)\,K^2\int_0^t e^{2m(2m+1)K^2(t-s)} \\ \times[\mathbb{M}(\eta(0))^{2m}+s\,m(2m+1)\,K^2]\,ds,$$

whence follows the inequality (7) for $\eta_N(t)$ with $C=2m(m+1)K^2$. Proceeding to the limit for $N\to\infty$ we obtain (7) for $\eta(t)$.

We now establish (8). We have

$$\mathbb{M}[\eta(t)-\eta(0)]^{2m}=\mathbb{M}\left[\int_0^t a(s,\eta(s))\,ds+\int_0^t \sigma(s,\eta(s))\,dw(s)\right]^{2m} \\ \leqq 2^{2m-1}\left\{\mathbb{M}\left[\int_0^t a(s,\eta(s))\,ds\right]^{2m}\right. \\ \left.+\mathbb{M}\left[\int_0^t \sigma(s,\eta(s))\,dw(s)\right]^{2m}\right\}.$$

Using Hölder's inequality and Theorem 6 § 3 we find

$$\mathbb{M}[\eta(t)-\eta(0)]^{2m} \leqq 2^{2m-1}\Big[t^{2m-1}\int_0^t \mathbb{M}[a(s,\eta(s))]^{2m}\,ds$$

$$+t^{m-1}[m(2m-1)]^m\,\mathbb{M}\int_0^t \sigma^{2m}(s,\eta(s))\,ds\Big]$$

consequently, there exists a K_1 such that

$$\begin{aligned}\mathbb{M}[\eta(t)-\eta(0)]^{2m} &\leqq K_1 t^{m-1}\int_0^t \mathbb{M}(1+\eta^2(s))^m\,ds\\ &\leqq K_1 t^{m-1}\int_0^t [2^{m-1}+2^{m-1}\mathbb{M}\eta^{2m}(s)]\,ds\\ &\leqq 2^{m-1}K_1 t^{m-1}\int_0^t \big[1+(1+\mathbb{M}[\eta(0)]^{2m})\,e^{Cs}\big]\,ds\\ &=2^m K_1 t^m+2^{m-1}t^m K_1\mathbb{M}[\eta(0)]^{2m}\frac{e^{Ct}-1}{Ct}\\ &\leqq 2^m K_1 t^m\{1+\mathbb{M}[\eta(0)]^{2m}\}\,e^{Ct},\end{aligned}$$

since $\dfrac{e^x-1}{x}\leqq e^x$ for $x>0$. The theorem is proved.

§ 7. Stochastic Equations which Depend on a Parameter

In the study of the stochastic differential equations introduced in § 5 it is sometimes fitting to consider auxiliary equations of more general form

$$\eta(t)=\varphi(t)+\int_0^t A(s,\eta(s))\,ds+\int_0^t B(s,\eta(s))\,dw(s), \tag{1}$$

where $\varphi(t)$, $A(t,x)$ and $B(t,x)$ are already random functions. In order to list the conditions which these functions will satisfy we assume given a family of σ-algebras $\mathfrak{F}_t$, $0\leqq t\leqq T$ for which $\mathfrak{F}_{t_1}\subset\mathfrak{F}_{t_2}$ for $t_1<t_2$, that $w(t)$ is $\mathfrak{F}_t$-measurable, and that the process $w_h(t)=w(t+h)-w(t)$ does not depend on $\mathfrak{F}_h$, $0\leqq h\leqq T$. In this paragraph the functions $\varphi(t)$, $A(t,x)$ and $B(t,x)$ (with or without indices) are assumed defined for $t\in[0,T]$, $x\in(-\infty,\infty)$, measurable w.r.t. all their arguments (including the variable belonging to a probability space), and for fixed t and x, $\mathfrak{F}_t$-measurable. The solution $\eta(t)$ of (11) must be $\mathfrak{F}_t$-measurable for $t\in[0,T]$, the integral on the right in (1) must be defined, and (1) must be satisfied w.p. 1 for each $t\in[0,T]$. We will apply the methodology

used in the preceding paragraph to the existence and uniqueness problem for (1).

Theorem 1. *Let* $\sup_{0 \leqq t \leqq T} \mathbb{M}\varphi^2(t) < \infty$ *and assume there exists a constant* K *such that, w.p. 1*

$$|A(t,x)|^2 + |B(t,x)|^2 \leqq K^2(1+x^2),$$

$$|A(t,x) - A(t,y)| + |B(t,x) - B(t,y)| \leqq K|x-y|.$$

Then (1) *has a solution satisfying* $\sup_{0 \leqq t \leqq T} \mathbb{M}\eta^2(t) < \infty$ *and any two solutions* $\eta_1(t)$ *and* $\eta_2(t)$ *for which this is true are stochastically equivalent, i.e.,* $\mathbb{P}\{\eta_1(t) = \eta_2(t)\} = 1$ *for all* $t \in [0, T]$.

Proof. Set $\eta_0(t) = \varphi(t)$,

$$\eta_n(t) = \varphi(t) + \int_0^t A(s, \eta_{n-1}(s))\, ds + \int_0^t B(s, \eta_{n-1}(s))\, dw(s),$$

then

$$\begin{aligned} \mathbb{M}[\eta_n(t) - \eta_{n-1}(t)]^2 &\leqq 2\mathbb{M}\left\{\int_0^t [A(s, \eta_{n-1}(s)) - A(s, \eta_{n-2}(s))]\, ds\right\}^2 \\ &\quad + 2\mathbb{M}\left\{\int_0^T [B(s, \eta(s)) - B(s, \eta_{n-2}(s))]\, dw(s)\right\}^2 \\ &\leqq 2(t+1)K^2 \int_0^t \mathbb{M}[\eta_{n-1}(s) - \eta_{n-2}(s)]^2\, ds \end{aligned}$$

and

$$\begin{aligned} \mathbb{M}[\eta_1(s) - \eta_0(s)]^2 &\leqq 2tK^2 \int_0^t \mathbb{M}A^2(s, \eta_0(s))\, ds + 2K^2 \int_0^t \mathbb{M}B^2(s, \eta_0(s))\, ds \\ &\leqq 2(t+1)K^2 \int_0^t \mathbb{M}(1+\varphi^2(s))\, ds. \end{aligned}$$

Hence for some L

$$\mathbb{M}[\eta_n(t) - \eta_{n-1}(t)]^2 \leqq \frac{L^{n-1}}{(n-1)!}.$$

The remainder of the existence and uniqueness proof is carried out as in Theorem 1 § 6.

Remark 1. If $\varphi(t)$ is continuous w.p. 1, then there exists, also w.p. 1, a continuous solution of (1). If $\eta_1(t)$ and $\eta_2(t)$ are two such solutions, then $\mathbb{P}\{\sup_{0 \leqq t \leqq T} |\eta_1(t) - \eta_2(t)| > 0\} = 0$.

Remark 2. If $\varphi(t)$ is bounded w.p. 1, then it is sufficient for the theorem to hold that $A(s,x)$ and $B(s,x)$ satisfy $A^2(s,x) + B^2(s,x) \leqq K^2(1+x^2)$ for some K and that for each $N > 0$ there exist an L_N so that for $|x| \leqq N$

$$|A(s,x)-A(s,y)|+|B(s,x)-B(s,y)| \leqq L_N |x-y|.$$

In this case there exists w.p. 1 a unique bounded solution.

Remark 3. If $\sup_{0\leqq t\leqq T} \mathbb{M}|\varphi(t)|^m < C$ $(m>0)$ then it is possible to find a C_1, depending only on C, K, T and m, for which $\sup_{0\leqq t\leqq T} \mathbb{M}|\eta(t)|^m < C_1$.

We now establish a limit theorem for equations of the form (1).

Theorem 2. *Let $\eta_n(t)$, $n=0,1,2,\dots$ be solutions of*

$$\eta_n(t)=\varphi_n(t)+\int_0^t A_n(s,\eta_n(s))\,ds+\int_0^t B_n(s,\eta_n(s))\,dw(s), \tag{2}$$

where the coefficients $A_n(s,x)$ and $B_n(s,x)$ satisfy the conditions of Theorem 1 for a single K, $\varphi_n(t)$ is continuous w.p. 1, and $\sup_{0\leqq t\leqq T}\mathbb{M}\varphi_n^2(t)\leqq C$. If for each $N>0$, $s\in[0,T]$ and $\varepsilon>0$

$$\lim_{n\to\infty}\mathbb{P}\{\sup_{|x|\leqq N}(|A_n(s,x)-A_0(s,x)|+|B_n(s,x)-B_0(s,x)|)>\varepsilon\}=0$$

and

$$\lim_{n\to\infty}\sup_{0\leqq t\leqq T}\mathbb{M}(\varphi_n(t)-\varphi_0(t))^2=0,$$

then

$$\lim_{n\to\infty}\sup_{0\leqq t\leqq T}\mathbb{M}(\eta_n(t)-\eta_0(t))^2=0.$$

Proof. We write

$$\eta_n(t)-\eta_0(t)=\zeta_n(t)+\int_0^t[A_n(s,\eta_n(s))-A_n(s,\eta_0(s))]\,ds$$
$$+\int_0^t[B_n(s,\eta_n(s))-B_n(s,\eta_0(s))]\,dw(s),$$

where

$$\zeta_n(t)=\varphi_n(t)-\varphi_0(t)+\int_0^t[A_n(s,\eta_0(s))-A_0(s,\eta_0(s))]\,ds$$
$$+\int_0^t[B_n(s,\eta_0(s))-B_0(s,\eta_0(s))]\,dw(s).$$

Using the Lipschitz condition we find that for $L=3(T+1)K^2$

$$\mathbb{M}[\eta_n(t)-\eta_0(t)]^2\leqq 3\mathbb{M}\zeta_n^2(t)+L\int_0^t\mathbb{M}[\eta_n(s)-\eta_0(s)]^2\,ds.$$

On the basis of Lemma 1, §6

$$\mathbb{M}[\eta_n(t)-\eta_0(t)]^2\leqq 3\mathbb{M}\zeta_n^2(t)+L\int_0^t e^{L(t-s)}\mathbb{M}\zeta_n^2(s)\,ds.$$

Hence to prove the theorem it is sufficient to show that $\sup_t \zeta_n^2(t) \to 0$ and to use the fact that by assumption, $\sup_t \mathbb{M}[\varphi_n(t)-\varphi_0(t)]^2 \to 0$. Furthermore,

$$\mathbb{M}\int_0^t [A_n(s,\eta_0(s))-A_0(s,\eta_0(s))]\,ds \leqq t\mathbb{M}\int_0^t [A_n(s,\eta_0(s))-A_0(s,\eta_0(s))]^2\,ds.$$

The term under the integral sign on the right is bounded by $2K^2(1+\eta_0^2(s))$, for which $\mathbb{M}\int_0^t [1+\eta_0^2(s)]\,ds<\infty$ and w.p. 1 it tends to zero. Hence from Lebesque's theorem

$$\sup_{0\leqq t\leqq T} \mathbb{M}\left[\int_0^t (A_n(s,\eta_0(s))-A_0(s,\eta_0(s)))\,ds\right]^2$$
$$\leqq T\int_0^T \mathbb{M}[A_n(s,\eta_0(s))-A_0(s,\eta_0(s))]^2\,ds \to 0.$$

Using Theorem 1 § 3 we can show that

$$\mathbb{M}\sup_{0\leqq t\leqq T}\left\{\int_0^t [B_n(s,\eta_0(s))-B_0(s,\eta_0(s))]\,dw(s)\right\}^2$$
$$\leqq 4\mathbb{M}\int_0^T [B_n(s,\eta_0(s))-B_0(s,\eta_0(s))]^2\,ds \to 0.$$

This completes the proof of the theorem.

Remark 4. One can prove in a completely analogous manner that if $\mathbb{M}\sup_{0\leqq t\leqq T}|\varphi_n(t)-\varphi_0(t)|^2\to 0$, then $\mathbb{M}\sup_{0\leqq t\leqq T}|\eta_n(t)-\eta_0(t)|^2\to 0$.

Corollary 1. *Assume the coefficients in* (1) *depend on a parameter* α *which varies through some set of numbers* G*:*

$$\eta_\alpha(t)=\varphi_\alpha(t)+\int_0^t A_\alpha(s,\eta_\alpha(s))\,ds+\int_0^t B_\alpha(s,\eta_\alpha(s))\,dw(s), \tag{3}$$

where

$$\lim_{\alpha'\to\alpha}\sup_t \mathbb{M}[\varphi_{\alpha'}(t)-\varphi_\alpha(t)]^2=0,$$

and that for $\varepsilon>0$, $N>0$

$$\lim_{\alpha'\to\alpha}\mathbb{P}\{\sup_{|x|\leqq N}(|A_{\alpha'}(s,x)-A_\alpha(s,x)|+|B_{\alpha'}(s,x)-B_\alpha(s,x)|)>\varepsilon\}=0$$

and finally that the coefficients $A_\alpha(s,x)$ *and* $B_\alpha(s,x)$ *satisfy the assumptions of Theorem 1 for a single* K*. Then* $\eta_\alpha(t)$ *is continuous in mean square w.r.t.* α*; moreover,*

$$\lim_{\alpha'\to\alpha}\sup_{0\leqq t\leqq T}\mathbb{M}[\eta_{\alpha'}(t)-\eta_\alpha(t)]^2=0.$$

This follows immediately from an application of Theorem 2 to the sequence $\eta_{\alpha_n}(t)$, where $\alpha_n \to \alpha$.

We now investigate some consequences of Theorem 2 for solutions of stochastic differential equations of the form (1) in § 6.

Theorem 3. *Let $\eta_n(t)$, $n=0, 1, \ldots$, be solutions of the stochastic equations*

$$\eta_n(t)=\eta_n(0)+\int_0^t a_n(s, \eta_n(s))\, ds+\int_0^t \sigma_n(s, \eta_n(s))\, dw(s),$$

assume $\eta_n(0)$ converges in probability to $\eta_0(0)$, $a_n(s, x)$ and $\sigma_n(s, x)$ satisfy for each n the assumptions of the existence and uniqueness theorem, that for each $N>0$

$$\lim_{n\to\infty} \sup_{|x|\leqq N} \big(|a_n(s, x)-a_0(s, x)|+|\sigma_n(s, x)-\sigma_0(s, x)|\big)=0$$

and that there exists a K such that for all n with $s\in[0, T]$

$$|a_n(s, x)|^2+|\sigma_n(s, x)|^2 \leqq K^2(1+x^2). \tag{4}$$

Then $\sup_{0\leqq t\leqq T} |\eta_n(t)-\eta_0(t)| \to 0$ in probability for $n\to\infty$.

Proof. Set $g_N(x)=x$ for $|x|\leqq N$, $g_N(x)=N \operatorname{sign} x$ for $|x|>N$, $\eta_n^{(N)}(0)=g_N(\eta_n(0))$, $a_n^{(N)}(s, x)=a_n(s, g_N(x))$ and $\sigma_n^{(N)}(s, x)=\sigma_n(s, g_N(x))$. Let $\eta_n^{(N)}(t)$ be a solution of

$$\eta_n^{(N)}(t)=\eta_n^{(N)}(0)+\int_0^t a_n^{(N)}(s, \eta_n^{(N)}(s))\, ds+\int_0^t \sigma_n^{(N)}(s, \eta_n^{(N)}(s))\, dw(s).$$

From Theorem 2 § 6

$$\mathbb{P}\{\sup_{0\leqq t\leqq T} |\eta_n^{(N)}(t)-\eta_n(t)|>0\} \leqq \mathbb{P}\{\sup_{0\leqq t\leqq T} |\eta_n^{(N)}(t)|>N\}.$$

Thus,

$$\begin{aligned}\mathbb{P}\{\sup_{0\leqq t\leqq T} |\eta_n(t)-\eta_0(t)|>\varepsilon\} \leqq{}& \mathbb{P}\{\sup_{0\leqq t\leqq T} |\eta_n^{(N)}(t)-\eta_0^{(N)}(t)|>\varepsilon\}\\ &+\mathbb{P}\{\sup_{0\leqq t\leqq T} |\eta_n^{(N)}(t)|>N\}\\ &+\mathbb{P}\{\sup_{0\leqq t\leqq T} |\eta_0^{(N)}(t)|>N\}.\end{aligned} \tag{5}$$

The first summand on the right side of (5) tends to zero as a result of Remark 4. From the inequality at (4) and the boundedness in probability of $\eta_0(0)$ (this is consequence of the convergence in probability of $\eta_n(0)$ to $\eta_0(0)$) we can prove as in Theorem 3 § 6 that, uniformly w.r.t. n,

$$\lim_{N\to\infty} \mathbb{P}\{\sup_{0\leqq t\leqq T} |\eta_n^{(N)}(t)|>N\}=0.$$

Letting $n\to\infty$ in (5) and then $N\to\infty$, we obtain the theorem.

We now turn to the question of differentiability of the solution of a stochastic differential equation with respect to a parameter. We will consider mean square derivatives with respect to a parameter: if ξ_α is a family of random variables, then $\frac{\partial}{\partial\alpha}\xi_\alpha$ is a variable for which

$$\lim_{\Delta\alpha\to 0} \mathbb{M}\left|\frac{\xi_{\alpha+\Delta\alpha}-\xi_\alpha}{\Delta\alpha}-\frac{\partial}{\partial\alpha}\xi_\alpha\right|^2=0.$$

Theorem 4. *Let $\eta_\alpha(t)$ be solutions of (3), whereby the following conditions are satisfied:*

1) $\sup_{0\leq t\leq T}\mathbb{M}\varphi_\alpha^2(t)<\infty$, $\frac{\partial}{\partial\alpha}\varphi_\alpha(t)$ *exists, and*

$$\lim_{\Delta\alpha\to 0}\sup_{0\leq t\leq T}\mathbb{M}\left(\frac{\partial}{\partial\alpha}\varphi_\alpha(t)-\frac{1}{\Delta\alpha}[\varphi_{\alpha+\Delta\alpha}(t)-\varphi_\alpha(t)]\right)^2=0;$$

2) $\frac{\partial}{\partial\alpha}A_\alpha(s,x)$ *and* $\frac{\partial}{\partial\alpha}B_\alpha(s,x)$ *exist and*

$$\lim_{\Delta\alpha\to 0}\mathbb{M}\int_0^T\left\{\left[\frac{A_{\alpha+\Delta\alpha}(s,\eta_\alpha(s))-A_\alpha(s,\eta_\alpha(s))}{\Delta\alpha}-\frac{\partial}{\partial\alpha}A_\alpha(s,\eta_\alpha(s))\right]^2\right.$$
$$\left.+\left[\frac{B_{\alpha+\Delta\alpha}(s,\eta_\alpha(s))}{\Delta\alpha}-\frac{B_\alpha(s,\eta_\alpha(s))}{\Delta\alpha}-\frac{\partial}{\partial\alpha}B_\alpha(s,\eta_\alpha(s))\right]^2\right\}ds=0;$$

3) $\frac{\partial}{\partial x}A_\alpha(s,x)$ *and* $\frac{\partial}{\partial x}B_\alpha(s,x)$ *exist and are w.p. 1 continuous w.r.t. all arguments, and for some K*

$$\mathbb{P}\left\{\left|\frac{\partial}{\partial x}A_\alpha(s,x)\right|\leq K\right\}=1,\quad \mathbb{P}\left\{\left|\frac{\partial}{\partial x}B_\alpha(s,x)\right|\leq K\right\}=1.$$

Then $\eta_\alpha(t)$ is differentiable w.r.t. α.

Proof. Denote

$$\zeta_{\alpha,\Delta\alpha}(t)=\frac{1}{\Delta\alpha}(\eta_{\alpha+\Delta\alpha}(t)-\eta_\alpha(t)).$$

This random process satisfies

$$\zeta_{\alpha,\Delta\alpha}(t)=\frac{\varphi_{\alpha+\Delta\alpha}(t)-\varphi_\alpha(t)}{\Delta\alpha}$$
$$+\int_0^t\frac{A_{\alpha+\Delta\alpha}(s,\eta_{\alpha+\Delta\alpha}(s))-A_{\alpha+\Delta\alpha}(s,\eta_\alpha(s))}{\eta_{\alpha+\Delta\alpha}(s)-\eta_\alpha(s)}\zeta_{\alpha,\Delta\alpha}(s)\,ds$$
$$+\int_0^t\frac{B_{\alpha+\Delta\alpha}(s,\eta_{\alpha+\Delta\alpha}(s))-B_{\alpha+\Delta\alpha}(s,\eta_\alpha(s))}{\eta_{\alpha+\Delta\alpha}(s)-\eta_\alpha(s)}\zeta_{\alpha,\Delta\alpha}(s)\,dw(s)$$

$$+\int_0^t \frac{A_{\alpha+\Delta\alpha}(s,\eta_\alpha(s))-A_\alpha(s,\eta_\alpha(s))}{\Delta\alpha}\,ds$$
$$+\int_0^t \frac{B_{\alpha+\Delta\alpha}(s,\eta_\alpha(s))-B_\alpha(s,\eta_\alpha(s))}{\Delta\alpha}\,dw(s). \tag{6}$$

The symbol $\zeta_{\alpha,0}(t)$ will denote the solution of

$$\begin{aligned}\zeta_{\alpha,0}(t)=&\frac{\partial}{\partial\alpha}\varphi_\alpha(t)+\int_0^t\frac{\partial}{\partial\alpha}A_\alpha(s,\eta_\alpha(s))\,ds+\int_0^t\frac{\partial}{\partial\alpha}B_\alpha(s,\eta_\alpha(s))\,dw(s)\\&+\int_0^t\frac{\partial}{\partial x}A_\alpha(s,\eta_\alpha(s))\,\zeta_{\alpha,0}(s)\,ds+\int_0^t\frac{\partial}{\partial x}B_\alpha(s,\eta_\alpha(s))\,\zeta_{\alpha,0}(s)\,dw(s).\end{aligned}\tag{7}$$

We introduce the functions

$$\begin{aligned}\tilde\varphi_{\Delta\alpha}(t)=&\frac{\varphi_{\alpha+\Delta\alpha}(t)-\varphi_\alpha(t)}{\Delta\alpha}+\int_0^t\frac{A_{\alpha+\Delta\alpha}(s,\eta_\alpha(s))-A_\alpha(s,\eta_\alpha(s))}{\Delta\alpha}\,ds\\&+\int_0^t\frac{B_{\alpha+\Delta\alpha}(s,\eta_\alpha(s))-B_\alpha(s,\eta_\alpha(s))}{\Delta\alpha}\,dw(s),\end{aligned}$$

if $\Delta\alpha\neq 0$;

$$\tilde\varphi_{\Delta\alpha}(t)=\frac{\partial}{\partial\alpha}\varphi_\alpha(t)+\int_0^t\frac{\partial}{\partial\alpha}A_\alpha(s,\eta_\alpha(s))\,ds+\int_0^t\frac{\partial}{\partial\alpha}B_\alpha(s,\eta_\alpha(s))\,dw(s),$$

if $\Delta\alpha=0$;

$$\tilde A_{\Delta\alpha}(s,x)=\frac{A_{\alpha+\Delta\alpha}(s,\eta_{\alpha+\Delta\alpha}(s))-A_{\alpha+\Delta\alpha}(s,\eta_\alpha(s))}{\eta_{\alpha+\Delta\alpha}(s)-\eta_\alpha(s)}\,x,$$

if $\Delta\alpha\neq 0$;

$$\tilde A_{\Delta\alpha}(s,x)=\frac{\partial}{\partial x}A_\alpha(s,\eta_\alpha(s))\,x,$$

if $\Delta\alpha=0$;

$$\tilde B_{\Delta\alpha}(s,x)=\frac{B_{\alpha+\Delta\alpha}(s,\eta_{\alpha+\Delta\alpha}(s))-B_{\alpha+\Delta\alpha}(s,\eta_\alpha(s))}{\eta_{\alpha+\Delta\alpha}(s)-\eta_\alpha(s)}\,x,$$

if $\Delta\alpha\neq 0$; and

$$\tilde B_{\Delta\alpha}(s,x)=\frac{\partial}{\partial x}B_\alpha(s,\eta_\alpha(s))\,x,$$

if $\Delta\alpha=0$.

Using these functions we can express (6) and (7) in the form

$$\zeta_{\alpha,\Delta\alpha}(s)=\tilde\varphi_{\Delta\alpha}(t)+\int_0^t\tilde A_{\Delta\alpha}(s,\zeta_{\alpha,\Delta\alpha}(s))\,ds+\int_0^t\tilde B_{\Delta\alpha}(s,\zeta_{\alpha,\Delta\alpha}(s))\,dw(s). \tag{8}$$

To prove the theorem it is sufficient to show that

$$\mathbb{M}[\zeta_{\alpha,\Delta\alpha}(t)-\zeta_{\alpha,0}]^2\to 0$$

for $\Delta\alpha\to 0$. To do this it is sufficient, due to Corollary 1, to verify that

$$\lim_{\Delta\alpha\to 0}\sup_{0\leq t\leq T}\mathbb{M}|\tilde{\varphi}_{\Delta\alpha}(t)-\tilde{\varphi}_0(t)|^2=0, \tag{9}$$

and that for each $N>0$, $\varepsilon>0$,

$$\lim_{n\to\infty}\mathbb{P}\{\sup_{|x|\leq N}(|\tilde{A}_{\Delta\alpha}(s,x)-\tilde{A}_0(s,x)|+|\tilde{B}_{\Delta\alpha}(s,x)-\tilde{B}_0(s,x)|)>\varepsilon\}=0. \tag{10}$$

Relation (10) follows from Condition 3 and the linear dependence of $\tilde{A}_{\Delta\alpha}(s,x)$ and $\tilde{B}_{\Delta\alpha}(s,x)$ on x; (9) follows from Conditions 1 and 2. The theorem is completed.

Remark 5. In order to find $\frac{\partial}{\partial\alpha}\eta_\alpha(t)$ it is necessary to differentiate (3) w.r.t. α. One obtains thereby a linear stochastic equation of the form (7).

Remark 6. If the coefficients of (7) are such that the conditions of Theorem (4) are satisfied, then $\frac{\partial^2}{\partial\alpha^2}\eta_\alpha(t)$ exists.

Remark 7. If $\mathbb{M}(\eta_\alpha(t))^m$ is bounded for each $m>0$, then in order to satisfy Condition 2 of Theorem 4 it is sufficient to require that for some $m>0$ and $r>2$ there exists an L for which

$$\mathbb{M}\sup_x\left(\frac{|A_{\alpha+\Delta\alpha}(s,x)-A_\alpha(s,x)|+|B_{\alpha+\Delta\alpha}(s,x)-B_\alpha(s,x)|}{\Delta\alpha(1+|x|^m)}\right)^r\leq L$$

and that for each $N>0$

$$\sup_{|x|\leq N}\left[\left|\frac{A_{\alpha+\Delta\alpha}(s,x)-A_\alpha(s,x)}{\Delta\alpha}-\frac{\partial}{\partial\alpha}A_\alpha(s,x)\right|\right.$$
$$\left.+\left|\frac{B_{\alpha+\Delta\alpha}(s,x)-B_\alpha(s,x)}{\Delta\alpha}-\frac{\partial}{\partial\alpha}B_\alpha(s,x)\right|\right]\to 0$$

in probability. In fact, we then have

$$\mathbb{M}\int_0^T\left[\frac{A_{\alpha+\Delta\alpha}(s,\eta_\alpha(s))-A_\alpha(s,\eta_\alpha(s))}{\Delta\alpha}-\frac{\partial}{\partial\alpha}A_\alpha(s,\eta_\alpha(s))\right]^2ds\to 0,$$

since the function under the integral sign converges to zero in probability and since from Hölder's inequality for $2<r_1<r$

$$\mathbb{M}\int_0^T\left[\frac{A_{\alpha+\Delta\alpha}(s,\eta_\alpha(s))-A_\alpha(s,\eta_\alpha(s))}{\Delta\alpha}-\frac{\partial}{\partial\alpha}A_\alpha(s,\eta_\alpha(s))\right]^{r_1}ds$$

$$\leqq\left\{\mathbb{M}\int_0^T\left[\frac{A_{\alpha+\Delta\alpha}(s,\eta_\alpha(s))-A_\alpha(s,\eta_\alpha(s))-\Delta\alpha\frac{\partial}{\partial\alpha}A_\alpha(s,\eta_\alpha(s))}{\Delta\alpha(1+|\eta_\alpha(s)|^m)}\right]^{r}ds\right\}^{\frac{r_1}{r}}$$

$$\times\left\{\mathbb{M}\int_0^T(1+|\eta_\alpha(s)|^m)^{\frac{r}{r_1-r}}ds\right\}^{1-\frac{r_1}{r}}.$$

Consequently, we can use the theorem on passage to the limit under the integral sign. Analogous considerations apply to $B_\alpha(s,x)$.

Derivatives of the solutions of stochastic differential equations can be used to construct asymptotic expansions for small parameter by means of Taylor's theorem. We consider for the sake of illustration two simple examples.

Example 1. Let

$$d\eta_\varepsilon(t)=\varepsilon\, a(t,\eta_\varepsilon(t))\,dt+dw(t),$$

where ε is a small parameter. The solution of this equation for $\varepsilon=0$ is $\eta_0(t)=\eta(0)+w(t)$. For $\frac{\partial}{\partial\varepsilon}\eta_\varepsilon(t)$ we obtain the equation

$$\frac{\partial}{\partial\varepsilon}\eta_\varepsilon(t)=\int_0^t a(s,\eta_\varepsilon(s))\,ds+\varepsilon\int_0^t a'_x(s,\eta_\varepsilon(s))\frac{\partial}{\partial\varepsilon}\eta_\varepsilon(s)\,ds,$$

from which we find

$$\frac{\partial}{\partial\varepsilon}\eta_\varepsilon(t)|_{\varepsilon=0}=\int_0^t a(s,\eta(0)+w(s))\,ds.$$

This means that

$$\eta_\varepsilon(t)=\eta(0)+w(t)+\varepsilon\int_0^t a(s,\eta(0)+w(s))\,ds+o(\varepsilon).$$

Example 2. $d\eta_\varepsilon(t)=a(t,\eta_\varepsilon(t))\,dt+\varepsilon\,dw(t)$. Then $\eta_0(t)$ will be a solution of the ordinary differential equation (possibly with random initial condition) $\frac{\partial\eta_0(t)}{\partial t}=a(t,\eta_0(t))$.

The process $\eta^{(1)}(t)=\frac{\partial}{\partial\varepsilon}\eta_\varepsilon(t)|_{\varepsilon=0}$ satisfies

$$d\eta^{(1)}(t)=a'_x(t,\eta_0(t))\,\eta^{(1)}(t)\,dt+dw(t)$$

with initial condition $\eta^{(1)}(0)=0$. The solution of this equation is

$$\eta^{(1)}(t)=\exp\left\{\int_0^t a'_x(s,\eta_0(s))\,ds\right\}\int_0^t\exp\left\{-\int_0^s a'_x(u,\eta_0(u))\,du\right\}dw(s).$$

We can again write down the solution of the initial equation with accuracy up to $o(\varepsilon)$.

§ 8. Dependence of the Solutions of Stochastic Differential Equations on the Initial Data

We again consider

$$d\eta(t)=a(t,\eta(t))\,dt+\sigma(t,\eta(t))\,dw(t). \tag{1}$$

We assume that the coefficients of this equation satisfy the assumptions of Theorem 1 § 6. Let $\eta_{z,s}(t)$ be the solution of

$$\eta_{z,s}(t)=z+\int_s^t a(u,\eta_{z,s}(u))\,du+\int_0^t \sigma(u,\eta_{z,s}(u))\,dw(u), \tag{2}$$

i.e., the solution of (1) on the interval $[s,T]$ satisfying the initial condition $\eta_{z,s}(s)=z$. We consider this equation as one of the form (3) in § 7, regarding z as a parameter. The coefficients of the equation are independent of the parameter and so if the derivatives $\partial a(s,x)/\partial x$ and $\partial\sigma(s,x)/\partial x$ exist and are continuous and bounded, we can show on the basis of Theorem 4 § 7 that the mean square derivative $\frac{\partial}{\partial z}\eta_{z,s}(t)$ also exists and satisfies

$$\begin{aligned}\frac{\partial}{\partial z}\eta_{z,s}(t)=1&+\int_s^t a'_x(u,\eta_{z,s}(u))\frac{\partial}{\partial z}\eta_{z,s}(u)\,du\\&+\int_s^t \sigma'_x(u,\eta_{z,s}(u))\frac{\partial}{\partial z}\eta_{z,s}(u)\,dw(u).\end{aligned} \tag{3}$$

It follows from Remark 3 § 7 that for each $m>0$ there exists a constant C (in addition to m, this constant depends also on the K entering into the Lipschitz condition and on T) for which

$$\sup_{s\leq t\leq T}\mathsf{M}\left|\frac{\partial}{\partial z}\eta_{z,s}(t)\right|^m<C.$$

Assume further that $\frac{\partial^2}{\partial x^2}a(s,x)$ and $\frac{\partial^2}{\partial x^2}\sigma(s,x)$ exist, are continuous, and satisfy for some m_2 and L_2 the inequality

$$\left|\frac{\partial^2}{\partial x^2}a(s,x)\right|+\left|\frac{\partial^2}{\partial x^2}\sigma(s,x)\right|\leq L_2(1+|x|^{m_2}).$$

We now show that when these assumptions hold $\frac{\partial^2}{\partial z^2}\eta_{z,s}(t)$ exists and satisfies

$$\begin{aligned}\frac{\partial^2}{\partial z^2}\eta_{z,s}(t)=&\int_s^t a''_{xx}(u,\eta_{z,s}(u))\left(\frac{\partial}{\partial z}\eta_{z,s}(u)\right)^2 du\\&+\int_s^t \sigma''_{xx}(u,\eta_{z,s}(u))\left(\frac{\partial}{\partial z}\eta_{z,s}(u)\right)^2 dw(u)\\&+\int_s^t a'_x(u,\eta_{z,s}(u))\frac{\partial^2}{\partial z^2}\eta_{z,s}(u)\,du\\&+\int_s^t \sigma'_x(u,\eta_{z,s}(u))\frac{\partial^2}{\partial z^2}\eta_{z,s}(u)\,dw(u).\end{aligned}\tag{4}$$

To show this we note that (4) reduces to (3) § 7 if we set $\alpha=z$, $\eta_\alpha(u)=\frac{\partial}{\partial z}\eta_{z,s}(u)$, $\varphi_\alpha(t)=1$, $A_\alpha(u,x)=a'_x(u,\eta_{z,s}(u))\,x$, $B_\alpha(u,x)=\sigma'_x(u,\eta_{z,s}(u))\,x$.

We verify that the assumptions of Theorem 4 and Remark 7 in § 7 are fulfilled. Since

$$\frac{A_{\alpha+\Delta\alpha}(u,x)-A_\alpha(u,x)}{\Delta\alpha}=x\,\frac{a'_x(u,\eta_{z+\Delta z,s}(u))-a'_x(u,\eta_{z,s}(u))}{\eta_{z+\Delta z,s}(u)-\eta_{z,s}(u)}\times\frac{\eta_{z+\Delta z,s}(u)-\eta_{z,s}(u)}{\Delta z},$$

we have

$$\left|\frac{A_{\alpha+\Delta\alpha}(u,x)-A_\alpha(u,x)}{\Delta\alpha(1+|x|)}\right|\leq L_2\left(1+|\eta_{z+\Delta z,s}(u)|^{m_2}+|\eta_{z,s}(u)|^{m_2}\right)\times\left|\frac{\eta_{z+\Delta z,s}(u)-\eta_{z,s}(u)}{\Delta z}\right|$$

and all moments (of positive power) on the right side of this inequality are bounded in u (for $\eta_{z,s}(u)$ this follows from Theorem 4 § 6). Also,

$$\zeta_{\Delta z}(u)=\frac{\eta_{z+\Delta z,s}(u)-\eta_{z,s}(u)}{\Delta z}$$

satisfies

$$\begin{aligned}\zeta_{\Delta z}(t)=1&+\int_0^t\frac{a(u,\eta_{z+\Delta z,s}(u))-a(u,\eta_{z,s}(u))}{\eta_{z+\Delta z,s}(u)-\eta_{z,s}(u)}\zeta_{\Delta z}(u)\,du\\&+\int_0^t\frac{\sigma(u,\eta_{z+\Delta z,s}(u))-\sigma(u,\eta_{z,s}(u))}{\eta_{z+\Delta z,s}(u)-\eta_{z,s}(u)}\zeta_{\Delta z}(u)\,dw(u),\end{aligned}$$

so that the boundedness of all moments $\zeta_{\Delta z}(u)$ results from Remark 3 § 7.

Using reasoning similar to the preceding we can establish the existence of higher order derivatives with corresponding assumptions on the coefficients. Typical is the following

Theorem 1. *Assume the coefficients of* (1) $a(t, x)$ *and* $\sigma(t, x)$ *satisfy the conditions of Theorem 1 § 6, that the derivatives* $\frac{\partial^k}{\partial x^k} a(t, x)$, $\frac{\partial^k}{\partial x^k}\sigma(t, x)$, $k=1, 2, \dots, r$ *exist, are continuous and satisfy for some* $m_k>0$ *and* C_k

$$\left|\frac{\partial^k}{\partial x^k} a(t, x)\right| + \left|\frac{\partial^k}{\partial x^k}\sigma(t, x)\right| \leqq C_k(1+|x|^{m_k}), \qquad k=2, \dots, r.$$

Then the solution of (2) $\eta_{z,s}(t)$ *is r-times differentiable w.r.t. z, whereby for each* $m>0$

$$\lim_{\Delta z\to 0} \mathbb{M}\left|\frac{1}{\Delta z}\left[\frac{\partial^{k-1}}{\partial z^{k-1}}\eta_{z+\Delta z, s}(t) - \frac{\partial^{k-1}}{\partial z^{k-1}}\eta_{z,s}(t)\right] - \frac{\partial^k}{\partial z^k}\eta_{z,s}(t)\right|^m = 0,$$

and for each $m>0$ *and* $k\leqq r$ *there exists a constant* $C_{m,k}$ *for which*

$$\mathbb{M}\left|\frac{\partial^k}{\partial z^k}\eta_{z,s}(t)\right|^m \leqq C_{m,k},$$

moreover, $\frac{\partial^k}{\partial z^k}\eta_{z,s}(t)$ *is continuous w.r.t. z in mean square.*

Remark 1. We now investigate the dependence of the solution of (2) on s. Let $s\leqq s_1\leqq t\leqq T$. Then

$$\eta_{z,s}(t) = \eta_{z,s}(s_1) + \int_{s_1}^{t} a(u, \eta_{z,s}(u))\, du + \int_{s_1}^{t} \sigma(u, \eta_{z,s}(u))\, dw(u),$$

$$\eta_{z_1,s_1}(t) = z_1 + \int_{s_1}^{t} a(u, \eta_{z_1,s_1}(u))\, du + \int_{s_1}^{t} \sigma(u, \eta_{z_1,s_1}(u))\, dw(u).$$

Then for some L

$$\mathbb{M}|\eta_{z_1,s_1}(t) - \eta_{z,s}(t)|^2 \leqq L\Big(\mathbb{M}|\eta_{z,s}(s_1) - z_1|^2 + \int_{s_1}^{t} (|\eta_{z_1,s_1}(u) - \eta_{z,s}(u)|^2\, du\Big)$$

and from Lemma 1 § 6 for some C we have

$$\begin{aligned}\mathbb{M}|\eta_{z_1,s_1}(t) - \eta_{z,s}(t)|^2 &\leqq C\mathbb{M}|\eta_{z,s}(s_1) - z_1|^2\\ &\leqq 2C\mathbb{M}|\eta_{z,s}(s_1) - z|^2 + 2C|z - z_1|^2\\ &= O\big((s - s_1)^2 + (z - z_1)^2\big)\end{aligned}$$

(we used Theorem 4 § 6 here). One establishes analogously that the derivatives $\frac{\partial^k}{\partial z^k}\eta_{z,s}(t)$ are also mean square continuous w.r.t. z and s.

Corollary 1. *Let $f(x)$ be r-times differentiable and for some $m>0$ satisfy $|f^{(k)}(x)| \leqq L(1+|x|^m)$. Then under the conditions of the preceding theorem the function $\Phi_s(z)=\mathbb{M} f(\eta_{z,s}(t))$ will be r-times differentiable w.r.t. z.*

In fact,

$$\frac{\Phi_s(z+\Delta z)-\Phi_s(z)}{\Delta z}=\mathbb{M}\, \frac{f(\eta_{z+\Delta z,s}(t))-f(\eta_{t,s}(t))}{\eta_{z+\Delta z,s}(t)-\eta_{z,s}(t)} \cdot \frac{\eta_{z+\Delta z,s}(t)-\eta_{z,s}(t)}{\Delta z}.$$

Hence,

$$\begin{aligned}
&\left|\frac{\Phi_s(z+\Delta z)-\Phi_s(z)}{\Delta z}-\mathbb{M} f'(\eta_{z,s}(t)) \frac{\partial}{\partial z} \eta_{z,s}(t)\right| \\
&\quad \leqq \mathbb{M}\left|\frac{f(\eta_{z+\Delta z,s}(t))-f(\eta_{z,s}(t))}{\eta_{z+\Delta z,s}(t)-\eta_{z,s}(t)}\left[\frac{\eta_{z+\Delta z,s}(t)-\eta_{z,s}(t)}{\Delta z}-\frac{\partial}{\partial z} \eta_{z,s}(t)\right]\right| \\
&\qquad +\mathbb{M}\left|\frac{\partial}{\partial z} \eta_{z,s}(t)\left[\frac{f(\eta_{z+\Delta z,s}(t))-f(\eta_{z,s}(t))}{\eta_{z+\Delta z,s}(t)-\eta_{z,s}(t)}-f'(\eta_{z,s}(t))\right]\right| \\
&\quad \leqq\left\{\mathbb{M}\left|\frac{f(\eta_{z+\Delta z,s}(t))-f(\eta_{z,s}(t))}{\eta_{z+\Delta z,s}(t)-\eta_{z,s}(t)}\right|^2\right\}^{\frac{1}{2}} \\
&\qquad \times\left\{\mathbb{M}\left(\frac{\eta_{z+\Delta z,s}(t)-\eta_{z,s}(t)}{\Delta z}-\frac{\partial}{\partial z} \eta_{z,s}(t)\right)^2\right\}^{\frac{1}{2}} \\
&\qquad +\left\{\mathbb{M}\left(\frac{\partial}{\partial z} \eta_{z,s}(t)\right)^2\right\}^{\frac{1}{2}}\left\{\mathbb{M}\left[\frac{f(\eta_{z+\Delta z,s}(t))-f(\eta_{z,s}(t))}{\eta_{z+\Delta z,s}(t)-\eta_{z,s}(t)}-f'(\eta_{z,s}(t))\right]^2\right\}^{\frac{1}{2}}.
\end{aligned}$$

Therefore

$$\frac{\partial}{\partial z} \Phi_s(z)=\mathbb{M} f'(\eta_{z,s}(t)) \frac{\partial}{\partial z} \eta_{z,s}(t).$$

One can show that the expression for $\frac{\partial}{\partial z} \Phi_s(z)$ under the expectation sign can again be differentiated, etc. Using Remark 1, we can prove that $\Phi_s(z)$ and its derivatives w.r.t. z are continuous in all their arguments.

Chapter 3. Solutions of Stochastic Differential Equations and Markov Diffusion Processes

§ 9. Markov Processes. Diffusion Processes

In various problems in the natural sciences it is often fitting to consider deterministic systems whose present state determines their future evolution. A natural generalization of these are stochastically determined systems which evolve in a random fashion, for which the current state completely determines the probability of occupying the various possible ones at all future times. Such systems are called *Markov* and can be described by means of Markov processes.

Definition 1. Let $\xi(t)$ be a random process defined on $[0, T]$ and $\mathfrak{F}_t$ the minimal σ-algebra with respect to which $\xi(s)$ are measurable for $s \leqq t$. We assume that there exists a function $P(t, x, s, A)$, defined for all x, $0 \leqq t \leqq s \leqq T$ and Borel sets A for which there holds w.p. 1

$$\mathbb{P}\{\xi(s) \in A / \mathfrak{F}_t\} = P(t, \xi(t), s, A) = \mathbb{P}\{\xi(s) \in A / \xi(t)\},$$

for any $0 \leqq t < s < T$ and Borel set A. Then the process $\xi(t)$ will be called a *Markov process* and the function $P(t, x, s, A)$ the *transition probability* for $\xi(t)$.

We will assume in what follows that the transition probability $P(t, x, s, A)$ for the process $\xi(t)$ satisfies the following conditions:

I. $P(t, x, s, A)$ is a Borel function of x;

II. $P(t, x, s, A)$ is a measure w.r.t. A for fixed t, x, s;

III. $\int P(t, x, s, dy)\, P(s, y, \tau, A) = P(t, x, \tau, A)$ for all x and $0 \leqq t < s < \tau \leqq T$.

Regarding II and III, we note that since $P(t, \xi(t), s, A)$ is a conditional distribution w.r.t. A, it is countably additive w.r.t. A. Moreover, using the properties of conditional probabilities and of the conditional expectation, we can write

$$\begin{aligned} P(t, \xi(t), \tau, A) &= \mathbb{P}\{\xi(\tau) \in A / \mathfrak{F}_t\} = \mathbb{M}(\mathbb{P}\{\xi(\tau) \in A / \mathfrak{F}_s\} / \mathfrak{F}_t) \\ &= \mathbb{M}(P(s, \xi(s), \tau, A) / \mathfrak{F}_t) = \mathbb{M}(P(s, \xi(s), \tau, A) / \xi(t)). \end{aligned}$$

Since

$$\mathbb{M}(\varphi(\xi(s)) / \xi(t)) = \int \varphi(y)\, P(t, \xi(t), s, dy),$$

we have
$$P(t, \xi(t), \tau, A) = \int P(s, y, \tau, A) P(t, \xi(t), s, dy).$$

Thus, Properties II and III follow from the definition of a Markov process if we weaken them somewhat: they are satisfied w.p. 1 if x is replaced by $\xi(t)$.

With the help of the transition probabilities of the process we can write down all its finite distributions if we know the distribution of $\xi(0)$. Denote $\mathbb{P}\{\xi(0)<x\}$ by $F_0(x)$. Let $0<t_1<t_2<\dots<t_n=T$ and $A_0, \dots, A_n$ be Borel sets on the real line. Then

$$\begin{aligned}\mathbb{P}\{&\xi(0)\in A_0, \xi(t_1)\in A_1, \dots, \xi(t_n)\in A_n\}\\ &= \int_{A_0} dF_0(x_0) \int_{A_1} P(0, x_0, t_1, dx_1) \dots \int_{A_n} P(t_{n-1}, x_{n-1}, t_n, dx_n).\end{aligned} \tag{1}$$

To show this we use the relations

$$\begin{aligned}\mathbb{P}\{&\xi(0)\in A_0, \dots, \xi(t_{n-1})\in A_{n-1}, \xi(t_n)\in A_n\}\\ &= \mathbb{M}\,\mathbb{P}\{\xi(t_n)\in A_n/\xi(t_{n-1})\} \prod_{i=0}^{n-1} \chi_{A_i}(\xi(t_i)),\end{aligned}$$

$$\chi_A(x)=1, \quad \text{if } x\in A, \quad \chi_A(x)=0, \quad \text{if } x\notin A.$$

If follows from (1) that for any Borel function $f(x_0, x_1, \dots, x_n)$

$$\begin{aligned}\mathbb{M} f(\xi(0), \xi(t_1), \dots, \xi(t_n)) &= \int \dots \int f(x_0, x_1, \dots, x_n)\\ &\times dF_0(x_0)\, P(0, x_0, t_1, dx_1) \dots P(t_{n-1}, x_{n-1}, t_n, dx_n),\end{aligned} \tag{2}$$

if the right side makes sense.

We assume that $\xi(t)$ is a Markov process on $[0, T]$ and $g(t, x)$ is defined for $t\in[0, T]$, $x\in(-\infty, \infty)$ and is for each $t\in[0, T]$ monotone in x. Then the process $\eta(t)=g(t, \xi(t))$ is also Markov with transition function $\tilde{P}(t, x, s, A)$ expressed in terms of the transition function $P(t, x, s, A)$ of $\xi(t)$ by means of

$$\tilde{P}(t, x, s, A) = P(t, g^{-1}(t, x), s, g^{-1}(s, A)),$$

where $g^{-1}(t, x)$ is the inverse w.r.t. x of $g(t, x)$ and $g^{-1}(s, A)$ denotes the set of y's for which $g(s, y)\in A$.

A basic role will be played in what follows by *Markov diffusions* (or symply *diffusions*).

Definition 2. A Markov process $\xi(t)$, $t\in[0, T]$ will be called a *diffusion* if its transition function $P(t, x, s, A)$ possesses the following properties:

1) for any $\varepsilon>0$ and $t\in[0, T]$, $x\in(-\infty, \infty)$

$$\lim_{\Delta\to 0} \frac{1}{\Delta} \int_{|x-y|>\varepsilon} P(t, x, t+\Delta, dy) = 0;$$

2) there exist functions $a(t,x)$ and $b(t,x)$ such that for all $\varepsilon>0$, $t\in[0,T]$ and $x\in(-\infty,\infty)$

a) $\lim_{\Delta\to 0}\frac{1}{\Delta}\int_{|x-y|\leqq\varepsilon}(y-x)P(t,x,t+\Delta,dy)=a(t,x)$,

b) $\lim_{\Delta\to 0}\frac{1}{\Delta}\int_{|x-y|<\varepsilon}(y-x)^2 P(t,x,t+\Delta,dy)=b(t,x)$.

$b(t,x)$ and $a(t,x)$ will be called, resp., the *coefficients of diffusion* and *displacement (drift)*.

The designation "diffusion process" refers to the fact that under certain conditions one can describe the movement of a diffusing particle rather exactly with their help. The displacement of such a particle during the time Δ, assuming it was at x at time t, can be represented as the sum of two displacements $a(t,x)\Delta+\delta x+o(\Delta)$, where $a(t,x)$ is the macroscopic velocity of the surrounding fluid and δx is a random displacement caused by collisions of the particle with the molecules undergoing the chaotic thermal movement of the fluid; Furthermore, $\mathbb{M}(\delta x)^2=b(t,x)\Delta+o(\Delta)$ ($b(t,x)$ is proportional to the mean energy of the molecules of the fluid in the vicinity of the point x at time t).

Remark 1. For $\xi(t)$ to be a diffusion it is sufficient that its transition probability satisfy the following assumptions:

1*) for some $\delta>0$

$$\lim_{\Delta\to 0}\frac{1}{\Delta}\int|x-y|^{2+\delta}P(t,x,t+\Delta,dy)=0,$$

2*) there exist functions $a(t,x)$ and $b(t,x)$ such that for all t, x

a) $\lim_{\Delta\to 0}\frac{1}{\Delta}\int(y-x)P(t,x,t+\Delta,dy)=a(t,x)$,

b) $\lim_{\Delta\to 0}\frac{1}{\Delta}\int(y-x)^2 P(t,x,t+\Delta,dy)=b(t,x)$.

In fact, in this case

$$\int_{|y-x|>\varepsilon}P(t,x,t+\Delta,dy)\leqq\frac{1}{\varepsilon^{2+\delta}}\int|y-x|^{2+\delta}P(t,x,t+\Delta,dy)=o(\Delta),$$

$$\left|\int_{|y-x|>\varepsilon}(y-x)P(t,x,t+\Delta,dy)\right|\leqq\frac{1}{\varepsilon^{1+\delta}}\int|y-x|^{2+\delta}P(t,x,t+\Delta,dy)=o(\Delta),$$

$$\int_{|y-x|>\varepsilon}(y-x)^2 P(t,x,t+\Delta,dy)=\frac{1}{\varepsilon^{\delta}}\int|y-x|^{2+\delta}P(t,x,t+\Delta,dy)=o(\Delta).$$

Conditions 1* and 2* are in many cases more convenient to verify.

Let $g(t,x)$ be twice continuously differentiable and monotone in x and continuously differentiable in t and let $\xi(t)$ be a diffusion. We will show that $\eta(t)=g(t,\xi(t))$ is also a diffusion and will calculate its coefficients. We have shown above that $\eta(t)$ is a Markov process. Furthermore,

$$\lim_{\Delta\to 0}\frac{1}{\Delta}\int_{|x-y|>\varepsilon} P(t,g^{-1}(t,x),t+\Delta,g^{-1}(t+\Delta,dy))$$
$$=\lim_{\Delta\to 0}\frac{1}{\Delta}\int_{|g(t,u)-g(t+\Delta,v)|>\varepsilon} P(t,u,t+\Delta,dv)=0,$$

$$\frac{1}{\Delta}\int_{|x-y|\leqq\varepsilon} P(t,g^{-1}(t,x),t+\Delta,g^{-1}(t+\Delta,dy))(y-x)$$
$$=\frac{1}{\Delta}\int_{|g(t,u)-g(t+\Delta,v)|\leqq\varepsilon}[g(t+\Delta,v)-g(t,u)]\,P(t,u,t+\Delta,dv)$$
$$=\frac{\partial}{\partial t}g(t,u)+\frac{1}{\Delta}\int_{|g(t,u)-g(t+\Delta,v)|\leqq\varepsilon}\frac{\partial}{\partial u}g(t,u)(v-u)\,P(t,u,t+\Delta,dv)$$
$$+\frac{1+o(\varepsilon)}{\Delta}\int_{|g(t,u)-g(t+\Delta,v)|\leqq\varepsilon}\frac{1}{2}\frac{\partial^2 g(t,u)}{\partial u^2}(v-u)^2\,P(t,u,t+\Delta,dv)+o(\Delta)$$
$$=\frac{\partial}{\partial t}g(t,u)+a(t,u)\frac{\partial g(t,u)}{\partial u}+\frac{1}{2}b(t,u)\frac{\partial^2 g(t,u)}{\partial u^2}+o(\Delta),$$

where $u=g^{-1}(t,x)$. Analogously

$$\frac{1}{\Delta}\int_{|x-y|\leqq\varepsilon}(y-x)^2\,P(t,g^{-1}(t,x),t+\Delta,g^{-1}(t+\Delta,dy))$$
$$=b(t,u)\left(\frac{\partial}{\partial u}g(t,u)\right)^2+o(\Delta).$$

Thus the coefficients for $\eta(t)=g(t,\xi(t))$ are defined by

$$\begin{aligned}\bar a(t,x)=&\frac{\partial}{\partial t}g(t,g^{-1}(t,x))+a(t,g^{-1}(t,x))\frac{\partial}{\partial x}g(t,g^{-1}(t,x))\\&+\frac{1}{2}b(t,g^{-1}(t,x))\frac{\partial^2}{\partial x^2}g(t,g^{-1}(t,x)),\end{aligned}\tag{3}$$

$$\bar b(t,x)=b(t,g^{-1}(t,x))\left[\frac{\partial}{\partial x}g(t,g^{-1}(t,x))\right]^2.\tag{4}$$

It is interesting to note the analogy between these formulas and (3), (4) § 5 (it is merely necessary to set $b=\sigma^2$). The preceding formulas show that when $b(t,x)$ is positive and continuously differentiable in both arguments the diffusion coefficient of the process $\eta(t)=g(t,\xi(t))$, where

$$g(t,x)=\int_0^x \frac{du}{\sqrt{b(t,u)}},$$

is identically equal to one.

§ 10. Diffusion Processes as Solutions of Stochastic Equations

Theorem 1. *Assume that $\eta(t)$ is a solution of*

$$d\eta(t)=a(t,\eta(t))\,dt+\sigma(t,\eta(t))\,dw(t), \tag{1}$$

whose coefficients satisfy the conditions of the existence and uniqueness theorem. Then $\eta(t)$ will be a Markov process whose transition probability is defined by

$$P(t,x,s,A)=\mathbb{P}(\eta_{x,t}(s)\in A), \tag{2}$$

where $\eta_{x,t}(s)$ is a solution of

$$\eta_{x,t}(s)=x+\int_t^s a(u,\eta_{x,t}(u))\,du+\int_t^s \sigma(u,\eta_{x,t}(u))\,dw(u) \tag{3}$$

on the interval $[t,T]$.

Proof. Denote by $\mathfrak{F}_t$ the minimal σ-algebra of events relative to which $\eta(0)$ and $w(s)$ for $s\leqq t$ are measurable, and G^t the σ-algebra generated by $w(s)-w(t)$ for $s\geqq t$. It is obvious that the events of the σ-algebra G^t are independent of those of $\mathfrak{F}_t$. The value of $\eta_{x,t}(s)$ is completely determined by the increments $w(u)-w(t)$ for $u\geqq t$ and is measurable w.r.t. G^t. We note that $\eta(s)=\eta_{\eta(t),t}(s)$ since for $s>t$ $\eta(s)$ and $\eta_{\eta(t),t}(s)$ satisfy

$$\eta(s)=\eta(t)+\int_t^s a(u,\eta(u))\,du+\int_t^s \sigma(u,\eta(u))\,dw(u), \tag{4}$$

whose solution is unique. Therefore, $\eta(s)=f(\eta(t),\omega)$, where $f(x,\omega)$ is a random function independent of the events of $\mathfrak{F}_t$. We now establish an auxiliary

Lemma 1. *Let $f(x,\omega)$ be a bounded, measurable random function of x, independent of events in $\mathfrak{F}_t$ and ζ a $\mathfrak{F}_t$-measurable random variable. Then*

$$\mathbb{M}(f(\zeta,\omega)/\mathfrak{F}_t)=g(\zeta), \quad \text{where } g(x)=Mf(x,\omega).$$

Proof. Assume $f(x,\omega)=\sum\psi_k(x)\,\lambda_k(\omega)$ (ψ_k a nonrandom function). Then for any random variable ζ_1, measurable w.r.t. $\mathfrak{F}_t$, we have

$$\mathbb{M}f(\zeta,\omega)\,\zeta_1=\mathbb{M}\sum\psi_k(\zeta)\,\lambda_k(\omega)\,\zeta_1=\mathbb{M}\sum\psi_k(\zeta)\,\zeta_1\,\mathbb{M}\,\lambda_k(\omega)=\mathbb{M}\,g(\zeta)\,\zeta_1,$$

since $g(x)$ is in this case equal to $\sum \psi_k(x) \mathbb{M} \lambda_k(\omega)$. Since such a sum can be approximated by an arbitrary measurable bounded function, we can prove the lemma in the general case by passage to the limit.

We return to the proof of the theorem. Using the lemma we find

$$\mathbb{M}(\chi_A(\eta(s))/\mathfrak{F}_t) = \mathbb{M}\,\chi_A(\eta_{x,t}(s))|_{x=\eta(t)} = P(t, x, s, A)|_{x=\eta(t)},$$

if $P(t, x, s, A)$ is defined by (2). The theorem is proved.

Theorem 2. *Let $a(t, x)$ and $\sigma(t, x)$ be continuous in both arguments and assume that for some K*

$$|a(t, x)|^2 + |\sigma(t, x)|^2 \leqq K(1 + |x|^2)$$

and that for each N there exists an L_N with $|x| \leqq N$, $|y| \leqq N$ for which

$$|a(t, x) - a(t, y)| + |\sigma(t, x) - \sigma(t, y)| \leqq L_N |x - y|.$$

Then the process $\eta(t)$ as a solution of (1) will be a diffusion with diffusion coefficient $b(t, x) = \sigma^2(t, x)$ and displacement coefficient $a(t, x)$.

Proof. It suffices to show that

$$\mathbb{M}(\eta_{x,t}(t+\Delta) - x)^4 = \int (y - x)^4 P(t, x, t+\Delta, dy) = o(\Delta),$$

$$\lim_{\Delta \to 0} \frac{1}{\Delta} (\eta_{x,t}(t+\Delta) - x) = \lim_{\Delta \to 0} \frac{1}{\Delta} \int (y - x) P(t, x, t+\Delta, dy) = a(t, x),$$

$$\lim_{\Delta \to 0} \frac{1}{\Delta} (\eta_{x,t}(t+\Delta) - x)^2 = \lim_{\Delta \to 0} \frac{1}{\Delta} \int (y - x)^2 P(t, x, t+\Delta, dy) = b(t, x).$$

From Theorem 4 § 6 there exists a constant K, independent of t and x, such that

$$\mathbb{M}[\eta_{x,t}(t+\Delta) - \eta_{x,t}(t)]^4 = \mathbb{M}(\eta_{x,t}(t+\Delta) - x)^4 \leqq K_1 \Delta^2 (1 + x^4).$$

Moreover,

$$\frac{1}{\Delta} \mathbb{M}(\eta_{x,t}(t+\Delta) - x) = \frac{1}{\Delta} \int_t^{t+\Delta} \mathbb{M}\, a(u, \eta_{x,t}(u))\, du$$

$$= \int_0^1 \mathbb{M}\, a(t + s\Delta, \eta_{x,t}(t + s\Delta))\, ds.$$

Since for $\Delta \to 0$ $a(t + s\Delta, \eta_{x,t}(t + s\Delta)) \to a(t, x)$ w.p. 1 and

$$|a(t + s\Delta, \eta_{x,t}(t + s\Delta))|^2 \leqq K_1(1 + |\eta_{x,t}(t + s\Delta)|^2),$$

whereby $\mathbb{M}\int_0^1 (1+|\eta_{x,t}(t+s\Delta)|^2)\,ds<\infty$, we find using Lebesque's theorem on exchange of limit and integration signs that

$$\lim_{\Delta\to 0}\int_0^1 \mathbb{M}\,a(t+s\Delta,\eta_{x,t}(t+s\Delta))\,ds=a(t,x).$$

Using Itô's formula, we write

$$\begin{aligned}\mathbb{M}(\eta_{x,t}(t+\Delta)-x)^2&=\mathbb{M}[\eta_{x,t}(t+\Delta)]^2-x^2-2x[\mathbb{M}\,\eta_{x,t}(t+\Delta)-x]\\&=\mathbb{M}\left[\int_t^{t+\Delta}[2\eta_{x,t}(u)\,a(u,\eta_{x,t}(u))+\sigma^2(u,\eta_{x,t}(u))]\,du\right.\\&\qquad\left.+\int_t^{t+\Delta}2\eta_{x,t}(u)\,\sigma(u,\eta_{x,t}(u))\,dw(u)\right]\\&\quad-2x(a(t,x)\Delta+o(\Delta)).\end{aligned}$$

In an analogous manner one can justify the following limit passage:

$$\begin{aligned}&\lim_{\Delta\to 0}\frac{1}{\Delta}\mathbb{M}(\eta_{x,t}(t+\Delta)-x)^2\\&\quad=\lim_{\Delta\to 0}\int_0^1[2\eta_{x,t}(t+s\Delta)\,a(t+s\Delta,\eta_{x,t}(t+s\Delta))+\sigma^2(t+s\Delta,\eta_{x,t}(t+s\Delta))]\,ds\\&\qquad-2x\,a(t,x)=\sigma^2(t,x).\end{aligned}$$

The theorem is proved.

Remark 1. Assume the conditions of Theorem 2 are fulfilled and that $f(t,x)$ is continuous in both arguments with, for some K and $m>0$ $|f(t,x)|\leqq K(1+|x|^m)$. Using the same considerations as in the proofs of Theorem 2 and Theorem 4 § 6, we can show that

$$\lim_{\Delta\to 0}\frac{1}{\Delta}\int_t^{t+\Delta}\mathbb{M}f(s,\eta_{x,t}(s))\,ds=\lim_{\Delta\to 0}\frac{1}{\Delta}\int_{t-\Delta}^{t}\mathbb{M}f(s,\eta_{x,t}(s))\,ds=f(t,x).$$

We now prove that, under certain conditions, a diffusion process satisfies a stochastic differential equation.

Lemma 2. *Assume that the process $\zeta(t)$ is defined on $[0,T]$ and $\mathfrak{F}_t$ is the σ-algebra generated by the variables $\zeta(s)$ for $s\leqq t$. If the following conditions are satisfied: a) $\zeta(t)$ is continuous w.p.1 and $\zeta(0)=0$; b) there exist random variables ξ_1 and ξ_2 for which*

$$\left|\frac{1}{\Delta}\mathbb{M}(\zeta(t+\Delta)-\zeta(t)/\mathfrak{F}_t)\right|\leqq\xi_1,\qquad\frac{1}{\Delta}\mathbb{M}([\zeta(t+\Delta)-\zeta(t)]^2/\mathfrak{F}_t)\leqq\xi_2$$

and $\mathbb{M}\,\xi_k<\infty$*; c) for each* $t\in[0,T]$, *w.p. 1*

$$\lim_{\Delta\to 0}\frac{1}{\Delta}\mathbb{M}(\zeta(t+\Delta)-\zeta(t)/\mathfrak{F}_t)=0,$$

$$\lim_{\Delta\to 0}\frac{1}{\Delta}\mathbb{M}([\zeta(t+\Delta)-\zeta(t)]^2/\mathfrak{F}_t)=1,$$

then $\zeta(t)$ *is a Wiener process.*

Proof. We show that under these assumptions the conditions of Theorem 1 § 1 are fulfilled. For $t>s$ set

$$\varphi(t)=\mathbb{M}(\zeta(t)-\zeta(s)/\mathfrak{F}_s).$$

Then

$$\lim_{\Delta\downarrow 0}\frac{\varphi(t+\Delta)-\varphi(t)}{\Delta}=\lim_{\Delta\downarrow 0}\mathbb{M}\left(\frac{\zeta(t+\Delta)-\zeta(t)}{\Delta}/\mathfrak{F}_s\right)$$
$$=\lim_{\Delta\downarrow 0}\mathbb{M}\left(\mathbb{M}\left\{\frac{\zeta(t+\Delta)-\zeta(t)}{\Delta}/\mathfrak{F}_t\right\}/\mathfrak{F}_s\right).$$

Because of b) it is possible to pass to the limit under the expectation sign, thus

$$\lim_{\Delta\downarrow 0}\frac{\varphi(t+\Delta)-\varphi(t)}{\Delta}=0.$$

It is easy to see that $\varphi(t)$ is continuous w.p. 1. But every continuous function whose right-sided derivative is equal to zero at each point is a constant. Since $\varphi(s)=0$, $\varphi(t)=0$ for $s<t$. Thus

$$\mathbb{M}(\zeta(t)-\zeta(s)/\mathfrak{F}_s)=0\qquad(t\geqq s).$$

Analogous considerations show that for the function

$$\psi(t)=\mathbb{M}([\zeta(t)-\zeta(s)]^2/\mathfrak{F}_t)$$

we have

$$\lim_{\Delta\downarrow 0}\frac{\psi(t+\Delta)-\psi(t)}{\Delta}=1$$

for $t\geqq s$ and $\psi(t)=t-s$. The proof of the lemma then follows from Theorem 1 § 1.

Theorem 3. *Let* $\xi(t)$ *be w.p. 1 a diffusion process on* $[0,T]$ *with coefficients* $a(t,x)$ *and* $b(t,x)$ *which satisfy*

1) $a(t,x)$ *is continuous in both arguments and for some* K *satisfies* $|a(t,x)|\leqq K(1+|x|)$;

2) *$b(t, x)$ is continuous in both arguments and has continuous, bounded derivatives $\frac{\partial}{\partial t} b(t, x)$ and $\frac{\partial}{\partial x} b(t, x)$, also $\frac{1}{b(t, x)}$ is bounded;*

3) *there exists a function $\psi(x)$, independent of t and Δ, for which*

$$\psi(x) > 1 + |x|, \qquad \sup_{0 \leqq t \leqq T} \mathbb{M}\, \psi(\xi(t)) < \infty$$

and

$$\left|\int (y-x)\, P(t, x, t+\Delta, dy)\right| + \int (y-x)^2\, P(t, x, t+\Delta, dy) \leqq \psi(x)\, \Delta,$$

$$\int (|y| + y^2)\, P(t, x, t+\Delta, dy) \leqq \psi(x).$$

Then there exists a Wiener process $w(t)$ for which $\xi(t)$ satisfies the stochastic differential equation

$$d\xi(t) = a(t, \xi(t))\, dt + \sqrt{b(t, \xi(t))}\, dw(t). \tag{5}$$

Proof. Since the diffusion coefficients and the coefficients of (5) are transformed by the substitution $\eta(t) = g(t, \xi(t))$ according to the same formula, it suffices to show that $\eta(t) = g(t, \xi(t))$ satisfies

$$d\eta(t) = a(t, \eta(t))\, dt + dw(t), \tag{6}$$

where $\bar{a}(t, x)$ is defined by formula (3) § 9 and

$$g(t, x) = \int_0^x \frac{dz}{\sqrt{b(t, z)}}.$$

We have

$$\begin{aligned}
\mathbb{M}(\eta(t+\Delta) - \eta(t)/\eta(t)) &= \mathbb{M}(g(t+\Delta, \xi(t+\Delta)) - g(t, \xi(t))/\xi(t)) \\
&= \mathbb{M}(g'_t(t+\theta\Delta, \xi(t+\Delta))/\xi(t))\, \Delta \\
&\quad + g'_x(t, \xi(t))\, \mathbb{M}(\xi(t+\Delta) - \xi(t)/\xi(t)) \\
&\quad + \tfrac{1}{2} \mathbb{M}(g''_{xx}(t, \xi(t) + \theta'[\xi(t+\Delta) - \xi(t)]) \\
&\qquad \times (\xi(t+\Delta) - \xi(t))^2/\xi(t))
\end{aligned}$$

$$(\theta \in (0, 1),\ \theta' \in (0, 1)).$$

By assumption, $|g_t| = K|x|$ and g'_x and g''_{xx} are bounded, thus

$$|\mathbb{M}[\eta(t+\Delta) - \eta(t)/\eta(t)]| \leqq L\psi(t, g^{-1}(t, \eta(t)))\, \Delta = L\bar{\psi}(t, \eta(t))\, \Delta,$$

where $\bar{\psi}(t, x) = \psi(t, g^{-1}(t, x))$ and $\sup_t \mathbb{M}\, \bar{\psi}(t, \eta(t)) = \sup_t \mathbb{M}\, \psi(\xi(t)) < \infty$. Analogously, we conclude that

$$\mathbb{M}([\eta(t+\Delta) - \eta(t)]^2/\eta(t)) \leqq L\bar{\psi}(t, \eta(t))\, \Delta.$$

We introduce the process

$$\zeta(t)=\eta(t)-\eta(0)-\int_0^t \bar{a}(s,\eta(s))\,ds.$$

Then $\zeta(t)$ is measurable w.r.t. the σ-algebra $\mathfrak{F}_t$ generated by the variables $\eta(s)$ for $s\leqq t$ and

$$\begin{aligned}|\mathbb{M}(\zeta(t+\Delta)-\zeta(t)/\mathfrak{F}_t)| &= \left|\mathbb{M}(\eta(t+\Delta)-\eta(t)/\eta(t))-\int_t^{t+\Delta}(\mathbb{M}\,\bar{a}(s,\eta(s))/\mathfrak{F}_t)\,ds\right| \\ &\leqq L\bar{\psi}(t,\eta(t))\,\Delta+K_1\int_t^{t+\Delta}\mathbb{M}(1+|g(s,\xi(s))|/\mathfrak{F}_t)\,ds \\ &\leqq L_1\bar{\psi}(t,\eta(t))\,\Delta,\end{aligned}$$

where L_1 is some constant. It is also clear that w.p. 1

$$\lim_{\Delta\to 0}\frac{1}{\Delta}\int_t^{t+\Delta}\mathbb{M}(\bar{a}(s,\eta(s))/\mathfrak{F}_t)\,ds=\bar{a}(t,\eta(t)),$$

thus

$$\lim_{\Delta\to 0}\frac{1}{\Delta}\mathbb{M}(\zeta(t+\Delta)-\zeta(t)/\mathfrak{F}_t)=0.$$

Analogously,

$$\mathbb{M}([\zeta(t+\Delta)-\zeta(t)]^2/\mathfrak{F}_t)\leqq L_1\bar{\psi}(t,\eta(t))\,\Delta$$

and

$$\lim_{\Delta\to 0}\frac{1}{\Delta}\mathbb{M}([\zeta(t+\Delta)-\zeta(t)]^2/\mathfrak{F}_t)=\lim_{\Delta\to 0}\frac{1}{\Delta}\mathbb{M}([\eta(t+\Delta)-\eta(t)]^2/\mathfrak{F}_t)=1.$$

By construction $\zeta(t)$ is continuous w.p. 1. Thus $\zeta(t)$ is a Wiener process. Setting $w(t)=\zeta(t)$ we find that for $\eta(t)$

$$\eta(t)=\eta(0)+\int_0^t \bar{a}(s,\eta(s))\,ds+w(t),$$

which is equivalent to (6). The theorem is proved.

We have thus established that, under certain assumptions, the solutions of stochastic differential equations and diffusions comprise one and the same class of processes.

As a consequence of Theorem 3 we can deduce that when $a(t,x)$ and $\sqrt{b(t,x)}$ satisfy the conditions of the existence and uniqueness theorem for stochastic differential equations, the transition probability of the diffusion process $\xi(t)$ is completely determined by the diffusion coefficients. This fact is not trivial since $a(t,x)$ and $b(t,x)$ are determined by means of the first two moments of a conditional distribution and a distribution is not in general determined by two of its moments.

§ 11. Kolmogorov's Equation

To determine the transition probability of the process $\eta(t)$, which is the solution of Eq. (1) § 10 it is sufficient to find the mathematical expectation $\mathbb{M} f(\eta_{x,t}(s))$ for some set of functions $f(x)$ which is everywhere dense in the space of all continuous functions; $\eta_{x,t}(s)$ is the solution of (3) § 10. For the function

$$u(t,x)=\mathbb{M} f(\eta_{x,t}(s)),$$

defined for $0 \leqq t \leqq s$ and $x \in (-\infty, \infty)$ we introduce a partial differential equation which is called Kolmogorov's backward equation. We first establish an auxiliary

Lemma 1. *If the function $f(x)$ is twice continuously differentiable and satisfies for some $C>0$ and $m>0$ $|f(x)|+|f'(x)|+|f''(x)| \leqq C(1+|x|^m)$, and if the coefficients $a(t,x)$ and $\sigma(t,x)$ fulfill the conditions of Theorem 2 § 10, then*

$$\lim_{\Delta \to 0} \frac{1}{\Delta}(\mathbb{M} f(\eta_{x,t-\Delta}(t+\Delta))-f(x))=a(t,x) f'(x)+\tfrac{1}{2}\sigma^2(t,x) f''(x).$$

Proof. From Itô's formula we have

$$\begin{aligned} f(\eta_{x,t-\Delta}(t))-f(x)= \int_{t-\Delta}^{t} \big[a(s,\eta_{x,t-\Delta}(s)) f'(\eta_{x,t-\Delta}(s)) \\ +\tfrac{1}{2}\sigma^2(s,\eta_{x,t-\Delta}(s)) f''(\eta_{x,t-\Delta}(s))\big]\, ds \\ + \int_{t-\Delta}^{t} \sigma(s,\eta_{x,t-\Delta}(s)) f'(\eta_{x,t-\Delta}(s))\, dw(s). \end{aligned}$$

Taking mathematical expectations on both sides and using Remark 1 § 10 we convince ourselves of the validity of the lemma.

Theorem 1. *Let the coefficients $a(t,x)$ and $\sigma(t,x)$ be continuous and have continuous partial derivatives $a'_x(t,x)$, $a''_{xx}(t,x)$, $\sigma'(t,x)$ and $\sigma''_{xx}(t,x)$. Assume also that for some K and $m>0$*

$$|a(t,x)|+|\sigma(t,x)| \leqq K(1+|x|),$$

$$|a'_x(t,x)|+|a''_{xx}(t,x)|+|\sigma'_x(t,x)|+|\sigma''_{xx}(t,x)| \leqq K(1+|x|^m),$$

and that the function $f(x)$ is twice continuously differentiable with

$$|f(x)|+|f'(x)|+|f''(x)| \leqq K(1+|x|^m).$$

Then the function $u(t,x)=\mathbb{M}f(\eta_{x,t}(s))$ *in the region* $t\in(0,s)$, $x\in(-\infty,\infty)$ *satisfies*

$$\frac{\partial u(t,x)}{\partial t}+a(t,x)\frac{\partial u(t,x)}{\partial x}+\frac{1}{2}\sigma^2(t,x)\frac{\partial^2 u(t,x)}{\partial x^2}=0 \tag{1}$$

with boundary condition $\lim\limits_{t\uparrow s} u(t,x)=f(x)$.

Proof. From the relation $\eta_{x,t-\Delta}(s)=\eta_{\eta_{x,t-\Delta}(t),t}(s)$ it follows that

$$u(t-\Delta,x)=\mathbb{M}\mathbb{M}\big(f(\eta_{x,t-\Delta}(s))/\eta_{x,t-\Delta}(t)\big)=\mathbb{M}\,u(t,\eta_{x,t-\Delta}(t)).$$

By the corollary in § 8, the function $u(t,x)$ is twice continuously differentiable in x. Since from Lemma 1

$$\lim_{\Delta\to 0}\frac{1}{\Delta}\big[\mathbb{M}\,u(t,\eta_{x,t-\Delta}(t))-u(t,x)\big]=a(t,x)\frac{\partial u(t,x)}{\partial x}+\frac{1}{2}\sigma^2(t,x)\frac{\partial^2 u(t,x)}{\partial x^2}$$

and
$$\frac{u(t,x)-u(t-\Delta,x)}{\Delta}+\frac{\mathbb{M}\,u(t,\eta_{x,t-\Delta})-u(t,x)}{\Delta}=0,$$

the limit

$$\lim_{\Delta\to 0}\frac{1}{\Delta}\big(u(t,x)-u(t-\Delta,x)\big)=-a(t,x)\frac{\partial u(t,x)}{\partial x}-\frac{1}{2}\sigma^2(t,x)\frac{\partial^2 u(t,x)}{\partial x^2}$$

exists. We will now show that the right side of the last relation is continuous in t. For this purpose we will need the formula

$$\lim_{t_1-t_2\downarrow 0}\mathbb{M}\big(\eta_{x,t_1}(s)-\eta_{x,t_2}(s)\big)^2=0.$$

Since $\eta_{x,t_1}(s)=\eta_{\eta_{x,t_1}(t_2),t_2}(s)$ we can find constants L_1 and L_2 for which

$$\mathbb{M}\big(\eta_{x,t_1}(s)-\eta_{x,t_2}(s)\big)^2\leqq L_1\mathbb{M}\big(\eta_{x,t_1}(t_2)-x\big)^2\leqq L_2(1+|x|^2)(t_2-t_1),$$

consequently,

$$\begin{aligned}&\mathbb{M}\,|f(\eta_{x,t_1}(s))-f(\eta_{x,t_2}(s))|\\&\quad\leqq L_3\mathbb{M}\,|\eta_{x,t_1}(s)-\eta_{x,t_2}(s)|\,(1+|\eta_{x,t_1}(s)|^m+|\eta_{x,t_2}(s)|^m)\\&\quad\leqq L_3\sqrt{\mathbb{M}\,|\eta_{x,t_1}(s)-\eta_{x,t_2}(s)|^2}\,\sqrt{\mathbb{M}(1+|\eta_{x,t_1}(s)|^m+|\eta_{x,t_2}(s)|^m)^2}\to 0\end{aligned}$$

(we have used the fact that for some $\theta\in(0,1)$ $|f(x)-f(y)|\leqq|x-y|\cdot|f'(x+\theta(y-x))|$ which implies that for some $L_3\,|f(x)-f(y)|\leqq L_3|x-y|(1+|x|^m+|y|^m)$). From these considerations follows the continuity of $u(t,x)$ w.r.t. t. It can also be shown that $u'_x(t,x)$ and $u''_{xx}(t,x)$ are continuous in t. Consequently, $u(t,x)$ is continuous in t and has a continuous left derivative w.r.t. t, i.e., its derivative in t exists and

$$\lim_{\Delta \downarrow 0} \frac{u(t,x)-u(t-\Delta,x)}{\Delta} = \frac{\partial u(t,x)}{\partial t}.$$

That $\lim_{t \uparrow s} u(t,x) = f(x)$ follows from the fact that $\eta_{x,t}(s) \to x$ for $t \uparrow s$ in probability and the possibility of proceeding to the limit under the expectation sign in $\lim_{t \uparrow s} u(t,x) = \lim_{t \uparrow s} \mathbb{M} f(\eta_{x,t}(s))$. The theorem is finished.

Let L denote the differential operator $a(t,x)\frac{\partial}{\partial x} + \frac{1}{2}\sigma^2(t,x)\frac{\partial^2}{\partial x^2}$, which appears in Kolmogorov's equation. We shall see that with the help of equations containing this operator we can determine many other characteristics of the process. We consider the equation satisfied by the distribution of the variable

$$\int_0^T g(s, \eta(s))\, ds, \tag{2}$$

where $\eta(t)$ is the solution of a stochastic equation. We establish an equation for the joint characteristic function of the variables

$$\int_t^T g(s, \eta_{x,t}(s))\, ds \quad \text{and} \quad f(\eta_{x,t}(T)):$$

$$V_{\lambda,\mu}(t,x) = \mathbb{M} \exp\left\{ i\lambda \int_t^T g(s, \eta_{x,t}(s))\, ds + i\mu f(\eta_{x,t}(T)) \right\}.$$

Knowing $V_{\lambda,\mu}(t,x)$, we can determine the characteristic function of the variable (2) by integrating $V_{\lambda,0}(0,x)$ w.r.t. the distribution of $\eta(0)$.

Theorem 2. *If the assumptions of Theorem 1 are fulfilled and the function $g(t,x)$ is twice continuously differentiable in x and for some $K>0$ and $m>0$*

$$|g(t,x)| + |g'_x(t,x)| + |g''_{xx}(t,x)| \leq K(1+|x|^m),$$

then the function $V_{\lambda,\mu}(t,x)$ for $t \in [0,T]$, $x \in (-\infty, \infty)$ satisfies

$$\frac{\partial}{\partial t} V_{\lambda,\mu}(t,x) + LV_{\lambda,\mu}(t,x) + i\lambda g(t,x) V_{\lambda,\mu}(t,x) = 0 \tag{3}$$

with boundary condition $\lim_{t \uparrow T} V_{\lambda,\mu}(t,x) = e^{i\mu f(x)}$.

Proof. The validity of the boundary condition is obvious. The twice differentiability of $V_{\lambda,\mu}(t,x)$ w.r.t. x follows from the possibility of differentiating w.r.t. x under the expectation sign which follows from the differentiability of $g(t,x)$ and $f(x)$, the mean square differentiability w.r.t. x of the process $\eta_{x,t}(s)$ and Corollary 1 §8. We have

$$\begin{aligned}
&V_{\lambda,\mu}(t,x)-V_{\lambda,\mu}(t-\Delta,x)\\
&\quad=V_{\lambda,\mu}(t,x)-\mathbb{M}\exp\left\{i\lambda\int_{t-\Delta}^{T}g(s,\eta_{x,t-\Delta}(s))\,ds+i\mu f(\eta_{x,t-\Delta}(T))\right\}\\
&\quad=V_{\lambda,\mu}(t,x)-\mathbb{M}\exp\left\{i\lambda\int_{t}^{T}g(s,\eta_{\eta_{x,t-\Delta}(t),t}(s))\,ds+i\mu f(\eta_{\eta_{x,t-\Delta}(t),t}(T))\right\}\\
&\qquad+\mathbb{M}\exp\left\{i\lambda\int_{t}^{T}g(s,\eta_{x,t-\Delta}(s))\,ds+i\mu f(\eta_{x,t}(T))\right\}\\
&\qquad\times\left[1-\exp\left\{i\lambda\int_{t-\Delta}^{t}g(s,\eta_{x,t-\Delta}(s))\,ds\right\}\right]\\
&\quad=V_{\lambda,\mu}(t,x)-\mathbb{M}V_{\lambda,\mu}(t,\eta_{x,t-\Delta}(t))\\
&\qquad+\mathbb{M}\exp\left\{i\lambda\int_{t}^{T}g(s,\eta_{x,t-\Delta}(s))\,ds+i\mu f(\eta_{x,t}(T))\right\}\\
&\qquad\times\left[1-\exp\left\{i\lambda\int_{t-\Delta}^{t}g(s,\eta_{x,t-\Delta}(s))\,ds\right\}\right].
\end{aligned}$$

From Lemma 1

$$\lim_{\Delta\downarrow 0}\frac{1}{\Delta}\left[V_{\lambda,\mu}(t,x)-V_{\lambda,\mu}(t,\eta_{x,t-\Delta}(t))\right]=-LV_{\lambda,\mu}(t,x).$$

Since, in probability,

$$\begin{aligned}
\lim_{\Delta\to 0}\frac{1}{\Delta}\left[1-\exp\left\{i\lambda\int_{t-\Delta}^{t}g(s,\eta_{x,t-\Delta}(s))\,ds\right\}\right]&=-\lim_{\Delta\to 0}\frac{i\lambda}{\Delta}\int_{t-\Delta}^{t}g(s,\eta_{x,t-\Delta}(s))\,ds\\
&=-i\lambda g(t,x),
\end{aligned}$$

$$\begin{aligned}
&\lim_{\Delta\to 0}\frac{1}{\Delta}\mathbb{M}\exp\left\{i\lambda\int_{t}^{T}g(s,\eta_{x,t-\Delta}(s))\,ds+i\mu f(\eta_{x,t-\Delta}(T))\right\}\\
&\qquad\times\left[1-\exp\left\{i\lambda\int_{t-\Delta}^{t}g(s,\eta_{x,t-\Delta}(s))\,ds\right\}\right]\\
&\quad=-\lim_{\Delta\to 0}\frac{1}{\Delta}\mathbb{M}\exp\left\{i\lambda\int_{t}^{T}g(s,\eta_{x,t-\Delta}(s))\,ds+i\mu f(\eta_{x,t-\Delta}(T))\right\}i\lambda g(t,x)\\
&\quad=-i\lambda g(t,x)V_{\lambda,\mu}(t,x).
\end{aligned}$$

Thus,

$$\lim_{\Delta\downarrow 0}\frac{V_{\lambda,\mu}(t,x)-V_{\lambda,\mu}(t-\Delta,x)}{\Delta}=-LV_{\lambda,\mu}(t,x)-i\lambda g(t,x)V_{\lambda,\mu}(t,x).$$

From the existence of the left derivative w.r.t. t of $V_{\lambda,\mu}(t,x)$ follows as in Theorem 1 the existence of $\frac{\partial}{\partial t} V_{\lambda,\mu}(t,x)$, so that Eq. (3) is a consequence of the above considerations.

Remark 1. One can show that under the conditions of Theorem 2 the function

$$V_\lambda(t,x)=\mathbf{M} f(\eta_{x,t}(T)) \exp\left\{i\lambda \int_t^T g(s,\eta_{x,t}(s))\,ds\right\}$$

also satisfies (3) with boundary condition $\lim_{t\uparrow T} V_\lambda(t,x)=f(x)$. This result combines those of Theorem 1 and Theorem 2.

Remark 2. In order to find the distribution of the variable (2), it is necessary to solve Eq. (3) with the boundary condition $V_\lambda(T,x)=1$.

Remark 3. Let us establish an equation for the moments

$$M_k(t,x)=\mathbf{M}\left[\int_t^T g(s,\eta_{x,t}(s))\,ds\right]^k.$$

We use the expansion

$$V_{\lambda,0}(t,x)=1+\sum_{k=1}^{\infty}\frac{(i\lambda)^k}{k!}M_k(t,x).$$

Substituting this into (3) and comparing corresponding powers of λ, we obtain

$$\frac{\partial}{\partial t} M_k(t,x)+LM_k(t,x)+k\,g(t,x)\,M_{k-1}(t,x)=0. \tag{4}$$

The boundary condition for $M_k(t,x)$ will be $M_k(t,x)=0$ for $k\geqq 1$, $M_0(t,x)\equiv 1$. We now turn to equations for the distribution of stochastic integrals of the form $\int_t^T h(s,\eta_{x,t}(s))\,dw(s)$.

Let us solve a more general problem: find an equation satisfied by the joint distribution of the variables

$$\eta_{x,t}(T),\quad \int_t^T g(s,\eta_{x,t}(s))\,ds,\quad \int_t^T h(s,\eta_{x,t}(s))\,dw(s).$$

Let

$$\psi(t,x)=\mathbf{M}\exp\left(i\lambda\int_t^T g(s,\eta_{x,t}(s))\,ds+i\mu\int_t^T h(s,\eta_{x,t}(s))\,dw(s)\right\} f(\eta_{x,t}(T)). \tag{5}$$

Theorem 3. *Assume the assumptions of Theorem 2 are satisfied and that for some $K>0$ and $m>0$ the twice continuously differentiable function $h(t,x)$ satisfies*

$$|h(t,x)|+|h'_x(t,x)|+|h''_{xx}(t,x)|\leqq K(1+|x|^m).$$

Then the function $\psi(t,x)$ *satisfies for* $t\in(0,T)$, $x\in(-\infty,\infty)$ *the equation*

$$\frac{\partial\psi(t,x)}{\partial t}+L\psi(t,x)+i\mu h(t,x)\frac{\partial\psi(t,x)}{\partial x}+\left[i\lambda g(t,x)-\frac{\mu^2}{2}h^2(t,x)\right]\psi(t,x)=0 \tag{6}$$

with boundary condition $\lim_{t\uparrow T}\psi(t,x)=f(x)$.

Proof. Let $H(t,x)$ be a function for which

$$H'_x(t,x)=\frac{h(t,x)}{\sigma(t,x)}.$$

Then

$$\begin{aligned}H(T,\eta_{x,t}(T))-H(t,x)&=\int_t^T H'_x(s,\eta_{x,t}(s))\,\sigma(s,\eta_{x,t}(s))\,dw(s)\\&+\int_t^T\big[H'_s(s,\eta_{x,t}(s))+H'_x(s,\eta_{x,t}(s))\,a(s,\eta_{x,t}(s))\\&\quad+\tfrac{1}{2}H''_{xx}(s,\eta_{x,t}(s))\,\sigma^2(s,\eta_{x,t}(s))\big]\,ds.\end{aligned}$$

Taking into account the value of H'_x, we find

$$\begin{aligned}\int_t^T h(s,\eta_{x,t}(s))\,dw(s)&=H(T,\eta_{x,t}(T))-H(t,x)\\&-\int_t^T\big[H'_s(s,\eta_{x,t}(s))-LH(s,\eta_{x,t}(s))\big]\,ds.\end{aligned}$$

Thus

$$\begin{aligned}\psi(t,x)&=\exp\{-i\mu H(t,x)\}\\&\times\mathbb{M}\exp\left\{i\int_t^T\big[\lambda g(s,\eta_{x,t}(s))-\mu H'_s(s,\eta_{x,t}(s))-\mu LH(s,\eta_{x,t}(s))\big]\,ds\right\}\\&\times e^{i\mu H(T,\eta_{x,t}(T))}f(\eta_{x,t}(T)).\end{aligned}$$

Assume that $H'_s(s,x)+LH(s,x)$ is twice continuously differentiable in x and that

$$\frac{\partial^2}{\partial x^2}(H'_s(s,x)+LH(s,x))\leqq K(1+|x|^m).$$

Then from Theorem 2 the function $\psi(t,x)\,e^{i\mu H(t,x)}$ satisfies

$$\begin{aligned}&\frac{\partial}{\partial t}[\psi(t,x)\,e^{i\mu H(t,x)}]+L[\psi(t,x)\,e^{i\mu H(t,x)}]\\&\quad+\big[i\lambda g(t,x)-i\mu(H'_t(t,x)+LH(t,x))\big]\,\psi(t,x)\,e^{i\mu H(t,x)}=0.\end{aligned}$$

Differentiating and cancelling out $e^{i\mu H(t,x)}$ we obtain, after simplification of like terms, Eq. (6). In order to free the proof from the special assumptions made above it remains to note that under the assumptions of the theorem the function $\psi(t,x)\,e^{i\mu H(t,x)}$ is twice continuously differentiable in x and this was all we used in the proof of Theorem 2. The theorem is proved.

Remark 4. Let $\sigma(t,x)=1$ and $\Phi(x)$ be an arbitrary twice continuously differentiable function vanishing outside some finite interval. We denote the set of all such functions by $C_f^{(2)}$. Multiplying (5) by $\Phi(x)$ and integrating w.r.t. x over the whole line, we obtain (using the formula for integration by parts)

$$\frac{\partial}{\partial t}\int \psi(t,x)\,\Phi(x)\,dx+\tfrac{1}{2}\int \psi(t,x)\,\Phi''(x)\,dx$$
$$-\int \psi(t,x)\frac{\partial}{\partial x}\left[(a(t,x)+i\mu h(t,x))\,\Phi(x)\right]dx$$
$$+\int\left(i\lambda g(t,x)-\frac{\mu^2}{2}h^2(t,x)\right)\psi(t,x)\,\Phi(x)\,dx=0.$$

Denote by $L_1^*(t)$ the differential operator

$$L_1^*(t)[U]=\frac{1}{2}\cdot\frac{\partial^2}{\partial x^2}U-\frac{\partial}{\partial x}\left[(a(t,x)+i\mu h(t,x))\,U\right]$$
$$+\left(i\lambda g(t,x)-\frac{\mu^2}{2}h^2(t,x)\right)U.$$

Then from (6)

$$\frac{\partial}{\partial t}\int \psi(t,x)\,\Phi(x)\,dx+\int \psi(t,x)\,L_1^*(t)\,\Phi(x)\,dx=0. \tag{7}$$

If (7) holds for all functions Φ from $C_f^{(2)}$ and $\psi(t,x)$ is twice continuously differentiable in x, then (6) follows from (7).

We call $\psi(t,x)$ a generalized solution of Eq. (6) if it satisfies (7) for arbitrary Φ from $C_f^{(2)}$. We will now show that the solution $\psi(t,x)$ of (5) determined above is a generalized solution of (5) if $\sigma(t,x)=1$, $a_x'(t,x)$ is continuous and bounded and the functions $g(t,x)$ and $h(t,x)$ are continuous. For this purpose we consider the process $\eta_{x,t}^{(n)}(s)$ as solution of

$$\eta_{x,t}^{(n)}(s)=x+\int_t^s a_n(u,\eta_{x,t}^{(n)}(u)\,du)+w(s)-w(t),$$

where $a_n(t,x)$ has bounded, continuous derivatives

$$\frac{\partial}{\partial x}a_n(t,x),\qquad \frac{\partial^2}{\partial x^2}a_n(t,x)\quad\text{and}\quad \frac{\partial}{\partial x}a_n(t,x)$$

converges uniformly to $\frac{\partial}{\partial x} a(t,x)$ and $a_n(t,x)$ to $a(t,x)$. Let the functions $h_n(t,x)$, $g_n(t,x)$ and $f_n(x)$ for each n satisfy the conditions of Theorem 3 and for $n\to\infty$ converge uniformly to $h(t,x)$, $g(t,x)$ and $f(x)$, resp. Set

$$\psi^{(n)}(t,x)=\mathbb{M}\exp\left\{i\lambda\int_t^T g_n(s,\eta_{x,t}^{(n)}(s))\,ds+i\mu\int_t^T h_n(s,\eta_{x,t}^{(n)}(s))\,dw(s)\right\} f_n(\eta_{x,t}^{(n)}(T)).$$

From Theorem 3 and (7) we can write

$$\int[\psi^{(n)}(T,x)-\psi^{(n)}(t,x)]\,\Phi(x)\,dx+\int_t^T\int\psi^{(n)}(s,x)\,L_1^{*(n)}(s)\,\Phi(x)\,dx\,ds=0, \tag{8}$$

where $L_1^{*(n)}(s)$ is the operator resulting from $L_1^*(s)$ when $a_n(s,x)$ is substituted for $a(s,x)$. Using Theorem 2 §7 we can show that $\psi^{(n)}(t,x)$ converges to $\psi(t,x)$ when $n\to\infty$. Proceeding to the limit in (8) and differentiating the result w.r.t. t, we obtain (7).

Remark 5. Using Theorem 2 §7 we can show that if the functions $a(t,x)$, $\sigma(t,x)$, $f(x)$, $g(t,x)$ and $h(t,x)$ are continuous and satisfy the assumptions of the existence and uniqueness theorem, then $\psi(t,x)$ defined by formula (5), satisfies (6) in the following generalized sense: $\psi(t,x)=\lim_{n\to\infty}\psi_n(t,x)$, where $\psi_n(t,x)$ satisfies

$$\begin{aligned}\frac{\partial\psi_n(t,x)}{\partial t}&+\frac{1}{2}\sigma_n(t,x)\frac{\partial^2\psi_n(t,x)}{\partial x^2}\\&+(a_n(t,x)+i\mu h_n(t,x)\,\sigma_n(t,x))\frac{\partial\psi_n(t,x)}{\partial x}\\&+\left(i\lambda g_n(t,x)-\frac{\mu^2}{2}h_n^2(t,x)\right)\psi_n(t,x)=0,\end{aligned}$$

the functions $a_n(t,x)$, $\sigma_n(t,x)$, $f_n(x)$, $g_n(t,x)$ and $h_n(t,x)$ fulfill the conditions of Theorem 3 and converge uniformly for $n\to\infty$ to $a(t,x)$, $\sigma(t,x)$, $f(x)$, $g(t,x)$ and $h(t,x)$.

§ 12. Measures in Function Space Induced by Diffusion Processes

We recall the general definition of a random process. Let $\{\Omega,\mathfrak{s},\mathbb{P}\}$ be some probability space, where Ω is a set of elementary events ω, $\mathfrak{s}$ is a σ-algebra of events (subsets of Ω) and $\mathbb{P}$ is a probability measure on $\mathfrak{s}$. A random process $\xi(t)$ on $[0,T]$ is determined by the probability space $\{\Omega,\mathfrak{s},\mathbb{P}\}$ and the function $\xi(t,\omega)$ is defined for $t\in[0,T]$ and $\omega\in\Omega$ and is $\mathfrak{s}$-measurable in ω for each $t\in[0,T]$.

Denote by $F[0, T]$ the space of real functions $x(t)$ defined on $[0, T]$ and by $C_s(A)$, where $s \in [0, T]$ and A is a Borel set on the line, the set of functions $x(t)$ for which $x(s) \in A$. We will call cylinder sets those subsets of $F[0, T]$ of the form $\prod_{k=1}^{n} C_{t_k}(A_k)$, where $t_k \in [0, T]$ and A_k are Borel sets. Let $\mathfrak{F}[0, T]$ be the minimal σ-algebra of subsets of $F[0, T]$ containing all cylinder sets. With each process $\xi(t)$ defined on $[0, T]$ is associated a probability measure μ on the σ-algebra $\mathfrak{F}[0, T]$. For cylinder sets the value of the measure μ is defined by the relation

$$\mu\left(\prod_{k=1}^{n} C_{t_k}(A_k)\right) = \mathbb{P}\{\xi(t_1) \in A_1, \ldots, \xi(t_n) \in A_n\},$$

and μ is then extended to all of $\mathfrak{F}[0, T]$ by means of additivity and continuity. The measure constructed in this way will be called the measure corresponding to the process $\xi(t)$. These measures are useful in the study of many questions in the theory of random processes.

Let us consider the following transformation of a random process. Let $\rho(\omega)$ be a real $\mathfrak{s}$-measurable function defined on Ω for which the measure

$$\mathbb{P}_1(E) = \int_E \rho(\omega)\, \mathbb{P}(d\omega) \qquad (E \in \mathfrak{s})$$

is also a probability. We treat a random process which is also defined by means of $\xi(t) = \xi(t, \omega)$, but is considered on the probability space $\{\Omega, \mathfrak{s}, \mathbb{P}_1\}$. In order to distinguish the two processes we use the notation $\{\xi(t), \mathbb{P}\}$ and $\{\xi(t), \mathbb{P}_1\}$ referring along with the process to the probability measure appearing in the definition of the probability space. It is not difficult to see that the measure $\mathbb{P}_1$ will be a probability if and only if $\rho(\omega) \geqq 0$ almost everywhere w.r.t. $\mathbb{P}$ and $\int \rho(\omega)\, \mathbb{P}(d\omega) = 1$. If the process $\{\xi(t), \mathbb{P}_1\}$ corresponds to the measure μ_1 then the latter will be absolutely continuous w.r.t. the measure μ which corresponds to the process $\{\xi(t), \mathbb{P}\}$; moreover, its density w.r.t. μ (which we will designate by $\frac{d\mu_1}{d\mu}(x(\cdot))$) is related to $\rho(\omega)$ through

$$\frac{d\mu_1}{d\mu}(\xi(\cdot, \omega)) = \rho(\omega).$$

We will be interested in the form of $\rho(\omega)$ when $\{\xi(t), \mathbb{P}\}$ and $\{\xi(t), \mathbb{P}_1\}$ are diffusions.

We first establish a lemma which, for a certain class of random variables ρ, allows as to conclude that $\mathbb{M}\rho = 1$. Let $H_2[0, T]$ be the space of random functions introduced in §2 and $\mathfrak{F}_t$ the family of σ-algebras appearing in the definition of $H_2[0, T]$.

Lemma 1. *If* $\alpha(t)\in H_2[0,T]$ *and for some* $\delta>0$

$$\mathbb{M}\exp\left\{(1+\delta)\int_0^T \alpha^2(s)\,ds\right\}<\infty,$$

then

$$\mathbb{M}\exp\left\{\int_0^T \alpha(s)\,dw(s)-\tfrac{1}{2}\int_0^T \alpha^2(s)\,ds\right\}=1.$$

Proof. We assume first that $\alpha(s)$ is a bounded step function. It is easy to see that in this case

$$\mathbb{M}\exp\left\{2\int_0^T \alpha(s)\,dw(s)\right\}<\infty.$$

From Itô's formula

$$\exp\left\{\int_{t_1}^{t_2} \alpha(s)\,dw(s)-\tfrac{1}{2}\int_{t_1}^{t_2} \alpha^2(s)\,ds\right\}-1$$
$$=\int_{t_1}^{t_2}\exp\left\{\int_{t_1}^{t} \alpha(s)\,dw(s)-\tfrac{1}{2}\int_{t_1}^{t} \alpha^2(s)\,ds\right\}\alpha(t)\,dw(t).$$

Since under our assumptions

$$\int_{t_1}^{t_2}\mathbb{M}\left[\exp\left\{\int_{t_1}^{t} \alpha(s)\,dw(s)-\tfrac{1}{2}\int_{t_1}^{t} \alpha^2(s)\,ds\right\}\right]^2\alpha^2(t)\,dt<\infty,$$

we have

$$\mathbb{M}\exp\left\{\int_{t_1}^{t_2} \alpha(s)\,dw(s)-\tfrac{1}{2}\int_{t_1}^{t_2} \alpha^2(s)\,ds\right\}=1,\qquad 0\leqq t_1<t_2\leqq T.$$

Going to the limit, we find that for all $\alpha(t)\in H_2[0,T]$

$$\mathbb{M}\exp\left\{\int_{t_1}^{t_2} \alpha(s)\,dw(s)-\tfrac{1}{2}\int_{t_1}^{t_2} \alpha^2(s)\,ds\right\}\leqq 1.$$

Now let $\alpha_n(t)$ be a sequence of bounded step functions for which $\alpha_n^2(t)\leqq\alpha^2(t)$ and w.p. 1

$$\lim_{n\to\infty}\int_0^T(\alpha(t)-\alpha_n(t))^2\,dt=0.$$

Then

$$\mathbb{M}\left[\exp\left\{\int_{t_1}^{t_2} \alpha_n(s)\,dw(s)-\tfrac{1}{2}\int_{t_1}^{t_2} \alpha_n^2(s)\,ds\right\}\right]^{1+\varepsilon}$$
$$=\mathbb{M}\exp\left\{(1+\varepsilon)\int_{t_1}^{t_2} \alpha_n(s)\,dw(s)-\frac{(1+\varepsilon)^3}{2}\int_{t_1}^{t_2} \alpha_n^2(s)\,ds\right\}$$
$$\times\exp\left\{\frac{(1+\varepsilon)(2\varepsilon+\varepsilon^2)}{2}\int_{t_1}^{t_2} \alpha_n^2(s)\,ds\right\}$$

$$\leqq\left[\mathbb{M}\exp\left\{(1+\varepsilon)^2\int_{t_1}^{t_2}\alpha_n(s)\,dw(s)-\tfrac{1}{2}(1+\varepsilon)^4\int_{t_1}^{t_2}\alpha_n^2(s)\,ds\right\}\right]^{\frac{1}{1+\varepsilon}}$$

$$\times\left[\mathbb{M}\exp\left\{\frac{(1+\varepsilon)^2(2+\varepsilon)}{2}\int_{t_1}^{t_2}\alpha_n^2(s)\,ds\right\}\right]^{\frac{\varepsilon}{1+\varepsilon}}$$

$$\leqq\left[\mathbb{M}\exp\left\{\frac{(1+\varepsilon)^2(2+\varepsilon)}{2}\int_{t_1}^{t_2}\alpha_n^2(s)\,ds\right\}\right]^{\frac{\varepsilon}{1+\varepsilon}}$$

$$\leqq\left[\mathbb{M}\exp\left\{\left(1+\frac{\varepsilon(2+\varepsilon)^2}{2}\right)\int_{t_1}^{t_2}\alpha^2(s)\,ds\right\}\right]^{\frac{\varepsilon}{1+\varepsilon}}<\infty \quad \text{for } \frac{\varepsilon(2+\varepsilon)^3}{2}<\delta,$$

hence, using the theorem on passage to the limit under the integral sign we have

$$1=\lim_{n\to\infty}\mathbb{M}\exp\left\{\int_{t_1}^{t_2}\alpha_n(s)\,dw(s)-\tfrac{1}{2}\int_{t_1}^{t_2}\alpha_n^2(s)\,ds\right\}$$

$$=\mathbb{M}\exp\left\{\int_{t_1}^{t_2}\alpha(s)\,dw(s)-\tfrac{1}{2}\int_{t_1}^{t_2}\alpha^2(s)\,ds\right\},\qquad 0\leqq t_1<t_2\leqq T.$$

The proof is finished.

Remark 1. Without at all changing the proof of the lemma – in place of the usual we consider the conditional expectation – we can show that under the conditions of the lemma

$$\mathbb{M}\left(\exp\left\{\int_{t_1}^{t_2}\alpha(s)\,dw(s)-\tfrac{1}{2}\int_{t_1}^{t_2}\alpha^2(s)\,ds\right\}/\mathfrak{F}_{t_1}\right)=1.$$

Remark 2. In order for Lemma 1 to hold it is sufficient that for some $\delta>0$ there exist a $C>0$ such that for $s\in[0,T]$, $\mathbb{M}\exp\{\delta\,\alpha^2(s)\}\leqq C$.

In fact from Jensen's inequality

$$e^{\lambda\int_{t_1}^{t_2}\alpha^2(s)\,ds}=\exp\left\{\lambda(t_2-t_1)\frac{1}{t_2-t_1}\int_{t_1}^{t_2}\alpha^2(s)\,ds\right\}\leqq\frac{1}{t_2-t_1}\int_{t_1}^{t_2}e^{\lambda(t_2-t_1)\alpha^2(s)}\,ds$$

(the function $e^{\lambda x}$ is convex), i.e., for each $\lambda>1$ we can find a $\varepsilon>0$ such that for $0<t''-t'<\varepsilon$

$$\mathbb{M}\exp\left\{\lambda\int_{t'}^{t''}\alpha^2(s)\,ds\right\}\leqq\frac{1}{t''-t'}\int_{t'}^{t''}\mathbb{M}\,e^{\lambda(t''-t')\alpha^2(s)}\,ds\leqq C.$$

Thus from Remark 1

$$\mathbb{M}\left(\exp\left\{\int_{t'}^{t''}\alpha(s)\,dw(s)-\tfrac{1}{2}\int_{t'}^{t''}\alpha^2(s)\,ds\right\}/\mathfrak{F}_{t'}\right)=1.$$

Partitioning the arbitrary interval $[t_1, t_2]$ by means of the points $t_1 = s_0 < \cdots < s_n = t_2$ in such a way that $s_{k+1} - s_k < \varepsilon$, we find that

$$\begin{aligned}
&\mathbb{M} \exp\left\{\int_{t_1}^{t_2} \alpha(s)\, dw(s) - \tfrac{1}{2} \int_{t_1}^{t_2} \alpha^2(s)\, ds\right\} \\
&= \mathbb{M} \exp\left\{\int_{s_0}^{s_{n-1}} \alpha(s)\, dw(s) - \tfrac{1}{2} \int_{s_0}^{s_{n-1}} \alpha^2(s)\, ds\right\} \\
&\quad \times \exp\left\{\int_{s_{n-1}}^{s_n} \alpha(s)\, dw(s) - \tfrac{1}{2} \int_{s_{n-1}}^{s_n} \alpha^2(s)\, ds\right\} \\
&= \mathbb{M}\left[\mathbb{M}\left(\exp\left\{\int_{s_{n-1}}^{s_n} \alpha(s)\, dw(s) - \tfrac{1}{2} \int_{s_{n-1}}^{s_n} \alpha^2(s)\, ds\right\} / \mathfrak{F}_{n-1}\right)\right. \\
&\quad \left. \times \exp\left\{\int_{s_0}^{s_{n-1}} \alpha(s)\, dw(s) - \tfrac{1}{2} \int_{s_0}^{s_{n-1}} \alpha^2(s)\, ds\right\}\right] \\
&= \mathbb{M} \exp\left\{\int_{s_0}^{s_{n-1}} \alpha(s)\, dw(s) - \tfrac{1}{2} \int_{s_0}^{s_{n-1}} \alpha^2(s)\, ds\right\} = \cdots \\
&= \mathbb{M} \exp\left\{\int_{s_0}^{s_1} \alpha(s)\, dw(s) - \tfrac{1}{2} \int_{s_0}^{s_1} \alpha^2(s)\, ds\right\} = 1.
\end{aligned}$$

Theorem 1. *Let $\eta(t)$, $0 \leqq t \leqq T$ be a solution of the stochastic equation (1) § 10, and $\eta_{x,t}(s)$ that of (3) § 10. Let the Borel function $\alpha(s, x)$, defined for $s \in [0, T]$, $x \in (-\infty, \infty)$ be such that for some $\delta > 0$ there exists a function $C(x)$ with $\sup_{0 \leqq t \leqq T} \mathbb{M}\, C(\eta(t)) < \infty$ and $\mathbb{M} \exp\{\delta \alpha^2(s, \eta_{x,t}))\} \leqq C(x)$. Set*

$$\mathbb{P}^*(E) = \int_E \exp\left\{\int_0^T \alpha(s, \eta(s))\, dw(s) - \tfrac{1}{2} \int_0^T \alpha^2(s, \eta(s))\, ds\right\} \mathbb{P}(d\omega), \tag{1}$$

then the process $\{\eta(t), \mathbb{P}^*\}$ is a Markov process with transition probability $P^*(t, x, s, A)$ satisfying

$$\begin{aligned}
&P^*(t, x, s, A) \\
&= \mathbb{M} \exp\left\{\int_t^s \alpha(u, \eta_{x,t}(u))\, dw(u) - \tfrac{1}{2} \int_t^s \alpha^2(u, \eta_{x,t}(u))\, du\right\} \chi_A(\eta_{x,t}(s))
\end{aligned} \tag{2}$$

($\chi_A(x)$ is the indicator of the set A).

Proof. Set

$$\rho(x, t_1, t_2) = \exp\left\{\int_{t_1}^{t_2} \alpha(s, \eta_{x,t_1}(s))\, dw(s) - \tfrac{1}{2} \int_{t_1}^{t_2} \alpha^2(s, \eta_{x,t_1}(s))\, ds\right\}.$$

The following relations are obvious:

$$\exp\left\{\int_{t_1}^{t_2} \alpha(s,\eta(s))\,dw(s)-\tfrac{1}{2}\int_{t_1}^{t_2}\alpha^2(s,\eta(s))\,ds\right\}=\rho(\eta(t_1),t_1,t_2),$$

since $\eta_{\eta(t_1),t_1}(s)=\eta(s)$ for $s>t_1$; and

$$\rho(\eta(t_1),t_1,t_3)=\rho(\eta(t_1),t_1,t_2)\,\rho(\eta(t_2),t_2,t_3),\quad \text{if } t_1<t_2<t_3. \tag{3}$$

We denote by $\mathbb{M}(\cdot)$ and $\mathbb{M}^*(\cdot)$ the mathematical expectations corresponding to the measures $\mathbb{P}$ and $\mathbb{P}^*$. To prove the theorem it is sufficient to show that for any bounded Borel function $\varphi(x_1,x_2,\dots,x_n)$ and arbitrary Borel set A

$$\begin{aligned}\mathbb{M}^*\varphi(\eta(t_1),\dots,\eta(t_k))\,\mathbb{P}^*(t_k,\eta(t_k),t_{k+1},A)\\ =\mathbb{M}^*\varphi(\eta(t_1),\dots,\eta(t_k))\,\chi_A(\eta(t_{k+1})),\end{aligned}\tag{4}$$

if $t_1<\dots<t_k<t_{k+1}$. The right side of (4) can be rewritten as

$$\begin{aligned}&\mathbb{M}\,\varphi(\eta(t_1),\dots,\eta(t_k))\,\chi_A(\eta(t_{k+1}))\,\rho(\eta(0),0,T)\\ &\quad=\mathbb{M}\,\varphi(\eta(t_1),\dots,\eta(t_k))\,\chi_A(\eta(t_{k+1}))\,\rho(\eta(0),0,t_{k+1})\,\rho(\eta(t_{k+1}),t_{k+1},T)\\ &\quad=\mathbb{M}\,\varphi(\eta(t_1),\dots,\eta(t_k))\,\chi_A(\eta(t_{k+1}))\,\rho(\eta(0),0,t_{k+1}),\end{aligned}\tag{5}$$

since from Remark 2

$$\mathbb{M}(\rho(\eta(t_{k+1}),t_{k+1},T)/\mathfrak{F}_{t_{k+1}})=1.$$

The left side of (4) can be written, taking account of (2), as

$$\begin{aligned}&\mathbb{M}\,\varphi(\eta(t_1),\dots,\eta(t_k))\,P^*(t_k,\eta(t_k),t_{k+1},A)\,\rho(\eta(0),0,t_k)\\ &\quad=\mathbb{M}\,\varphi(\eta(t_1),\dots,\eta(t_k))\,\rho(\eta(0),0,t_k)\\ &\qquad\times\mathbb{M}(\rho(\eta(t_k),t_k,t_{k+1})\,\chi_A(\eta(t_{k+1}))/\eta(t_k))\\ &\quad=\mathbb{M}\,\varphi(\eta(t_1),\dots,\eta(t_k))\,\rho(\eta(0),0,t_k)\\ &\qquad\times\mathbb{M}(\rho(\eta(t_k),t_k,t_{k+1})\,\chi_A(\eta(t_{k+1}))/\mathfrak{F}_{t_k})\\ &\quad=\mathbb{M}\,\varphi(\eta(t_1),\dots,\eta(t_k))\,\rho(\eta(0),0,t_k)\,\rho(\eta(t_k),t_k,t_{k+1})\,\chi_A(\eta(t_{k+1}))\\ &\quad=\mathbb{M}\,\varphi(\eta(t_1),\dots,\eta(t_k))\,\rho(\eta(0),0,t_{k+1})\,\chi_A(\eta(t_{k+1})).\end{aligned}$$

Comparing this result with (5), we complete the proof.

We can convince ourselves that $\{\eta(t),\mathbb{P}^*\}$ is also a diffusion and find its coefficients. For this purpose we need some auxiliary propositions.

Lemma 2. *If the assumptions of Theorem 1 are fulfilled, then*

$$\mathbb{M}\exp\left\{\int_t^s \alpha(u,\eta_{x,t}(u))\,dw(u)\right\}\le\mathbb{M}\exp\left\{2\int_t^s\alpha^2(u,\eta_{x,t}(u))\,du\right\}.$$

Proof. Since

$$\mathbb{M}\exp\left\{2\int_t^s \alpha(u,\eta_{x,t}(u))\,dw(u) - 2\int_t^s \alpha^2(u,\eta_{x,t}(u))\,du\right\} \leqq 1,$$

$$\begin{aligned}\mathbb{M}\exp\left\{\int_t^s \alpha(u,\eta_{x,t}(u))\,dw(u)\right\} &\leqq \sqrt{\mathbb{M}\exp\left\{2\int_t^s \alpha(u,\eta_{x,t}(u))\,dw(u) - 2\int_t^s \alpha^2(u,\eta_{x,t}(u))\,du\right\}}\\ &\quad\times\sqrt{\mathbb{M}\exp\left\{2\int_t^s \alpha^2(u,\eta_{x,t}(u))\,du\right\}}\\ &\leqq \mathbb{M}\exp\left\{2\int_t^s \alpha^2(u,\eta_{x,t}(u))\,du\right\}\end{aligned}$$

(we used here the inequality $\sqrt{a}<a$ for $a>1$). The lemma is proved.

Corollary 1. *Under the assumptions of Theorem 1 for $m>0$ we can find a $\delta_1>0$ such that for $0<s-t<\delta_1$*

$$\mathbb{M}[\rho(x,t,s)]^m \leqq C(x)$$

($C(x)$ is introduced in Theorem 1).

Indeed,

$$\begin{aligned}\mathbb{M}[\rho(x,t,s)]^m &\leqq \mathbb{M}\exp\left\{m\int_t^s \alpha(u,\eta_{x,t}(u))\,dw(u)\right\}\\ &\leqq \mathbb{M}\exp\left\{2m^2\int_t^s \alpha^2(u,\eta_{x,t}(u))\,du\right\}\\ &\leqq \frac{1}{s-t}\int_t^s \mathbb{M}\exp\{2m^2(s-t)\,\alpha^2(u,\eta_{x,t}(u))\}\,du < C(x),\end{aligned}$$

if $(s-t)\,2m^2<\delta$.

Corollary 2. *Under the assumptions of Theorem 1 there exists a constant L (depending on x) such that for sufficiently small δ*

$$\mathbb{M}(\rho(x,s,t)-1)^2 \leqq L(s-t).$$

Indeed, by Itô's formula

$$\begin{aligned}\rho(x,s,t)-1 &= \exp\left\{\int_t^s \alpha(u,\eta_{x,t}(u))\,dw(u) - \tfrac{1}{2}\int_t^s \alpha^2(u,\eta_{x,t}(u))\,du\right\} - 1\\ &= \int_t^s \rho(x,t,u)\,\alpha(u,\eta_{x,t}(u))\,dw(u).\end{aligned}$$

Hence

$$\begin{aligned}\mathbb{M}(\rho(x,t,u)-1)^2 &= \int_t^s \mathbb{M}\,\rho^2(x,t,u)\,\alpha^2(u,\eta_{x,t}(u))\,du\\ &\leqq \left[\int_t^s \mathbb{M}\,\rho^4(x,t,u)\,du\right]^{\frac{1}{2}}\left[\int_t^s \mathbb{M}\,\alpha^4(u,\eta_{x,t}(u))\,du\right]^{\frac{1}{2}}\\ &\leqq [C(x)(s-t)]^{\frac{1}{2}}\left[\int_t^s \mathbb{M}\frac{2}{\delta^2}\,e^{\delta\alpha^2(u,\eta_{x,t}(u))}\,du\right]^{\frac{1}{2}}\\ &\leqq \frac{C(x)\sqrt{2}\,(s-t)}{\delta},\end{aligned}$$

if $8(s-t)<\delta$.

Theorem 2. *Assume the assumptions of Theorem 1 are fulfilled, and that the function* $a^*(t,x)=a(t,x)+\alpha(t,x)\,\sigma(t,x)$ *is continuous in all its arguments and for some* $K>0$ *satisfies* $|a^*(t,x)|^2\leqq K(1+x^2)$. *Then the process* $\{\eta(t),\mathbb{P}^*\}$ *is a diffusion with coefficients* $b^*(t,x)=\sigma^2(t,x)$ *and* $a^*(t,x)$.

Proof. Since on the basis of Corollary 1 and Theorem 4 § 6

$$\begin{aligned}\mathbb{M}^*(\eta_{x,t}(s)-x)^4 &= \mathbb{M}(\eta_{x,t}(s)-x)^4\,\rho(x,t,s)\\ &\leqq (\mathbb{M}(\eta_{x,t}(s)-x)^8)^{\frac{1}{2}}\,(\mathbb{M}\,\rho^2(x,t,s))^{\frac{1}{2}} = O(s-t)^2\end{aligned}$$

it is sufficient to show that

$$\begin{aligned}\mathbb{M}^*(\eta_{x,t}(s)-x) &= a^*(x,t)(s-t)+o(s-t),\\ \mathbb{M}^*(\eta_{x,t}(s)-x)^2 &= \sigma^2(t,x)(s-t)+o(s-t).\end{aligned}\tag{6}$$

The second of these relations follows from the inequality

$$\begin{aligned}|\mathbb{M}^*(\eta_{x,t}(s)-x)^2-\mathbb{M}(\eta_{x,t}(s)-x)^2| &= |\mathbb{M}(\eta_{x,t}(s)-x)^2(\rho(x,t,s)-1)|\\ &\leqq [\mathbb{M}(\eta_{x,t}(s)-x)^4]^{\frac{1}{2}}(\mathbb{M}(\rho(x,t,s)-1)^2)^{\frac{1}{2}}\\ &= O(s-t)^{\frac{3}{2}}\end{aligned}$$

and Theorem 2 § 10. We now show that the first is also true. We have

$$\begin{aligned}\mathbb{M}^*(\eta_{x,t}(s)-x) &= \mathbb{M}(\eta_{x,t}(s)-x)\,\rho(x,t,s)\\ &= \mathbb{M}\left(\int_t^s(\sigma(u,\eta_{x,t}(u)))\,dw(u)+\int_t^s a(u,\eta_{x,t}(u))\,du\right)\\ &\quad\times\left(1+\int_t^s \rho(x,t,u)\,\alpha(u,\eta_{x,t}(u))\,dw(u)\right)\end{aligned}$$

$$=\mathbb{M}\int_t^s \left(a(u,\eta_{x,t}(u))+\sigma(u,\eta_{x,t}(u))\,\alpha(u,\eta_{x,t}(u))\right)du$$
$$+\mathbb{M}\int_t^s a(u,\eta_{x,t}(u))\,du(\rho(x,t,s)-1)$$
$$+\mathbb{M}\int_t^s \sigma(u,\eta_{x,t}(u))\,\alpha(u,\eta_{x,t}(u))(\rho(x,t,u)-1)\,du.$$

Moreover,

$$\left|\mathbb{M}\int_t^s a(u,\eta_{x,t}(u))\,du(\rho(x,t,s)-1)\right.$$
$$\leqq\left\{\mathbb{M}\left[\int_t^s a(u,\eta_{x,t}(u))\,du\right]^2\right\}^{\frac{1}{2}}(\mathbb{M}\,|\rho(x,t,s)-1|^2)^{\frac{1}{2}}=O(s-t)^{\frac{3}{2}}$$

and

$$\left|\mathbb{M}\int_t^s \sigma(u,\eta_{x,t}(u))\,\alpha(u,\eta_{x,t}(u))(\rho(x,t,u)-1)\,du\right|$$
$$\leqq\left(\int_t^s \mathbb{M}\,\sigma^4(u,\eta_{x,t}(u))\,du\right)^{\frac{1}{4}}\left(\int_t^s \mathbb{M}(\rho(x,t,u)-1)^2\,du\right)^{\frac{1}{2}}\left(\int_t^s \mathbb{M}\,\alpha^4(u,\eta_{x,t}(u))\,du\right)^{\frac{1}{4}}$$
$$=O(s-t)^{\frac{1}{4}+1+\frac{1}{4}}.$$

Therefore

$$\mathbb{M}^*(\eta_{x,t}(s)-x)=\mathbb{M}\int_t^s a^*(u,\eta_{x,t}(u))\,du+o(s-t)^{\frac{3}{2}}.$$

From this, as in Theorem 2 § 10, there follows (6). The theorem is finished.

Lemma 3. *Set* $w^*(t)=w(t)+\int_0^t \alpha(s,\eta(s))\,ds$. *Then under the conditions of Theorem 1, the process* $\{w^*(t),\mathbb{P}^*\}$ *is Wiener.*

Proof. Since $w^*(t)$ is continuous w.p. 1 it is sufficient from Theorem 1 to show that

$$\mathbb{M}^*[w^*(t+h)-w^*(t)/\mathfrak{F}_t]=0,\qquad \mathbb{M}^*\left[(w^*(t+h)-w^*(t))^2/\mathfrak{F}_t\right]=h. \tag{7}$$

But

$$\mathbb{M}^*[w^*(t+h)-w^*(t)/\mathfrak{F}_t]$$
$$=\mathbb{M}\left[(w(t+h)-w(t))\,\rho(\eta(t),t,t+h)-\int_t^{t+h}\alpha(s,\eta(s))\,ds\,\rho(\eta(t),t,t+h)/\mathfrak{F}_t\right]$$

$$=\mathbb{M}\left[\int_t^{t+h} dw(s)\int_t^{t+h}\rho(\eta(t),t,s)\,\alpha(s,\eta(s))\,dw(s)\right.$$
$$\left.-\int_t^{t+h}\alpha(s,\eta(s))\,\rho(\eta(t),t,t+h)\,ds/\mathfrak{F}_t\right]$$
$$=\mathbb{M}\left[\int_t^{t+h}\alpha(s,\eta(s))\,\rho(\eta(t),t,s)\,ds\right.$$
$$\left.-\int_t^{t+h}\alpha(s,\eta(s))\,\rho(\eta(t),t,s)\,\rho(\eta(s),s,t+h)\,ds/\mathfrak{F}_t\right]$$
$$=\mathbb{M}\left[\int_t^{t+h}\alpha(s,\eta(s))\,\rho(\eta(t),t,s)\,\mathbb{M}(1-\rho(\eta(s),s,t+h)/\mathfrak{F}_s)\,ds/\mathfrak{F}_t\right]=0.$$

Moreover,

$$\mathbb{M}^*[(w^*(t+h)-w^*(t))^2/\mathfrak{F}_t]=\mathbb{M}^*\left(\sum_{k=1}^{n}[w^*(t_k)-w^*(t_{k-1})]^2/\mathfrak{F}_t\right),$$

where $t=t_0<\cdots<t_n=t+h$. Thus,

$$\mathbb{M}^*[(w^*(t+h)-w^*(t))^2/\mathfrak{F}_t]$$
$$=\mathbb{M}\left(\sum_{k=1}^{n}(w(t_k)-w(t_{k-1}))^2\,\rho(\eta(t),t,t+h)/\mathfrak{F}_t\right)$$
$$+2\mathbb{M}\left(\sum_{k=1}^{n}(w(t_k)-w(t_{k-1}))\int_{t_{k-1}}^{t_k}\alpha(u,\eta(u))\,du\,\rho(\eta(t),t,t+h)/\mathfrak{F}_t\right)$$
$$+\mathbb{M}\left(\sum_{k=1}^{n}\left(\int_{t_{k-1}}^{t_k}\alpha(u,\eta(u))\,du\right)^2\rho(\eta(t),t,t+h)/\mathfrak{F}_t\right).$$

Since

$$\left|\mathbb{M}\left(\sum_{k=1}^{n}(w(t_k)-w(t_{k-1}))\int_{t_{k-1}}^{t_k}\alpha(u,\eta(u))\,du\,\rho(\eta(t),t,t+h)/\mathfrak{F}_t\right)\right|$$
$$\leqq\left[\mathbb{M}\left(\sum_{k=1}^{n}(w(t_k)-w(t_{k-1}))^2\,\rho(\eta(t),t,t+h)/\mathfrak{F}_t\right)\right]^{\frac{1}{2}}$$
$$\times\left[\mathbb{M}\left(\sum_{k=1}^{n}\left[\int_{t_{k-1}}^{t_k}\alpha(u,\eta(u))\,du\right]^2\rho(\eta(t),t,t+h)/\mathfrak{F}\right)\right]^{\frac{1}{2}},$$

it is sufficient for the proof of the second part of (7) to show that

$$\lim_{\lambda\to 0}\mathbb{M}\left(\sum_{k=1}^{n}\mathbb{M}(w(t_k)-w(t_{k-1}))^2\,\rho(\eta(t),t,h+t)/\mathfrak{F}_t\right)=h, \tag{8}$$

$$\lim_{\lambda\to 0}\mathbb{M}\left(\sum_{k=1}^{n}\left(\int_{t_{k-1}}^{t_k}\alpha(u,\eta(u))\,du\right)^2\rho(\eta(t),t,t+h)/\mathfrak{F}_t\right)=0, \tag{9}$$

$$\lambda=\max_k(t_k-t_{k-1}).$$

But

$$\left|\mathbb{M}\left(\left[\sum_{k=1}^{n}(w(t_k)-w(t_{k-1}))^2-h\right]\rho(\eta(t),t,t+h)/\mathfrak{F}_t\right)\right|$$
$$\leq\left[\mathbb{M}\left(\left[\sum_{k=1}^{n}(w(t_k)-w(t_{k-1}))^2-h\right]^2/\mathfrak{F}_t\right)\right]^{\frac{1}{2}}[\mathbb{M}(\rho^2(\eta(t),t,t+h)/\mathfrak{F}_t)]^{\frac{1}{2}}\to 0,$$

if h is so small that $\mathbb{M}(\rho^2(\eta(t),t,t+h)/\mathfrak{F}_t)\leq C(\eta(t))$. Using the fact that $\alpha^4\leq\frac{2}{\delta^2}e^{\delta\alpha^2}$, we can also establish (9). The proof is finished.

Theorem 3. *Let $\xi_1(t)$ and $\xi_2(t)$ be solutions of the stochastic differential equation*

$$d\xi_i(t)=a^{(i)}(t,\xi_i(t))\,dt+\sigma(t,\xi_i(t))\,dw(t)$$

on the interval $[0,T]$ with the single initial condition $\xi_i(0)=\xi(0)$ and μ_i $(i=1,2)$ measures corresponding to the two processes on $\mathfrak{F}[0,T]$. If the coefficients $a^{(1)}(t,x)$, $a^{(2)}(t,x)$ and $\sigma(t,x)$ are such that the assumptions of the existence and uniqueness theorem are satisfied and if for the function

$$\alpha(t,x)=\frac{a^{(2)}(t,x)-a^{(1)}(t,x)}{\sigma(t,x)}$$

there exists a $\delta>0$ such that $\sup_t\mathbb{M}\exp\{\delta\alpha^2(t,\xi_1(t))\}<\infty$, then the measure μ_2 will be absolutely continuous w.r.t. the measure μ_1 and

$$\frac{d\mu_2}{d\mu_1}(\xi_1(\cdot))=\exp\left\{\int_0^T\alpha(s,\xi_1(s))\,dw(s)-\tfrac{1}{2}\int_0^T\alpha^2(s,\xi_1(s))\,ds\right\}. \tag{10}$$

Proof. From Lemma 3 the process $\{w^*(t),\mathbb{P}^*\}$, where

$$w^*(t)=w(t)+\int_0^t\alpha(u,\xi_1(u))\,du,$$

is a Wiener process. Since $\xi_1(t)$ is a solution of

$$d\xi_1(t)=a^{(2)}(t,\xi_1(t))\,dt+\sigma(t,\xi_1(t))\,dw^*(t),$$

the process $\{\xi_1(x),\mathbb{P}^*\}$ has the same distribution as $\{\xi_2(t),\mathbb{P}\}$. Thus the same measure μ_2 corresponds to both $\{\xi_1(t),\mathbb{P}^*\}$ and $\{\xi_2(t),\mathbb{P}\}$ on $\mathfrak{F}[0,T]$. Since the measures μ_1 and μ_2 correspond to processes which can be obtained from one another by means of on absolutely continuous transformation of the measure on the probability space, we have

$$\frac{d\mu_2}{d\mu_1}(\xi_1(\cdot))=\frac{d\mathbb{P}^*}{d\mathbb{P}}(\omega).$$

The proof is finished.

§ 13. Formulas for Transition Density Functions

We will use the results of the preceding paragraph to prove the existence of and derive formulas for the transition density function of a diffusion process. We assume initially that the process has translation (drift) coefficient $a(t, x)$ and diffusion coefficient equal to unity. It follows from Theorem 3 § 10 that such a process satisfies the stochastic differential equation

$$d\xi(t) = a(t, \xi(t))\,dt + dw(t). \tag{1}$$

Let $a(t, x)$ be continuous and satisfy the conditions of the existence and uniqueness theorem. Denote by $\xi_{x,t}(s)$ the solution of (1) for $s>t$ with initial condition $\xi_{x,t}(t)=x$. From Theorem 1 § 10 the transition probability for the process $\xi(t)$ coincides with the function

$$P(t, x, s, A) = \mathbb{P}\{\xi_{x,t}(s) \in A\}.$$

Let $\bar{\xi}_{x,t}(s) = x + w(s) - w(t)$. This process is a solution of the equation $d\bar{\xi}_{x,t}(s) = dw(s)$ for $s>t$ with unitial condition $\bar{\xi}_{x,t}(t)=x$. We conclude from Theorem 3 § 12 that

$$P(t, x, s, A) = \mathbb{M}\,\chi_A(\bar{\xi}_{x,t}(s)) \exp\left\{\int_t^s \alpha(u, \bar{\xi}_{x,t}(u))\,dw(u) - \tfrac{1}{2}\int_t^s a^2(u, \bar{\xi}_{x,t}(u))\,du\right\}$$

(since $|a(s,x)|^2 \leqq K(1+x^2)$, we have

$$\mathbb{M}\exp\{\delta\,|a(s, \bar{\xi}_{x,t}(s))|^2\} = O(\mathbb{M}\exp\{\delta K\,|w(s)-w(t)|^2\}) < \infty,$$

provided that $2K\delta(s-t)<1$). Setting

$$\begin{aligned}\Phi&(t, s, x, \bar{\xi}_{x,t}(s))\\ &= \mathbb{M}\left(\exp\left\{\int_t^s a(u, \bar{\xi}_{x,t}(u))\,dw(u) - \tfrac{1}{2}\int_t^s a^2(u, \bar{\xi}_{x,t}(u))\,du\right\} / \bar{\xi}_{x,t}(s)\right),\end{aligned}$$

we obtain

$$\begin{aligned}P(t, x, s, A) &= \mathbb{M}\,\chi_A(\bar{\xi}_{x,t}(s))\,\Phi(t, s, x, \bar{\xi}_{x,t}(s))\\ &= \int \chi_A(y)\,\Phi(t, s, x, y)\,\mathbb{P}\{\bar{\xi}_{x,t}(s) \in dy\}\\ &= \int_A \frac{1}{\sqrt{2\pi(s-t)}}\,\Phi(t, s, x, y)\,e^{-\frac{(y-x)^2}{2(s-t)}}\,dy.\end{aligned}$$

The latter expression shows that there exists a density $p(t, x, s, y)$ corresponding to the transition probability, i.e., the function $P(t, x, s, A)$ can be represented by $P(t, x, s, A) = \int_A p(t, x, s, y)\,dy$, with

$$p(t, x, s, y) = \frac{1}{\sqrt{2\pi(s-t)}} \exp\left\{-\frac{(y-x)^2}{2(s-t)}\right\} \Phi(t, s, x, y). \tag{2}$$

The function $\Phi(t, s, x, y)$ is expressed by means of a certain conditional expectation and is somewhat inconvienient in this form.

Assuming additional smoothness properties for the coefficient $a(t, x)$, we will find an expression for $\Phi(t, s, x, y)$ in terms of the unconditional mathematical expectation of a certain expression. With this goal in mind we prove several auxiliary lemmas.

Lemma 1. *If $w(t)$ is a Wiener process, then the random process*

$$\eta(u)=w(u)-w(t)-\frac{u-t}{s-t}[w(s)-w(t)],$$

defined for $u\in(t, s)$, is independent of $w(t)$ and $w(s)$.

Proof. That $w(s)-w(t)$ and $\eta(u)$ are independent of $w(t)$ follows from the fact that the Wiener process has independent increments. Since $w(s), \eta(u_1), \ldots, \eta(u_k); u_1, \ldots, u_k\in[t, s]$ have a joint Gaussian distribution and $\mathbb{M}\, w(s)=\mathbb{M}\,\eta(u)=0$, it is sufficient to show that $w(s)$ is uncorrelated with $\eta(u)$, i.e., that $\mathbb{M}\, w(s)\,\eta(u)=0$ for $u\in[t, s]$. But

$$\begin{aligned}\mathbb{M}\, w(s)\,\eta(u)&=\mathbb{M}\, w(s)\,[w(u)-w(t)]-\frac{u-t}{s-t}\mathbb{M}\, w(s)\,[w(s)-w(t)]\\&=\mathbb{M}\,[w(s)-w(u)]\,[w(u)-w(t)]+\mathbb{M}\,[w(u)-w(t)]^2\\&\quad+\mathbb{M}\, w(t)\,[w(u)-w(t)]-\frac{u-t}{s-t}\mathbb{M}\,[w(s)-w(t)]^2\\&\quad-\frac{u-t}{s-t}\mathbb{M}\, w(t)\,[w(s)-w(t)]\\&=u-t-\frac{u-t}{s-t}(s-t)=0\end{aligned}$$

(again we have used the independence of the increments of $w(u)$). The lemma is proved.

Lemma 2. *Let the random variable ζ have the form*

$$\zeta=g\left(\int_t^s K\big(u, w(t), w(s), \eta(u)\big)\, du\right),$$

where $g(x)$ is continuous, $K(u, x, y, z)$ is an integrable function of its arguments and $\eta(u)$ is the process defined in Lemma 1. Then

$$\mathbb{M}\,(\zeta/w(t), w(s))=r\big(w(t), w(s)\big), \tag{3}$$

where

$$r(x, y)=\mathbb{M}\, g\left(\int_t^s K\big(u, x, y, \eta(u)\big)\, du\right).$$

Proof. To show (3) it is sufficient to prove that for any measurable function $\lambda(x, y)$

$$\mathbb{M}\zeta\lambda(w(t), w(s)) = \mathbb{M}r(w(t), w(s))\lambda(w(t), w(s)). \tag{4}$$

Consider the random variable $\zeta_n = \psi(w(t), w(s), \eta(u_1), \ldots, \eta(u_n))$, where $\psi(x, y, z_1 \ldots z_n)$ is a measurable, bounded function of all its arguments and $r_n(x, y) = \mathbb{M}\psi(x, y, \eta(u_1), \ldots, \eta(u_n))$. Then

$$\begin{aligned}\mathbb{M}\zeta_n\lambda(w(t), w(s)) &= \mathbb{M}\psi(w(t), w(s)), \eta(u_1), \ldots, \eta(u_n)\lambda(w(t), w(s))\\ &= \int \psi(x, y, z_1, \ldots, z_n)\lambda(x, y)\, g_n(dz_1, \ldots, dz_n)\, P(dx, dy)\\ &= \int \lambda(x, y)\left[\int \psi(x, y, z_1, \ldots, z_n)\, g_n(dz_1, \ldots, dz_n)\right] P(dx, dy),\end{aligned}$$

where $g_n(dz_1, \ldots, dz_n)$ is the joint distribution of $\eta(u_1), \ldots, \eta(u_n)$, and $P(dx, dy)$ is the joint distribution of $w(t)$ and $w(s)$. Since

$$r_n(x, y) = \int \psi(x, y, z_1, \ldots, z_n)\, g_n(dz_1, \ldots, dz_n),$$

we have

$$\begin{aligned}\mathbb{M}\zeta_n\lambda(w(t), w(s)) &= \int \lambda(x, y)\, r_n(x, y)\, P(dx, dy)\\ &= \mathbb{M}r_n(w(t), w(s))\lambda(w(t), w(s)).\end{aligned}$$

Consequently, (4) holds if ζ and r are replaced in it by ζ_n and r_n. We now note that we can choose a sequence ζ_n converging in probability to ζ in such a way that $r_n \to r$ and that it is possible to go to the limit under the expectation sign. The proof is finished.

We will transform the expression

$$\exp\left\{\int_t^s a(u, \bar{\xi}_{x,t}(u))\, dw(u) - \tfrac{1}{2}\int_t^s a^2(u, \bar{\xi}_{x,t}(u))\, du\right\} \tag{5}$$

in such a way that Lemma 2 can be applied. Assume that $a(u, x)$ is differentiable in u and x. Then, letting

$$A(u, x) = \int_0^x a(u, y)\, dy,$$

we find from Itô's formula

$$\begin{aligned}A(s, \bar{\xi}_{x,t}(s)) - A(t, \xi_{x,t}(t)) &= \int_t^s \frac{\partial A}{\partial x}(u, \bar{\xi}_{x,t}(u))\, dw(u)\\ &\quad + \int_t^s \left[\frac{\partial A}{\partial u}(u, \bar{\xi}_{x,t}(u)) + \frac{1}{2}\cdot\frac{\partial^2 A}{\partial x^2}(u, \bar{\xi}_{x,t}(u))\right] du\end{aligned}$$

$(d\bar{\xi}_{x,t}(u)=dw(u))$. Since $\dfrac{\partial A}{\partial x}=a(u,x)$, $\bar{\xi}_{x,t}(t)=x$, we have

$$\int_t^s a(u,\bar{\xi}_{x,t}(u))\,dw(u)=A(s,\bar{\xi}_{x,t}(s))-A(t,x)$$
$$-\int_t^s\left[\frac{\partial A}{\partial u}(u,\bar{\xi}_{x,t}(u))+\frac{1}{2}\cdot\frac{\partial a}{\partial x}(u,\bar{\xi}_{x,t}(u))\right]du.$$

Hence (5) can be rewritten as

$$\exp\left\{A(s,\bar{\xi}_{x,t}(s))-A(t,x)+\int_t^s B(u,\bar{\xi}_{x,t}(u))\,du\right\}, \tag{6}$$

where

$$B(u,x)=-\tfrac{1}{2}a^2(u,x)-\tfrac{1}{2}a'_x(u,x)-\int_0^x a'_u(u,y)\,dy.$$

Furthermore,

$$\bar{\xi}_{x,t}(u)=x+w(u)-w(t)=x+\eta(u)+\frac{u-t}{s-t}[w(s)-w(t)].$$

Substituting this expression in (6) we obtain

$$\exp\left\{A(s,x+w(s)-w(t))-A(t,x)\right.$$
$$\left.+\int_t^s B\left(u,x+\eta(u)+\frac{u-t}{s-t}[w(s)-w(t)]\right)du\right\},$$

thus from Lemma 2

$$\Phi(t,s,x,\bar{\xi}_{x,t}(s))=\mathbb{M}\left(\exp\left\{A(s,x+w(s)-w(t))-A(t,x)\right.\right.$$
$$\left.\left.+\int_t^s B\left(u,x+\eta(u)+\frac{u-t}{s-t}[w(s)-w(t)]\right)du\right\}/\bar{\xi}_{x,t}(s)\right)$$
$$=\exp\{A(s,\bar{\xi}_{x,t}(s))-A(t,x)\}$$
$$\times\mathbb{M}\left(\exp\left\{\int_t^s B\left(u,x+\eta(u)+\frac{u-t}{s-t}[\bar{\xi}_{x,t}(s)-x]\right)du\right\}/\bar{\xi}_{x,t}(s)\right)$$
$$=\exp\{A(s,\bar{\xi}_{x,t}(s))-A(t,x)\}\,G(\bar{\xi}_{x,t}(s)),$$

where

$$G(y)=\mathbb{M}\exp\left\{\int_t^s B\left(u,x+\eta(u)+\frac{u-t}{s-t}(y-x)\right)du\right\}.$$

Substituting the value of $\Phi(t, s, x, y)$ into (2), we obtain the following formula for the transition density

$$p(t, x, s, y) = \frac{1}{\sqrt{2\pi(s-t)}} \exp\left\{-\frac{(y-x)^2}{2(s-t)} + A(s, y) - A(t, x)\right\}$$
$$\times \mathbb{M} \exp\left\{\int_t^s B\left(u, x + \eta(u) + \frac{u-t}{s-t}(y-x)\right) du\right\}.$$

This formula is inconvienient because $\eta(u)$ depends on t and s. Let us perform the substitution $\frac{u-t}{s-t} = v$, $u = t + (s-t)v$, $0 \leqq v \leqq 1$, $\eta(t + (s-t)v) = w(t + (s-t)v) - w(t) - [w(s) - w(t)]$. The process

$$w^*(v) = \frac{w(t + (s-t)v) - w(t)}{\sqrt{s-t}}$$

is also Wiener. Setting $\eta^*(u) = w^*(u) - u w^*(1)$ for $0 \leqq u \leqq 1$, we have the following (in its final form) formula for the transition probability density:

$$\begin{aligned} p(t, x, s, y) &= \frac{1}{\sqrt{2\pi(s-t)}} \exp\left\{-\frac{(y-x)^2}{2(s-t)} + A(s, y) - A(t, x)\right\} \\ &\quad \times \mathbb{M} \exp\left\{(s-t) \int_0^1 B\left(t + u(s-t), x + \sqrt{s-t}\,\eta^*(u) + u(y-x)\right) du\right\}. \end{aligned} \tag{7}$$

We now find conditions which guarantee the finiteness of the expectation on the right side of (7).

Lemma 3. *For the measurable function $C(u, x)$ let there exist a $\delta < 2$ and a $K > 0$ such that $C(u, x) < K + \delta x^2$. Then*

$$\mathbb{M} \exp\left\{\int_0^1 C(u, \eta^*(u))\, du\right\} < \infty.$$

To prove this it suffices to show that

$$\mathbb{M} \exp\left\{\delta \int_0^1 [\eta^*(u)]^2\, du\right\} < \infty.$$

But by Jensen's theorem

$$\exp\left\{\delta \int_0^1 [\eta^*(u)]^2\, du\right\} \leqq \int_0^1 e^{\delta[\eta^*(u)]^2}\, du,$$

further, $\eta^*(u)$ is normally distributed with variance $u(1-u)$ and mean zero, so that

$$\mathbb{M}\, e^{\delta[\eta^*(u)]^2} \leqq \mathbb{M}\, e^{\delta u(1-u)\xi^2} \leqq \mathbb{M}\, e^{\frac{\delta}{4}\xi^2},$$

where ξ is normally distributed, $\mathbb{M}\xi=0$, $\operatorname{Var}\xi=1$. Consequently,

$$\mathbb{M}\, e^{\frac{\delta}{4}\xi^2} = \frac{1}{\sqrt{2\pi}} \int_{-\infty}^{\infty} e^{\left(\frac{\delta}{4}-\frac{1}{2}\right)x^2}\, dx = \sqrt{\frac{2}{2-\delta}}.$$

Finally,

$$\mathbb{M}\exp\left\{\delta \int_0^1 [\eta^*(u)]^2\, du\right\} \leqq \sqrt{\frac{2}{2-\delta}}.$$

The lemma is proved.

Theorem 1. *Assume that the derivatives $a'_x(t,x)$, $a'_t(t,x)$ of the coefficient $a(t,x)$ of the diffusion $\xi(t)$ exist, that $\sigma(t,x)=1$ and that the function*

$$B(t,x) = -\tfrac{1}{2}a^2(t,x) - \tfrac{1}{2}a'_x(t,x) - \int_0^x a'_t(t,y)\, dy$$

satisfies

$$\overline{\lim_{|x|\to\infty}} \frac{\sup_{0\leqq t\leqq T} B(t,x)}{1+x^2} = 0.$$

Then for all $0\leqq t<s\leqq T$ the transition probability density of the diffusion is given by (7).

Remark 1. If in place of the condition imposed above on $B(t,x)$ we require that there exist a K for which $B(t,x)\leqq K(1+x^2)$, then formula (7) still holds when $K(s-t)^2<1$.

Theorem 2. *Let $\xi(t)$ be a diffusion on $[0,T]$ with coefficients $a(t,x)$ and $\sigma^2(t,x)>0$ satisfying the conditions:*

1) $a'_x(t,x)$ and $\sigma'_x(t,x)$ exist and are bounded; the derivatives σ''_{xx}, $\sigma''_{x,t}$, $\sigma''_{t,t}$ and σ'_t exist;

2) $g(t,x)$ is defined by $x=\int_0^{g(t,x)} \frac{dy}{\sigma(t,y)}$;

$$\bar{a}(t,x) = \int_0^{g(t,x)} \frac{\sigma'_t(t,y)}{\sigma^2(t,y)}\, dy + \frac{a(t,g(t,x))}{\sigma(t,g(t,x))} - \frac{1}{2}\sigma'_x(t,g(t,x)),$$

$$\bar{B}(t,x) = -\tfrac{1}{2}\bar{a}^2(t,x) - \tfrac{1}{2}\bar{a}'_x(t,x) - \int_0^x \bar{a}'_t(t,y)\, dy,$$

$\bar{B}(t,x)$ satisfies

$$\overline{\lim_{|x|\to\infty}} \frac{1}{1+x^2} \sup_{0\leqq t\leqq T} \bar{B}(t,x) \leqq 0.$$

Then under these assumptions the transition probability density of the process exists and is defined by the formula

$$\begin{aligned}
p(t,x,s,y)=&\frac{1}{\sqrt{2\pi(s-t)}\,\sigma(s,y)}\\
&\times\exp\left\{-\frac{(f(s,y)-f(t,x))^2}{2(s-t)}+\bar{A}(s,f(s,y))-\bar{A}(t,f(t,x))\right\}\\
&\times\mathbb{M}\exp\left\{(s-t)\int_0^1\bar{B}\big(t+u(s-t),f(t,x)\right.\\
&\qquad\left.+\sqrt{s-t}\,\eta^*(u)+u[f(s,y)-f(t,x)]\big)\,du\right\},
\end{aligned}\tag{8}$$

where

$$\bar{A}(s,x)=\int_0^x\bar{a}(s,y)\,dy,\qquad f(t,x)=\int_0^x\frac{dy}{\sigma(t,y)}.$$

Proof. We introduce the diffusion process $\eta(t)=f(t,\xi(t))$ with coefficients $\bar{b}(t,x)=1$ and $\bar{a}(t,x)$ defined in Condition 2 of the theorem. The transition probability density $\bar{p}(t,x,s,y)$ for $\eta(t)$ is given by (7) if we replace in it $A(t,x)$ by $\bar{A}(t,x)$ and $B(t,x)$ by $\bar{B}(t,x)$. Finally we have

$$p(t,x,s,y)=\bar{p}(t,f(t,x),s,f(s,y))\,f_y'(s,y)=\frac{1}{\sigma(s,y)}\,\bar{p}(t,f(t,x),s,f(s,y)).$$

The theorem is proved.

Formula (8) can be simplified considerably if the coefficients are independent of time, i.e., $a(t,x)=a(x)$ and $b(t,x)=\sigma^2(x)=b(x)$. In this case,

$$\bar{a}(x)=\frac{a(g(x))}{\sigma(g(x))}-\frac{1}{2}\sigma'(g(x)),$$

where $g(x)$ is a solution of $x=\int_0^{g(x)}\frac{dy}{\sigma(y)}$,

$$f(x)=\int_0^x\frac{dy}{\sigma(y)},\qquad (f(g(x))=x),$$

$$\bar{A}(f(x))=\int_0^{f(x)}\left[\frac{a(g(y))}{\sigma(g(y))}-\frac{1}{2}\sigma'(g(y))\right]dy.$$

Performing in the integral the substitution $g(y)=z$, we obtain

$$\bar{A}(f(x))=\int_0^x\left(\frac{a(z)}{\sigma(z)}-\frac{1}{2}\sigma'(z)\right)\frac{dz}{\sigma(z)}=\int_0^x\frac{a(z)}{b(z)}\,dz-\frac{1}{2}\ln\frac{\sigma(x)}{\sigma(0)}.$$

Setting $\bar{B}(x) = -\frac{1}{2}\bar{a}^2(x) - \frac{1}{2}a'_x(x)$, we find

$$\begin{aligned} &p(t,x,s,y)\\ &\quad= \frac{1}{\sqrt{2\pi(s-t)\,b(y)}}\left(\frac{b(x)}{b(y)}\right)^{\frac{1}{4}}\\ &\quad\times \exp\left\{-\frac{1}{2(s-t)}\left(\int_x^y \frac{dz}{\sqrt{b(z)}}\right)^2 + \int_x^y \frac{a(z)}{b(z)}\,dz\right\} \qquad (9)\\ &\quad\times \mathbb{M}\exp\left\{(s-t)\int_0^1 \bar{B}\left(f(x)+\sqrt{s-t}\,\eta^*(u)+u[f(y)-f(x)]\right)du\right\}. \end{aligned}$$

We now treat the question of the existence of derivatives of the transition density given by (8). To this end we investigate the possibility of differentiating under the expectation sign in formula (8).

Lemma 4. *Assume the function $c(\alpha,t,x)$ satisfies the conditions*

1) $\displaystyle\lim_{|x|\to\infty}\ \sup_{\substack{0\le t\le T\\ a\le\alpha\le b}} \frac{c(\alpha,t,x)}{1+x^2} \le 0;$

2) *$c'_\alpha(\alpha,t,x)$ exists and for all $\delta>0$*

$$\lim_{|x|\to\infty}\ \sup_{\substack{0\le t\le T\\ \alpha\in[a,b]}} |c'_\alpha(\alpha,t,x)|\,e^{-\delta x^2} = 0.$$

Then the function

$$\psi(\alpha) = \mathbb{M}\exp\left\{\int_0^1 c(\alpha,u,\eta^*(u))\,du\right\}$$

is differentiable in α for $\alpha\in[a,b]$; if, moreover, $c(\alpha,t,x)$ and $c'_\alpha(\alpha,t,x)$ are continuous in α, then $\psi'(\alpha)$ will also be continuous.

Proof. The formal derivative w.r.t. α has the form

$$\mathbb{M}\exp\left\{\int_0^1 c(\alpha,u,\eta^*(u))\,du\right\}\int_0^1 c'_\alpha(\alpha,u,\eta^*(u))\,du.$$

The indicated mathematical expectation converges uniformly with respect to α since

$$\exp\left\{\int_0^1 c(\alpha,u,\eta^*(u))\,du\right\} \le K_1\int_0^1 e^{\delta[\eta^*(u)]^2}\,du,$$

$$\left|\int_0^1 c'_\alpha(\alpha,u,\eta^*(u))\,du\right| \le K_2\int_0^1 e^{\delta[\eta^*(u)]^2}\,du$$

and, consequently, the expression under the expectation sign is bounded by a variable, having (for sufficiently small δ) a finite expectation. We can then differentiate the expectation w.r.t. the parameter. If the function under the expectation sign depends continuously on α, then from uniform convergence there follows the continuity of the expectation as function of the parameter. The lemma is proved.

Applying Lemma 4 to the transition probability density at (8), we can obtain the following

Theorem 3. *Assume that the assumptions of Theorem 2 are fulfilled and that $a(t,x)$ and $\sigma(t,x)$ have continuous derivatives up to 3rd order satisfying, for $\delta>0$*

$$\lim_{|x|\to\infty}\sup_{0\leqq t\leqq T}\left\{\left|\frac{\partial \bar{B}}{\partial t}(t,x)\right|+\left|\frac{\partial \bar{B}}{\partial x}(t,x)\right|+\left|\frac{\partial^2 B}{\partial x^2}(t,x)\right|\right\}e^{-\delta x^2}=0.$$

Then for $t<s$ the following partial derivatives of the transition probability density (t.p.d.) exist and are continuous:

$$\frac{\partial p(t,x,s,y)}{\partial t},\quad \frac{\partial p(t,x,s,y)}{\partial s},\quad \frac{\partial p(t,x,s,y)}{\partial x},\quad \frac{\partial p(t,x,s,y)}{\partial y},$$

$$\frac{\partial^2 p(t,x,s,y)}{\partial x^2},\quad \frac{\partial^2 p(t,x,s,y)}{\partial y^2}.$$

Remark 2. We consider the question of the boundedness of the t.p.d. From (7) it follows that in case $\sigma(t,x)=1$, $p(t,x,s,y)$ will be bounded for $s-t>\delta$, when the following conditions are satisfied:

1) $\dfrac{\partial}{\partial s}\displaystyle\int_0^x a(s,y)\,dy=\dfrac{\partial}{\partial s}A(s,y)$ is a bounded function,

2) $\displaystyle\lim_{|y-x|\to\infty}\frac{A(s,y)-A(s,x)}{1+(y-x)^2}=0,$

3) $B(s,x)$ a bounded function.

If in place of 3) we require that $\displaystyle\lim_{|x|\to\infty}\sup_{0\leqq s\leqq T}\frac{B(s,x)}{1+x^2}=0$, then $p(t,x,s,y)$ will be bounded in y for x varying through an arbitrary finite interval and $s-t>\delta$; moreover, $p(t,x,s,y)$ will also be bounded in x for y in such an interval and $s-t>\delta$.

§ 14. Kolmogorov's Equation for the Transition Probability Density

Let $\xi(t)$ be a Markov process on $[0,T]$ whose transition density $p(t,x,s,y)$ exists and is measurable in x and for $t<s$.

We take $t_1 < t_2 < t_3$ and consider the double integral

$$\int\int p(t_1, x, t_2, y)\, p(t_2, y, t_3, z)\, dy\, dz. \tag{1}$$

Since the integrated function is nonnegative and

$$\int\left[\int p(t_1, x, t_2, y)\, p(t_2, y, t_3, z)\, dz\right] dy = \int p(t_1, x, t_2, y)\left[\int p(t_2, y, t_3, z)\, dz\right] dy = 1$$

($\int p(t, x, s, y)\, dy = 1$ since $p(t, x, s, y)$ is a probability density w.r.t. y), we conclude from Fubini's theorem that for almost all z the following integral exists

$$\psi(z) = \int p(t_1, x, t_2, y)\, p(t_2, y, t_3, z)\, dy$$

and that $\psi(z)$ is integrable in z. If A is an arbitrary measurable set, then

$$\begin{aligned}\int_A \psi(z)\, dz &= \int_A \left[\int p(t_1, x, t_2, y)\, p(t_2, y, t_3, z)\, dy\right] dz \\ &= \int p(t_1, x, t_2, y)\left[\int_A p(t_2, y, t_3, z)\, dz\right] dy \\ &= \int p(t_1, x, t_2, y)\, P(t_2, y, t_3, A)\, dy \\ &= \int P(t_2, y, t_3, A)\, P(t_1, x, t_2, dy),\end{aligned}$$

where $P(t, x, s, A)$ is the transition probability of the Markov process. The exchange of the order of integration is possible because of Fubini's theorem. Thus, taking into account Property III §9 of the transition probability of a Markov process, we obtain

$$\int_A \psi(z)\, dz = P(t_1, x, t_3, A) = \int_A p(t_1, x, t_3, z)\, dz.$$

This relation is valid for arbitrary measurable sets A. Thus $\psi(z)$ coincides almost everywhere (w.r.t. Lebesgue measure) with $p(t, x, t_3, z)$.

Theorem 1. *If the transition probability density $p(t, x, s, y)$ of a Markov process is measurable in all its arguments, then for each x, $t_1 < t_2 < t_3$ and for almost all z*

$$\int p(t_1, x, t_2, y)\, p(t_2, y, t_3, z)\, dy = p(t_1, x, t_3, z). \tag{2}$$

This relation is called the Chapman-Kolmogorov equation for the transition density.

Remark 1. If the density $p(t, x, s, y)$ is continuous in x and y and the integral in (2) converges uniformly w.r.t. z in each bounded interval (for example, if in each interval $[a, b]$ $\sup_y \sup_{z \in [a, b]} p(t_2, y, t_3, z) < \infty$ is satisfied, we have the conditions of Remark 2 § 13), then (2) holds for all z and two continuous solutions of (2) which coincide almost everywhere are identical.

We consider a diffusion on $[0, T]$ with coefficients $a(t,x)$ and $b(t,x)$ satisfying the assumptions of Theorem 3 § 13. Then the continuous partial derivatives $\frac{\partial p}{\partial t}, \frac{\partial p}{\partial s}, \frac{\partial p}{\partial x}, \frac{\partial p}{\partial y}, \frac{\partial^2 p}{\partial x^2}$ and $\frac{\partial^2 p}{\partial y^2}$ of the density $p(t,x,s,y)$ exist and are continuous.

Under these assumptions we will introduce two partial differential equations for $p(t,x,s,y)$. We first prove the following

Lemma 1. *Let $g(y)$ be a twice continuously differentiable function. Then*

$$\lim_{h\downarrow 0} \frac{1}{h} \int [g(y)-g(x)]\, p(t,x,t+h,y)\, dy = a(t,x)\, g'(x) + \frac{b(t,x)}{2} g''(x). \quad (3)$$

Proof. For each $\varepsilon > 0$

$$\int_{|x-y| \geqq \varepsilon} |g(y)-g(x)|\, p(t,x,t+h,y)\, dy \leqq 2 \sup_y |g(y)| \int_{|x-y| \geqq \varepsilon} P(t,x,t+h,dy) = o(h).$$

Moreover, using Taylor's formula, we find

$$\int_{|x-y| \leqq \varepsilon} (g(y)-g(x))\, p(t,x,t+h,y)\, dy = \int_{|x-y| \leqq \varepsilon} [g'(x)(y-x) + \tfrac{1}{2} g''(x)(y-x)^2 + r_\varepsilon(x,y)(y-x)^2] \times p(t,x,t+h,y)\, dy,$$

where $\lim_{\varepsilon\to 0} \sup_{|x-y|<\varepsilon} |r_\varepsilon(x,y)| = 0$. Taking account of the relations

$$\int_{|y-x| \leqq \varepsilon} (y-x)\, p(t,x,t+h,dy) = a(t,x)\, h + o(h),$$

$$\int_{|y-x| \leqq \varepsilon} (y-x)^2\, p(t,x,t+h,dy) = b(t,x)\, h + o(h),$$

we obtain

$$\frac{1}{h} \int [g(y)-g(x)]\, p(t,x,t+h,y)\, dy - a(t,x)\, g'(x) - \tfrac{1}{2} b(t,x)\, g''(x) = \frac{o(h)}{h} + o\Big(\sup_{|y-x| \leqq \varepsilon} r_\varepsilon(x,y)\Big).$$

Hence

$$\overline{\lim_{h\downarrow 0}} \left| \frac{1}{h} \int [g(y)-g(x)]\, p(t,x,t+h,y)\, dy - a(t,x)\, g'(x) - \tfrac{1}{2} b(t,x)\, g''(x) \right| \leqq O\Big(\sup_{|y-x| \leqq \varepsilon} r_\varepsilon(x,y)\Big).$$

Going to the limit as $\varepsilon \to 0$ we obtain the proof of the lemma.

Remark 2. One proves in exactly the same way that

$$\lim_{h\downarrow 0} \frac{1}{h} \int [g(h,y)-g(h,x)]\, p(t,x,t+h,y)\, dy$$
$$= a(t,x)\, g'_x(0,x) + \tfrac{1}{2} b(t,x)\, g''_{xx}(0,x),$$

provided that $g'_x(h,x)$ and $g''_{xx}(h,x)$ are continuous in all arguments and $g(h,x)$ is bounded.

Theorem 1. *Assume the partial derivatives*

$$\frac{\partial}{\partial x} a(t,x), \quad \frac{\partial}{\partial x} b(t,x), \quad \frac{\partial^2 b(t,x)}{\partial x^2}, \quad \frac{\partial p(t,x,s,y)}{ds},$$

$$\frac{\partial p(t,x,s,y)}{\partial y} \quad \textit{and} \quad \frac{\partial^2 p(t,x,s,y)}{\partial y^2}$$

exist. Then for $s>t$, $p(t,x,s,y)$ satisfies

$$\begin{aligned}\frac{\partial p(t,x,s,y)}{\partial s} = &-\frac{\partial}{\partial y}(a(s,y)\, p(t,x,s,y)) \\ &+\frac{1}{2}\cdot\frac{\partial^2}{\partial y^2}(b(s,y)\, p(t,x,s,y)).\end{aligned} \tag{4}$$

This equation is *Kolmogorov's first equation* or the *Fokker-Planck equation.*

Proof. We take a twice continuously differentiable function $g(z)$ which vanishes outside some finite interval. From (2) follows

$$\int p(t,x,s+h,z)\, g(z)\, dz = \int p(t,x,s,y) \left[\int p(s,y,s+h,z)\, g(z)\, dz\right] dy,$$

consequently,

$$\begin{aligned}&\int [p(t,x,s+h,z)-p(t,x,s,z)]\, g(z)\, dz \\ &\quad = \int p(t,x,s,y)\left[\int p(s,y,s+h,z)[g(z)-g(y)]\, dz\right] dy.\end{aligned} \tag{5}$$

Dividing both sides of (5) by h and letting $h \to 0$ we obtain on the left

$$\int \frac{\partial p}{\partial s}(t,x,s,z)\, g(z)\, dz.$$

The admissability of this limiting procedure follows from the continuity of $\frac{\partial p}{\partial s}$ and the fact that $g(z)$ is different from zero on only a finite interval.

We now show that

$$\lim_{h\to 0} \int p(t,x,s,y)\left[\frac{1}{h}\int p(s,y,s+h,z)\,[g(z)-g(y)]\,dz\right]dy \\ = \int p(t,x,s,y)\,[a(s,y)\,g'(y)+\tfrac{1}{2}b(s,y)\,g''(y)]\,dy. \tag{6}$$

Applying Lemma 1 we see that it is sufficient to justify a limit passage under the integral sign in (6) for $h\to 0$. Let $\xi_{x,s}(u)$ denote a solution of the stochastic equation

$$d\xi_{x,s}(u)=a(u,\xi_{x,s}(u))\,du+\sqrt{b(u,\xi_{x,s}(u))}\,dw(u).$$

It follows from Theorem 1 § 10 that the distribution of $\xi_{x,s}(s+h)$ coincides with the transition probability $P(s,x,s+h,A)$ of the process $\xi(t)$,

$$\int p(s,y,s+h,z)\,[g(z)-g(y)]\,dz=\mathbb{M}\,g(\xi_{y,s}(s+h))-g(y).$$

From Itô's formula

$$\mathbb{M}\,g(\xi_{y,s}(s+h))-g(y) \\ =\mathbb{M}\int_s^{s+h}[a(u,\xi_{y,s}(u))\,g'_x(\xi_{y,s}(u))+\tfrac{1}{2}b(u,\xi_{y,s}(u))\,g''_{xx}(\xi_{y,s}(u))]\,du=O(h),$$

since $a(u,z)\,g'(z)+\frac{1}{2}b(u,z)\,g''(z)$ is continuous and different from zero on only a finite interval and is thus bounded. Hence, for some $C>0$

$$\left|\frac{1}{h}\int p(s,y,s+h,z)\,[g(z)-g(y)]\,dz\right|\leq C$$

and a limit passage in (6) is possible due to Lebesgue's theorem. We have thus shown that

$$\int \frac{\partial p}{\partial s}(t,x,s,z)\,g(z)\,dz=\int p(t,x,s,z)\,(a(s,z)\,g'(z)+\tfrac{1}{2}g''(z))\,dz.$$

A partial integration yields

$$\int p(t,x,s,z)\,a(s,z)\,g'(z)\,dz=-\int g(z)\frac{\partial}{\partial z}(a(s,z)\,p(t,x,s,z))\,dz,$$

$$\int p(t,x,s,z)\,b(s,z)\,g''(z)\,dz=\int g(z)\frac{\partial^2}{\partial z^2}(b(s,z)\,p(t,x,s,z))\,dz,$$

whence

$$\int\left\{\frac{\partial p}{\partial s}(t,x,s,z)+\frac{\partial}{\partial z}[a(s,z)\,p(t,x,s,z)] \\ -\frac{1}{2}\cdot\frac{\partial^2}{\partial z^2}[b(t,z)\,p(t,x,s,z)]\right\}g(z)\,dz=0. \tag{7}$$

Since the expression in curly brackets in (7) is continuous and $g(z)$ is an arbitrary twice continuously differentiable function vanishing outside a finite interval, the integrand in (7) is equal to zero, i.e., (4) holds. The theorem is proved.

Corollary 1. *Let the distribution function $F(y)$ be such that the integrals*

$$\int dF(x)\,p(0,x,t,y),\qquad \int dF(x)\,p'_y(0,x,t,y),\qquad \int dF(x)\,p'_t(0,x,t,y)$$

converge uniformly on each finite interval of variation of y and $t\in[\delta,T]$ with $\delta>0$. We denote

$$F(t,y)=\int_{-\infty}^{\infty} dF(x)\int_{-\infty}^{y} p(0,x,t,z)\,dz$$

($F(t,y)$ is the distribution function of $\xi(t)$ if $\xi(0)$ has the distribution $F(x)$). If

$$\lim_{y\to-\infty}\frac{\partial}{\partial t}F(t,y)+a(t,y)\frac{\partial F(t,y)}{\partial y}-\frac{1}{2}\cdot\frac{\partial}{\partial y}\left(b(t,y)\frac{\partial F(t,y)}{\partial y}\right)=0,$$

then $F(t,x)$ satisfies the following (also called the Fokker-Planck) equation:

$$\frac{\partial F(t,y)}{\partial t}=-a(t,y)\frac{\partial F(t,y)}{\partial y}+\frac{1}{2}\cdot\frac{\partial}{\partial y}\left(b(t,y)\frac{\partial F(t,y)}{\partial y}\right)\tag{8}$$

with initial condition $\lim_{t\to 0}F(t,y)=F(y)$. Eq. (8) *is obtained from* (4) *by integrating w.r.t. $dF(x)$ from $-\infty$ to $+\infty$ and w.r.t. y from $-\infty$ to y.*

Theorem 2. *Assume that the transition density $p(t,x,s,y)$ of a diffusion process on $[0,T]$ with coefficients $a(t,x)$ and $b(t,x)$ satisfies:*

1) *for $s-t>\delta>0$ $p(t,x,s,y)$ is continuous and bounded in x, s and t;*
2) *the partial derivatives*

$$\frac{\partial p}{\partial t}(t,x,s,y),\qquad \frac{\partial p}{\partial x}(t,x,s,y),\qquad \frac{\partial^2 p}{\partial x^2}(t,x,s,y)$$

exist. Then $p(t,x,s,y)$ for $0<t<s$ satisfies

$$\frac{\partial p(t,x,s,y)}{\partial t}=-a(t,x)\frac{\partial p(t,x,s,y)}{\partial x}-\frac{1}{2}b(t,x)\frac{\partial^2 p(t,x,s,y)}{\partial x^2}.\tag{9}$$

This is the backward equation of A. N. Kolmogorov for the transition density.

Proof. From (2) we have

$$p(t+h, x, s, y) - p(t, x, s, y)$$
$$= \int p(t, x, t+h, z)\,[p(t+h, x, s, y) - p(t+h, z, s, y)]\, dz.$$

Dividing both sides by h and letting $h \to 0$ we complete the proof using Remark 2.

Remark 3. If $f(x)$ is a continuous bounded function vanishing outside some bounded interval and we set $u_s(t, x) = \int f(y)\, p(t, x, s, y)\, dy$, then under the assumptions of Theorem 2

$$\frac{\partial u_s(t, x)}{\partial t} = -a(t, x)\frac{\partial u_s(t, x)}{\partial x} - \frac{1}{2} b(t, x)\frac{\partial p^2 u_s(t, x)}{\partial x^2} \tag{10}$$

and $u_s(t, x) \to f(x)$ for $t \uparrow s$. Eq. (10) follows from differentiating within the integral $\int f(y)\, p(t, x, s, y)\, dy$ w.r.t. t and twice w.r.t. x. We establish analogously that if $v_t(s, y) = \int g(x)\, p(t, x, s, y)\, dx$, where $g(y)$ is a continuous function vanishing outside a finite interval, then under the assumptions of Theorem 1

$$\frac{\partial v_t(s, y)}{\partial s} = -\frac{\partial}{\partial y}(a(s, y)\, v_t(s, y)) + \frac{1}{2} \cdot \frac{\partial^2}{\partial y^2}(b(s, y)\, v_t(s, y))$$

and $\lim_{s \downarrow t} v(s, y) = g(y)$. Hence $p(t, x, s, y)$ for fixed s is a fundamental solution of

$$\frac{\partial u}{\partial t} + a(t, x)\frac{\partial u}{\partial x} + \frac{1}{2} b(t, x)\frac{\partial^2 u}{\partial x^2} = 0, \qquad 0 \leqq t \leqq s \tag{11}$$

and $p(s, y, t, x)$ for fixed s is a fundamental solution of

$$-\frac{\partial v}{\partial t} - \frac{\partial}{\partial x}(a(t, x)\, u) + \frac{1}{2} \cdot \frac{\partial^2}{\partial x^2}(b(t, x)\, u) = 0, \qquad s \leqq t \leqq T. \tag{12}$$

We remark that the differential operators on the left sides of (11) and (12) are formal adjoints of one another. Using the formulas of the preceding paragraph for the transition density one can study properties of the fundamental solutions of (11) and (12).

§ 15. Time-homogeneous Solutions of Stochastic Differential Equations

We treat stochastic differential equations whose coefficients do not depend explicitly on time, i.e., equations of the form

$$d\xi(t) = a(\xi(t))\, dt + \sigma(\xi(t))\, dw(t), \tag{1}$$

where $a(x)$ and $\sigma(x)$ are functions defined on $(-\infty, \infty)$. If these functions satisfy the following conditions:

a) for some K

$$|a(x)|+|\sigma(x)| \leqq K(1+|x|);$$

b) for each $C>0$ there exists an L_C such that for $|x| \leqq C$ and

$$|a(x)-a(y)|+|\sigma(x)-\sigma(y)| \leqq L_C|x-y|,$$

then for each random variable $\xi(0)$ not dependent on $w(t)$ there exists a unique solution of (1) on an arbitrary interval $[0, T]$, $T>0$ satisfying the initial condition $\xi(t)=\xi(0)$ for $t=0$. Thus, in this case we can consider the solution of (1) on the semi-infinite interval $[0, \infty)$. In § 10 we proved that $\xi(t)$ is Markov process whose transition probability $P(t, x, s, A)$ coincides with $\mathbb{P}\{\xi_{x,t}(s) \in A\}$, where $\xi_{x,t}(s)$ for $s \geqq t$ satisfies

$$d\xi_{x,t}(s)=a(\xi_{x,t}(s))\, ds+\sigma(\xi_{x,t}(s))\, dw(s) \tag{2}$$

with initial condition $\xi_{x,t}(t)=x$. Set $\tilde{\xi}_x(s)=\xi_{x,t}(t+s)$ and $\tilde{w}(s)=w(t+s)-w(t)$. The process $\tilde{w}(s)$ is also a Wiener process and $\tilde{\xi}_x(s)$ satisfies

$$d\tilde{\xi}_x(s)=a(\tilde{\xi}_x(s))\, ds+\sigma(\tilde{\xi}_x(s))\, d\tilde{w}(s) \tag{3}$$

with initial condition $\tilde{\xi}_x(0)=x$. It is obvious that (3) can be obtained from (2) if we set $t=0$ in (2) and substitute $\tilde{w}(s)$ for $w(s)$. From the uniqueness of the solution and the fact that $w(s)$ and $\tilde{w}(s)$ are identically distributed it follows that the distributions of $\xi_x(s)=\xi_{x,0}(s)$ and $\tilde{\xi}_x(s)$ are also identical. In other words, the distribution of $\xi_{x,t}(t+s)$ is independent of t and the transition probability $P(t, x, s, A)$ of the process $\xi(t)$, as a solution of (1), depends merely on the difference $s-t$: $P(t, x, s, A)=P(0, x, s-t, A)$. A Markov process whose transition probability satisfies the last relation is called homogeneous. The transition probability time at t for such a process will be denoted by

$$P(t, x, A)=P(0, x, t, A).$$

We have thus established that the solution of (1) is a homogeneous Markov process.

It was proved above that the distribution of $\xi(\tau+s)$ for $s>0$ (τ is a nonrandom variable) coincides with that of $\check{\xi}_{\xi(\tau)}(s)$, where $\check{\xi}_x(s)$ is a solution of $d\check{\xi}_x(s)=a(\check{\xi}_x(s))\, ds+\sigma(\check{\xi}_x(s))\, d\check{w}(s)$ with initial condition $\check{\xi}_x(0)=x$ and $\check{w}(s)$ is a Wiener process independent of $\xi(t)$. We will show that this property is preserved if τ is a Markov time (see § 4), i.e., for all $t>0$ the event $\{\tau>t\}$ belongs to the σ-algebra $\mathfrak{F}_t$, the minimum σ-algebra with respect to which $\xi(0)$ and $w(s)$ are measurable for $s \leqq t$. It follows from Theorem 2 § 4 that $w(\tau+s)-w(\tau)$ is a Wiener process independent of τ and $w(t)$ for $t \leqq \tau$. Since $\xi(t)$ is expressed by means of $\xi(0)$ and $w(t)$ for $t \leqq \tau$, $\xi(0)$ is independent of $w(t)$ and $w(t)$ for $t \leqq \tau$ is independent of $w(\tau+s)-w(\tau)$ for $s \geqq 0$, the process $\hat{w}(t)=w(t+\tau)-w(\tau)$ is likewise independent of τ and $\xi(\tau)$. Moreover,

$$\xi(\tau+t)=\xi(\tau)+\int_{\tau}^{\tau+t} a(\xi(s))\,ds+\int_{\tau}^{\tau+t} \sigma(\xi(s))\,dw(s)$$

$$=\xi(\tau)+\int_0^t a(\xi(s+\tau))\,ds+\int_0^t \sigma(\xi(s+\tau))\,d\hat{w}(s),$$

consequently, $\xi(t+\tau)=\hat{\xi}_{\xi(\tau)}(t)$, where $\hat{\xi}_x(t)$ is a solution of

$$\hat{\xi}_x(t)=x+\int_0^t a(\hat{\xi}_x(s))\,ds+\int_0^t \sigma(\hat{\xi}_x(s))\,d\hat{w}(s).$$

Let $\mathfrak{F}_\tau$ be the σ-algebra generated by events of the form $\mathfrak{A}_t \cap \{\tau>t\}$, where $\mathfrak{A}_t \in \mathfrak{F}_t$. From the considerations above follows the proof of

Theorem 1. *For any Markov time τ and measurable bounded function $f(x)$ we have*

$$\mathbb{M}[f(\xi(\tau+t))/\mathfrak{F}_\tau]=\mathbb{M}_{\xi(\tau)} f(\xi(t)),$$

where

$$\mathbb{M}_x f(\xi(t))=\mathbb{M} f(\xi_x(t))=\int f(y)\,P(t,x,dy).$$

Processes for which the conclusion of Theorem 1 holds will be called *strong Markov*.

Remark 1. If $0<t_1<t_2<\cdots<t_n$ and $f(x_1,\ldots,x_n)$ is a bounded Borel function of all its arguments, then

$$\mathbb{M}[f(\xi(\tau+t_1),\ldots,\xi(\tau+t_n))/\mathfrak{F}_t]=\mathbb{M}_{\xi(\tau)} f(\xi(t_1),\ldots,\xi(t_n)).$$

Remark 2. Let τ_x be the time $\xi(t)$ first attains the point x: $\tau_x=\inf\{t;\,\xi(t)=x\}$, $=+\infty$ if the set in brackets is empty. Then, if $\mathbb{P}\{\tau_x<+\infty\}=1$, the distribution of the process $\xi(\tau_x+t)$ coincides with that of $\xi_x(t)$ and $\xi(\tau_x+t)$ is independent of events of the σ-algebra $\mathfrak{F}_{\tau_x}$. This is true for any Markov time τ for which $\mathbb{P}\{\xi(\tau)=x\}=1$.

Corollary 1. *Let $\xi(0)=x$ and τ_0 be a Markov time for which $\xi(\tau_0)=x$ w.p.1. We set $\xi_1(t)=\xi(\tau_0+t)$ and designate by τ_1 the random time with the same meaning for $\xi_1(t)$ as τ_0 has for $\xi(t)$. We construct inductively a sequence of processes $\xi_k(t)$ and random times τ_k connected by the relation $\xi_k(t)=\xi_{k+1}(\tau_{k-1}+t)$ in such a way that the variable τ_k is defined w.r.t. the process $\xi_k(t)$ in the same manner as τ_0 w.r.t. $\xi(t)$. We assume that for the function $f(x)$ the integral $\int_0^{\tau_0} f(\xi(s))\,ds$ is defined. Then the random variables*

$$\int_0^{\tau_0} f(\xi(s))\,ds,\quad \int_0^{\tau_1} f(\xi_1(s))\,ds,\quad \ldots,\quad \int_0^{\tau_k} f(\xi_k(s))\,ds$$

are independent and identically distributed. In particular (if $f=1$), the random variables $\tau_0, \tau_1, \ldots, \tau_k$ are independent and identically distributed.

The truth of this statement follows from the fact that $\tau_0+\tau_1+\cdots+\tau_{k-1}$ is a Markov time for which $\xi(\tau_0+\tau_1+\cdots+\tau_{k-1})=x$, hence $\xi(\tau_0+\tau_1+\cdots+\tau_{k-1}+t)$ is independent of $w(s)$ for $s\leqq\tau_0+\cdots+\tau_{k-1}$ and $\tau_0, \ldots, \tau_{k-1}$, and the distributions of this process coincide with those of the process $\xi(t)$; finally

$$\int_0^{\tau_k} f(\xi_k(s))\,ds=\int_{\tau_0+\cdots+\tau_{k-1}}^{\tau_0+\cdots+\tau_k} f(\xi(s))\,ds.$$

We now study certain properties of first passage times for a process $\xi_x(t)$ to the boundaries of some interval. We write $\tau_x[a,b]=\inf\{t\colon \xi(t)\notin(a,b)\}$; $=+\infty$ if the set in brackets is empty.

Theorem 2. *If $\sigma(x)>0$ for $x\in[a,b]$, then the variable $\tau_x[a,b]$ is finite w.p.1 for $x\in(a,b)$ and $\mathbb{M}\tau_x[a,b]=v(x)$, where $v(x)$ is a solution of the differential equation*

$$\tfrac{1}{2}\sigma^2(x)\,v''(x)+a(x)\,v'(x)=-1, \tag{4}$$

with $v(a)\geqq v(b)=0$.

Proof. Using Itô's formula we write

$$\begin{aligned} v(\xi_x(t))-v(x)=&\int_0^t v'(\xi_x(s))\,\sigma(\xi_x(s))\,dw(s)\\ &+\int_0^t\left[a(\xi_x(s))\,v'(\xi_x(s))+\tfrac{1}{2}\sigma^2(\xi_x(s))\,v''(\xi_x(s))\right]ds. \end{aligned} \tag{5}$$

Set $\tau_T=\inf\{T,\tau_x[a,b]\}$. Then for $s<\tau_T$ the variable $\xi_x(s)$ is contained in the interval (a,b), and so, w.p.1

$$a(\xi_x(s))\,v'(\xi_x(s))+\tfrac{1}{2}\sigma^2(\xi_x(s))\,v''(\xi_x(s))=-1.$$

Setting $t=\tau_T$ in (5) we obtain after some simple manipulations

$$\tau_T=\int_0^{\tau_T} v'(\xi_x(s))\,\sigma(\xi_x(s))\,dw(s)+v(x)-v(\xi_x(\tau_T)), \tag{6}$$

whence $\mathbb{M}\tau_T=v(x)-\mathbb{M}v(\xi_x(\tau_T))$. Since τ_T increases monotonically to $\tau_x[a,b]$ for $T\to\infty$ and $\mathbb{M}\tau_T$ is uniformly bounded in T, we have $\mathbb{M}\tau_T\to\mathbb{M}\tau_x[a,b]<\infty$. Consequently, $\tau_x[a,b]$ is finite w.p. 1, $\xi_x(\tau_T)\to\xi_x(\tau_x[a,b])$ w.p.1 and $v(\xi_x(\tau_T))\to v(\xi_x(\tau_x[a,b]))=0$ since $\xi_x[\tau_x[a,b]]$ is equal to either a or b. Hence, from Lebesgue's theorem on passage to the limit under the integral sign (the function v is bounded in $[a,b]$), we have

$\mathbb{M}\, v(\xi_x(\tau_T)) \to 0$ for $T \to \infty$, so that

$$\mathbb{M}\,\tau_x[a,b] = \lim_{T\to\infty} \mathbb{M}\,\tau_T = v(x) - \lim_{T\to\infty} \mathbb{M}\, v(\xi_x(\tau_T)) = v(x).$$

The proof is completed.

Corollary 2. *Set*

$$\Phi(x) = \exp\left\{-\int_a^x \frac{2a(z)}{\sigma^2(z)}\,dz\right\}.$$

Solving above for $v(x)$, we find

$$\begin{aligned}\mathbb{M}\,\tau_x[a,b] = &-\int_a^x 2\Phi(y)\int_a^y \frac{dz}{\sigma^2(z)\,\Phi(z)}\,dy\\ &+\int_a^b 2\Phi(y)\int_a^y \frac{dz}{\sigma^2(z)\,\Phi(z)}\,dy\,\frac{\int_a^x \Phi(z)\,dz}{\int_a^b \Phi(z)\,dz}.\end{aligned} \tag{7}$$

Theorem 3. *Under the assumptions of Theorem 2, $\mathbb{M}(\tau_x[a,b])^2 = v_1(x)$, where $v_1(x)$ is a solution of*

$$\tfrac{1}{2}\sigma^2(x)\,v_1''(x) + a(x)\,v_1'(x) = -2v(x), \tag{8}$$

with $v_1(a) = v_1(b) = 0$ and $v(x)$ is as defined in Theorem 2.

Proof. Using the finiteness of $\tau_x[a,b]$ and proceeding to the limit as $T\to\infty$ in (6) we obtain

$$\tau_x[a,b] = \int_0^{\tau_x[a,b]} v'(\xi_x(s))\,\sigma(\xi_x(s))\,dw(s) + v(x).$$

From this follows

$$\mathbb{M}(\tau_x[a,b])^2 = \mathbb{M}\int_0^{\tau_x[a,b]} \{v'(\xi_x(s))\,\sigma(\xi_x(s))\}^2\,ds + v^2(x)$$

(the finiteness of the expectation of the integral follows from the boundedness of $v'(x)\,\sigma(x)$ on $[a,b]$ and the finiteness of $\mathbb{M}\,\tau_x[a,b]$).

Let $z(x)$ be a solution of

$$\tfrac{1}{2}\sigma^2(x)\,z''(x) + a(x)\,z'(x) = -\{v'(x)\,\sigma(x)\}^2,$$

satisfying the boundary conditions $z(a) = z(b) = 0$. Using Itô's formula we can write

$$\begin{aligned}z(\xi_x(\tau_x[a,b]) - z(x)) = &-\int_0^{\tau_x[a,b]} [v'(\xi_x(s))\,\sigma(\xi_x(s))]^2\,ds\\ &+\int_0^{\tau_x[a,b]} z'(\xi_x(s))\,\sigma(\xi_x(s))\,dw(s).\end{aligned}$$

Since $z(\xi_x(\tau_x[a,b]))=0$, we have

$$\mathbf{M}\int_0^{\tau_x[a,b]}\{v'(\xi_x(s))\,\sigma(\xi_x(s))\}^2\,ds=z(x),$$

consequently, $\mathbf{M}(\tau_x[a,b])^2=z(x)-v^2(x)$. If we set $v_1(x)=z(x)-v^2(x)$ then $v_1(a)=v_1(b)=0$ and

$$\begin{aligned}\tfrac{1}{2}\sigma^2(x)\,v_1''(x)+a(x)\,v_1'(x)=\tfrac{1}{2}\sigma^2(x)\,z''(x)&+a(x)\,z'(x)\\&+\tfrac{1}{2}\sigma^2(x)\,[2v(x)\,v''(x)+2[v'(x)]^2]\\&+2a(x)\,v(x)\,v'(x)\\&=[\sigma^2(x)\,v''(x)+2a(x)\,v'(x)]\,v(x)=-2v(x).\end{aligned}$$

The theorem is proved.

We introduce the probability $p_a(x;b)$ — that $\xi_x(t)$ hits the point a before the point b. If x does not lie between a and b, then it is obvious that $\xi_x(t)$ hits the point closest to x first w.p. 1 since $\xi_x(t)$ is w.p. 1 continuous in t. The case of interest is thus that in which x lies between a and b.

Theorem 4. *Let $a<x<b$, $\sigma(x)>0$ for $x\in[a,b]$ and $u(x)$ be a not identically constant solution of $\frac{1}{2}\sigma^2(x)\,u''(x)+a(x)\,u'(x)=0$ on the interval $[a,b]$. Then $p_a(x;b)=\dfrac{u(x)-u(b)}{u(a)-u(b)}$.*

Proof. The function $\psi(x)=\dfrac{u(x)-u(b)}{u(a)-u(b)}$ satisfies the equation $\frac{1}{2}\sigma^2(x)\,\psi''(x)+a(x)\,\psi'(x)=0$ with $\psi(a)=1$, $\psi(b)=0$. Hence,

$$\begin{aligned}\psi(\xi_x(\tau_x[a,b]))-\psi(x)&=\int_0^{\tau_x[a,b]}\sigma(\xi_x(s))\,\psi'(\xi_x(s))\,dw(s)\\&\quad+\int_0^{\tau_x[a,b]}\left\{\frac{\sigma^2(\xi_x(s))}{2}\psi''(\xi_x(s))+a(\xi_x(s))\,\psi'(\xi_x(s))\right\}ds\\&=\int_0^{\tau_x[a,b]}\sigma(\xi_x(s))\,\psi'(\xi_x(s))\,dw(s),\end{aligned}$$

and consequently, $\mathbf{M}\psi(\xi_x(\tau_x[a,b]))=\psi(x)$. But $\xi_x(\tau_x[a,b])=a$ with probability $p_a(x;b)$ and $\xi_x(\tau_x[a,b])=b$ with probability $p_b(x;a)$, thus

$$\psi(x)=\psi(\mathbf{M}\,\xi_x(\tau_x[a,b]))=\psi(a)\,p_a(x;b)+\psi(b)\,p_b(x;a)=p_a(x;b).$$

The theorem is proved.

Corollary 3. *If we use the notation of Corollary 2 we find that*

$$p_a(x;b)=\frac{\int_x^b \Phi(z)\,dz}{\int_a^b \Phi(z)\,dz}; \qquad p_b(x;a)=\frac{\int_a^x \Phi(z)\,dz}{\int_a^b \Phi(z)\,dz}. \tag{9}$$

To establish the first of these results it is sufficient to show that the function $u(x)=\int_a^x \Phi(z)\,dz$ satisfies the assumptions of Theorem 4. The second formula at (9) can be obtained from considerations analogous to those in Theorem 4.

Remark 3. If $a(x)\equiv 0$ for $x\in[a,b]$, then the formulas for $p_a(x;b)$ and $p_b(x;a)$ have the simple forms

$$p_a(x;b)=\frac{b-x}{b-a}; \qquad p_b(x;a)=\frac{x-a}{b-a}. \tag{10}$$

We now consider the question of time-substitutions in stochastic differential equations whose coefficients do not depend explicitly on time. Let $\xi(t)$ be a solution of (1) defined for $t\geqq 0$ and satisfying the initial condition $\xi(t)=\xi(0)$ for $t=0$. Let $g(x)$ be a positive continuous function for which $\int_0^\infty g(\xi(s))\,ds=+\infty$ w.p. 1. We define a family τ_t of random variables by

$$t=\int_0^{\tau_t} g(\xi(s))\,ds. \tag{11}$$

For each t the variable τ_t is a Markov time and is w.p. 1 a monotone and continuous function of t. Introduce the random process $\eta(t)=\xi(\tau_t)$. Then

$$\eta(t)=\xi(\tau_t)=\xi(0)+\int_0^{\tau_t} a(\xi(s))\,ds+\int_0^{\tau_t}\sigma(\xi(s))\,dw(s).$$

We make the substitution of variable of integration $s=\tau_u$ in the first integral. Then

$$ds=\frac{d\tau_u}{du}\,du=\frac{du}{g(\xi(\tau_u))},$$

since from (1) it follows that $1=g(\xi(\tau_t))\dfrac{d\tau_t}{dt}$. Carrying through the substitution, we find

$$\int_0^{\tau_t} a(\xi(s))\,ds=\int_0^t \frac{a(\xi(\tau_u))}{g(\xi(\tau_u))}\,du=\int_0^t \frac{a(\eta(u))}{g(\eta(u))}\,du.$$

We perform a similar substitution in the stochastic integral

$$\int_0^{\tau_t} \sigma(\xi(s))\, dw(s) = \int_0^t \frac{\sigma(\xi(\tau_u))}{\sqrt{g(\xi(\tau_u))}} \sqrt{g(\xi(\tau_u))}\, dw(\tau_u) = \int_0^t \frac{\sigma(\eta(u))}{\sqrt{g(\eta(u))}}\, dw_1(u),$$

where $w_1(u) = \int_0^{\tau_u} \sqrt{g(\xi(s))}\, dw(s)$. We note that the integral w.r.t. $dw_1(u)$ exists as limit in mean square of the corresponding Riemann sum. We will prove that $w_1(t)$ is also a Wiener process. To this end it is sufficient to show that

$$\mathbb{M}(w_1(u+h) - w_1(u)/\mathfrak{F}_{\tau_u}) = 0, \qquad \mathbb{M}([w_1(u+h) - w_1(u)]^2/\mathfrak{F}_{\tau_u}) = h$$

for all $u > 0$ and $h > 0$ (see Theorem 1 § 1). But on the basis of Remark 2 §4,

$$\mathbb{M}\left(\int_{\tau_u}^{\tau_{u+h}} \sqrt{g(\xi(s))}\, dw(s)/\mathfrak{F}_{\tau_u}\right) = 0.$$

Furthermore,

$$\begin{aligned}
&\mathbb{M}\left[(w_1(u+h) - w_1(u))^2/\mathfrak{F}_{\tau_u}\right] \\
&= \mathbb{M}\left[\left(\int_{\tau_u}^{\tau_{u+h}} \sqrt{g(\xi(s))}\, dw(s)\right)^2/\mathfrak{F}_{\tau_u}\right] = \mathbb{M}\left[\int_{\tau_u}^{\tau_{u+h}} g(\xi(s))\, ds/\mathfrak{F}_{\tau_u}\right] \\
&= \mathbb{M}\left[\int_0^{\tau_{u+h}} g(\xi(s))\, ds - \int_0^{\tau_u} g(\xi(s))\, ds/\mathfrak{F}_{\tau_u}\right] = \mathbb{M}[u+h-u/\mathfrak{F}_{\tau_u}] = h.
\end{aligned}$$

Hence, $w_1(t)$ is a Wiener process and because of Theorem 1 § 1 $w_1(t+h) - w_1(t)$ is independent of events of the σ-algebra $\mathfrak{F}_{\tau_t}$ and of $\xi(s)$ for $s \leqq \tau_t$, i.e., of $\eta(u)$ for $u \leqq t$. Substituting the transformed integrals into the expression for $\eta(t)$ we obtain

$$\eta(t) = \xi(0) + \int_0^t \frac{a(\eta(s))}{g(\eta(s))}\, ds + \int_0^t \frac{\sigma(\eta(s))}{\sqrt{g(\eta(s))}}\, dw_1(s).$$

Using the fact that $\xi(0) = \eta(0)$, we can formulate the following

Theorem 5. *Let $\xi(t)$ be a solution of* (1) *and $\eta(t) = \xi(\tau_t)$, where τ_t is defined as in* (11) *with $g(x)$ a continuous positive function and*

$$\mathbb{P}\left\{\int_0^\infty g(\xi(s))\, ds = +\infty\right\} = 1.$$

Then $\eta(t)$ will also be a solution of the stochastic differential equation

$$d\eta(t) = \frac{a(\eta(t))}{g(\eta(t))}\, dt + \frac{\sigma(\eta(t))}{\sqrt{g(\eta(t))}}\, dw_1(t),$$

where $w_1(t)$ is some Wiener process and $\eta(0)$ is independent of $w_1(t)$.

Corollary 4. *If $w(t)$ is a Wiener process and τ_t is defined by*

$$t=\int_0^{\tau_t}\frac{1}{\sigma^2(w(s))}\,ds,$$

where $\sigma(x)>0$ is such that $\mathbb{P}\left\{\int_0^{\infty}\frac{1}{\sigma^2(w(s))}\,ds=+\infty\right\}=1$, *then the process $\eta(t)=w(\tau_t)$ will be a solution of the stochastic equation $d\eta(t)=\sigma(\eta(t))\,dw_1(t)$, where $w_1(t)$ is likewise a Wiener process.*

We remark that if we take a twice continuously differentiable, increasing function $f(x)$ and consider the process $\xi(t)=f(\eta(t))$ we can show that $\xi(s)$ satisfies Eq. (1) with the previously given coefficients. Consequently, any homogeneous Markov process which is a solution of (1) can be obtained from a Wiener process by means of a random time-substitution and a state transformation.

Chapter 4. Asymptotic Behavior of the Solutions of Stochastic Equations

§ 16. Bounded and Unbounded Solutions of Stochastic Equations

We study here conditions for one-sided boundedness and unboundedness of the solutions of stochastic differential equations on the entire interval of their definition. In the case of homogeneous processes (coefficients of the equation independent of time) such an interval will be $[0, \infty)$. In the general case we observe that from the continuity of the solution follows its boundedness on an arbitrary closed finite interval. Thus, in the inhomogeneous case, the main interest is in the study of the solution's behavior in a neighborhood of the interval $[0, T)$ and such an interval can be always be transformed into $[0, \infty)$ by a time substitution. We therefore consider equations with solutions defined on $[0, \infty)$. We first treat the case of time-independent coefficients: We study the one-sided boundedness of solutions of the equation

$$d\xi(t) = a(\xi(t))\,dt + \sigma(\xi(t))\,dw(t). \tag{1}$$

Assume that (1) has a unique solution for arbitrary initial condition $\xi(0)$, and that $\sigma(x) > 0$ for all x. Equations for which $\sigma(x)$ vanishes will be considered later in this chapter. An important role in investigations of boundedness of solutions of (1) will be played by the function $\varphi(x)$, defined by $a(x)\varphi'(x) + \frac{1}{2}\sigma^2(x)\varphi''(x) = 0$. For the sake of convenience, we designate the differential operator $a(x)\dfrac{d}{dx}[\,] + \dfrac{1}{2}\sigma^2(x)\dfrac{d^2}{dx^2}[\,]$ by $L[\,]$. A function φ satisfying $L[\varphi] = 0$ will be called *L-harmonic*. It is easy to see that for any arbitrary L-harmonic function

$$\varphi'(x) = C\exp\left\{-\int_0^x \frac{2a(z)}{\sigma^2(z)}\,dz\right\},$$

hence, if φ is not identically constant, then $\varphi'(x)$ does not vanish and, consequently, φ is monotone. We now prove several lemmas on the connection between the behavior of L-harmonic functions and that of solutions of Eq. (1).

Lemma 1. *If there exists an L-harmonic function* $\varphi(x)$ *for which* $\varphi(x) \to +\infty$ *when* $x \to +\infty$ *and* $\varphi(x) \to -\infty$ *when* $x \to -\infty$, *then*

$$\mathbb{P}\{\sup_{t>0} \xi(t) = +\infty\} = \mathbb{P}\{\inf_{t>0} \xi(t) = -\infty\} = 1. \tag{2}$$

Proof. Since $\xi(t) = \xi_{\xi(0)}(t)$, where $\xi_x(t)$ is a solution of (1) with boundary condition $\xi_x(0) = x$, it suffices to prove that the lemma is true for $\xi_x(t)$ with arbitrary x.

Let $x_1 < x < x_2$. Let $\tau_x[x_1, x_2]$ be the first time that $\xi_x(t)$ reaches one of the boundaries of the interval (x_1, x_2). Then $\xi_x(\tau_x[x_1, x_2])$ is equal to either x_1 or x_2. From Theorem 4 § 15 we can write

$$\mathbb{P}\{\xi_x(\tau_x[x_1, x_2]) = x_1\} = \frac{\varphi(x_2) - \varphi(x)}{\varphi(x_2) - \varphi(x_1)},$$

$$\mathbb{P}\{\xi_x(\tau_x[x_1, x_2]) = x_2\} = \frac{\varphi(x) - \varphi(x_1)}{\varphi(x_2) - \varphi(x_1)}.$$

Obviously,

$$\mathbb{P}\{\sup_{t>0} \xi_x(t) \geqq x_2\} \geqq \mathbb{P}\{\xi_x(\tau_x[x_1, x_2]) = x_2\},$$

for any $x_1 < x$. Thus

$$\mathbb{P}\{\sup_{t>0} \xi_x(t) \geqq x_2\} \geqq \frac{\varphi(x) - \varphi(x_1)}{\varphi(x_2) - \varphi(x_1)}.$$

Proceeding to the limit as $x_1 \to -\infty$ and recalling that $\varphi(x_1) \to -\infty$, we find that

$$\mathbb{P}\{\sup_{t>0} \xi_x(t) \geqq x_2\} = 1.$$

But then

$$\mathbb{P}\{\sup_{t>0} \xi_x(t) = +\infty\} = \lim_{x_2 \to +\infty} \mathbb{P}\{\sup_{t>0} \xi_x(t) \geqq x_2\} = 1.$$

One establishes analogously that $\mathbb{P}\{\inf_{t>0} \xi_x(t) = -\infty\} = 1$. The lemma is proved.

Lemma 1 gives sufficient conditions for the unboundedness from both sides of a solution of (1). One can show that these conditions are also necessary: under the conditions of Lemma 1 $\xi(t)$ has neither finite nor infinite limit as $t \to +\infty$, but oscillates with ever growing amplitude.

We now treat the case in which there exists a one-sided bounded L-harmonic function.

Lemma 2. *Assume there exists an L-harmonic function* $\varphi(x)$ *for which* $\varphi(x) \to +\infty$ *when* $x \to +\infty$ *and* $\varphi(x) \geqq C$ *for all* x. *Then*

$$\mathbb{P}\{\sup_{t>0} \xi(t) < +\infty\} = \mathbb{P}\{\inf_{t>0} \xi(t) = -\infty\} = \mathbb{P}\{\lim_{t \to +\infty} \xi(t) = -\infty\} = 1.$$

Proof. We use again the relation

$$\mathbb{P}\{\xi_x(\tau_x[x_1, x_2]) = x_2\} = \frac{\varphi(x) - \varphi(x_1)}{\varphi(x_2) - \varphi(x_1)}.$$

It is clear that

$$\lim_{x_1 \to -\infty} \mathbb{P}\{\xi_x(\tau_x[x_1, x_2]) = x_2\} = \mathbb{P}\{\sup_{t>0} \xi_x(t) \geqq x_2\}$$

(if for some $s>0$ $\xi_x(s) = x_2$, then $x_1 < \inf_{t<s} \xi(s)$ implies $\xi_x(\tau_x[x_1, x_2]) = x_2$). Since $\varphi(x)$ is increasing and bounded from below, $\lim_{x_1 \to -\infty} \varphi(x) = \varphi(-\infty)$ exists. Consequently,

$$\begin{aligned} \mathbb{P}\{\sup_{t>0} \xi_x(t) \geqq x_2\} &= \frac{\varphi(x) - \varphi(-\infty)}{\varphi(x_2) - \varphi(-\infty)}, \\ \mathbb{P}\{\sup_{t>0} \xi_x(t) < x_2\} &= \frac{\varphi(x_2) - \varphi(x)}{\varphi(x_2) - \varphi(-\infty)}. \end{aligned} \tag{3}$$

Going to the limit as $x_2 \to +\infty$ and using the fact that $\lim_{x_2 \to +\infty} \varphi(x_2) = +\infty$, we find that $\mathbb{P}\{\sup_{t>0} \xi(t) < +\infty\} = 1$. The equality $\mathbb{P}\{\inf_{t>0} \xi(t) = -\infty\} = 1$ is proved as in Lemma 1 since for its proof we need only use the fact that $\varphi(x) \to +\infty$ when $x \to +\infty$.

We now show that $\mathbb{P}\{\lim_{t\to\infty} \xi(t) = -\infty\} = 1$. Let τ_y be the first passage time to the point y. Since $\mathbb{P}\{\inf_{t>0} \xi_x(t) = -\infty\} = 1$, τ_y is finite w.p. 1 for all $y < x$. Because of Remark 2 § 15 we can write

$$\mathbb{P}\{\sup_{t>0} \xi_x(t+\tau_y) \geqq x_2\} = \mathbb{P}\{\sup_{t>0} \xi_y(t) \geqq x_2\} = \frac{\varphi(y) - \varphi(-\infty)}{\varphi(x_2) - \varphi(-\infty)},$$

provided that $y < x_2$. But

$$\mathbb{P}\{\sup_{t>0} \xi_x(t+\tau_y) \geqq x_2\} = \mathbb{P}\{\sup_{t>\tau_y} \xi_x(t) \geqq x_2\} \geqq \mathbb{P}\{\overline{\lim_{t\to\infty}}\, \xi_x(t) \geqq x_2\}$$

and

$$\mathbb{P}\{\overline{\lim_{t\to\infty}}\, \xi_x(t) \geqq x_2\} \leqq \frac{\varphi(y) - \varphi(-\infty)}{\varphi(x_2) - \varphi(-\infty)}$$

for any y. Letting $y \to -\infty$, we obtain $\mathbb{P}\{\overline{\lim_{t\to\infty}}\, \xi_x(t) \geqq x_2\} = 0$ for all x_2. This means that $\overline{\lim_{t\to\infty}}\, \xi_x(t) = -\infty$ w.p. 1. But then $\lim_{t\to\infty} \xi_x(t) = -\infty$, also w.p. 1. The lemma is finished.

Remark 1. If there exists an L-harmonic function $\varphi(x)$, bounded from above, for which $\lim_{x\to-\infty}\varphi(x)=-\infty$, then

$$\mathbb{P}\{\lim_{t\to\infty}\xi(t)=+\infty\}=\mathbb{P}\{\sup_{t>0}\xi(t)=+\infty\}=\mathbb{P}\{\inf_{t>0}\xi(t)>-\infty\}=1.$$

The proof of this statement proceeds analogously to that of Lemma 2.

We turn finally to the case for which there exists an L-harmonic function bounded from both sides.

Lemma 3. *Assume there exists an increasing, L-harmonic function φ for which $\lim_{x\to-\infty}\varphi(x)=\varphi(-\infty)$ and $\lim_{x\to+\infty}\varphi(x)=\varphi(+\infty)$ are finite. Then*

$$\mathbb{P}\{\sup_{t>0}\xi_x(t)<\infty\}=\mathbb{P}\{\inf_{t>0}\xi_x(t)=-\infty\}=\frac{\varphi(+\infty)-\varphi(x)}{\varphi(+\infty)-\varphi(-\infty)}, \tag{4}$$

$$\mathbb{P}\{\inf_{t>0}\xi_x(t)>-\infty\}=\mathbb{P}\{\sup_{t>0}\xi_x(t)=\infty\}=\frac{\varphi(x)-\varphi(-\infty)}{\varphi(+\infty)-\varphi(-\infty)} \tag{5}$$

with

$$\mathbb{P}\{\lim_{t\to\infty}\xi_x(t)=-\infty\}=\mathbb{P}\{\sup\xi_x(t)<\infty\},$$

$$\mathbb{P}\{\lim_{t\to\infty}\xi_x(t)=+\infty\}=\mathbb{P}\{\inf\xi_x(t)>-\infty\}.$$

Proof. Relation (4) follows from (3) if we let $x_2\to+\infty$. (5) can be obtained from

$$\mathbb{P}\{\inf_{t>0}\xi_x(t)>x_1\}=\frac{\varphi(x)-\varphi(x_1)}{\varphi(+\infty)-\varphi(x_1)} \tag{6}$$

by letting $x_1\to-\infty$. (6) is obtained as (3). That

$$\mathbb{P}\{\sup_{t>0}\xi_x(t)<\infty\}=\mathbb{P}\{\inf_{t>0}\xi_x(t)=-\infty\}$$

and

$$\mathbb{P}\{\inf_{t>0}\xi_x(t)>-\infty\}=\mathbb{P}\{\sup_{t>0}\xi_x(t)=+\infty\}$$

follows from the fact that $\xi_x(t)$ is unbounded from at least one side, since for all $x_1<x<x_2$, the variable $\tau_x[x_1,x_2]$ is, because of Theorem 2 § 15, bounded w.p. 1.

We now establish that

$$\mathbb{P}\{\lim_{t\to\infty}\xi_x(t)=+\infty\}=P\{\inf_{t>0}\xi_x(t)>-\infty\}.$$

To do this it is sufficient to demonstrate that for each C we can find a T such that for $t\geqq T$, $\xi_x(t)\notin[-C,C]$. Indeed, in this case $|\xi_x(t)|\to\infty$ for $t\to\infty$ and since $\xi_x(t)$ is a continuous function w.p. 1 either $\xi_x(t)\to+\infty$ or $\xi_x(t)\to-\infty$ for $t\to\infty$, the probability of any other possibilities being equal to zero. But then

$$\mathbb{P}\{\xi_x(t)\to+\infty\}=\mathbb{P}\{\sup_{t>0}\xi_x(t)=+\infty\};$$

$$\mathbb{P}\{\xi_x(t)\to-\infty\}=\mathbb{P}\{\inf_{t>0}\xi_x(t)=-\infty\}.$$

We choose $x_1<-C$ and $x_2>C$ in such a way that

$$\mathbb{P}\{\sup_{t>0}\xi_{x_1}(t)<-C\}=\frac{\varphi(-C)-\varphi(x_1)}{\varphi(-C)-\varphi(-\infty)}=1-\varepsilon,$$

$$\mathbb{P}\{\inf_{t>0}\xi_{x_2}(t)>C\}=\frac{\varphi(x_2)-\varphi(C)}{\varphi(+\infty)-\varphi(C)}=1-\varepsilon,$$

where $\varepsilon>0$ is a number chosen in advance. Let $\tau=\tau_x[x_1,x_2]$. Then

$$\begin{aligned}&\mathbb{P}\{\xi(t)\notin[-C,C]\text{ for } t>\tau\}\\&\quad=\mathbb{P}\{\inf_{t>0}\xi_x(t+\tau)>C/\xi(\tau)=x_2\}\,\mathbb{P}\{\xi(\tau)=x_2\}\\&\qquad+\mathbb{P}\{\sup_{t>0}\xi_x(t+\tau)<-C/\xi(\tau)=x_1\}\,\mathbb{P}\{\xi(\tau)=x_1\}.\end{aligned}$$

On the basis of Remark 2 §15

$$\mathbb{P}\{\inf_{t>0}\xi(t+\tau)>C/\xi(\tau)=x_2\}=\mathbb{P}\{\inf_{t>0}\xi_{x_2}(t)>C\}=1-\varepsilon,$$

$$\mathbb{P}\{\sup_{t>0}\xi(t+\tau)<-C/\xi(\tau)=x_1\}=\mathbb{P}\{\sup_{t>0}\xi_{x_1}(t)<-C\}=1-\varepsilon.$$

Since $\mathbb{P}\{\xi(\tau)=x_2\}+\mathbb{P}\{\xi(\tau)=x_1\}=1$, we have $\mathbb{P}\{\inf_{t>\tau}|\xi(t)|\geqq C\}=1-\varepsilon$. But

$$\mathbb{P}\{\varliminf_{t\to\infty}|\xi(t)\geqq C\}\geqq\mathbb{P}\{\inf_{t>\tau}|\xi(t)|\geqq C\}=1-\varepsilon.$$

Since $\mathbb{P}\{\varliminf_{t\to\infty}|\xi(t)|\geqq C\}$ does not depend on ε, we get on letting $\varepsilon\to 0$

$$\mathbb{P}\{\varliminf_{t\to\infty}|\xi(t)|\geqq C\}=1.$$

From the last relation, as already mentioned above, we obtain the complete proof of the lemma.

Remark 2. In Lemmas 1, 2, 3 and Remark 1 we treated all possible and mutually exclusive combinations of behavior of L-harmonic functions and listed the corresponding mutually exclusive possibilities for the behavior of the solutions of stochastic differential equations as $t\to+\infty$. It follows that a knowledge of the behavior of the corresponding L-harmonic function is not only sufficient but also necessary for determining that of the solutions referred to in the lemmas.

Let us introduce the integrals (which can assume the value $+\infty$)

$$I_1(x)=\int_{-\infty}^{x}\exp\left\{-\int_0^z\frac{2a(u)}{\sigma^2(u)}du\right\}dz,$$

$$I_2(x)=\int_x^{\infty}\exp\left\{-\int_0^z\frac{2a(u)}{\sigma^2(u)}du\right\}dz.$$

Combining the results of Lemmas 1, 2, 3 and Remark 1, we obtain

Theorem 1. *Assume that the coefficients of Eq. (1) satisfy the conditions of the existence and uniqueness theorem, $\sigma(x)>0$ and that $\xi(t)$ is a solution of (1). Then*

1) *if $I_1(x)=+\infty$ and $I_2(x)=+\infty$, then*

$$\mathbb{P}\{\sup_{t>0}\xi(t)=+\infty\}=\mathbb{P}\{\inf_{t>0}\xi(t)=-\infty\}=1;$$

2) *if $I_1(x)<\infty$ and $I_2(x)<\infty$, then*

$$\mathbb{P}\{\sup_{t>0}\xi(t)=+\infty\}=\mathbb{P}\{\lim_{t\to\infty}\xi(t)=+\infty\}=\mathbb{M}\frac{I_1(\xi(0))}{I_1(\xi(0))+I_2(\xi(0))};$$

$$\mathbb{P}\{\inf_{t>0}\xi(t)=-\infty\}=\mathbb{P}\{\lim_{t\to\infty}\xi(t)=-\infty\}=\mathbb{M}\frac{I_2(\xi(0))}{I_1(\xi(0))+I_2(\xi(0))};$$

3) *if $I_1(x)<+\infty$, $I_2(x)=+\infty$, then*

$$\mathbb{P}\{\sup_{t>0}\xi(t)<\infty\}=\mathbb{P}\{\inf_{t>0}\xi(t)=-\infty\}=\mathbb{P}\{\lim_{t\to\infty}\xi(t)=-\infty\}=1;$$

4) *if $I_1(x)=\infty$, $I_2(x)<+\infty$, then*

$$\mathbb{P}\{\sup_{t>0}\xi(t)=+\infty\}=\mathbb{P}\{\inf\xi(t)>-\infty\}=\mathbb{P}\{\lim_{t\to\infty}\xi(t)=+\infty\}=1.$$

Proof. From the definitions of the integrals $I_k(x)$ is seen that $I_k(x)$ is either finite for all x or identically equal to $+\infty$. If

$$\varphi(x)=\int_0^x\exp\left\{-\int_0^z\frac{2a(u)}{\sigma^2(u)}du\right\}dz,$$

then in case 1 the conditions of Lemma 1 are fulfilled, in case 4 those of Lemma 2, in case 3 those of Remark 1 and in case 2 those of Lemma 3. It remains to note that in case 2

$$\mathbb{P}\{\lim_{t\to\infty}\xi(t)=+\infty/\xi(0)\}=\mathbb{P}\{\lim_{t\to\infty}\xi_{\xi(0)}(t)=+\infty/\xi(0)\}$$

$$=\frac{I_1(\xi(0))}{I_1(\xi(0))+I_2(\xi(0))}.$$

Taking expectations on both sides of this equality, we get the first formula of claim 2. The second follows analogously. The theorem is proved.

When the coefficients depend on t it is not possible to obtain such complete results. We turn to the consideration of some sufficient conditions for the one-sided boundedness of solutions for which $\sigma(t,x)=1$. For the purpose of estimating these solutions the following lemma will prove useful.

Lemma 4. *Let $\xi_1(t)$ and $\xi_2(t)$ be solutions of the equations*

$$d\xi_i(t)=a_i(t,\xi_i(t))\,dt+dw(t) \qquad (i=1,2),$$

satisfying the same boundary condition $\xi_i(0)=x$. If for all $t\geqq 0$ and x we have $a_1(t,x)<a_2(t,x)$, then w.p.1 $\xi_1(t)<\xi_2(t)$ for all $t>0$.

Proof. Since

$$\xi_2(t)-\xi_1(t)=\int_0^t [a_2(s,\xi_2(s))-a_1(s,\xi_1(s))]\,ds,$$

$\Delta(t)=\xi_2(t)-\xi_1(t)$ is a differentiable function and at all points t where $\Delta(t)=0$, $\Delta'(t)=a_2(t,\xi_2(t))-a_1(t,\xi_1(t))>0$ because $\xi_1(t)=\xi_2(t)$. In particular, $\Delta(0)=0$, $\Delta'(0)>0$. Thus for some $\delta>0$ $\Delta(t)>0$ for $0<t<\delta$. Set $t_1=\inf\{t;t>0,\Delta(t)=0\}$ if the set in brackets is not empty. Then $\Delta(t_1)=0$, $\Delta'(t_1)>0$. Consequently, we can find a δ_1 for which $\Delta(t)<0$ with $t\in[t_1-\delta_1,t_1]$. But then on the interval $[\delta,t-\delta_1]$ the function $\Delta(t)$ changes signs, i.e., it takes on the value zero there, which is a contradiction of the definition of t_1. Thus, the set $\{t: t>0, \Delta(t)=0\}$ is empty w.p.1 and since $\Delta(t)>0$ for sufficiently small t, $\Delta(t)>0$ for all $t>0$. The lemma is proved.

We use this lemma to prove

Theorem 2. *Let $\xi(t)$ be a solution of*

$$d\xi(t)=a(t,\xi(t))\,dt+dw(t) \tag{7}$$

with initial condition $\xi(0)=\xi_0$. We set

$$\alpha_1(t)=\inf_{x\in R_1} a(t,x), \qquad \alpha_2(t)=\sup_{x\in R_1} a(t,x),$$

$$a_1(x)=\inf_{t>0} a(t,x), \qquad a_2(x)=\sup_{t>0} a(t,x).$$

Then the following relations hold:

1) *if*

$$\varliminf_{T\to\infty}\frac{1}{\sqrt{2T\log\log T}}\int_0^T \alpha_1(t)\,dt>1,$$

we have

$$\mathbb{P}\{\lim_{t\to\infty}\xi(t)=\infty\}=1;$$

if

$$\overline{\lim_{T\to\infty}}\frac{1}{\sqrt{2T\log\log T}}\int_0^T\alpha_2(t)\,dt<-1,$$

then

$$\mathbb{P}\{\lim_{t\to\infty}\xi(t)=-\infty\}=1;$$

2) *if*

$$\int_{-\infty}^0 e^{-\int_0^x 2a_1(z)\,dz}dx=+\infty,\qquad \int_0^\infty e^{-\int_0^x 2a_1(z)\,dz}dx<+\infty,$$

then

$$\mathbb{P}\{\lim_{t\to\infty}\xi(t)=+\infty\}=1;$$

if, however,

$$\int_{-\infty}^0 e^{-\int_0^x 2a_2(z)\,dz}dx<+\infty,\qquad \int_0^\infty e^{-\int_0^x 2a_2(z)\,dz}dx=+\infty,$$

then

$$\mathbb{P}\{\lim_{t\to\infty}\xi(t)=-\infty\}=1.$$

Proof. Let $\varepsilon(t)$ be a positive function for which $\int_0^\infty\varepsilon(t)\,dt<\infty$. Denote by $\xi_1(t)$ a solution of

$$d\xi_1(t)=[\alpha_1(t)-\varepsilon(t)]\,dt+dw(t).$$

From Lemma 4, $\xi_1(t)<\xi(t)$, since $\alpha_1(t)-\varepsilon(t)<a(t,x)$ for all $t>0$ and x. Consequently,

$$\xi(t)>\xi_0+w(t)+\int_0^t[\alpha_1(s)-\varepsilon(s)]\,ds.$$

On the basis of the law of the integrated logarithm for the Wiener process (see [20])

$$\underline{\lim}_{t\to\infty}\frac{w(t)}{\sqrt{2t\log\log t}}=-1,$$

w.p. 1, thus with the same probability

$$\underline{\lim}_{t\to\infty}\frac{\xi(t)}{\sqrt{2t\log\log t}}\geqq\underline{\lim}_{t\to\infty}\frac{w(t)}{\sqrt{2t\log\log t}}+\underline{\lim}_{t\to\infty}\frac{1}{\sqrt{2t\log\log t}}\int_0^t[\alpha_1(s)-\varepsilon(s)]\,ds>0.$$

This last inequality implies the first claim at 1) in the theorem. The second is proved analogously.

To prove the first statement at 2) we consider a process $\eta_1(t)$ satisfying

$$d\eta_1(t)=\bar{a}_1(\eta_1(t))\,dt+dw(t)$$

(with the same initial condition as for $\xi(t)$), where $\bar{a}_1(x)$ satisfies the assumptions of the existence and uniqueness theorem,

$$\bar{a}_1(x) < a_1(x) \quad \text{and} \quad \int_{-\infty}^{0} \exp\left\{-\int_0^x 2\bar{a}_1(z)\,dz\right\} dx = +\infty,$$

$$\int_0^{\infty} \exp\left\{-\int_0^x 2\bar{a}_1(z)\,dz\right\} dx < +\infty.$$

Then from Lemma 4, $\eta_1(t) < \xi(t)$ for $t > 0$ and from the third statement of Theorem 1 $\mathbb{P}\{\lim_{t\to\infty} \eta_1(t) = +\infty\} = 1$ and $\mathbb{P}\{\lim_{t\to\infty} \xi(t) = +\infty\} = 1$. The second claim at 2) is proved analogously. □

We will now establish a general theorem on the boundedness of the solutions of (7). In this regard the following lemma will prove to be useful.

Lemma 5. *Let $\xi(t)$, defined on $[0, T]$ be continuous and nonnegative w.p.1. Assume that for $t \in [0, T]$ $\mathbb{M}\,\xi(t) < \infty$ and that for arbitrary $0 \leqq t_1 < t_2 < \cdots < t_n \leqq T$ $\mathbb{M}(\xi(t)/\xi(t_1) \ldots \xi(t_{n-1})) \leqq \xi(t_{n-1})$. Then for $C > 0$*

$$\mathbb{P}\{\sup_{0 \leqq t \leqq T} \xi(t) > C\} \leqq \frac{1}{C}\,\mathbb{M}\,\xi(0).$$

Proof. Let $\xi_k = \xi(t_k)$, where $0 = t_0 < t_1 < \cdots < t_n = T$. Then for $k = 0, 1, 2, \ldots$, $\mathbb{M}(\xi_{k+1}/\xi_0, \ldots, \xi_k) \leqq \xi_k$. Let $C > 0$ and introduce

$$\xi_k^* = \begin{cases} \xi_k, & \text{if } \sup_{i \leqq k} \xi_i < C; \\ C, & \text{if } \sup_{i \leqq k} \xi_i \geqq C. \end{cases}$$

Denote by $\mathfrak{B}_k$ the event $\{\sup_{i \leqq k} \xi_i < C\}$; $g_C(x) = x$ if $x < C$; $g_C(x) = C$ if $x \geqq C$. Assume the event $\mathfrak{B}_k$ occurs. Then $\xi_i = \xi_i^*$ for $i \leqq k$ and

$$\begin{aligned} \mathbb{M}(\xi_{k+1}^*/\xi_0^*, \ldots, \xi_k^*) &= \mathbb{M}(\xi_{k+1}^*/\xi_0, \ldots, \xi_k) \\ &= \mathbb{M}(g_C(\xi_{k+1})/\xi_0, \ldots, \xi_k) \leqq \mathbb{M}(\xi_{k+1}/\xi_0, \ldots, \xi_k) \leqq \xi_k = \xi_k^*. \end{aligned}$$

If the event $\mathfrak{B}_k$ does not occur and $\xi_{k+1}^* = \xi_k^* = C$, then $\mathbb{M}(\xi_{k+1}^*/\xi_0^*, \ldots, \xi_k^*) = \xi_k^*$. Hence, $\mathbb{M}(\xi_{k+1}^*/\xi_0^*, \ldots, \xi_k^*) \leqq \xi_k^*$ and so $\mathbb{M}\,\xi_{k+1}^* \leqq \mathbb{M}\,\xi_k^*$ and $\mathbb{M}\,\xi_n^* \leqq \mathbb{M}\,\xi_0^*$. Since

$$C\,\mathbb{P}\{\sup_{0 \leqq k \leqq n} \xi_k > C\} \leqq \mathbb{M}\,\xi_n^*,$$

we have

$$\mathbb{P}\{\sup_{0 \leqq k \leqq n} \xi_k \leqq C\} \leqq \frac{\mathbb{M}\,\xi_0}{C}.$$

If we set $t_k = \frac{k}{n} T$, then we obtain

$$\mathbb{P}\left\{\sup_{0 \leqq k \leqq n} \xi\left(\frac{kT}{n}\right) \geqq C\right\} \leqq \frac{\mathbb{M}\,\xi(0)}{C}.$$

Letting $n \to \infty$ and considering the continuity of $\xi(t)$, we finish the proof of the lemma.

Theorem 3. *Suppose there exists a nonnegative function $\varphi(t, x)$, twice differentiable w.r.t. x and once w.r.t. t, which satisfies*

$$\frac{\partial \varphi(t, x)}{\partial t} + a(t, x) \frac{\partial \varphi(t, x)}{\partial x} + \frac{1}{2} \cdot \frac{\partial^2 \varphi(t, x)}{\partial x^2} \leqq 0$$

for all $t > 0$ and $x \in R$, for which $\lim\limits_{\substack{x \to +\infty \\ T \to +\infty}} \inf\limits_{t > T} \varphi(t, x) = +\infty$. Then

$$\mathbb{P}\{\sup_{t > 0} \xi(t) < \infty\} = 1.$$

Proof. Using Itô's formula, we can write

$$\begin{aligned} &\varphi(t, \xi(t)) - \varphi(t_1, \xi(t_1)) \\ &= \int_{t_1}^{t} \left[\frac{\partial \varphi}{\partial s}(s, \xi(s)) + a(s, \xi(s)) \frac{\partial \varphi}{\partial x}(s, \xi(s)) + \frac{1}{2} \cdot \frac{\partial^2 \varphi}{\partial x^2}(s, \xi(s))\right] ds \\ &\quad + \int_{t_1}^{t} \frac{\partial \varphi}{\partial x}(s, \xi(s))\, dw(s) \leqq \int_{t_1}^{t} \frac{\partial \varphi}{\partial x}(s, \xi(s))\, dw(s). \end{aligned}$$

From this we find that for $t_0 < t_1 < \cdots < t_n$

$$\begin{aligned} &\mathbb{M}\{\varphi(t_{k+1}, \xi(t_{k+1})) - \varphi(t_k, \xi(t_k)) / \varphi(t_0, \xi(t_0)), \ldots, \varphi(t_k, \xi(t_k))\} \\ &\quad \leqq \mathbb{M}\left\{\int_{t_k}^{t_{k+1}} \frac{\partial \varphi(s, \xi(s))}{\partial x}\, dw(s) / \varphi(t_0, \xi(t_0)), \ldots, \varphi(t_k, \xi(t_k))\right\} = 0, \end{aligned}$$

thus from Lemma 5

$$\mathbb{P}\{\sup_{0 < t < T} \varphi(t, \xi(t)) > C / \xi(0)\} \leqq \frac{\varphi(0, \xi(0))}{C}.$$

Letting $T \to \infty$, we get

$$\mathbb{P}\{\sup_{0 < t} \varphi(t, \xi(t)) > C / \xi(0)\} \leqq \frac{\varphi(0, \xi(0))}{C}. \tag{8}$$

From (8) it follows that

$$\mathbb{P}\{\sup_{t > 0} \varphi(t, \xi(t)) < +\infty\} = 1. \tag{9}$$

If there existed a sequence $t_n \to \infty$ for which $\xi(t_n) \to +\infty$, then from the properties of $\varphi(t, x)$ we would have $\varphi(t_n, \xi(t_n)) \to +\infty$ for $n \to \infty$. Thus the proof of the theorem follows from (9).

Corollary 1. *Assume that $a(t, x)$ is such that for all $t, x > 0$ the derivative $\frac{\partial}{\partial t} \int_0^x a(t, z)\, dz$ exists and is $\geqq 0$ and that*

$$v(t, x) = \int_{-\infty}^{x} \exp\left\{-\int_0^y 2a(t, z)\, dz\right\} dy$$

converges.

Assume further that $\lim_{t \to +\infty} v(t, x) = v(x)$ exists and that $\lim_{x \to +\infty} v(x) = +\infty$. Then $\mathbb{P}\{\sup_t \xi(t) < +\infty\} = 1$.

In fact, since

$$\frac{\partial v(t, x)}{\partial t} = -\int_{-\infty}^{x} \exp\left\{-\int_0^y 2a(t, z)\, dz\right\} \frac{\partial}{\partial t} \int_0^y 2a(t, z)\, dz\, dy \leqq 0,$$

and

$$a(t, x) \frac{\partial v(t, x)}{\partial x} + \frac{1}{2} \cdot \frac{\partial^2 v(t, x)}{\partial x^2} = 0,$$

we have

$$\frac{\partial v(t, x)}{\partial t} + a(t, x) \frac{\partial v(t, x)}{\partial x} + \frac{1}{2} \cdot \frac{\partial^2 v(t, x)}{\partial x^2} \leqq 0.$$

Moreover, $v(t, x)$ is monotone decreasing in t which means that $v(t, x) \geqq v(x)$. Thus from $\lim_{x \to \infty} v(x) = +\infty$ it follows that $\lim_{x \to \infty} \inf_{t > 0} v(t, x) = +\infty$. To complete the proof we apply Theorem 3.

§ 17. Theorems on the Asymptotic Behavior of Solutions of Stochastic Equations

We treat here solutions of equations whose coefficients do not depend on time. When $\xi(t) \to +\infty$ w.p. 1 we will look for a function $\varphi(t)$, tending to infinity for $t \to \infty$, which satisfies

$$\mathbb{P}\left\{\lim_{t \to \infty} \frac{\xi(t)}{\varphi(t)} = 1\right\} = 1. \tag{1}$$

It is natural to call such a function $\varphi(t)$ *the order of growth* of $\xi(t)$ for $t \to \infty$. We prove a very simple theorem on the existence of an order of growth.

Theorem 1. *Let $\xi(t)$ satisfy*

$$d\xi(t)=a(\xi(t))\,dt+\sigma(\xi(t))\,dw(t), \tag{2}$$

whose coefficients fulfill the following conditions: 1) $\lim_{x\to+\infty} a(x)=a_0>0$; 2) $\sigma(x)$ *is bounded and positive;* 3) $a(x)$ *and* $\sigma(x)$ *are such that*

$$\mathbb{P}\{\lim_{t\to\infty}\xi(t)=+\infty\}=1. \tag{3}$$

Then

$$\mathbb{P}\left\{\lim_{t\to\infty}\frac{\xi(t)}{a_0 t}=1\right\}=1. \tag{4}$$

Proof. Since

$$\xi(t)=\xi(0)+\int_0^t a(\xi(s))\,ds+\int_0^t \sigma(\xi(s))\,dw(s),$$

it is sufficient to show that

$$\mathbb{P}\left\{\lim_{t\to\infty}\frac{1}{t}\int_0^t a(\xi(s))\,ds=a_0\right\}=1, \tag{5}$$

and

$$\mathbb{P}\left\{\lim_{t\to\infty}\frac{1}{t}\int_0^t \sigma(\xi(s))\,dw(s)=0\right\}=1. \tag{6}$$

Eq. (5) follows from the relation $\mathbb{P}\{\lim_{t\to\infty} a(\xi(t))=a_0\}=1$ which is a consequence of assumption 1) of the theorem and Eq. (3).

We now prove (6). From Theorem 1 § 3

$$\begin{aligned}
&\mathbb{P}\left\{\sup_{A\le t\le A_1}\frac{1}{t}\left|\int_0^t \sigma(\xi(u))\,dw(u)\right|>\varepsilon\right\}\\
&\le\mathbb{P}\left\{\sup_{A\le t\le A_1}\frac{1}{A}\left|\int_0^t \sigma(\xi(u))\,dw(u)\right|>\varepsilon\right\}\\
&\le\frac{1}{A^2\varepsilon^2}\,\mathbb{M}\left[\int_0^{A_1}\sigma(\xi(u))\,dw(u)\right]^2\le\frac{CA_1}{\varepsilon^2A^2},
\end{aligned}$$

where $C=\sup_x \sigma(x)$. Thus,

$$\begin{aligned}
&\mathbb{P}\left\{\sup_{A\le t}\frac{1}{t}\left|\int_0^t \sigma(\xi(u))\,dw(u)\right|>\varepsilon\right\}\\
&\le\sum_{k=0}^{\infty}\mathbb{P}\left\{\sup_{A\cdot 2^k\le t\le A\cdot 2^{k+1}}\frac{1}{t}\left|\int_0^t \sigma(\xi(u))\,dw(u)\right|>\varepsilon\right\}\le\sum_{k=0}^{\infty}\frac{CA\cdot 2^{k+1}}{\varepsilon^2A^2\cdot 2^{2k}}=\frac{4C}{\varepsilon^2A}.
\end{aligned}$$

The variable $\alpha(A)=\sup_{t\geq A}\frac{1}{t}\left|\int_0^t \sigma(\xi(u))\,dw(u)\right|$ falls monotonically as $A\to\infty$ which implies the existence of the limit $\alpha=\lim_{A\to\infty}\alpha(A)$. Then,

$$\mathbb{P}\{\alpha>\varepsilon\}=\lim_{A\to\infty}\mathbb{P}\{\alpha(A)>\varepsilon\}=0.$$

Formula (6) and thus the theorem are proved.

Theorem 1 can be used to prove the existence of a nonlinear order of growth. In many cases the following theorem is useful for this purpose. It is an easy consequence of Theorem 1.

Theorem 2. *Let $\xi(t)$ satisfy* (2) *whose coefficients fulfill the following conditions:*

1) *$a(x)$ and $\sigma(x)$ are such that $\mathbb{P}\{\lim_{t\to\infty}\xi(t)=+\infty\}=1$;* 2) *there exists an increasing, twice continuously differentiable function $f(x)$ for which $f(x)\to+\infty$ when $x\to+\infty$, $\lim_{x\to+\infty}[a(x)f'(x)+\frac{1}{2}f''(x)\sigma^2(x)]=C>0$ and $f'(x)\sigma(x)$ is bounded. Then*

$$\mathbb{P}\left\{\lim_{t\to\infty}\frac{f(\xi(t))}{t}=C\right\}=1. \tag{7}$$

Proof. Let $\varphi(x)$ be the inverse of f. Set $\eta(t)=f(\xi(t))$. It is clear that

$$\mathbb{P}\{\lim_{t\to\infty}\eta(t)=+\infty\}=1.$$

Moreover, $\eta(t)$ satisfies $d\eta(t)=\bar{a}(\eta(t))\,dt+\bar{\sigma}(\eta(t))\,dw(t)$ with coefficients

$$\bar{a}(x)=a(\varphi(x))f'(\varphi(x))+\tfrac{1}{2}\sigma(\varphi(x))f''(\varphi(x)),$$

and

$$\bar{\sigma}(x)=\sigma(\varphi(x))f'(\varphi(x)).$$

From the assumptions of the theorem it follows that $\lim_{x\to+\infty}\bar{a}(x)=C>0$ and that $\bar{\sigma}(x)$ is bounded. Thus, from Theorem 1

$$\mathbb{P}\left\{\lim_{t\to\infty}\frac{\eta(t)}{Ct}=1\right\}=1.$$

Since $\eta(t)=f(\xi(t))$, the last relation implies the proof of the theorem.

Remark 1. Conditions which f must satisfy in order to fulfill the assertions of the theorem are for example the following: 1) $\lim_{x\to+\infty}a(x)f'(x)=C>0$; 2) $\lim_{x\to+\infty}f''(x)\sigma^2(x)=0$; 3) the function $f'(x)\sigma(x)$ is bounded.

We use Remark 1 to obtain a simple consequence of Theorem 2.

Corollary 1. *Assume $a(x)\sim Cx^{\alpha}$ for $x\to+\infty$, $0<\alpha<1$, $C>0$. Take $f(x)=x^{1-\alpha}$ for sufficiently large x. Then $f'(x)=(1-\alpha)x^{-\alpha}$ and $f''(x)=$*

$-\alpha(1-\alpha)\,x^{-1-\alpha}$. *Since* $f'(x)\,a(x)\to C(1-\alpha)$, *we have*

$$\mathbb{P}\left\{\lim_{t\to\infty}\frac{\xi(t)^{1-\alpha}}{t}=C(1-\alpha)\right\}=1,$$

provided that $\lim\limits_{x\to\infty}\frac{\sigma^2(x)}{x^{1+\alpha}}=0$ *and* $\lim\limits_{x\to\infty}\frac{\sigma(x)}{x^{\alpha}}=0$. *Then if the listed conditions are satisfied, we have*

$$\mathbb{P}\left\{\lim_{t\to\infty}\frac{\xi(t)}{t^{\frac{1}{1-\alpha}}}=[C(1-\alpha)]^{\frac{1}{1-\alpha}}\right\}=1. \tag{8}$$

Corollary 2. *Let* $a(x)\sim Cx$ *for* $x\to\infty$, $C>0$. *Assume* $f(x)=\ln x$ *for sufficiently large* x. *Since* $a(x)\,f'(x)\to C$ *for* $x\to+\infty$, *we can conclude that if* $\frac{\sigma^2(x)}{x^2}\to 0$ *for* $x\to\infty$ *and* $\mathbb{P}\{\lim\limits_{t\to\infty}\xi(t)=+\infty\}=1$, *then*

$$\mathbb{P}\left\{\lim_{t\to\infty}\frac{\ln\xi(t)}{t}=C\right\}=1.$$

From the preceding formula it is not possible to derive statements on the order of growth of $\xi(t)$. The equation $d\xi(t)=C\,\xi(t)\,dt+dw(t)$ with solution

$$\xi(t)=\xi(0)\,e^{Ct}+e^{Ct}\int_0^t e^{-Cu}\,dw(u)$$

shows that the expression $\xi(t)\,e^{-Ct}$ can converge to the random variable η $\left(\text{here } \eta=\xi(0)+\int_0^\infty e^{-Cu}\,dw(u)\right)$.

We turn now to the case in which $\sigma(x)=1$ and introduce for it more general conditions for power-law growth.

Theorem 3. *Set*

$$A(z)=2\int_0^z a(y)\,dy.$$

Assume that for some $\alpha\in(1,2)$ *and for all* $z>0$

$$A(z)=C\,z^{\alpha}+\beta(z)+\gamma(z),$$

where $C>0$, $\beta(z)=o(z^{\alpha})$, $\beta'(z)=o(z^{\alpha-1})$ *for* $z\to+\infty$ *and* $\gamma(z)$ *is a bounded and uniformly continuous function. If* $\int_{-\infty}^{0} e^{-A(z)}\,dz=+\infty$, *then the solution of*

$$d\xi(t)=a(\xi(t))\,dt+dw(t) \tag{9}$$

satisfies

$$\mathbb{P}\left\{\lim_{t\to\infty}\frac{\xi(t)}{t^{\frac{1}{2-\alpha}}}=\left[\frac{C(2-\alpha)}{2}\right]^{\frac{1}{2-\alpha}}\right\}=1. \tag{10}$$

Proof. Since $\int_0^\infty e^{-A(z)}dz<\infty$, from statement 4 of Theorem 1 § 16 we have $\mathbb{P}\{\lim_{t\to\infty}\xi(t)=+\infty\}=1$. Assume $f(x)$ satisfies

$$a(x)f'(x)+\tfrac{1}{2}f''(x)=\chi(x),$$

where $\chi(x)$ is a continuous function equal to one for $x>0$. Then for $x>0$

$$f'(x)=2e^{-A(x)}\left[\int_0^x e^{A(z)}dz+C_1\right],$$

where C_1 is some constant. Taking into account the representation of $A(x)$ we can write down the asymptotic formula

$$f'(x)=2e^{-Cx^\alpha-\beta(x)}\int_a^x\frac{de^{Cz^\alpha+\beta(z)}}{\alpha Cz^{\alpha-1}+\beta'(z)}e^{\gamma(z)-\gamma(x)}+o(e^{-Cx^\alpha-\beta(x)})$$

(for $x\to+\infty$), where a is chosen large enough so that $\alpha Cz^{\alpha-1}+\beta'(z)>1$ when $z>a$. Since

$$e^{-Cx^\alpha-\beta(x)}\int_a^{x/2}\frac{de^{Cz^\alpha+\beta(z)}}{\alpha Cz^{\alpha-1}+\beta'(z)}e^{\gamma(z)-\gamma(x)}=o(e^{-\delta x^\alpha})$$

for some $\delta>0$, we have for $\varepsilon>0$

$$\begin{aligned}
&e^{-Cx^\alpha-\beta(x)}\int_{x/2}^{x-\varepsilon}\frac{de^{Cz^\alpha+\beta(z)}}{\alpha Cz^{\alpha-1}+\beta'(z)}e^{\gamma(z)-\gamma(x)}\\
&\quad=O(\exp\{-C[x^\alpha-(x-\varepsilon)^\alpha-\beta(x)+\beta(x-\varepsilon)]\})\\
&\quad=O(\exp\{-\alpha C(x-\varepsilon)^{\alpha-1}\varepsilon+|\beta'(x+\theta)|\,\varepsilon\})\\
&\quad=O(\exp\{-\alpha C(x-\varepsilon)^{\alpha-1}\varepsilon+o(x^{\alpha-1})\})=o(e^{-\delta_1\varepsilon x^{\alpha-1}}),
\end{aligned}$$

where δ_1 is some constant. Hence, for arbitrary $\varepsilon>0$

$$f'(x)=2e^{-Cx^\alpha-\beta(x)}\int_{x-\varepsilon}^x\frac{de^{Cz^\alpha+\beta(z)}}{\alpha Cz^{\alpha-1}+\beta'(z)}e^{\gamma(z)-\gamma(x)}+o(e^{-2\delta_1x^{\alpha-1}}).$$

From the latter relation it is easy to conclude, using the uniform continuity of $\gamma(z)$, that

$$\lim_{x\to+\infty}f'(x)\frac{\alpha Cx^{\alpha-1}}{2}=1,$$

consequently, $f'(x) \sim \frac{2}{\alpha C x^{\alpha-1}}$ for $x \to +\infty$, and so

$$f(x) \sim \frac{2x^{2-\alpha}}{\alpha C(2-\alpha)} \quad \text{for } x \to +\infty.$$

It is clear that we can extend $f(x)$ for $x<0$ in such a way that $f'(x)$ will be bounded for all x. Then we conclude from Theorem 2 that

$$\mathbb{P}\left\{\lim_{t \to +\infty} \frac{f(\xi(t))}{t} = 1\right\} = 1,$$

and since $\mathbb{P}\{\xi(t) \to +\infty\} = 1$, we have

$$\mathbb{P}\left\{\lim_{t \to +\infty} \frac{f(\xi(t))}{\frac{2\xi(t)^{2-\alpha}}{\alpha C(2-\alpha)}} = 1\right\} = 1 \quad \text{and} \quad \mathbb{P}\left\{\lim_{t \to \infty} \frac{2\xi(t)^{2-\alpha}}{\alpha C(2-\alpha)t} = 1\right\} = 1.$$

The last equality completes the proof.

A very simple example of an equation for which $a(x)$ does not have a power-law asymptote for $x \to \infty$ but $\xi(t)$ grows as a power at infinity is given by an equation with $a(x) = x^\alpha + \lambda x^\beta \cos x^\gamma$ for $x>0$ with $\beta > \alpha > 0$ and γ such that $\int_0^\infty x^\beta \cos x^\gamma \, dx$ converges ($\gamma > \beta + 1$).

We now use Theorem 2 to find conditions under which the order of growth $\varphi(t)$ of a solution $\xi(t)$ of (2) can be chosen to be a solution of an ordinary differential equation obtained from (2) by neglecting the term containing the stochastic differential.

In other words, we are interested in the question of when the order of growth can be taken as the solution of

$$d\varphi(t) = a(\varphi(t))\, dt. \tag{11}$$

We remark that is quite natural to look for a (nonrandom) function which a) grows as $t \to \infty$ in the same way as the solution of a stochastic equation and b) is a solution of the latter without random term. We make precise conditions under which such an approach to the investigation of the asymptotic behavior is legitimate. It is obviously reasonable to consider merely the case in which the solution of (11) grows without bound for $t \to +\infty$. Such a solution will exist provided that $a(x) > 0$ for all sufficiently large x.

Theorem 4. *Assume that $a(x)$ is such that* (2) *and* (11) *have a unique solution with arbitrary unitial data for all sufficiently large $x>0$, that $a(x)>0$, $\frac{\sigma(x)}{a(x)}$ is bounded, $a'(x)$ exists and that $a'(x) \to 0$. If $\varphi(t)$ is a* *solution of* (11) *for which* $\lim_{t\to\infty}\varphi(t)=+\infty$ *and if for some* $C>0$

$$\lim_{\varepsilon\to 0}\sup_{\left|\frac{z}{u}-1\right|\leqq\varepsilon,\, z>C}\left|\frac{\varphi(z)}{\varphi(u)}-1\right|=0, \tag{12}$$

then

$$\mathbb{P}\left\{\lim_{t\to\infty}\frac{\xi(t)}{\varphi(t)}=1\right\}=1.$$

Proof. Assume $f'(x)=\frac{1}{a(x)}$ for sufficiently large x. Extend $f'(x)$ for the remaining real x in such a way that $f'(x)\sigma(x)$ is positive and bounded and that $f''(x)$ exists. Since

$$\lim_{x\to+\infty}a(x)f'(x)=1,\qquad \lim_{x\to+\infty}\sigma^2(x)f''(x)=\lim_{x\to+\infty}\frac{\sigma^2(x)a'(x)}{a^2(x)}=0,$$

and $f'(x)\sigma(x)=\frac{\sigma(x)}{a(x)}$ is bounded, all the assumptions of Remark 1 are met. Hence

$$\mathbb{P}\left\{\lim_{t\to\infty}\frac{f(\xi(t))}{t}=1\right\}=1.$$

We solve the equation $d\varphi(t)=a(\varphi(t))\,dt$ with initial condition $\varphi(0)=x_0$ which is such that $f'(x)=\frac{1}{a(x)}>0$ for $x\geqq x_0$. Then $f'(\varphi(t))\,d\varphi(t)=dt$ and $f(\varphi(t))-f(x_0)=t$. It follows from the properties of $\varphi(t)$ that

$$\mathbb{P}\left\{\lim_{t\to\infty}\frac{\varphi(f(\xi(t)))}{\varphi(t)}=1\right\}=1. \tag{13}$$

We will show that $\lim_{x\to\infty}\frac{\varphi(f(x))}{x}=1$. Since $\varphi(t)=+\infty$ for $t\to+\infty$ it is sufficient to prove that $\lim_{t\to\infty}\frac{\varphi(f(\varphi(t)))}{\varphi(t)}=1$. But from (12) we have

$$\frac{\varphi(f(\varphi(t)))}{\varphi(t)}=\frac{\varphi(f(x_0)+t)}{\varphi(t)}\to 1$$

because $\frac{t+f(x_0)}{t}\to 1$. Since $\xi(t)\to+\infty$ w.p. 1, $\lim_{t\to\infty}\frac{\varphi(f(\xi(t)))}{\xi(t)}=1$ w.p. 1. The proof now follows from (13).

We investigate the case in which $a(x)\to 0$ for $x\to+\infty$. We will need

Lemma 1. *Assume* $\xi(t)$ *satisfies*

$$d\xi(t)=dt+\sigma(\xi(t))\,dw(t), \tag{14}$$

where $\sigma(x)$ is such that $\sigma^2(x) \leqq C(1+|x|^\alpha)$ for some $\alpha<1$. Then

$$\mathbb{P}\left\{\lim_{t\to\infty}\frac{\xi(t)}{t}=1\right\}=1.$$

Proof. It suffices to prove the lemma for the case where $\xi(0)$ is deterministic and $\xi(0)=x$. From (14)

$$\xi(t)=x+t+\int_0^t \sigma(\xi(s))\,dw(s),$$

so that

$$\mathbb{M}\,\xi(t)^2=(t+x)^2+\int_0^t \mathbb{M}\,\sigma^2(\xi(s))\,ds \leqq (t+x)^2+Ct+C\int_0^t \mathbb{M}\,|\xi(s)|^\alpha\,ds$$

$$\leqq (t+x)^2+Ct+C\int_0^t [\mathbb{M}\,\xi(s)^2]^{\alpha/2}\,ds$$

and for some x_0

$$\mathbb{M}\,\xi(t)^2 \leqq (t+x_0)^2+C\int_0^t (\mathbb{M}\,\xi(s)^2)^{\alpha/2}\,ds.$$

Set $\sup\limits_{0\leqq t\leqq T}\mathbb{M}\,\xi(t)^2=g(T)$. Then $g(T)\leqq(T+x_0)^2+CTg(T)^{\alpha/2}$. Obviously, the solution of the inequality $x<a+k\,x^\beta$, where $\beta<1$, will also satisfy $x<a+k\,\alpha^\beta+\beta\,k\,a^{\beta-1}(x-a)$ since for $x>a$,

$$k\,x^\beta<k\,\alpha^\beta+\frac{d}{dx}(k\,x^\beta)|_{x=a}(x-a).$$

Then

$$g(t)\leqq(t+x_0)^2+Ct(t+x_0)^\alpha+\frac{\alpha}{2}Ct(t+x_0)^{\alpha-2}[g(t)-(t+x_0)^2],$$

i.e.,

$$g(t)\leqq\frac{(t+x_0)^2+Ct\left(1-\dfrac{\alpha}{2}\right)(t_0+x_0)^\alpha}{1-C\dfrac{\alpha}{2}\cdot\dfrac{t}{(t_0+x_0)^{2-\alpha}}}.$$

This implies that $\overline{\lim\limits_{t\to\infty}}\dfrac{g(t)}{t^2}\leqq 1$ and for some A and B $\mathbb{M}\,\xi(t)^2\leqq A+Bt^2$. But then

$$\mathbb{M}\left[\frac{1}{t}\int_0^t \sigma(\xi(s))\,dw(s)\right]^2\leqq\frac{1}{t^2}\int_0^t \mathbb{M}\,\sigma(\xi(s))^2\,ds\leqq\frac{1}{t^2}C\int_0^t(1+\mathbb{M}\,|\xi(s)|^\alpha)\,ds$$

$$\leqq\frac{C}{t^2}\left[t+\int_0^t [M(\xi(s))^2]^{\alpha/2}\,ds\right]$$

$$\leqq \frac{C}{t^2}\left[t+\int_0^t (A+Bs^2)^{\alpha/2}\,ds\right]$$

$$\leqq \frac{C}{t}+O\left(\frac{1}{t^{1-\alpha}}\right)\leqq \frac{L}{t^{1-\alpha}} \quad \text{for } t>1,$$

where L is some constant. From Theorem 1 §3

$$\mathbb{P}\left\{\sup_{N\cdot 2^k\leqq t\leqq N\cdot 2^{k+1}}\left|\frac{1}{t}\int_0^t \sigma(\xi(u))\,dw(u)\right|>\varepsilon\right\}$$

$$\leqq \mathbb{P}\left\{\sup_{N\cdot 2^k\leqq t\leqq N\cdot 2^{k+1}}\left|\frac{1}{N\cdot 2^k}\int_0^t \sigma(\xi(u))\,dw(u)\right|>\varepsilon\right\}$$

$$\leqq \frac{1}{\varepsilon^2 N^2\cdot 2^{2k}}\int_0^{N\cdot 2^{k+1}} M\sigma^2(\xi(u))\,du \leqq \frac{4L}{\varepsilon^2 N^{1-\alpha}\cdot 2^{(1-\alpha)(k+1)}}.$$

From this we can (as in Theorem 1) conclude that

$$\mathbb{P}\left\{\lim_{t\to\infty}\frac{1}{t}\int_0^t \sigma(\xi(u))\,dw(u)=0\right\}=1,$$

which finishes the proof.

Remark 2. With only slight changes in the proof we can establish the claim of Lemma 1 for the solution of (2), where $a(x)$ is bounded, $\sigma(x)$ meets the conditions of Lemma 1, $\lim_{x\to\infty} a(x)=1$ and

$$\mathbb{P}\{\lim_{t\to\infty}\xi(t)=\infty\}=1.$$

We apply this remark to the investigation of the growth of solutions of stochastic equations for which $a(x)$ has a power-law asymptote with negative exponent for $x\to+\infty$.

Theorem 5. *Assume $a(x)\sim Cx^\alpha$ $(C>0)$ for $x\to+\infty$, $-1<\alpha<0$. If $\xi(t)$ is a solution of* (2) *with bounded $\sigma(x)$ and $\mathbb{P}\{\lim_{t\to\infty}\xi(t)=+\infty\}=1$, then*

$$\mathbb{P}\left\{\lim_{t\to\infty}\frac{\xi(t)}{t^{\frac{1}{1-\alpha}}}=[C(1-\alpha)]^{\frac{1}{1-\alpha}}\right\}=1.$$

Proof. Let $f(x)$ be an increasing function equal to $\dfrac{1}{C}\dfrac{x^{1-\alpha}}{1-\alpha}$ for sufficiently large x. We extend f in such a way that it is monotone increasing on the rest of the line, that $f'(x)$ and $f''(x)$ exist and the function $a(x)f'(x)+\frac{1}{2}\sigma^2(x)f''(x)$ is bounded. Then the process $\eta(t)=$

$f(\xi(t))$ will satisfy

$$d\eta(t)=\bar{a}(\eta(t))\,dt+\bar{\sigma}(\eta(t))\,dw(t),$$

where

$$\lim_{x\to+\infty}\bar{a}(x)=\lim_{x\to+\infty}\left[a(x)f'(x)+\tfrac{1}{2}\sigma^2(x)f''(x)\right]=1,$$

$$\bar{\sigma}(x)=\sigma(g(x))\,f'(g(x)),$$

and $g(x)$ is the inverse function to $f(x)$. Hence, for sufficiently large x

$$g(x)=[C(1-\alpha)\,x]^{\frac{1}{1-\alpha}},$$

$$\bar{\sigma}(x)=\sigma(g(x))\left\{\frac{1}{C}[C(1-\alpha)\,x]^{\frac{1}{1-\alpha}}\right\}^{-\alpha}\leqq L_1\,x^{-\frac{\alpha}{1-\alpha}}\left(0<\frac{-\alpha}{1-\alpha}<\frac{1}{2}\right).$$

This means that the assumptions of Remark 2 are met for $\eta(t)$ and

$$\mathbb{P}\left\{\lim_{t\to\infty}\frac{\eta(t)}{t}=1\right\}=1 \quad\text{or}\quad \mathbb{P}\left\{\lim_{t\to\infty}\frac{\xi(t)^{1-\alpha}}{C(1-\alpha)\,t}=1\right\}=1.$$

This completes the proof of the theorem.

Remark 3. Proceeding analogously to the proof of Theorem 3, we can show that if $a(x)$ in (9) is such that $A(x)$ (defined in Theorem 3) satisfies $A(x)=C\,x^{\alpha}+\beta(x)$, where $C>0$ and $0<\alpha\leqq 1$, $\lim_{x\to\infty}\beta(x)$ exists and $\int_{-\infty}^{0}\exp\{-A(z)\}\,dz=+\infty$, then the claim of Theorem 3 still holds.

This follows from the fact that the process $\eta(t)=f(\xi(t))$, where f is the function introduced in Theorem 3, satisfies

$$d\eta(t)=\bar{a}(\eta(t))\,dt+\bar{\sigma}(\eta(t))\,dw(t),$$

where $\bar{a}(x)=1$ for $x>0$ and

$$\bar{\sigma}(x)\sim\frac{1}{C}[C(2-\alpha)]^{\frac{1-\alpha}{2-\alpha}}\,x^{\frac{1-\alpha}{2-\alpha}},$$

with $0\leqq\dfrac{1-\alpha}{2-\alpha}<\dfrac{1}{2}$. Thus we can use Lemma 1.

Remark 4. If $a(x)\sim\dfrac{C}{x}$ as $x\to\infty$ (i.e., in Theorem 5, $\alpha=-1$), then the conclusion of Theorem 5 no longer holds.

Indeed, assume there exists an $a>0$ such that

$$\mathbb{P}\left\{\lim_{t\to\infty}\frac{\xi(t)}{\sqrt{t}}=a\right\}=1.$$

Then

$$\mathbb{P}\left\{\lim_{t\to+\infty}\frac{1}{\sqrt{t}}\int_0^t a(\xi(s))\,ds=\lim_{t\to\infty}\frac{1}{\sqrt{t}}\int_0^t \frac{C}{\xi(s)}\,ds\right.$$

$$\left.=\lim_{t\to\infty}\frac{1}{\sqrt{t}}\int_0^t \frac{C}{a\sqrt{s}}\,ds=\frac{2C}{a}\right\}=1.$$

Since

$$\xi(t)=\xi(0)+\int_0^t a(\xi(s))\,ds+w(t),$$

we have

$$\mathbb{P}\left\{\lim_{t\to\infty}\frac{w(t)}{\sqrt{t}}=a-\frac{2C}{a}\right\}=1$$

which is impossible since $\frac{w(t)}{\sqrt{t}}$ is normally distributed with mean zero and variance one.

§ 18. Ergodic Theorems

We will deal with two kinds of ergodic theorems: a) those for the processes themselves, such as a law of large numbers for the averages $\frac{1}{T}\int_0^T f(\xi(s))\,ds$ $(T\to+\infty)$, where $\xi(t)$ is the process in question and f is a function belonging to a sufficiently wide class (for example, an arbitrary continuous function); b) ergodic theorems for the distributions, stating the existence of a limiting distribution for the variable $\xi(t)$ when $t\to+\infty$. The connection between these lies in the fact that in the most interesting cases we have, w.p. 1

$$\lim_{T\to\infty}\frac{1}{T}\int_0^T f(\xi(s))\,ds=\int f(x)\,dG(x), \tag{1}$$

where G is the limiting distribution of $\xi(t)$ for $t\to\infty$. Such a distribution G is called *ergodic for the process* $\xi(t)$.

We will study conditions which guarantee the existence of an ergodic distribution for the solution of stochastic equation and will find the form of such a distribution.

Let $\xi(t)$ be a solution of

$$d\xi(t)=a(\xi(t))\,dt+\sigma(\xi(t))\,dw(t),$$

whose coefficients $a(x)$ and $\sigma(x)$ satisfy a Lipschitz condition and $\sigma(x)>0$. It is clear that an ergodic distribution can only exist when $|\xi(t)|$

stays finite w.p. 1 which will be the case when the function

$$f(x)=\int_0^x \exp\left\{-\int_0^z \frac{2a(y)}{\sigma^2(y)}\,dy\right\}dz$$

satisfies $\lim_{x\to\infty} f(x)=+\infty$ and $\lim_{x\to-\infty} f(x)=-\infty$.

Set $\eta(t)=f(\xi(t))$. The process $\eta(t)$ satisfies an equation of the form $d\eta(t)=\bar{\sigma}(\eta(t))\,dw(t)$. It follows from the properties of f that the limiting distributions of $\eta(t)$ and $\xi(t)$ exist simultaneously. Thus, in the sequel we will study the limit distribution of $\xi(t)$ which satisfies

$$d\xi(t)=\sigma(\xi(t))\,dw(t). \tag{2}$$

As a preliminary we consider the random time $\tau_{x,y}=\inf\{s, \xi_x(s)=y\}$ required by the process $\xi_x(s)$, which satisfies (2) with $\xi_x(0)=x$, to reach the point y.

Since here

$$\mathbb{P}\{\sup_{t>0}\xi(t)=+\infty\}=1; \quad \mathbb{P}\{\inf_{t>0}\xi(t)=-\infty\}=1,$$

for an arbitrary pair x and y the time $\tau_{x,y}$ is defined (finite) w.p. 1. Let $a<x<b$ and let $\tau_x[a,b]$ be the time $\xi_x(t)$ first attains the boundary of (a,b). Using Corollary 2 § 15 and the fact that $\Phi(z)=1$ here (Φ is the function appearing in that corollary), we obtain

$$\mathbb{M}\tau_x[a,b]=\frac{x-a}{b-a}\int_a^b \frac{2(b-z)}{\sigma^2(z)}\,dz-\int_a^x \frac{2(x-z)}{\sigma^2(z)}\,dz. \tag{3}$$

By means of simple transformations we can reduce (3) to the following (more symmetric) form:

$$\mathbb{M}\tau_x[a,b]=\frac{x-a}{b-a}\int_x^b \frac{2(b-z)}{\sigma^2(z)}\,dz+\frac{b-x}{b-a}\int_a^x \frac{2(z-a)}{\sigma^2(z)}\,dz. \tag{4}$$

We apply this formula to the proof of the following

Lemma 1. *If* $\int_0^\infty \frac{1}{\sigma^2(z)}\,dz<\infty$, *then for* $y<x$

$$\mathbb{M}\tau_{x,y}=(x-y)\int_x^\infty \frac{2}{\sigma^2(z)}\,dz+\int_y^x (z-y)\frac{2}{\sigma^2(z)}\,dz, \tag{5}$$

if $\int_{-\infty}^0 \frac{1}{\sigma^2(z)}\,dz<\infty$, *then for* $x<y$

$$\mathbb{M}\tau_{x,y}=(y-x)\int_{-\infty}^x \frac{2}{\sigma^2(z)}\,dz+\int_x^y (y-z)\frac{2}{\sigma^2(z)}\,dz. \tag{6}$$

Proof. Since both (5) and (6) are proved identically we will only do (5). Let $C>x$ and consider the variable $\tau_x[y, C]$. It is clear that $\tau_x[y, C]\uparrow\tau_{x,y}$ when $C\to+\infty$. Now we use (4) and let $C\to+\infty$:

$$\mathbb{M}\tau_x[y, C]=\frac{C-x}{C-y}\int_y^x(z-y)\frac{2}{\sigma^2(z)}dz+(x-y)\int_x^C\frac{C-z}{C-y}\cdot\frac{2}{\sigma^2(z)}dz.$$

If $C\to+\infty$, then

$$\frac{C-x}{C-y}\to 1,\quad 0\leqq\frac{C-z}{C-y}\leqq 1,\quad \frac{C-z}{C-y}\to 1.$$

The proof is complete.

Lemma 2. *Let $\sigma(x)$ be such that $\int_{-\infty}^{\infty}\frac{1}{\sigma^2(x)}dx<\infty$ and $\xi_x(t)$ satisfy* (2) *with initial condition $\xi_x(0)=x$. Then, for arbitrary x and y and bounded, measurable Borel function $\varphi(z)$, there holds*

$$\lim_{T\to\infty}\mathbb{M}\left(\frac{1}{T}\int_0^T\varphi(\xi_x(t))\,dt-\frac{1}{T}\int_0^T\varphi(\xi_y(t))\,dt\right)=0. \tag{7}$$

Proof. Set $\bar{\xi}(t)=\xi_x(\tau_{x,y}+t)$. It follows from Remark 2 § 15 that $\bar{\xi}(t)$ and $\xi_y(t)$ are identically distributed. Furthermore,

$$\mathbb{M}\frac{1}{T}\int_0^T\varphi(\xi_x(t))\,dt=\mathbb{M}\frac{1}{T}\int_0^{\tau_{x,y}}\varphi(\xi_x(t))\,dt$$
$$+\mathbb{M}\frac{1}{T}\int_{\tau_{x,y}}^{T+\tau_{x,y}}\varphi(\xi_x(t))\,dt-\mathbb{M}\frac{1}{T}\int_T^{T+\tau_{x,y}}\varphi(\xi_x(t))\,dt,$$

clearly,

$$\mathbb{M}\frac{1}{T}\int_{\tau_{x,y}}^{T+\tau_{x,y}}\varphi(\xi_x(s))\,ds=\mathbb{M}\frac{1}{T}\int_0^T\varphi(\bar{\xi}(s))\,ds=\mathbb{M}\frac{1}{T}\int_0^T\varphi(\xi_y(s))\,ds$$

and

$$\left|\mathbb{M}\frac{1}{T}\int_0^{\tau_{x,y}}\varphi(\xi_x(t))\,dt\right|\leqq\frac{\mathbb{M}\tau_{x,y}}{T}\sup_z|\varphi(z)|.$$

A similar estimate holds for $\left|\mathbb{M}\frac{1}{T}\int_T^{\tau_{x,y}+T}\varphi(\xi_x(t))\,dt\right|$. Thus, for some $L>0$

$$\left|\mathbb{M}\frac{1}{T}\int_0^T\varphi(\xi_x(t))\,dt-\mathbb{M}\frac{1}{T}\int_0^T\varphi(\xi_y(t))\,dt\right|\leqq\frac{L}{T}. \tag{8}$$

From this the proof follows immediately.

Remark 1. Since L can be taken as

$$2\mathbb{M}\,\tau_{x,y}\sup_z \varphi(z),$$

we find, taking into account the estimate

$$\mathbb{M}\,\tau_{x,y} \leqq |x-y| \int_{-\infty}^{\infty} \frac{2}{\sigma^2(z)}\,dz,$$

that there exists a constant H such that

$$\left|\frac{1}{T}\int_0^T \mathbb{M}\,\varphi(\xi_x(t))\,dt - \frac{1}{T}\int_0^T \mathbb{M}\,\varphi(\xi_y(t))\,dt\right| \leqq \frac{H|x-y|\,\|\varphi\|}{T}, \tag{9}$$

where

$$\|\varphi\| = \sup_z |\varphi(z)|.$$

Lemma 3. *If* $\int_{-\infty}^{\infty} \frac{1}{\sigma^2(x)}\,dx < \infty$ *and* $F_{t,x}(y) = \mathbb{P}\{\xi_x(t) < y\}$ *is the distribution function of* $\xi_x(t)$, *then for all* $t > 0$

$$\int_{-\infty}^{\infty} \frac{1}{\sigma^2(x)} F_{t,x}(y)\,dx = \int_{-\infty}^{y} \frac{1}{\sigma^2(x)}\,dx. \tag{10}$$

Proof. We first assume that $\sigma(x)$ is such that the function

$$g(t, y) = \int_{-\infty}^{\infty} \frac{1}{\sigma^2(x)} F_{t,x}(y)\,dx$$

satisfies the Fokker-Planck equation (see Remark 3 § 14)

$$\frac{\partial g(t, y)}{\partial t} = \frac{1}{2} \cdot \frac{\partial}{\partial y}\left(\sigma^2(y)\frac{\partial g(t, y)}{\partial y}\right) \tag{11}$$

with initial condition $g(0, y) = \int_{-\infty}^{y} \frac{dx}{\sigma^2(x)}$. It's easy to see that $g(t, y) = \int_{-\infty}^{y} \frac{dx}{\sigma^2(x)}$ solves this Cauchy problem for (11). For sufficiently smooth $\sigma(x)$, the Eq. (11) has a unique solution of the posed Cauchy problem, so that under the assumptions made, (10) is valid. It remains to note that we can choose a sequence $\sigma_n(x)$ such that, for each n

$$\int_{-\infty}^{\infty} \frac{1}{\sigma_n^2(x)} F_{t,x}^{(n)}(y)\,dx = \int_{-\infty}^{y} \frac{1}{\sigma_n^2(z)}\,dz,$$

where $F_{t,x}^{(n)}(y)$ is the distribution function of the process $\xi_x^{(n)}(t)$, which satisfies $d\xi_x^{(n)}(t)=\sigma_n(\xi_x^{(n)}(t))\,dw(t)$ with $\xi_x^{(n)}(0)=x$ and also that

$$\frac{1}{\sigma_n^2(x)}\to\frac{1}{\sigma^2(x)}$$

uniformly and

$$\int_{-\infty}^{\infty}\left|\frac{1}{\sigma_n^2(y)}-\frac{1}{\sigma^2(y)}\right|dy\to 0.$$

Then the distribution of $\xi_x^{(n)}(t)$ will also converge uniformly to the distribution of $\xi_x(t)$ (e.g. on the basis of Theorem 2 § 7), which means that

$$\int_{-\infty}^{y}\frac{dz}{\sigma^2(z)}=\lim_{n\to\infty}\int_{-\infty}^{y}\frac{dz}{\sigma_n^2(z)}=\lim_{n\to\infty}\int_{-\infty}^{\infty}\frac{1}{\sigma_n^2(x)}F_{t,x}^{(n)}(y)\,dx$$
$$=\int_{-\infty}^{\infty}\frac{1}{\sigma^2(x)}F_{t,x}(y)\,dx.\quad\square$$

Corollary 1. *Under the assumptions of Lemma 3, an arbitrary Borel function* $\varphi(x)$ *for which* $\int_{-\infty}^{\infty}\frac{|\varphi(z)|}{\sigma^2(z)}dz<\infty$ *satisfies*

$$\int \mathbb{M}\varphi(\xi_y(t))\frac{1}{\sigma^2(y)}dy=\int\frac{\varphi(y)}{\sigma^2(y)}dy. \tag{12}$$

In fact, let $\xi(0)$ have the distribution function $D\int_{-\infty}^{y}\frac{1}{\sigma^2(z)}dz$, where $D^{-1}=\int_{-\infty}^{\infty}\frac{dz}{\sigma^2(z)}$. Then it follows from Lemma 3 that the distribution of $\xi_{\xi(0)}(t)$ coincides with that of $\xi(0)$ for all $t>0$. Thus for each bounded continuous function φ (12) holds since

$$D\int_{-\infty}^{\infty}\frac{\varphi(z)}{\sigma^2(z)}dz=\mathbb{M}\varphi(\xi(0))=\mathbb{M}\varphi(\xi(t))$$
$$=\mathbb{M}(\mathbb{M}\varphi(\xi_{\xi(0)}(t))/\xi(0))=D\int\mathbb{M}\varphi(\xi_y(t))\frac{dy}{\sigma^2(y)}.$$

Clearly, by means of a passage to the limit, (12) can be extended to all functions φ for which its right side makes sense.

Theorem 1. *Let* φ *be a bounded Borel function, then for arbitrary* x

$$\lim_{T\to\infty}\frac{1}{T}\int_0^T\mathbb{M}\varphi(\xi_x(t))\,dt=D\int_{-\infty}^{\infty}\frac{\varphi(y)}{\sigma^2(y)}dy.$$

Proof. We first show that

$$\lim_{T\to\infty} \int_{-\infty}^{\infty} \left[\frac{1}{T} \mathbb{M} \int_0^T \varphi(\xi_x(t))\,dt - \frac{1}{T} \mathbb{M} \int_0^T \varphi(\xi_y(t))\,dt\right] \frac{dy}{\sigma^2(y)} = 0. \tag{13}$$

Indeed, the function $\psi_T(x,y) = \frac{1}{T}\int_0^T [\mathbb{M}\varphi(\xi_x(t)) - \mathbb{M}\varphi(\xi_y(t))]\,dt$ satisfies $|\psi_T(x,y)| \leqq 2K$ $(K = \sup_x |\varphi(x)|)$ and $\psi_T(x,y) \to 0$ for each x and y if $T\to\infty$ (Lemma 2). Then (13) follows from Lebesgue's theorem on the interchange of limit and integration operations. From Corollary 1

$$\int_{-\infty}^{\infty} \frac{1}{T} \mathbb{M} \int_0^T \varphi(\xi_y(t)) \frac{dt}{\sigma^2(y)}\,dy = \frac{1}{T}\int_0^T \left[\int_{-\infty}^{\infty} \mathbb{M}\varphi(\xi_y(t)) \frac{dy}{\sigma^2(y)}\right] dt$$
$$= \int \frac{\varphi(z)}{\sigma^2(z)}\,dz,$$

so that (13) can be rewritten as

$$\lim_{T\to\infty} \left[\frac{1}{T}\int_0^T \mathbb{M}\varphi(\xi_x(t))\,dt \int_{-\infty}^{\infty} \frac{dy}{\sigma^2(y)} - \int_{-\infty}^{\infty} \frac{\varphi(z)}{\sigma^2(z)}\,dz\right] = 0.$$

The proof is finished.

Remark 2. Using (9) we can obtain the following estimate, more precise than Theorem 1:

$$\left|\frac{1}{T}\int_0^T \mathbb{M}\varphi(\xi_x(t))\,dt - D\int_{-\infty}^{\infty} \frac{\varphi(z)}{\sigma^2(z)}\,dz\right|$$
$$\leqq D\|\varphi\| \left\{\int_{|x-y|\geqq T} \frac{dy}{\sigma^2(y)} + \frac{H}{T}\int_{|x-y|\leqq T} |x-y| \frac{dy}{\sigma^2(y)}\right\}.$$

Lemma 4. *Under the assumptions of Theorem 1*

$$\lim_{T\to\infty} \mathbb{M}\left\{\frac{1}{T}\int_0^T \varphi(\xi_x(t))\,dt - D\int \frac{\varphi(z)}{\sigma^2(z)}\,dz\right\}^2 = 0,$$

i.e., $\lim_{T\to\infty} \frac{1}{T}\int_0^T \varphi(\xi_x(t))\,dt = D\int \frac{\varphi(z)}{\sigma^2(z)}\,dz$, *in mean square.*

Proof. Without loss of generality we can assume that $\int \frac{\varphi(z)}{\sigma^2(z)}\,dz = 0$ (this condition can be met by subtracting a suitable constant from φ). It follows from Remark 2 that we can determine a function $\psi(T,x)$, bounded and tending to zero for $T\to+\infty$ uniformly w.r.t. x on bounded intervals, for which

$$\left|\frac{1}{T}\int_0^T \mathbb{M}\varphi(\xi_x(t))\,dt\right| \leqq \psi(T, x).$$

Then

$$\begin{aligned}
\mathbb{M}\left(\frac{1}{T}\int_0^T \varphi(\xi_x(t))\,dt\right)^2 &= \frac{2}{T^2}\,\mathbb{M}\int_0^T \varphi(\xi_x(t))\int_t^T \varphi(\xi_x(s))\,ds\,dt \\
&= \frac{2}{T^2}\,\mathbb{M}\int_0^T \varphi(\xi_x(t))\,\mathbb{M}\left[\int_t^T \varphi(\xi_x(s))\,ds/\xi_x(t)\right]dt \\
&\leqq \frac{2\|\varphi\|}{T}\,\mathbb{M}\int_0^T \frac{T-t}{T}\,\psi(T-t, \xi_x(t))\,dt,
\end{aligned}$$

thus,

$$\begin{aligned}
\int_{-\infty}^{\infty} \frac{1}{\sigma^2(x)}\,\mathbb{M}\left(\frac{1}{T}\int_0^T \varphi(\xi_x(t))\,dt\right)^2 &\leqq \frac{2\|\varphi\|}{T}\int_0^T\int_{-\infty}^{\infty} \mathbb{M}\psi(T-t, \xi_x(t))\,\frac{dx}{\sigma^2(x)}\,dt \\
&= \frac{2\|\varphi\|}{T}\int_0^T\int_{-\infty}^{\infty} \psi(T-t, y)\,\frac{dy}{\sigma^2(y)}\,dt \to 0
\end{aligned}$$

for $T\to\infty$ and $\mathbb{M}\left(\frac{1}{T}\int_0^T \varphi(\xi_x(t))\,dt\right)^2 \to 0$ for almost all x. Furthermore, this expression is bounded by $\|\varphi\|^2$. This means that for $h>0$,

$$\int p(h, x, y)\,\mathbb{M}\left(\frac{1}{T-h}\int_0^{T-h} \varphi(\xi_y(t))\,dt\right)^2 dy \to 0,$$

where $p(h, x, y)$ is the density of the distribution of $\xi_x(t)$ (the existence of which density follows from (9) § 13). But

$$\begin{aligned}
\mathbb{M}\left(\frac{1}{T}\int_0^T \varphi(\xi_x(t))\,dt\right)^2 &= \frac{1}{T^2}\,\mathbb{M}\left[\int_0^h \varphi(\xi_x(s))\,ds\right]^2 \\
&\quad + \frac{2}{T^2}\,\mathbb{M}\int_0^h \varphi(\xi_x(s))\,ds\int_h^T \varphi(\xi_x(s))\,ds + \frac{1}{T^2}\,\mathbb{M}\left(\int_h^T \varphi(\xi_x(s))\,ds\right)^2 \\
&\leqq \frac{h^2\|\varphi\|^2}{T^2} + \frac{2h\|\varphi\|^2}{T} + \frac{1}{T^2}\int_{-\infty}^{\infty} p(h, x, y)\,\mathbb{M}\left\{\int_0^{T-h} \varphi(\xi_y(s))\,ds\right\}^2 dy \to 0
\end{aligned}$$

for $T\to 0$. The lemma is proved.

As already mentioned, it follows from Lemma 4 that

$$\lim_{T\to\infty} \frac{1}{T}\int_0^T \varphi(\xi_x(t))\,dt = D\int_{-\infty}^{\infty} \frac{\varphi(z)}{\sigma^2(z)}\,dz \tag{14}$$

in the sense of mean square convergence. We will show that (14) holds w.p.1, i.e., that the ergodic theorem holds for the process $\xi_x(t)$.

Theorem 2. *If* $\int_{-\infty}^{\infty} \frac{dz}{\sigma^2(z)} < \infty$ *and* $\varphi(z)$ *is a Borel function for which*

$$\int_{-\infty}^{\infty} \frac{|\varphi(z)|}{\sigma^2(z)} dz < \infty,$$

then

$$\mathbb{P}\left\{\lim_{T\to\infty} \frac{1}{T}\int_0^T \varphi(\xi_x(t))\,dt = D\int \frac{\varphi(z)}{\sigma^2(z)}\,dz\right\} = 1.$$

Proof. Let $a<x$ and set

$$\tau = \inf\{s;\ \xi_x(s)=x,\ \inf_{0\leqq u\leqq s} \xi_x(u) \leqq a\}.$$

That is, τ is the shortest time required for $\xi_x(t)$ to leave the interval (a,∞) and then return to the point x. Obviously, $\tau < \tau_{x,a} + \tau'_{a,x}$, where $\tau'_{a,x} = \inf\{s;\ \xi'_a(s)=x\}$ and $\xi'_a(s) = \xi_x(\tau_{x,a}+s)$. It follows from Remark 2 § 15 that $\xi'_a(s)$ is distributed as $\xi_a(s)$, so that $\mathbb{M}\,\tau < \mathbb{M}\,\tau'_{a,x} + \mathbb{M}\,\tau_{x,a} < \infty$. Set $\tau^{(1)} = \tau$, $\xi_x^{(1)}(s) = \xi_x(\tau+s) = \xi_x^{(0)}(\tau^{(1)}+s)$. Let $\tau^{(2)}$ be the corresponding variable for the process $\xi_x^{(1)}(s)$. Further, we define $\xi_x^{(k)}(s)$ as $\xi_x^{(k-1)}(\tau^{(k)}+s)$ and $\tau^{(k+1)}$ as $\inf\{s:\ \xi_x^{(k)}(s)=x,\ \inf_{0\leqq u\leqq s} \xi_x^{(k)}(u) \leqq a\}$. The times $\tau^{(1)}, \tau^{(2)}, \ldots, \tau^{(k)}, \ldots$ are identically distributed and independent, as are the variables

$$\eta_{k+1} = \int_{\tau^{(k)}}^{\tau^{(k+1)}} \varphi(\xi_x^{(k)}(s))\,ds,$$

because of Corollary 1 § 15. We assume initially that the function φ is bounded and nonnegative. Then $\mathbb{M}\,\eta_k \leqq \|\varphi\|\,\mathbb{M}\,\tau^{(k)}$. From the strong law of large numbers

$$\mathbb{P}\left\{\lim_{k\to\infty} \frac{1}{k}\sum_{i=1}^k \eta_i = \mathbb{M}\,\eta_1\right\} = \mathbb{P}\left\{\lim_{k\to\infty} \frac{1}{k}\sum_{i=1}^k \tau^{(1)} = \mathbb{M}\,\tau^{(1)}\right\} = 1$$

and

$$\mathbb{P}\left\{\lim_{k\to\infty} \frac{\sum_{i=1}^k \eta_i}{\sum_{i=1}^k \tau^{(i)}} = \frac{\mathbb{M}\,\eta_1}{\mathbb{M}\,\tau^{(1)}}\right\} = 1. \tag{15}$$

We note that

$$\sum_{i=1}^k \eta_i = \int_0^{\tau^{(1)}+\cdots+\tau^{(k)}} \varphi(\xi_x(s))\,ds.$$

Let ν_T be such that $\tau^{(1)}+\cdots+\tau^{(\nu_T)} < T \leqq \tau^{(1)}+\cdots+\tau^{(\nu_T+1)}$. Then

$$\frac{\sum_{i=1}^{\nu_T} \eta_i}{\sum_{i=1}^{\nu_T+1} \tau^{(i)}} \leqq \frac{1}{T}\int_0^T \varphi(\xi_x(s))\,ds \leqq \frac{\sum_{i=1}^{\nu_T+1} \eta_i}{\sum_{i=1}^{\nu_T} \tau^{(i)}}. \tag{16}$$

Clearly, $\nu_T \to +\infty$ w.p. 1 and

$$\mathbb{P}\left\{\lim_{k\to\infty}\frac{\sum_{i=1}^{k}\tau^{(i)}}{\sum_{i=1}^{k+1}\tau^{(i)}}=1\right\}=1.$$

Hence the two outer terms in (16) tend to the same limit $\frac{\mathbb{M}\eta_1}{\mathbb{M}\tau^{(1)}}$ for $T\to\infty$ w.p. 1, whence

$$\mathbb{P}\left\{\lim\frac{1}{T}\int_0^T\varphi(\xi_x(s))\,ds=\frac{\mathbb{M}\eta_1}{\mathbb{M}\tau^{(1)}}\right\}=1. \tag{17}$$

From (14) we have

$$\frac{\mathbb{M}\eta_1}{\mathbb{M}\tau^{(1)}}=D\int\frac{\varphi(z)}{\sigma^2(z)}\,dz,$$

i.e., for all bounded nonnegative functions

$$\mathbb{M}\int_0^{\tau^{(1)}}\varphi(\xi_x(s))\,ds=D\mathbb{M}\tau^{(1)}\int\frac{\varphi(z)}{\sigma^2(z)}\,dz. \tag{18}$$

A limit passage shows that (18) holds for all nonnegative φ for which $\int\frac{\varphi(z)}{\sigma^2(z)}\,dz<\infty$. But then $\mathbb{M}\int_0^{\tau^{(1)}}\varphi(\xi_x(s))\,ds<\infty$ which implies that (15) and (17) are valid. The theorem is thus proved for nonnegative functions. The proof in the general case follows from the fact that an arbitrary function φ satisfying the conditions of the theorem can be represented as the difference of such functions.

We now prove an ergodic theorem for the distribution of the process $\xi_x(t)$.

Theorem 3. *Let $\sigma(x)$ satisfy a Lipschitz condition and let*

$$\int_{-\infty}^{\infty}\frac{1}{\sigma^2(y)}\,dy<\infty,\quad \textit{and}\quad F_t(y)=\mathbb{P}\{\xi_x(t)<y\}.$$

Then

$$\lim_{t\to\infty}F_t(y)=D\int_{-\infty}^{y}\frac{1}{\sigma^2(z)}\,dz.$$

Proof. Assume $x<y$. Then from the uniqueness of the solution of (2) we conclude that $\xi_y(t)-\xi_x(t)\geqq 0$ for all $t>0$. Set $\Delta_{x,y}(t)=\xi_y(t)-\xi_x(t)$. Since

$$\Delta_{x,y}(t_{n+1})-\Delta_{x,y}(t_n)=\int_{t_n}^{t_{n+1}}\left[\sigma(\xi_y(s))-\sigma(\xi_x(s))\right]dw(s),$$

we have

$$\mathbb{M}(\Delta_{x,y}(t_{n+1})-\Delta_{x,y}(t_n)/\Delta_{x,y}(t_1),\dots,\Delta_{x,y}(t_n))=0,$$

provided that $t_1<t_2<\cdots<t_n<t_{n+1}$. Thus from Lemma 5 § 16 we obtain

$$\mathbb{P}\{\sup_{t>0}\Delta_{x,y}(t)>a\}\leqq\frac{y-x}{a}.$$

But from Corollary 1 §4

$$\mathbb{P}\left\{\int_0^\infty(\sigma(\xi_y(t))-\sigma(\xi_x(t)))^2\,dt<C\right\}\geqq 2\mathbb{P}\{w(C)>a\}-\frac{y-x}{a}. \quad (19)$$

By choosing $a>0$ and $C>0$ sufficiently large we can make the right side of (19) arbitrarily close to unity, so that

$$\mathbb{P}\left\{\int_0^\infty[\sigma(\xi_y(t))-\sigma(\xi_x(t))]^2\,dt<\infty\right\}=1. \quad (20)$$

It follows from (20) that

$$\Delta_\infty=\lim_{t\to\infty}\Delta_{x,y}(t)=\int_0^\infty[\sigma(\xi_y(t))-\sigma(\xi_x(t))]\,dw(t).$$

exists w.p. 1. It is easy to see that

$$\lim_{T\to\infty}\frac{1}{T}\int_0^T[\sigma(\xi_x(t)+\Delta_{x,y}(t))-\sigma(\xi_x(t))]^2\,dt$$
$$=\lim_{T\to\infty}\frac{1}{T}\int_0^T[\sigma(\xi_x(t)+\Delta_\infty)-\sigma(\xi_x(t))]^2\,dt,$$

thus,

$$\lim_{T\to\infty}\frac{1}{T}\int_0^T[\sigma(\xi_x(t)+\Delta_\infty)-\sigma(\xi_x(t))]^2\,dt=0. \quad (21)$$

Since for any nonrandom Δ the function $(\sigma(y+\Delta)-\sigma(y))^2$ is bounded (σ satisfies a Lipschitz condition) we have using Theorem 2

$$\lim_{T\to\infty}\frac{1}{T}\int_0^T[\sigma(\xi_x(t)+\Delta)-\sigma(\xi_x(t))]^2\,dt=D\int_{-\infty}^{\infty}\frac{[\sigma(y+\Delta)-\sigma(y)]^2}{\sigma^2(y)}\,dy.$$

Moreover, the function

$$\psi_T(\Delta)=\frac{1}{T}\int_0^T[\sigma(\xi_x(t)+\Delta)-\sigma(\xi_x(t))]^2\,dt$$

satisfies

$$|\psi_T(\Delta)-\psi_T(\Delta_1)|\leqq\frac{1}{T}\int_0^T|\sigma(\xi_x(t)+\Delta)-\sigma(\xi_x(t)+\Delta_1)|$$
$$\times|\sigma(\xi_x(t)+\Delta)+\sigma(\xi_x(t)+\Delta_1)-2\sigma(\xi_x(t))|\,dt$$
$$\leqq 2K^2|\Delta-\Delta_1|(|\Delta|+|\Delta_1|),$$

where K is the Lipschitz constant for $\sigma(x)$. Hence the $\psi_T(\Delta)$ are equicontinuous in Δ on each finite interval from which it follows that $\psi_T(\Delta)$

converges uniformly on each finite interval to

$$D\int_{-\infty}^{\infty}\frac{[\sigma(y+\Delta)-\sigma(y)]^2}{\sigma^2(y)}dy$$

for $T\to+\infty$. Hence, if Δ_∞ is a random variable, then

$$\lim_{T\to\infty}\psi_T(\Delta_\infty)=D\int_{-\infty}^{\infty}\frac{[\sigma(y+\Delta_\infty)-\sigma(y)]^2\,dy}{\sigma^2(y)}.$$

But from (21), $\lim\limits_{T\to\infty}\psi_T(\Delta_\infty)=0$, i.e.,

$$\int_{-\infty}^{\infty}\frac{[\sigma(y+\Delta_\infty)-\sigma(y)]^2}{\sigma^2(y)}dy=0.$$

Since the integrand is continuous and nonnegative it is identically zero w.p. 1, i.e., $\mathbb{P}\{\sigma(y+\Delta_\infty)=\sigma(y),\ -\infty<y<\infty\}=1$. Thus it follows that either $\Delta_\infty=0$ w.p. 1 or that $\sigma(y)$ is a periodic function. The periodicity of $\sigma(y)$ contradicts the convergence of $\int\limits_{-\infty}^{\infty}\frac{1}{\sigma^2(y)}dy$. Hence $\mathbb{P}\{\Delta_\infty=0\}=1$ and we have proved that for arbitrary x and y $\lim\limits_{t\to\infty}\{\xi_y(t)-\xi_x(t)]=0$ w.p. 1. Thus for any uniformly continuous function $f(z)$

$$\mathbb{P}\{\lim_{t\to\infty}[f(\xi_y(t))-f(\xi_x(t))]=0\}=1.$$

Moreover, if f is bounded, then, applying Lebesgue's theorem, we obtain

$$\lim_{t\to\infty}[\mathbb{M}f(\xi_y(t))-\mathbb{M}f(\xi_x(t))]=0.$$

Multiplying this equation by $\frac{1}{\sigma^2(y)}$ and integrating w.r.t. y we get

$$\lim_{t\to\infty}\left[\int\frac{1}{\sigma^2(y)}\mathbb{M}f(\xi_y(t))\,dy-\int\frac{dy}{\sigma^2(y)}\mathbb{M}f(\xi_x(t))\right]=0.$$

Using Corollary 1 we find that

$$\lim_{t\to\infty}\mathbb{M}f(\xi_x(t))=D\int\frac{f(z)}{\sigma^2(z)}dz. \tag{22}$$

Let $f(z)=e^{i\lambda z}$. Then

$$\lim_{t\to\infty}\int_{-\infty}^{\infty}e^{i\lambda z}\,dF_t(z)=D\int_{-\infty}^{\infty}e^{i\lambda z}\frac{1}{\sigma^2(z)}dz, \tag{23}$$

i.e., the characteristic function of the distribution function $F_t(y)$ converges to that of $F(y)=D\int\limits_{-\infty}^{y}\frac{dz}{\sigma^2(z)}$. The theorem is proved.

§ 19. Stability of Solutions of Stochastic Differential Equations

Up to this point we have merely considered equations with coefficients independent of time and for which $\sigma(x)$ is nonvanishing. In the present paragraph we will allow the possibility that $\sigma(x)$ vanishes. We will study the behavior of a solution in the neighbourhood of a *stationary point*—i.e., a point x_0 for which $\sigma(x_0)=a(x_0)=0$. Assuming that the coefficients $a(x)$ and $\sigma(x)$ are such that the stochastic equation's solution is unique, we can convince ourselves that for $x<x_0<y$, $\xi_x(t)\leqq x_0\leqq \xi_y(t)$ since $\xi_{x_0}(t)=x_0$.

A stationary point x_0 will be called *stable* if for any $\varepsilon>0$ there exists a $\delta>0$ such that when $|x-x_0|<\delta$

$$\mathbb{P}\{\lim_{t\to\infty}\xi_x(t)=x_0\}\geqq 1-\varepsilon. \tag{1}$$

If for any $\varepsilon>0$ there exists a δ for which (1) holds for

$$x\in(x_0, x_0+\delta)\,(x\in(x_0-\delta, x_0)),$$

then the stationary point x_0 will be called *right-(left-)stable*.

We will study conditions for the right (or left) stability of stationary points (for stability to hold it is necessary and sufficient that right- and left-stability obtain). Assume that 0 is a stationary point. If $a(x)$ and $\sigma(x)$ fulfill the assumptions of the existence and uniqueness theorem, then the functions $\dfrac{a(x)}{x}$ and $\dfrac{\sigma(x)}{x}$ will be bounded. Let us assume that $\sigma(x)$ does not vanish for $x>0$. Let $\eta_x(t)$ satisfy

$$d\eta_x(t)=\bar{a}(\eta_x(t))\,dt+\bar{\sigma}(\eta_x(t))\,dw(t) \tag{2}$$

with initial condition $\eta_x(0)=x$, where

$$\bar{\sigma}(x)=\sigma(e^x)\,e^{-x} \quad\text{and}\quad \bar{a}(x)=a(e^x)\,e^{-x}+\tfrac{1}{2}\sigma^2(e^x)\,e^{-2x}.$$

These coefficients are bounded and satisfy Lipschitz conditions on each finite interval. It is easy to see that the process $\xi(t)=e^{\eta_x(t)}$ satisfies $d\xi(t)=a(\xi(t))\,dt+\sigma(\xi(t))\,dw(t)$ with $\xi(0)=e^x$ and so, for $x>0$

$$\xi_x(t)=\exp(\eta_{\log x}(t)),$$

whence the right-stability of the point 0 is equivalent to the statement: for any $\varepsilon>0$ there exists a C with $x<C$ for which

$$\mathbb{P}\{\lim_{t\to\infty}\eta_x(t)=-\infty\}\geqq 1-\varepsilon.$$

It follows from Theorem 1 § 16 that when $\bar{\sigma}>0$ (which is the case here) the event $\{\lim_{t\to\infty}\eta_x(t)=-\infty\}$ can occur under the following two sets of circumstances:

1) if

$$\int_0^{\infty} \exp\left\{-\int_0^y \frac{2\bar{a}(z)}{\bar{\sigma}^2(z)}\,dz\right\} dy = +\infty,$$

and

$$\int_{-\infty}^{0} \exp\left\{-\int_0^y \frac{2\bar{a}(z)}{\bar{\sigma}^2(z)}\,dz\right\} dy < \infty,$$

then $\mathbb{P}\{\lim_{t\to+\infty} \eta_x(t) = -\infty\} = 1$ for all x;

2) if

$$\int_{-\infty}^{\infty} \exp\left\{-\int_0^y \frac{2\bar{a}(z)}{\bar{\sigma}^2(z)}\,dz\right\} dy < \infty,$$

then

$$\mathbb{P}\{\lim_{t\to\infty} \eta_x(t) = -\infty\} = \frac{\int_x^{\infty} \exp\left\{-\int_0^y \frac{2\bar{a}(z)}{\bar{\sigma}^2(z)}\,dz\right\} dy}{\int_{-\infty}^{\infty} \exp\left\{-\int_0^y \frac{2\bar{a}(z)}{\bar{\sigma}^2(z)}\,dz\right\} dy}.$$

This expression tends to one for $x \to -\infty$. Otherwise,

$$\mathbb{P}\{\lim_{t\to\infty} \eta_x(t) = -\infty\} = 0$$

for all x. Therefore, in order that the point 0 be right stable it is necessary and sufficient that

$$\int_{-\infty}^{0} \exp\left\{-\int_0^y \frac{2\bar{a}(z)}{\bar{\sigma}^2(z)}\,dz\right\} dy < \infty.$$

Substituting the expressions for $\bar{a}$ and $\bar{\sigma}$, we find

$$\int_0^y \frac{2\bar{a}(z)}{\bar{\sigma}^2(z)}\,dz = \int_0^y \frac{2a(e^z)\,e^{-z} + \sigma^2(e^z)\,e^{-2z}}{\sigma^2(e^z)\,e^{-2z}}\,dz$$
$$= y + \int_0^y \frac{2a(e^z)}{\sigma^2(e^z)}\,e^z\,dz = y + \int_1^{e^y} \frac{2a(z)}{\sigma^2(z)}\,dz.$$

From which the following condition for right stability results

$$\int_{-\infty}^{0} \exp\left\{-y - \int_1^{e^y} \frac{2a(z)}{\sigma^2(z)}\,dz\right\} dy < \infty. \tag{3}$$

We perform the substitution $y = \ln u$ in the outer integral. Then (3) looks like

$$\int_0^1 \exp\left\{\int_u^1 \frac{2a(z)}{\sigma^2(z)}\,dz\right\} du < \infty. \tag{4}$$

Obviously (4) is equivalent to the following: for some $\delta > 0$

$$\int_0^\delta \exp\left\{\int_u^\delta \frac{2a(z)}{\sigma^2(z)}\, dz\right\} du < \infty.$$

An analogous condition can be established for the left stability of a stationary point. We then have

Lemma 1. *Let $a(x)$ and $\sigma(x)$ fulfill the conditions of the existence and uniqueness theorem, $a(0)=\sigma(0)=0$, $\sigma(x)>0$ for $x \neq 0$. Then the statements: the stationary point zero is a) right stable; b) left stable; c) stable are equivalent to the following: for some $\delta>0$ a) $I_1<\infty$; b) $I_2<\infty$; c) $I_1+I_2<\infty$, where*

$$I_1 = \int_0^\delta \exp\left\{\int_u^\delta \frac{2a(z)}{\sigma^2(z)}\, dz\right\} du, \qquad I_2 = \int_{-\delta}^0 \exp\left\{-\int_{-\delta}^u \frac{2a(z)}{\sigma^2(z)}\, dz\right\} du.$$

Corollary 1. *Assume $\dfrac{\sigma(x)}{x} \to \sigma$ and $\dfrac{a(x)}{x} \to a$ for $x \to 0$, $x>0$. Then for $\dfrac{a}{\sigma^2} < \dfrac{1}{2}$ the point zero is right stable and is not stable on the right for $\dfrac{a}{\sigma^2} > \dfrac{1}{2}$.*

Indeed, we have here

$$\int_u^\delta \frac{2a(z)}{\sigma^2(z)}\, dz = \int_u^\delta \frac{2a}{\sigma^2} \cdot \frac{1}{z}(1+o(z))\, dz = -\frac{2a}{\sigma^2} \ln u(1+o(1)) + o(1),$$

so that

$$I_1 = O\left\{\int_0^\delta \left(\frac{1}{u}\right)^{\frac{2a}{\sigma^2}(1+o(1))} du\right\}.$$

This integral converges for $\dfrac{2a}{\sigma^2} < 1$ and diverges for $\dfrac{2a}{\sigma^2} > 1$. An analogous condition obtains for left stability.

Lemma 1 allows one to investigate the stability of stationary points if there do not exist other points in which $\sigma(x)$ vanishes. At the same time, conditions for stability are expressed by means of the values of the coefficients in an arbitrarily small neighborhood of a stationary point 0. We note that the condition $\sigma(x) \neq 0$ for $x \neq 0$ in Lemma 1 is unnecessarily strong and that it can be replaced by the following condition: there exists a $\delta>0$ such that $\sigma(x) \neq 0$ for $0<|x|<\delta$. Consider, for example, the case of right stability. Let $\sigma(x)>0$ for $x \in (0, h)$. Let

$\sigma^*(x)$ be a function coinciding with $\sigma(x)$ on $\left(0, \frac{h}{2}\right)$ and equal to $\sigma\left(\frac{h}{2}\right)$ for $x \geqq \frac{h}{2}$. If for some $\delta < \frac{h}{2}$ $I_1 < \infty$ (I_1 is defined in Lemma 1), then for the process $\xi_x^*(t)$ satisfying

$$d\xi_x^*(t) = a(\xi_x^*(t))\,dt + \sigma^*(\xi_x^*(t))\,dw(t)$$

with $\xi_x^*(0) = x$, the point 0 is right stable. If we write $\eta_{\log x}^*(t)$ for $\log \xi_x^*(t)$ then by means of (3) § 16 we find that $\mathbb{P}\{\sup_{t \geqq 0} \eta_{\log x}^*(t) \geqq C\} \to 0$ for $\log x \to -\infty$, i.e., for $x \to 0$. Hence, for any $\varepsilon > 0$ we can find a $\delta > 0$ such that when $x < \delta$

$$\mathbb{P}\left\{\sup_{t \geqq 0} \eta_{\log x}^*(t) < \log \frac{h}{2},\ \lim_{t \to \infty} \eta_{\log x}^*(t) = -\infty\right\} \geqq 1 - \varepsilon.$$

Consequently, for $0 < x < \delta$

$$\mathbb{P}\left\{\sup_{t \geqq 0} \xi_x^*(t) < \frac{h}{2},\ \lim_{t \to \infty} \xi_x^*(t) = 0\right\} \geqq 1 - \varepsilon.$$

Let $\tau = \inf\left\{s, \xi_x^*(s) = \frac{h}{2}\right\}$, $\tau = +\infty$ if the set in brackets is empty. Then from Theorem 2 § 6 for $t < \tau$ we have $\xi_x^*(t) = \xi_x(t)$. Hence,

$$\begin{aligned}\mathbb{P}\{\lim_{t \to \infty} \xi_x(t) = 0\} &\geqq \mathbb{P}\left\{\lim_{t \to \infty} \xi_x(t) = 0,\ \sup_{t > 0} \xi_x(t) < \frac{h}{2}\right\} \\ &= \mathbb{P}\left\{\lim_{t \to \infty} \xi_x^*(t) = 0,\ \sup_{t \geqq 0} \xi_x^*(t) < \frac{h}{2}\right\} \geqq 1 - \varepsilon.\end{aligned}$$

From this it follows that zero is also a point of right stability for $\xi_x(t)$. The last relation (with the roles of $\xi_x(t)$ and $\xi_x^*(t)$ reversed) can also be used to prove the right stability of zero for $\xi_x^*(t)$ if it is right stable for $\xi_x(t)$. Hence we have proved the following theorem on the stability of a stationary point.

Theorem 1. *Assume the coefficients $a(x)$ and $\sigma(x)$ vanish for $x = 0$ and that there exists a $\delta > 0$ such that $\sigma(x) > 0$ for $0 < |x| < \delta$. For the stability of the stationary point 0 a) from the right b) from the left c) without modifier it is necessary and sufficient that for some $\delta > 0$, a) $I_1 < \infty$, b) $I_2 < \infty$, c) $I_1 + I_2 < \infty$ (I_1 and I_2 are defined as in Lemma 1).*

We now turn to the question of behavior of a solution in the neighborhood of a point in which $\sigma(x)$ equals zero and $a(x)$ is different from zero.

Theorem 2. *Assume* $\sigma(x_0)=0$, $a(x_0)>0$ *and that for some* $\delta>0$ $\sigma(x)>0$ *for* $0<|x-x_0|<\delta$. *Then for all* $t>0$ *and* $x\geqq x_0$ $\mathbb{P}\{\xi_x(t)>x_0\}=1$.

Proof. We show first that $\mathbb{P}\{\xi_x(t)>x_0\}=1$ for $x>x_0$. We denote for $x>x_0$, $x_1>x_0$

$$\varphi(x)=\int_{x_1}^{x}\exp\left\{-\int_{x_1}^{y}\frac{2a(z)}{\sigma^2(z)}\,dz\right\}dy.$$

Choose $\delta_1<\delta$ and $x\in(x_0+\delta_1,x_0+\delta)$. Let $\tau_x[x_0+\delta_1,x_0+\delta]$ be the first time $\xi_x(t)$ leaves the interval $(x_0+\delta_1,x_0+\delta)$ (this time was defined in § 15). From Corollary 2 § 15

$$\mathbb{P}\{\xi_x(\tau_x[x_0+\delta_1,x_0+\delta])=x_0+\delta_1\}=\frac{\varphi(x_0+\delta)-\varphi(x)}{\varphi(x_0+\delta)-\varphi(x_0+\delta_1)}.$$

We note that from the assumptions made

$$-\int_{x_1}^{y}\frac{2a(z)}{\sigma^2(z)}\,dz=\int_{y}^{x_1}\frac{2a(z)}{\sigma^2(z)}\,dz\geqq C\int_{y}^{x_1}\frac{dz}{(z-x_0)^2}=C\left[\frac{1}{(y-x_0)}-\frac{1}{x_1-x_0}\right]$$

for some $C>0$ since $a(z)>C_1>0$ and

$$\frac{1}{\sigma^2(z)}=\frac{1}{[\sigma(z)-\sigma(x_0)]}\geqq\frac{1}{h^2(z-x_0)^2}.$$

Hence,

$$\int_{x_0+\delta_1}^{x_0+\delta}\exp\left\{-\int_{x_1}^{y}\frac{2a(z)}{\sigma^2(z)}\,dz\right\}dy>C_2\int_{x_0+\delta_1}^{x_0+\delta}e^{\frac{C}{y-x_0}}\,dy\to+\infty$$

for $\delta_1\to 0$, so that $\mathbb{P}\{\xi_x(\tau)=\delta_1+x_0\}\to 0$ for $\delta_1\to 0$.

Hence, for all $x\in(x_0,x_0+\delta)$ the process $\xi_x(t)$ attains the point $x_0+\delta$ before x_0 w.p. 1. Let $\tau^{(1)}$ be a time for which $\tau^{(1)}>\tau_x[x_0,x_0+\delta]$, $\xi_x(\tau^{(1)})=x$ and $\xi_x(s)\neq x$ for $\tau<s<\tau^{(1)}$. Assume $\xi_x^{(1)}(t)=\xi_x(t+\tau^{(1)})$ and define $\tau^{(2)}$ for $\xi_x^{(1)}(t)$ in the same way as $\tau^{(1)}$ for $\xi_x(t)$. Continuing in this way, we define a sequence of times $\tau^{(k)}$ w.r.t. the processes $\xi_x^{(k)}(t)$. We conclude from Corollary 1 § 15 that the $\tau^{(k)}$ are identically distributed. Moreover, $\mathbb{P}\{\tau^{(k)}>0\}=1$. Thus,

$$\mathbb{P}\left\{\sum_{k=1}^{\infty}\tau^{(k)}=+\infty\right\}=1.$$

Since

$$\mathbb{P}\{\xi_x^{(k)}(t)>x_0\ \text{for}\ 0\leqq t\leqq\tau^{(k+1)}\}=1,$$

we have

$$\mathbb{P}\{\xi_x(t)>x_0\ \text{for}\ \tau^{(1)}+\cdots+\tau^{(k)}\leqq t\leqq\tau^{(1)}+\cdots+\tau^{(k+1)}\}=1$$

and

$$\mathbb{P}\{\xi_x(t)>x_0 \text{ for } t>0\}$$
$$=\mathbb{P}\left\{\prod_{n=0}^{\infty}\left\{\xi_x(t)>x_0 \text{ for } \sum_{k=1}^{n}\tau^{(k)}\leqq t\leqq\sum_{k=1}^{n+1}\tau^{(k)}\right\}\right\}=1,$$
$$\left(\sum_{k=1}^{0}\tau^{(k)}=0\right).$$

For $x>x_0$ the theorem is proved.

In order to show that $\mathbb{P}\{\xi_{x_0}(t)>x_0, t>0\}=1$ it is sufficient to verify that, w.p. 1, there exists an $\varepsilon>0$ such that $\xi_{x_0}(t)>x_0$ for $0<t<\varepsilon$. We now estimate $\int_0^t \sigma(\xi_{x_0}(s))\,dw(s)$ for small t.

Let

$$\varphi(t)=\mathbb{M}\left(\int_0^t \sigma(\xi_{x_0}(s))\,dw(s)\right)^2=\mathbb{M}\int_0^t \sigma^2(\xi_{x_0}(s))\,ds.$$

Using the Lipschitz condition we obtain

$$\mathbb{M}\sigma^2(\xi_{x_0}(s))=\mathbb{M}[\sigma(\xi_{x_0}(s))-\sigma(x_0)]^2\leqq \mathbb{M}K^2|\xi_{x_0}(s)-x_0|^2$$
$$=K^2\mathbb{M}\left|\int_0^s a(\xi_{x_0}(u))\,du+\int_0^s \sigma(\xi_{x_0}(u))\,dw(u)\right|^2\leqq K_1 s^2+K_2\varphi(s),$$

where K_1 and K_2 are some constants. Hence,

$$\varphi(t)\leqq K_3 t^3+K_2\int_0^t \varphi(s)\,ds.$$

Using Lemma 1 § 6 we have

$$\varphi(t)\leqq e^{K_2 t}\int_0^t e^{-K_2 s}\,3K_3 s^2\,ds\leqq K_3 e^{K_2 t} t^3,$$

and so there exists an H such that $\varphi(t)\leqq Ht^3$ for sufficiently small t. Furthermore, from Theorem 1 § 3 it follows that

$$\mathbb{P}\left\{\sup_{0\leqq s\leqq t}\left|\int_0^s \sigma(\xi_{x_0}(u))\,dw(u)\right|>\varepsilon t\right\}\leqq\frac{\varphi(t)}{\varepsilon^2 t^2}\leqq\frac{Ht}{\varepsilon^3},$$

thus

$$\mathbb{P}\left\{\sup_{2^{-k}\leqq s\leqq 2^{-k+1}}\frac{1}{s}\left|\int_0^s \sigma(\xi_{x_0}(u))\,dw(u)\right|>\frac{1}{k}\right\}$$
$$\leqq\mathbb{P}\left\{\sup_{0\leqq s\leqq 2^{-k+1}}\left|\int_0^s \sigma(\xi_{x_0}(u))\,dw(u)\right|>\frac{2^{-k}}{k}\right\}\leqq\frac{4Hk^2}{2^k}$$

and

$$\sum_{k=1}^{\infty} \mathbb{P}\left\{\sup_{2^{-k} \leq s \leq 2^{-k+1}} \frac{1}{s}\left|\int_0^s \sigma(\xi_{x_0}(u))\, dw(u)\right| > \frac{1}{k}\right\} \leq \sum_{k=1}^{\infty} \frac{4Hk^2}{2^k} < \infty,$$

so that from the Borel-Cantelli lemma for all sufficiently large k w.p. 1

$$\sup_{2^{-k} \leq s \leq 2^{-k+1}} \frac{1}{s}\left|\int_0^s \sigma(\xi_{x_0}(u))\, dw(u)\right| \leq \frac{1}{k},$$

i.e., w.p. 1

$$\lim_{t \to \infty} \frac{1}{t} \int_0 \sigma(\xi_{x_0}(u))\, dw(u) = 0.$$

Clearly, w.p. 1

$$\lim_{t \to \infty} \frac{1}{t} \int_0^t a(\xi_{x_0}(s))\, ds = a(x_0) > 0,$$

so that

$$\mathbb{P}\left\{\lim_{t \to \infty} \frac{1}{t} [\xi_{x_0}(t) - x_0] = a(x_0)\right\} = 1.$$

One can determine an $\varepsilon > 0$ such that for $0 < t < \varepsilon$ $\xi_{x_0}(t) - x_0 > \frac{a(x_0)}{2} t$. As already mentioned above, this implies the theorem.

Remark 1. Making use of the first passage time to the boundary of the interval $(x_0 - \delta, x_0)$ for $x < x_0$ we can show that

$$\mathbb{P}\{\xi_x(\tau_x[x_0 - \delta, x_0]) = x_0\} \to 1$$

when $x \uparrow x_0$. Hence x_0 is such that the a trajectory of the process $\xi(t)$ can pass through it from the left but not the right.

§ 20. Some Other Limit Theorems

In this section we will treat theorems on the existence of a limiting distribution for the variable $\frac{\xi(t)}{\beta(t)}$, where $\xi(t)$ is solution of a stochastic equation whose coefficients are independent of t and $\beta(t)$ is a suitable function tending to infinity for $t \to \infty$. These limit theorems do not pretend to give an exhaustive answer to the question of the existence of a limiting distribution for variables $\xi(t)$ normed in this way but provide rather examples of such theorems. The limiting distributions obtained here are either normal or simply expressed in terms of normal distributions.

We begin with the simplest case, for which $\sigma(x)=1$.

Lemma 1. *Assume $\xi(t)$ fulfills*

$$d\xi(t)=a(\xi(t))\,dt+dw(t).$$

If $a(x)$ is such that $\int\limits_{-\infty}^{\infty} a(x)\,dx=0$, then $\frac{\xi(t)-w(t)}{\sqrt{t}}\to 0$ in probability for $t\to\infty$.

Proof. Without loss of generality, we can assume $\mathbb{M}\,\xi(0)^2<\infty$. Since

$$\frac{\xi(t)-w(t)}{\sqrt{t}}=\frac{\xi(0)}{\sqrt{t}}+\frac{1}{\sqrt{t}}\int\limits_0^t a(\xi(s))\,ds,$$

it suffices to prove that $\lim\limits_{t\to 0}\frac{1}{\sqrt{t}}\int\limits_0^t a(\xi(s))\,ds=0$ in probability. If the function Φ satisfies $a(x)\,\Phi'(x)+\frac{1}{2}\Phi''(x)=a(x)$, then, according to Itô's formula

$$\Phi(\xi(t))-\Phi(\xi(0))=\int\limits_0^t \Phi'(\xi(s))\,dw(s)+\int\limits_0^t a(\xi(s))\,ds.$$

We show that Φ can be so chosen that the variables

$$\frac{\Phi(\xi(t))}{\sqrt{t}} \quad\text{and}\quad \frac{1}{\sqrt{t}}\int\limits_0^t \Phi'(\xi(s))\,dw(s)$$

will converge to zero in probability for $t\to\infty$. The general form of Φ is

$$\Phi(x)=C_1+\int\limits_0^x [C e^{A(z)}-1]\,dz,$$

where $A(x)=\int\limits_{-\infty}^x a(y)\,dy$. Choosing $C_1=0$ and $C=1$ we obtain $\Phi'(x)=e^{A(x)}-1\to 0$ for $|x|\to\infty$ ($A(x)\to 0$ for $|x|\to\infty$). This means that $\frac{\Phi(x)}{x}\to 0$ when $|x|\to\infty$. Since

$$\mathbb{M}\left[\frac{1}{\sqrt{t}}\int\limits_0^t \Phi'(\xi(s))\,dw(s)\right]^2=\frac{1}{t}\int\limits_0^t \mathbb{M}[\Phi'(\xi(s))]^2\,ds$$

is bounded, $\frac{1}{\sqrt{t}}\int\limits_0^t \Phi'(\xi(s))\,dw(s)$ is bounded in probability. Using the relation

$$\frac{\xi(t)-\Phi(\xi(t))}{\sqrt{t}}=\frac{\xi(0)-\Phi(\xi(0))}{\sqrt{t}}+\frac{w(t)}{\sqrt{t}}+\frac{1}{\sqrt{t}}\int\limits_0^t \Phi'(\xi(s))\,dw(s)$$

and the boundedness in probability of $\frac{w(t)}{\sqrt{t}}$ we can show that $\mathbb{M}\left[\frac{\xi(t)-\Phi(\xi(t))}{\sqrt{t}}\right]^2$ is bounded which implies that $\mathbb{M}\frac{\xi^2(t)}{t}$ is also bounded since $\sup\frac{x^2}{[x-\Phi(x)]^2+1}<\infty$. Consider the inequality

$$\frac{|\Phi(\xi(t))|}{\sqrt{t}}=\frac{|\Phi(\xi(t))|}{|\xi(t)|+1}\cdot\frac{|\xi(t)|+1}{\sqrt{t}}\leqq\sup_{|x|>\sqrt[3]{t}}\frac{|\Phi(x)|}{1+|x|}$$
$$\times\frac{|\xi(t)|+1}{\sqrt{t}}+B\frac{\sqrt[3]{t}+1}{\sqrt{t}},\qquad B=\sup_x\frac{\Phi(x)}{1+|x|}.$$

The second summand tends to zero for $t\to\infty$ and the first is the product of two factors the first of which also tends to zero; the second is bounded in probability. Hence $\frac{\Phi(\xi(t))}{\sqrt{t}}\to 0$ in probability. Using Property IV of stochastic integrals (§ 2), we find that in order to prove that $\frac{1}{\sqrt{t}}\int_0^t\Phi'(\xi(t))\,dw(s)$ converges to zero in probability it is sufficient to show that $\frac{1}{t}\int_0^t[\Phi'(\xi(s))]^2\,ds\to 0$ in probability. Let $\psi(x)$ be defined by

$$\psi(x)=\int_0^x e^{-2A(z)}\int_0^z e^{2A(y)}[\Phi'(y)]^2\,dy\,dz$$

$\left(\lim_{x\to\infty}\frac{\psi(x)}{x^2}=0\text{ and }\psi(x)\text{ satisfies }\tfrac{1}{2}\psi''(x)+a(x)\psi'(x)=[\Phi'(x)]^2\right)$. From Itô's formula

$$\psi(\xi(t))-\psi(\xi(0))=\int_0^t\psi'(\xi(s))\,dw(s)+\int_0^t[a(\xi(s))\psi'(\xi(s))+\tfrac{1}{2}\psi''(\xi(s))]\,ds$$
$$=\int_0^t\psi'(\xi(s))\,dw(s)+\int_0^t[\Phi'(\xi(s))]^2\,ds,$$

so that

$$\frac{1}{t}\mathbb{M}\int_0^t[\Phi'(\xi(s))]^2\,ds=\mathbb{M}\frac{\psi(\xi(t))-\psi(\xi(0))}{t}.$$

In addition,

$$\mathbb{M}\frac{\psi(\xi(t))}{t}=\mathbb{M}\frac{\psi(\xi(t))}{1+\xi^2(t)}\cdot\frac{1+\xi^2(t)}{t}\leqq\varepsilon_L\frac{1+\xi^2(t)}{t}+\frac{C_L}{t},$$

where

$$C_L=\sup_{|x|\leqq L}|\psi(x)|,\qquad\varepsilon_L=\sup_{x>L}\frac{|\psi(x)|}{1+|x|^2}.$$

By choosing L sufficiently large we can make ε_L as small as we like. It follows from what was proved earlier that $\mathbb{M}\dfrac{\xi^2(t)}{t}$ is bounded. Since $\dfrac{C_L}{t}\to 0$ for $t\to\infty$ and since ε_L can be made arbitrarily small, $\mathbb{M}\dfrac{\psi(\xi(t))}{t}\to 0$. The lemma is proved.

Corollary 1. *Under the conditions of the lemma* $\dfrac{\xi(t)}{\sqrt{t}}$ *is normally* (0, 1)*-distributed, i.e., for all* y

$$\lim_{t\to\infty}\mathbb{P}\{\xi(t)<y\sqrt{t}\}=\frac{1}{\sqrt{2\pi}}\int_{-\infty}^{y}e^{-z^2/2}\,dz.$$

From Lemma 1 we can deduce a more general condition for the asymptotic normality of $\xi(t)$ as solution of

$$d\xi(t)=a(\xi(t))\,dt+\sigma(\xi(t))\,dw(t). \tag{1}$$

Theorem 1. *Assume the coefficients of* (1) *fulfill the conditions*

a) $a(x)$, $\sigma(x)$ *and* $\sigma'(x)$ *satisfy a Lipschitz condition;*

b) $\sigma(x)>0$ *and* $\displaystyle\lim_{x\to\infty}\frac{1}{x}\int_0^x\frac{1}{\sigma(y)}\,dy=s_0$;

c) $\displaystyle\int_{-\infty}^{\infty}\frac{a(z)-\frac{1}{2}\sigma(z)\sigma'(z)}{\sigma^2(z)}\,dz=0.$

Then

$$\lim_{t\to\infty}\mathbb{P}\left\{\xi(t)<\frac{x\sqrt{t}}{s_0}\right\}=\frac{1}{\sqrt{2\pi}}\int_{-\infty}^{x}e^{-y^2/2}\,dy.$$

Proof. Let $\Phi(z)=\displaystyle\int_0^z\frac{1}{\sigma(y)}\,dy$, $\eta(t)=\Phi(\xi(t))$, and $g(z)=\Phi^{-1}(z)$, the inverse of Φ. The process $\eta(t)$ satisfies

$$d\eta(t)=\left[a(\xi(t))\,\Phi'(\xi(t))+\tfrac{1}{2}\sigma^2(\xi(t))\,\Phi''(\xi(t))\right]dt+\Phi'(\xi(t))\,\sigma(\xi(t))\,dw(t).$$

From the definition of Φ and the relation $\xi(t)=g(\eta(t))$ we obtain

$$d\eta(t)=\left(\frac{a(g(\eta(t)))}{\sigma(g(\eta(t)))}-\frac{1}{2}\sigma'(g(\eta(t)))\right)dt+dw(t).$$

Since

$$\int_{-\infty}^{\infty}\left[\frac{a(g(y))}{\sigma(g(y))}-\frac{1}{2}\sigma'(g(y))\right]dy=\int_{-\infty}^{\infty}\left[\frac{a(z)}{\sigma(z)}-\frac{1}{2}\sigma'(z)\right]\Phi'(z)\,dz$$

$$=\int_{-\infty}^{\infty}\frac{a(z)-\frac{1}{2}\sigma(z)\sigma'(z)}{\sigma^2(z)}\,dz=0,$$

we obtain from Corollary 1

$$\lim_{t\to\infty} \mathbb{P}\{\eta(t)<x\sqrt{t}\}=\frac{1}{\sqrt{2\pi}}\int_{-\infty}^{x} e^{-y^2/2}\,dy.$$

Assume $x \neq 0$. Then

$$\begin{aligned}\lim_{t\to\infty} \mathbb{P}\left\{\xi(t)<\frac{x\sqrt{t}}{s_0}\right\}&=\lim_{t\to\infty} \mathbb{P}\left\{\Phi(\xi(t))<\Phi\left(\frac{x\sqrt{t}}{s_0}\right)\right\}\\ &=\lim_{t\to\infty} \mathbb{P}\left\{\eta(t)<\frac{\Phi\left(\frac{x\sqrt{t}}{s_0}\right)\frac{x\sqrt{t}}{s_0}}{\frac{x\sqrt{t}}{s_0}}\right\}\\ &=\lim_{t\to\infty} \mathbb{P}\{\eta(t)<x\sqrt{t}\},\end{aligned} \tag{2}$$

since

$$\lim_{t\to\infty}\frac{\Phi\left(\frac{x\sqrt{t}}{s_0}\right)}{\frac{x\sqrt{t}}{s_0}}=s_0$$

from condition b). From the validity of (2) for all $x \neq 0$ it follows that it also holds for $x=0$. The proof is complete.

Remark 1. Conditions b) and c) can be replaced by the following, which are more rigid but are easy to verify:

b′) $\sigma(x)>0$ and $\lim_{|x|\to\infty}\sigma(x)=\frac{1}{s_0}>0$ exists;

c′) $\int_{-\infty}^{\infty}\frac{a(z)}{\sigma^2(z)}\,dz=0.$

Clearly, b′) implies b). If b′) and c′) are fulfilled, then

$$\int_{-\infty}^{\infty}\frac{a(z)-\frac{1}{2}\sigma(z)\sigma'(z)}{\sigma^2(z)}\,dz=-\frac{1}{2}\int_{-\infty}^{\infty}\frac{\sigma'(z)}{\sigma(z)}\,dz=-\frac{1}{2}\ln\frac{\sigma(+\infty)}{\sigma(-\infty)}=0.$$

Remark 2. If the fulfillment of b) is not required in Theorem 1, we can still show that $\Phi(\xi(t))$ is asymptotically (0, 1)-normal.

Assume that $\Phi(x)$ possesses the property of "regularity" for $x\to\infty$, i.e., $\lim_{k\to+\infty}\frac{\Phi(kx)}{\Phi(ky)}$ exists for all $x \neq 0$ and $y \neq 0$. Then, since $\Phi(x)$ is monotone increasing

$$g(x)=\lim_{k\to+\infty}\frac{\Phi(kx)}{\Phi(k)}=\begin{cases} x^{\alpha}, & x>0;\\ -C|x|^{\alpha}, & x\leqq 0,\end{cases}$$

where $C>0$, $\alpha>0$. Let $B(t)$ satisfy $\Phi(B(t))=\sqrt{t}$. From the properties of $\Phi(x)$ it follows that

$$\frac{\Phi(\xi(t))}{\sqrt{t}}=\frac{\Phi(\xi(t))}{\Phi(B(t))}\sim g\left(\frac{\xi(t)}{B(t)}\right),$$

where the asymptotic equivalence is to be understood in the sense that the ratio of two equivalent random variables tends to 1 in probability. Hence $g\left(\frac{\xi(t)}{B(t)}\right)$ has a limiting distribution with density $\frac{1}{\sqrt{2\pi}}\exp\left\{-\frac{x^2}{2}\right\}$, so that the density of the limiting distribution of $\frac{\xi(t)}{B(t)}$ will be

$$\varphi(x)=\begin{cases}\frac{1}{\sqrt{2\pi}}\cdot\frac{1}{\alpha}x^{1/\alpha-1}\exp\left\{-\frac{x^{2/\alpha}}{2}\right\}, & x>0,\\ \frac{1}{\sqrt{2\pi}}C^{-1/\alpha}\frac{1}{\alpha}|x|^{1/\alpha-1}\exp\left\{-\frac{|x|^{2/\alpha}}{2C^{2/\alpha}}\right\}, & x<0.\end{cases}\tag{3}$$

We have thus proved

Theorem 2. *Let there exist $\alpha>0$ and $C>0$ for which*

$$\lim_{k\to\infty}\frac{\int_0^{kx}\frac{1}{\sigma(y)}dy}{\int_0^{k}\frac{1}{\sigma(y)}dy}=\begin{cases}x^\alpha, & x>0;\\ -C|x|^\alpha, & x<0\end{cases}$$

and

$$\int_{-\infty}^{\infty}\left(a(y)-\tfrac{1}{2}\sigma(y)\sigma'(y)\right)\frac{1}{\sigma^2(y)}dy=0.$$

Denote by $B(t)$ the solution of $\sqrt{t}=\int_0^{B(t)}\frac{1}{\sigma(y)}dy$. Then $\frac{\xi(t)}{B(t)}$ will have a limiting distribution for $t\to\infty$ whose density is given by (3).

Chapter 5. Stochastics Differential Equations on a Finite Spatial Interval

§ 21. Boundary Conditions at the Ends of the Interval

Assume $a(t,x)$ and $\sigma(t,x)$ are defined and continuous for $t \geqq 0$, $r_1 < x < r_2$ and satisfy a Lipschitz condition on each closed interval in (r_1, r_2), i.e., for each interval $[\bar{r}_1, \bar{r}_2]$ from (r_1, r_2) there exists a constant K for which

$$|a(t,x) - a(t,y)| + |\sigma(t,x) - \sigma(t,y)| \leqq K|x-y|$$

when $x, y \in [\bar{r}_1, \bar{r}_2]$. We consider sequences of functions $a_n(t,x)$ and $\sigma_n(t,x)$ satisfying the conditions of the existence and uniqueness theorem and such that $a_n(t,x) = a(t,x)$, $\sigma_n(t,x) = \sigma(t,x)$ for $r_1^{(n)} \leqq x \leqq r_2^{(n)}$, where $r_1^{(n)} \downarrow r_1$, $r_2^{(n)} \uparrow r_2$. Let $\xi_n(t)$ be the solution of

$$d\xi_n(t) = a_n(t, \xi_n(t))\, dt + \sigma_n(t, \xi_n(t))\, dw(t)$$

with initial condition $\xi_n(0) = \xi(0) \in (r_1, r_2)$. Set

$$\tau_n = \inf\{s;\ \xi_n(s) \notin [r_1^{(n)}, r_2^{(n)}]\},$$

if the set in brackets is not empty and $+\infty$ otherwise. Then from Theorem 2 § 6 for $m > n$ we see that the processes $\xi_m(t)$ and $\xi_n(t)$ coincide on $[0, \tau_n]$ and that $\tau_m > \tau_n$. Furthermore, for $t < \tau_n$ we have

$$a_n(t, \xi_n(t)) = a(t, \xi_n(t)), \qquad \sigma_n(t, \xi_n(t)) = \sigma(t, \xi_n(t)).$$

Let $\tau = \sup_n \tau_n$, $\xi^{(0)}(t) = \xi_n(t)$ if $t \leqq \tau_n$ (the properties of $\xi_n(t)$ are such that for $m > n$ the function $\xi_m(t)$ is an extension of $\xi_n(t)$). It is clear that $\xi^{(0)}(t)$ satisfies

$$d\xi^{(0)}(t) = a(t, \xi^{(0)}(t))\, dt + \sigma(t, \xi^{(0)}(t))\, dw(t) \tag{1}$$

for $t < \tau$. In the sequel $\xi^{(0)}(t)$ will denote the solution of (1) and τ the first exit time of this solution at the boundary of (r_1, r_2). Thus we can construct the solution on a finite spatial interval up to the time of its first exit at the interval's boundary. If $\tau = +\infty$, then $\xi(t)$ is defined for all $t > 0$ and the presence of a boundary does not manifest itself in the solution. An example of such an equation would be one whose coefficients $a(x)$ and $\sigma(x)$ satisfy $\sigma(x) > 0$ for $x \in (r_1, r_2)$, $\sigma(r_1) = \sigma(r_2) = 0$, $a(r_1) > 0$, $a(r_2) < 0$.

Indeed (as follows from Theorems 1 and 2 § 19), in this case the solution $\xi_x(t)$ of

$$d\xi_x(t)=a(\xi_x(t))\,dt+\sigma(\xi_x(t))\,dw(t)$$

with initial condition $\xi_x(0)=x\in(r_1, r_2)$ satisfies $r_1<\xi_x(t)<r_2$ for all $t>0$. When $\tau=+\infty$ w.p. 1 for arbitrary initial value $\xi(0)\in(r_1, r_2)$, we will say that the boundary of the interval (r_1, r_2) is *natural* for Eq. (1).

Our goal is to construct a process $\xi(t)$ such that (1) is satisfied in the interior of (1); $\xi(t)$ is defined for all $t>0$ and $\xi(t)\in[r_1, r_2]$. When $\tau=+\infty$, then $\xi^{(0)}(t)$ is such a process. If, however, $\tau<+\infty$, it is necessary to supplement the definition of the solution on the boundary in such a way that $\xi(t)$ does not leave $[r_1, r_2]$ after attaining its boundary.

We enumerate some possible variants of the behavior of the process at the boundary.

1. The simplest way to extend the process after it reaches the boundary is to assign a constant value to it thereafter: $\xi(t)=r_i$ for $t>\tau$ if $\xi^{(0)}(\tau-0)=r_i$. The process obtained in this way will be called a *process with absorption at the boundary*.

2. A process with *reflection at the boundary*. After attaining the boundary the process returns continuously and immediately to the interior of the interval (r_1, r_2). In this case the process either spends zero time at the boundary or stays there some arbitrary positive length of time (the length of time spent at r_i is taken as the Lebesgue measure of the set $\{s\colon \xi(s)=r_i\}$). In the former case the process is said to be one with *instantaneous reflection*, in the latter, with *delayed reflection*.

3. Jump-type return to the interior of the interval. Here $\xi(\tau)$ is defined as some random variable dependent on the boundary point attained by the process and on the time the boundary is reached. The further course of the process is defined subsequently by Eq. (1).

We turn to the physical interpretation of the various boundary conditions. In the case of a process with absorption, once a particle attains the boundary it is no longer subject to perturbation and stays there forever. If, after hitting the boundary, the particle reenters the interior of the interval with finite velocity, positive time is required in order to attain finite displacement and we obtain a process with delayed reflection. If, however, the exit velocity at the boundary is infinite, the time spent at the boundary can only be equal to zero since otherwise an infinite displacement would occur in finite time. We recall that the velocity of a particle whose movement is described by a stochastic differential equation is infinite. Hence, in the case of a process with instantaneous reflection, the boundary can be considered to be completely elastic. The velocity on arrival at the boundary is the same as the exit velocity. For processes with delayed reflection, the velocity at reflection is partially damped; such boundaries are said to be *elastic*. It is natural to consider boundary conditions of type 3 within the framework of the following scheme of "restoration": a particle governed in (r_1, r_2) by Eq. (1) vanishes after reaching the boundary and is then replaced by a new one which initiates its movement from some interior point $\xi(\tau) \in (r_1, r_2)$ governed by the same law as before.

By combining boundary conditions of the types listed we can obtain much more complicated ones. Conditions 1 and 3 are simplest to study. Conditions of type 2 cannot be considered on every boundary. The boundary can have the property that it is impossible to exit in a continuous manner under such a condition.

We now investigate the dependence of the boundary on the coefficients in the case where the latter are independent of time. Let these coefficients be $a(x)$ and $\sigma(x)$ with $\sigma(x) > 0$ for $x \in (r_1, r_2)$. Set

$$\varphi(x) = \int_{x_1}^{x} \exp\left\{-\int_{x_0}^{y} \frac{2a(z)}{\sigma^2(z)}\,dz\right\} dy$$

for $x \in (r_1, r_2)$, x_0, x_1 points in (r_1, r_2). Let $\xi_x^{(0)}(t)$ be a solution of $d\xi_x^{(0)}(t) = a(\xi_x^{(0)}(t))\,dt + \sigma(\xi_x^{(0)}(t))\,dw(t)$ until exit from (r_1, r_2) with $\xi_x^{(0)}(0) = x$ and $\tau_x[\alpha, \beta]$ the first time $\xi_x^{(0)}(t)$ attains the boundary of $(\alpha, \beta) \subset (r_1, r_2)$ where $x \in (\alpha, \beta)$. From Theorem 4 § 15 we have

$$\mathbb{P}\{\xi_x^{(0)}(\tau_x[\alpha, \beta]) = \alpha\} = \frac{\varphi(\beta) - \varphi(x)}{\varphi(\beta) - \varphi(\alpha)}.$$

Assume $\lim_{\alpha \downarrow r_1} \varphi(\alpha) > -\infty$. Then

$$\mathbb{P}\{\xi_x^{(0)}(\tau_x[r_1, \beta] - 0) = r_1\} = \frac{\varphi(\beta) - \varphi(x)}{\varphi(\beta) - \varphi(r_1 + 0)}.$$

If, however, $\varphi(r_1 + 0) = -\infty$, then $\mathbb{P}\{\xi_x^{(0)}(\tau_x[r_1 + \varepsilon, \beta]) = r_1 + \varepsilon\}$ can be made arbitrarily small by suitable choice of $\varepsilon > 0$ for arbitrary x and β. In this case the boundary r_1 cannot be attained in finite time before the boundary β (i.e., it cannot be attained at all since after reaching β the process must return to the interior of the interval (r_1, β)). Moreover, in this case $\xi_x^{(0)}(t)$ cannot tend to r_1 with positive probability when $t \to +\infty$ since then $\xi_x^{(0)}(t)$ would, with the same probability, reach $r_1 + \varepsilon$ before β for each $\varepsilon > 0$, which is impossible. In order to study the case $\varphi(r_1 + 0) > -\infty$ we set $\eta^{(0)}(t) = \varphi(\xi^{(0)}(t))$, whereby in the definition of φ we assume $x_1 = r_1$. Then

$$d\eta^{(0)}(t) = \varphi'(\xi^{(0)}(t))\,\sigma(\xi^{(0)}(t))\,dw(t)$$

and the process $\eta^{(0)}(t)$ will be defined up to exit from the interval $(0, \bar{r})$,

$$\bar{r} = \int_{r_1}^{r_2} \exp\left\{-\int_{x_0}^{y} \frac{2a(z)}{\sigma^2(z)}\,dz\right\} dy.$$

Inside this interval it satisfies $d\eta^{(0)}(t) = \bar{\sigma}(\eta^{(0)}(t))\,dw(t)$, where

$$\bar{\sigma}(x) = \sigma(g(x))\,U(g(x)), \qquad U(x) = \exp\left\{-\int_{x_0}^{x} \frac{2a(z)}{\sigma^2(z)}\,dz\right\},$$

and $g(x)$ is the inverse of $\varphi(x)$.

Let $\eta_x(t)$ satisfy $d\eta_x(t) = \bar{\sigma}(\eta_x(t))\,dw(t)$ up to the time of exit from $(0, \bar{r})$ with initial condition $\eta_x(0) = x$ and let $\bar{\tau}_x[s_1, s_2]$ denote the first time $\eta_x(t)$ reaches the boundary of the interval (s_1, s_2), $0 < s_1 < s_2 < \bar{r}$. From Theorems 2 and 3 § 15 we have

$$\mathbb{M}\bar{\tau}_x[s_1, s_2] = 2\int_{s_1}^{x} \frac{(z-s_1)(s_2-x)}{(s_2-s_1)\bar{\sigma}^2(z)}\,dz + 2\int_{x}^{s_2} \frac{(x-s_1)(s_2-z)}{(s_2-s_1)\bar{\sigma}^2(z)}\,dz, \tag{2}$$

$$\mathbb{M}\bar{\tau}_x^2[s_1, s_2] = 4\int_{s_1}^{x} \frac{(z-s_1)(s_2-x)}{(s_2-s_1)\bar{\sigma}^2(z)}\,V(z)\,dz + 4\int_{x}^{s_2} \frac{(s_2-z)(x-s_1)}{(s_2-s_1)\bar{\sigma}^2(z)}\,V(z)\,dz, \tag{3}$$

where $V(x) = \mathbb{M}\bar{\tau}_x[s_1, s_2]$, $s_1 \leqq x \leqq s_2$.

Lemma 1. *If* $\lim_{s_1 \downarrow 0} \mathbb{M}\bar{\tau}_x[s_1, s_2] = +\infty$, *then*

$$\varlimsup_{s_1 \downarrow 0} \frac{\mathbb{M}\bar{\tau}_x^2[s_1, s_2]}{(\mathbb{M}\bar{\tau}_x[s_1, s_2])^2} \leqq \frac{s_2}{s_2 - x}.$$

Proof. Under our assumptions

$$\mathbb{M}\bar{\tau}_x[s_1, s_2] = 2\frac{s_2-x}{s_2-s_1}\int_{s_1}^{x} \frac{(z-s_1)}{\bar{\sigma}^2(z)}\,dz + O(1),$$

$$\begin{aligned}
\mathbb{M}\bar{\tau}_x^2[s_1, s_2] &= 8\frac{s_2-x}{(s_2-s_1)}\int_{s_1}^{x} \frac{(z-s_1)}{\bar{\sigma}^2(z)}\int_{s_1}^{z} \frac{(u-s_1)(s_2-z)}{\bar{\sigma}^2(u)}\,du\,dz + O(\mathbb{M}\bar{\tau}_x[s_1, s_2]) \\
&\leqq 8\frac{s_2-x}{s_2-s_1}\int_{s_1}^{x}\int_{s_1}^{u} \frac{(z-s_1)(u-s_1)}{\bar{\sigma}^2(z)\,\bar{\sigma}^2(u)}\,du\,dz + O(\mathbb{M}\bar{\tau}_x[s_1, s_2]) \\
&= 4\frac{s_2-x}{s_2-s_1}\left(\int_{s_1}^{x} \frac{z-s_1}{\bar{\sigma}^2(z)}\,dz\right)^2 + O(\mathbb{M}\bar{\tau}_x[s_1, s_2]) \\
&= \frac{s_2-s_1}{s_2-x}(\mathbb{M}\bar{\tau}_x[s_1, s_2])^2 + O(\mathbb{M}\bar{\tau}_x[s_1, s_2]).
\end{aligned}$$

The proof is complete.

Corollary 1. *If* $x<\frac{s_2}{2}$, *then from* $\mathbb{M}\bar{\tau}_x[0,s_2]=\infty$ *it follows that* $\mathbb{P}\{\bar{\tau}_x[0,s_2]=+\infty\}>0$.

Indeed, for $\varepsilon>0$

$$\overline{\lim_{s_1\downarrow 0}}\,\mathbb{P}\{|\bar{\tau}_x[s_1,s_2]-\mathbb{M}\bar{\tau}_x[s_1,s_2]|>(1-\varepsilon)\mathbb{M}\bar{\tau}_x[s_1,s_2]\}$$

$$\leqq\overline{\lim_{s_1\downarrow 0}}\frac{\mathbb{M}\bar{\tau}_x^2[s_1,s_2]-(\mathbb{M}\bar{\tau}_x[s_1,s_2])^2}{(1-\varepsilon)^2(\mathbb{M}\tau_x[s_1,s_2])^2}\leqq\frac{\frac{s_2}{s_2-x}-1}{(1-\varepsilon)^2}=\frac{x}{(1-\varepsilon)^2(s_2-x)},$$

hence

$$\overline{\lim_{s_1\downarrow 0}}\,\mathbb{P}\{\bar{\tau}_x[s_1,s_2]\geqq\varepsilon\mathbb{M}\bar{\tau}_x[s_1,s_2]\}\geqq 1-\frac{x}{(1-\varepsilon)^2(s_2-x)}>0$$

for sufficiently small $\varepsilon>0$. Therefore, for arbitrary $C>0$

$$\underline{\lim}_{s_1\downarrow 0}\,\mathbb{P}\{\bar{\tau}_x[s_1,s_2]\geqq C\}\geqq 1-\frac{x}{(1-\varepsilon)^2(s_2-x)}.$$

which proves our claim since $\tau_x[s_1,s_2]$ is nonincreasing for $s_1\downarrow 0$.

It follows from (2) that a necessary and sufficient condition that $\mathbb{M}\bar{\tau}_x[0,s_2]$ be finite is: for some $x_0>0$

$$\int_0^{x_0} z\frac{dz}{\bar{\sigma}^2(z)}<\infty.$$

We can thus obtain a condition for the finiteness of $\mathbb{M}\tau_x[r_1,\beta]$ ($\tau_x[r_1,\beta]$ is defined w.r.t. the process $\xi^{(0)}(t)$). Since $\bar{\sigma}(z)=\sigma(g(z))\,U(g(z))$, we can obtain, by making the substitution $g(z)=y$, $z=\varphi(y)$ in the integral $\int_0^{x_0}\frac{z\,dz}{\sigma^2(g(z))\,U^2(g(z))}$:

$$\int_{r_1}^{x_0}\frac{\varphi(y)}{\sigma^2(y)\,U(y)}\,dy<\infty. \tag{4}$$

We recall that under our assumptions $\varphi(r_1)=0$.

Assume that $\mathbb{M}\tau_x[r_1,\beta]=\infty$ for each $\beta>r_1$ if $x\in(r_1,\beta)$. We shall show that in this case either $\tau_x[r_1,\beta]>\infty$ and $\xi^{(0)}(\tau_x[r_1,\beta]=\beta$, or $\tau_x[r_1,\beta]=+\infty$. To this end we need

Lemma 2. *Let* $\tau_x[r_1,\beta]$ *be the time the process* $\xi_x^{(0)}(t)$ *attains the boundary of* (r_1,β). *If for some* $\delta>0$ *and* $C>0$ *we have* $\mathbb{P}\{\tau_x[r_1,\beta]\leqq C\}\geqq\delta$ *for all* $x\in(r_1,\beta)$, *then* $\mathbb{M}\tau_x(r_1,\beta)<\infty$.

Proof. For all $x \in (r_1, \beta)$ $\mathbb{P}\{\tau_x[r_1,\beta] > C\} \leqq 1-\delta$. Hence

$$\mathbb{P}\{\tau_x[r_1,\beta] > 2C\}$$

$$= \int_{r_1}^{\beta} \mathbb{P}\{\tau_y[r_1,\beta] > C\}\, \mathbb{P}\{\tau_x[r_1,\beta] > C,\ \xi_x^{(0)}(C) \leqq dy\} \leqq (1-\delta)^2.$$

We establish analogously that $\mathbb{P}\{\tau_x[r_1,\beta] > kC\} \leqq (1-\delta)^k$, so that

$$\mathbb{M}\tau_x[r_1,\beta] \leqq \sum_{k=0}^{\infty} (k+1)\, C(1-\delta)^k < \infty.$$

Lemma 3. *If for some $x \in (r_1, \beta)$ the probability $\mathbb{P}\{\tau_x[r_1,\beta] < \infty,$ $\xi_x^{(0)}(\tau_x[r_1,\beta]) = r_1\}$ is positive, then $\mathbb{M}\tau_x[r_1,\beta] < \infty$.*

Proof. For some $C > 0$ and $\delta > 0$

$$\mathbb{P}\{\tau_x[r_1,\beta] \leqq C, \xi_x^{(0)}(\tau_x[r_1,\beta]) = r_1\} \geqq \delta.$$

But then for all $\bar{x} \in (r_1, x)$

$$\mathbb{P}\{\tau_{\bar{x}}[r_1,\beta] \leqq C,\ \xi_{\bar{x}}^{(0)}(\tau_{\bar{x}}[r_1,\beta]) = r_1\}$$

$$\geqq \mathbb{P}\{\tau_x[r_1,\beta] \leqq C,\ \xi_x^{(0)}(\tau_x[r_1,\beta]) = r_1\} \geqq \delta.$$

Moreover, from Theorem 2 § 15 the function $\mathbb{M}\tau_y(x,\beta)$ is continuous for $y \in [x,\beta]$ and is consequently, bounded. Hence

$$\mathbb{P}\{\tau_y[r_1,\beta] \leqq C+Z\} \geqq \mathbb{P}\{\tau_y[x,\beta] \leqq Z\}\, \mathbb{P}\{\bar{\tau}_x \leqq C\}$$

$$\geqq \left(1 - \frac{\mathbb{M}\tau_y[x,\beta]}{Z}\right)\delta,$$

where $\bar{\tau}_x$ is the first time the process $\xi_y^{(0)}(\tau_y[x,\beta]+s)$ reaches the boundary of (r_1,β). Thus, we can pick $C_1 > 0$ and $\delta_1 > 0$ such that $\mathbb{P}\{\tau_y[r_1,\beta] \leqq C_1\} \geqq \delta_1$, for all $y \in [r_1,\beta]$. An application of Lemma 2 completes the proof.

Corollary 2. *If $\mathbb{M}\tau_x[r_1,\beta] = +\infty$ then for all $x \in (r_1,\beta)$*

$$\mathbb{P}\{\tau_x[r_1,\beta] < \infty,\ \xi_x^{(0)}(\tau_x[r_1,\beta]) = r_1\} = 0.$$

This also means that either $\tau_x[r_1,\beta] < \infty$ and $\xi_x(\tau_x[r_1,\beta]) = \beta$, or $\tau_x[r_1,\beta] = +\infty$.

From these lemmas we can obtain

Theorem 1. *Let*

$$L_1 = \int_{r_1}^{\beta} \exp\left\{-\int_{\beta}^{x} \frac{2a(z)}{\sigma^2(z)}\,dz\right\} dx.$$

If $L_1<\infty$ then we set

$$L_2=\int_{r_1}^{\beta}\frac{1}{\sigma^2(y)}\int_{r_1}^{y}\exp\left\{-\int_{\beta}^{x}\frac{2a(z)}{\sigma^2(z)}dz\right\}dx\exp\left\{\int_{\beta}^{y}\frac{2a(z)}{\sigma^2(z)}dz\right\}dy.$$

The following statements then hold:

1) *if $L_1=+\infty$, then $\xi_x^{(0)}(t)$ attains the point β before r_1, w.p.1 for any $x\in(r_1,\beta)$;*

2) *if $L_1<+\infty$, $L_2=+\infty$, then either $\tau_x[r_1,\beta]=\infty$ and $\xi_x^{(0)}(t)\to r_1$ for $t\to\infty$, or $\tau_x[r_1,\beta]<\infty$ and $\xi_x^{(0)}(\tau_x[r_1,\beta])=\beta$;*

3) *if $L_1<\infty$ and $L_2<\infty$, then for all $x\in(r_1,\beta)$ $\mathbb{M}\tau_x[r_1,\beta]<\infty$, which means that $\mathbb{P}\{\tau_x[r_1,\beta]<+\infty\}=1$ and $\mathbb{P}\{\xi_x^{(0)}(\tau_x[r_1,\beta]-0)=r_1\}>0$.*

We have proved all of these but 2) the fact that $\tau_x[r_1,\beta]=+\infty$ implies $\lim_{t\to\infty}\xi_x^{(0)}(t)=r_1$.

Assume we can determine an $s_1<s$ so that $\underline{\lim}_{t\to\infty}\xi_x^{(0)}(t)<s_1$ and $\overline{\lim}_{t\to\infty}\xi_x^{(0)}(t)>s_2$. Then we can choose an infinite sequence of times τ_n, where τ_n is the moment the point $s_0\in(s_1,s_2)$ is attained for the n-th time after the process has reached the boundary of (s_1,s_2), i.e., at time τ_n, $\xi_x^{(0)}(\tau_n)=s_0$; then $\xi_x^{(0)}$ reaches s_1 or s_2 and the first time that the process again attains s_0 is τ_{n+1}. Between τ_n and τ_{n+1}, $\xi_x^{(0)}(t)$ attains β with positive probability. Thus, the probability that after an infinite number of such cycles $\xi_x^{(0)}(t)$ has not hit β at all is equal to zero, so that $\lim_{t\to\infty}\xi_x^{(0)}(t)$ exists and must equal r_1 since $\sigma(x)>0$ for $x>r_1$.

We now study properties which ensure a continuous passage from the boundary r_1 into the interior of the interval (r_1,r_2) (we assume that $L_1<\infty$ and $L_2<\infty$ since otherwise the process will not hit the point r_1). Let $\varepsilon>0$ be arbitrary and let $\xi_\varepsilon^{(1)}(t)$ be a process on (r_1,β) constructed as follows: $\xi_\varepsilon^{(1)}(0)=r_1+\varepsilon$, $\xi_\varepsilon^{(1)}(t)$ satisfies (1) in (r_1,β), if τ is the exit time of this process from (r_1,β), then take $\xi_\varepsilon^{(1)}(\tau)=\varepsilon+r_1$ and let the further behavior of $\xi_\varepsilon^{(1)}(t)$ for $t>\tau$ be governed by (1) until it again attains the boundary of (r_1,β), etc. $\xi_\varepsilon^{(1)}(t)$ is defined in this way for all $t>0$. Naturally, the process can leave the boundary r_1 if for $\varepsilon\to 0$ the probability that $\xi_\varepsilon^{(1)}(t)$ hits the point β in finite time is greater than zero (β can be taken arbitrarily close to r_1, but this must be done independently of ε). Denote by τ_k the k-th time $\xi_\varepsilon^{(1)}(t)$ reaches the boundary of (r_1,β). The variables $\tau_{k+1}-\tau_k$ are identically distributed and are independent (see Corollary 1 § 15), also independent are the events $\mathfrak{A}_k=\{\xi_\varepsilon^{(1)}(\tau_k-0)=\beta\}$. Using the notation of this paragraph we can write

$$\mathbb{M}(\tau_{k+1}-\tau_k)=V(r_1+\varepsilon),\qquad \mathbb{P}\{\mathfrak{A}_k\}=\frac{\varphi(r_1+\varepsilon)-\varphi(r_1)}{\varphi(\beta)-\varphi(r_1)}$$

(according to the assumptions already made, the functions $\varphi(x)$ and $V(x)$ are defined and continuous in (r_1, β)) so that

$$\mathbb{P}\{\sup_{t\leqq\tau_n}\xi_\varepsilon^{(1)}(t)=\beta\}=1-\left(1-\frac{\varphi(r_1+\varepsilon)-\varphi(r_1)}{\varphi(\beta)-\varphi(r_1)}\right)^n. \tag{5}$$

Using (2) and (3) we can show that $\overline{\lim\limits_{\varepsilon\to 0}}\,\frac{\operatorname{Var}(\tau_k-\tau_{k-1})}{\mathbb{M}(\tau_k-\tau_{k-1})}$ can be made smaller than any pre-assigned number by choosing β sufficiently close to r_1. Let n_ε be the integral part of $\frac{1}{\mathbb{M}(\tau_k-\tau_{k-1})}$. Then

$$\begin{aligned}\mathbb{P}\{\tau_{n_\varepsilon}>t_1\}&=\mathbb{P}\left\{\sum_{k=1}^{n_\varepsilon}(\tau_k-\tau_{k-1})>t_1\right\}=1-\mathbb{P}\left\{\sum_{k=1}^{n_\varepsilon}(\tau_k-\tau_{k-1})\leqq t_1\right\}\\&\geqq 1-\mathbb{P}\left\{\left|\sum_{k=1}^{n_\varepsilon}[\tau_k-\tau_{k-1}-\mathbb{M}(\tau_k-\tau_{k-1})]\right|\leqq n_\varepsilon\mathbb{M}\tau_1-t_1\right\}\\&\geqq 1-\frac{n_\varepsilon\operatorname{Var}\tau_1}{(1-t_1)^2}\geqq 1-\frac{\operatorname{Var}\tau_1}{\mathbb{M}\tau_1(1-t_1)^2}>0\end{aligned}$$

for sufficiently small $\mathbb{M}\tau_1$, and

$$\mathbb{P}\{\tau_{n_\varepsilon}<t_2\}\geqq 1-\frac{\mathbb{M}\tau_{n_\varepsilon}}{t_2}$$

which is arbitrary close to unity for sufficiently large t_2. Thus, in order that $\mathbb{P}\{\sup_{t\leqq T}\xi_\varepsilon^{(1)}(t)=\beta\}$ be bounded from below by a positive constant it is necessary and sufficient that

$$\overline{\lim_{\varepsilon\to 0}}\left(1-\frac{\varphi(r_1+\varepsilon)-\varphi(r_1)}{\varphi(\beta)-\varphi(r_1)}\right)^{n_\varepsilon}<1,\quad \text{i.e.,}\quad \overline{\lim_{\varepsilon\to 0}}\, n_\varepsilon[\varphi(r_1+\varepsilon)-\varphi(r_1)]>0$$

$$\overline{\lim_{\varepsilon\to 0}}\,\frac{V(r_1+\varepsilon)}{\varphi(r_1+\varepsilon)-\varphi(r_1)}<\infty.$$

We note that $\lim\limits_{\varepsilon\to 0}\frac{V(r_1+\varepsilon)}{\varphi(r_1+\varepsilon)-\varphi(r_1)}$ exists (it can assume the value ∞). To show this we use the probabilistic significance of $V(x)$ and $\varphi(x)$:

$$\frac{V(r_1+\varepsilon)}{\varphi(r_1+\varepsilon)-\varphi(r_1)}=\frac{1}{\varphi(\beta)-\varphi(r_1)}\cdot\frac{\mathbb{M}\tau_{r_1+\varepsilon}[r_1,\beta]}{\mathbb{P}\{\xi_\varepsilon^{(0)}(\tau_{r_1+\varepsilon}[r_1,\beta]-0)=\beta\}}.$$

We now transform $\xi_x^{(0)}(t)$ in such a way that $a(x)$ takes the value zero. Then

$$\frac{V(r_1+\varepsilon)}{\varphi(r_1+\varepsilon)-\varphi(r_1)}=\frac{V_1(\bar r_1+\varepsilon)-V_1(\bar r_1)}{\varepsilon},$$

where $V_1(x)$ satisfies $\frac{\bar{\sigma}^2(x)}{2} V_1''(x) = -1$. $V_1(x)$ is concave and therefore has everywhere finite or infinite one-sided derivative. Moreover, the function $z(x) = \frac{V'(x)}{\varphi'(x)}$ satisfies $z'(x) = -\frac{2}{\sigma^2(x)\,\varphi'(x)}$, i.e., $z(x)$ is monotone so that $\lim_{x \downarrow r_1} z(x) = \lim_{x \downarrow r_1} \frac{V'(x)}{\varphi'(x)}$ always exists. Hence

$$\lim_{\varepsilon \downarrow 0} \frac{V(r_1+\varepsilon)}{\varphi(r_1+\varepsilon) - \varphi(r_1)} = \lim_{x \downarrow r_1} \frac{V'(x)}{\varphi'(x)}.$$

In order for the latter limit to be finite it is necessary and sufficient that

$$\int_{r_1}^{\beta} \frac{1}{\sigma^2(x)\,\varphi'(x)}\,dx < \infty,$$

i.e., that the constant

$$L_3 = \int_{r_1}^{\beta} \frac{1}{\sigma^2(z)} \exp\left\{\int_{x_0}^{z} \frac{2a(u)}{\sigma^2(u)}\,du\right\} dz.$$

be finite.

We will say that the boundary r_1 is *absorbing* if $L_3 = +\infty$, and *reflecting* otherwise. The notions of absorption and reflection for the boundary r_2 are defined analogously.

Depending on the values of the constants L_1, L_2 and L_3 the boundary r_1 will be called

1) natural, if $L_1 = +\infty$;
2) attracting, if $L_1 < +\infty$, $L_2 = +\infty$;
3) absorbing, if $L_1 < +\infty$, $L_2 < +\infty$, $L_3 = +\infty$;
4) regular, if $L_1 < +\infty$, $L_2 < +\infty$, $L_3 < +\infty$.

Boundary conditions are only be assigned to the boundaries in 3 and 4; those in 1 and 2 are inaccessible. All of the boundary conditions noted above can be assigned to 4, and to 3-merely absorption and jump exit from the boundary.

§ 22. Processes with Absorption at the Boundary

In this section we will consider processes of somewhat more general type than those we spoke of in § 21. We assume that the process is defined in some region of the (t, x)-plane bounded by the curves $x = g_1(t)$ and $x = g_2(t)$, where $g_i(t)$ is defined and continuous for $t \geqq 0$ and $g_1(t) < g_2(t)$. Let G denote the set of points (t, x) for which $t > 0$ and $g_1(t) < x < g_2(t)$. The boundary of the set G will be regarded as the set of points on the curves $x = g_i(t)$, $i = 1, 2$; the curve $x = g_2(t)$ will be called the *upper boundary* of the region G and $x=g_1(t)$ the *lower*. We assume that the functions $a(t,x)$ and $\sigma(t,x)$ are defined in the interior of G, are continuous in t and satisfy a Lipschitz condition in x: for some $K>0$

$$|a(t,x)-a(t,y)|+|\sigma(t,x)-\sigma(t,y)|\leqq K|x-y|, \tag{1}$$

for $x, y\in(g_1(t), g_2(t))$. In G, $\xi(t)$ satisfies

$$d\xi(t)=a(t,\xi(t))\,dt+\sigma(t,\xi(t))\,dw(t) \tag{2}$$

and is a process with absorption on the boundary of G provided that (2) holds for all t for which $g_1(t)<\xi(t)<g_2(t)$, $\xi(t)$ is continuous, and $\xi(t)=g_i(t)$ for $t>t'$ when $\xi(t')=g_i(t')$.

Theorem 1. *If G and the coefficients $a(t,x)$, $\sigma(t,x)$ satisfy the enumerated conditions and $\xi(0)$ is independent of $w(t)$ and*

$$\mathbb{P}\{\xi(0)\in[g_1(0), g_2(0)]\}=1,$$

then there exists a unique (up to stochastic equivalence) solution of (2) *with absorption on the boundary of G.*

Proof. In order to prove the existence we assume $\bar{a}(t,x)=a(t,g_1(t)+0)$, $\bar{\sigma}(t,x)=\sigma(t,g_1(t)+0)$ for $x\leqq g_1(t)$; $\bar{a}(t,x)=a(t,x)$, $\bar{\sigma}(t,x)=\sigma(t,x)$ for $g_1(t)<x<g_2(t)$; $\bar{a}(t,x)=a(t,g_2(t)-0)$, $\bar{\sigma}(t,x)=\sigma(t,g_2(t)-0)$ for $x>g_2(t)$ (the existence of limiting values for $a(t,x)$ and $\sigma(t,x)$ follows from (1)). Let $\bar{\xi}(t)$ satisfy $d\bar{\xi}(t)=\bar{a}(t,\bar{\xi}(t))\,dt+\bar{\sigma}(t,\bar{\xi}(t))\,dw(t)$ with initial condition $\bar{\xi}(0)=\xi(0)$. From Theorem 1 §6, such a solution exists. Set $\psi(t)$ if $g_1(s)<\bar{\xi}(s)<g_2(s)$ for $s<t$, $\psi(t)=0$ if for some $s<t$ $\bar{\xi}(s)\notin(g_1(s), g_2(s))$; let τ denote a variable for which $\psi(t)=1$ when $t<\tau$ and $\psi(t)=0$ when $t>\tau$ (it can happen that $\tau=+\infty$). If $\tau<+\infty$, then $\bar{\xi}(\tau)=g_i(\tau)$ for some i. Put $\xi(t)=\bar{\xi}(t)$ for $t<\tau$, $\xi(t)=g_i(t)$ if $t\geqq\tau$ and $\bar{\xi}(\tau)=g_i(\tau)$. It is easy to see that $\xi(t)$ is the desired process.

We demonstrate the uniqueness of the process $\xi(t)$ satisfying (2) in G with absorption at G's boundary. Let $\xi_1(t)$ and $\xi_2(t)$ be two such solutions and $\psi(t)=1$ if $\xi_1(t)\in(g_1(t), g_2(t))$ and $\xi_2(t)\in(g_1(t), g_2(t))$; $\psi(t)=0$ of either $\xi_1(t)\notin(g_1(t), g_2(t))$ or $\xi_2(t)\notin(g_1(t), g_2(t))$. Then

$$[\xi_1(t)-\xi_2(t)]\,\psi(t)=\psi(t)\left[\int_0^t[a(s,\xi_1(s))-a(s,\xi_2(s))]\,ds\right.$$
$$\left.+\int_0^t[\sigma(s,\xi_1(s))-\sigma(s,\xi_2(s))]\,dw(s)\right];$$

$$[\xi_1(t)-\xi_2(t)]^2\,\psi(t)\leqq 2\left(\int_0^t\psi(s)[a(s,\xi_1(s))-a(s,\xi_2(s))]\,ds\right)^2$$
$$+2\left(\int_0^t\psi(s)[\sigma(s,\xi_1(s))-\sigma(s,\xi_2(s))]\,dw(s)\right)^2.$$

Since

$$\psi(s)\{|a(s,\xi_1(s))-a(s,\xi_2(s))|^2+|\sigma(s,\xi_1(s))-\sigma(s,\xi_2(s))|^2\}$$
$$\leqq 2K^2\,|\xi_1(s)-\xi_2(s)|^2\,\psi(s),$$

we can establish as in Theorem 2 § 6 that for any $T>0$ there exists an L such that when $0\leqq t\leqq T$

$$\mathbb{M}\,|\xi_1(t)-\xi_2(t)|^2\,\psi(t)\leqq L\int_0^t \mathbb{M}\,|\xi_1(s)-\xi_2(s)|^2\,\psi(s)\,ds.$$

This implies with the help of Lemma 1 § 6 that $\mathbb{M}\,|\xi_1(t)-\xi_2(t)|^2\,\psi(t)=0$. Hence

$$\mathbb{P}\{\xi_1(t)=\xi_2(t)/\psi(t)=1\}=1.$$

From the continuity of $\xi_i(t)$ it follows that $\xi_1(t)$ and $\xi_2(t)$ coincide w.p. 1 up to the time τ^* for which $\psi(t)=1$ for $t<\tau^*$ and $\psi(t)=0$ for $t>\tau^*$. But then $\xi_1(\tau^*-0)=\xi_2(\tau^*-0)$. Since at least one of the processes $\xi_i(t)$ attains the boundary of G at time τ^*, we have for some i

$$\xi_1(\tau^*)=\xi_2(\tau^*)=g_i(\tau^*),$$

and since both ξ_1 and ξ_2 are processes with absorption on the boundary, $\xi_1(t)=\xi_2(t)$ for $t>\tau^*$. □

We now show that when $\sigma(t,x)>0$ and $a(t,x)$, $\sigma(t,x)$ possess certain smoothness properties along with $g_1(t)$ and $g_2(t)$, time and space variables can be transformed in such a way that G is mapped into the strip bounded by $x=0$ and $x=1$ with (2) assuming the form

$$d\tilde{\xi}(t)=\tilde{a}(t,\tilde{\xi}(t))\,dt+d\tilde{w}(t), \tag{3}$$

where $\tilde{w}(t)$ is also a Wiener process.

For $x\in[g_1(t),g_2(t)]$ let

$$\bar{g}_2(t)=\int_{g_1(t)}^{g_2(t)}\frac{dy}{\sigma(t,y)};\qquad \bar{g}_1(t)=0;\qquad f(t,x)=\int_{g_1(t)}^{z}\frac{dy}{\sigma(t,y)};$$

$\varphi(t,x)$ is the inverse of $f(t,x)$: $f(t,\varphi(t,x))=x$ for $0\leqq x\leqq\bar{g}_2(t)$; $\eta(t)=f(t,\xi(t))$, where $\xi(t)$ satisfies (2) with absorption on the boundary of G. Let $\bar{G}$ denote the region $\{(t,x)\colon t>0,\ x\in(0,\bar{g}_2(t))\}$ in the $(t;x)$-plane and assume

$$\begin{aligned}a(t,x)=f_t'(t,\varphi(t,x))+f_x'(t,\varphi(t,x))\,a(t,\varphi(t,x))\\+\tfrac{1}{2}f_{xx}''(t,\varphi(t,x))\,\sigma^2(t,\varphi(t,x)).\end{aligned}\tag{4}$$

Then $\eta(t)$ will be a process with absorption the boundary of $\bar{G}$ satisfying

$$d\eta(t)=\bar{a}(t,\eta(t))\,dt+dw(t)$$

(which follows from Itô's formula). Now let $\tilde{\xi}(t)=\frac{\eta(\lambda(t))}{\bar{g}_2(\lambda(t))}$, where $\lambda(t)$ in some increasing, continuously differentiable function which will be chosen later. Then

$$\begin{aligned} d\tilde{\xi}(t) &= -\frac{\eta(\lambda(t))}{[\bar{g}_2(\lambda(t))]^2}\lambda'(t)\,dt+\frac{d\eta(\lambda(t))}{\bar{g}_2(\lambda(t))} \\ &= -\frac{\tilde{\xi}(t)\lambda'(t)}{\bar{g}_2(\lambda(t))}dt+\frac{\bar{a}(\lambda(t),\eta(\lambda(t)))\lambda'(t)\,dt}{\bar{g}_2(\lambda(t))}+\frac{dw(\lambda(t))}{\bar{g}_2(\lambda(t))} \\ &= -\frac{\tilde{\xi}(t)+\bar{a}(\lambda(t),\bar{g}_2(\lambda(t))\,\tilde{\xi}(t))}{\bar{g}_2(\lambda(t))}\lambda'(t)\,dt+d\int_0^{\lambda(t)}\frac{dw(s)}{\bar{g}_2(s)}. \end{aligned}$$

The process $\zeta(u)=\int_0^u \frac{dw(s)}{\bar{g}_2(s)}$ is Gaussian with independent increments, for which $\mathbb{M}\zeta(u)=0$, $\operatorname{Var}\zeta(u)=\int_0^u \frac{ds}{(\bar{g}_2(s))^2}$. If $\lambda(t)$ is chosen so that $\int_0^{\lambda(t)} \frac{ds}{(\bar{g}_2(s))^2}=t$, then the process $\tilde{w}(t)=\zeta(\lambda(t))$ is a Wiener process. Hence $\tilde{\xi}(t)$ will be absorbed on the boundary of $G_1=\{(t,x)\colon t>0,\ x\in(0,1)\}$ and satisfies (3) inside this region, with

$$\tilde{a}(t,x)=-\frac{x+\bar{a}(\lambda(t),\bar{g}_2(\lambda(t))\,x)}{\bar{g}_2(\lambda(t))}\lambda'(t).$$

Since $\xi(t)$ and $\tilde{\xi}(t)$ are related to one another in a unique fashion by

$$\tilde{\xi}(t)=\frac{1}{\bar{g}_2(\lambda(t))}f(\lambda(t),\xi(\lambda(t))),$$

we can study $\xi(t)$ by means of $\tilde{\xi}(t)$.

Let $x\in(0,1)$ and assume $\tilde{\xi}_{x,s}(t)$ satisfies $d\tilde{\xi}_{x,s}(t)=\tilde{a}(t,\tilde{\xi}_{x,s}(t))\,dt+d\tilde{w}(t)$ for $t>s$ with absorption on the boundary of $(0,1)$ and initial condition $\tilde{\xi}_{x,s}(s)=x$. One can show that if $\tilde{\xi}(s)$ is a solution of (3), then in the region with absorption at the boundary of $(0,1)$, it is a Markov process whose transition probability $P(s,x,t,A)$ coincides with $\mathbb{P}\{\tilde{\xi}_{x,s}(t)\in A\}$.

We study the dependence of the solution of (3) on the initial data and will derive A.N. Kolmogorov's equation for the function $u(s,x)=\mathbb{M}f(\tilde{\xi}_{x,s}(t))$.

Let $\tilde{a}^*(t,x)$ be a function coinciding with $\tilde{a}(t,x)$ for $x\in(0,1)$, bounded and satisfying the assumptions of the existence and uniqueness theorem. $\xi^*_{x,s}(t)$ will denote the solution of

$$d\xi^*_{x,s}(t)=\tilde{a}^*(t,\xi^*_{x,s}(t))\,dt+d\tilde{w}(t)$$

for $t \geqq s$ with initial condition $\xi^*_{x,s}(s)=x$. Let $\tau_{x,s}$ be defined by

$$\tau_{x,s}=\inf\{t;\ \xi^*_{x,s}(t)\notin(0,1)\},$$

$=+\infty$ if the set in brackets is empty. Clearly, $\tilde{\xi}_{x,s}(t)$ is expressed by means of $\xi^*_{x,s}(t)$ as follows: $\tilde{\xi}_{x,s}(t)=\xi^*_{x,s}(t)$ for $t<\tau_{x,s}$, $\tilde{\xi}_{x,s}(t)=\xi^*_{x,s}(\tau_{x,s})$ for $t\geqq\tau_{x,s}$.

Lemma 1. *For any $T>0$ we can determine an L_T such that for $x_1, x_2\in(0,1)$, $0<s_1<s_2<T$*

$$\sup_{s_2\leqq t\leqq T}|\xi^*_{x_1,s_1}(t)-\xi^*_{x_2,s_2}(t)|\leqq L_T(|x_1-x_2|+|s_1-s_2|+|\tilde{w}(s_1)-\tilde{w}(s_2)|).$$

Proof. Since, for $t>s_2$

$$\xi^*_{x_1,s_1}(t)-\xi^*_{x_2,s_2}(t)=\xi^*_{x_1,s_1}(s_2)-x_2+\int_{s_2}^{t}[\tilde{a}^*(u,\xi^*_{x_1,s_1}(u))-\tilde{a}^*(u,\xi^*_{x_2,s_2}(u))]\,du,$$

on the basis of Lemma 1 § 6 we have

$$|\xi^*_{x_1,s_1}(t)-\xi^*_{x_2,s_2}(t)|\leqq|\xi^*_{x_1,s_1}(s_2)-x_2|\,e^{L(t-s_2)},$$

where L is such that $|\tilde{a}^*(t,x)-\tilde{a}^*(t,y)|\leqq L|x-y|$. We complete the proof by taking into consideration the fact that

$$\xi^*_{x_1,s_1}(s_2)=x_1+\tilde{w}(s_2)-\tilde{w}(s_1)+\int_{s_1}^{s_2}\tilde{a}^*(u,\xi^*_{x_1,s_1}(u))\,du$$

and that $\tilde{a}^*(u,x)$ is bounded.

Lemma 2. *For arbitrary $x\in(0,1)$ and $s>0$, $\tau_{x',s'}\to\tau_{x,s}$ for $x'\to x$ and $s'\to s$ w.p.1. If $x'\to 1\,(x'\to 0)$ and $s'\to s$, then $\tau_{x',s'}\to s$ w.p.1.*

Proof. Consider the first claim. Since $\xi^*_{x',s'}(t)$ converges uniformly (w.r.t. t) to $\xi^*_{x,s}(t)$ on any finite interval, it is sufficient for the proof of $\tau_{x',s'}\to\tau_{x,s}$ to show that for any $\delta>0$, $\sup\limits_{0<u<\delta}\xi^*_{x,s}(\tau_{x,s}+u)>1$ when $\xi^*_{x,s}(\tau_{x,s})=1$; $\inf\limits_{0<u<\delta}\xi^*_{x,s}(\tau_{x,s}+u)<0$ when $\xi^*_{x,s}(\tau_{x,s})=0$. Hence it suffices to show that the probability that $\xi^*_{x,s}(t)$ assumes a local maximum one or a local minimum zero, is equal to zero. Since the measure in the function space corresponding to $\xi^*_{x,s}(t)$ is absolutely continuous w.r.t. that corresponding to $w(t)-w(s)+x$ (see § 12), it is sufficient to show that for any rational interval (r_1,r_2) and arbitrary α, $\mathbb{P}\{\sup\limits_{r_1<t<r_2}w(t)=\alpha\}=0$. This is implied by the fact that $\sup\limits_{r_1<t<r_2}w(t)$ has a continuous distribution (see Lemma 1 § 1). An analogous statement holds for the infimum.

Now let $x'\to 1$ and $s'\to s$. It follows from the law of the iterated logarithm for Brownian motion that, w.p.1, we can find a sequence

$s_n \downarrow s$ for which $w(s_n)-w(s)>\sqrt{s_n-s}$. Thus, if $|\tilde{a}^*(t,x)|\leqq C$

$$\xi^*_{x',s'}(s_n)=x'+\int_{s'}^{s_n}\tilde{a}^*(u,\xi^*_{x',s'}(u)\,du+w(s_n)-w(s')$$
$$\geqq 1+\sqrt{s_n-s}-C(s_n-s')-|w(s)-w(s')|-(1-x'),$$

so that $s'\leqq\tau'_{x',s'}\leqq s_n$ provided that $|s-s'|$ and $|1-x'|$ are so small that $C|s-s'|+|w(s)-w(s')|+|1-x'|<\sqrt{s_n-s}-C|s_n-s|$. Since $w(s)$ is a continuous function and s_n arbitrarily close to s, the lemma is proved.

Corollary 1. *For any continuous function $f(x)$ for which $f(0)=f(1)=0$, the function $\mathbb{M}f(\tilde{\xi}_{x,s}(t))$ is continuous in all its arguments.*

This follows from the fact that $\tilde{\xi}_{x',s'}(t')\to\tilde{\xi}_{x,s}(t)$ in probability when $x'\to x$, $s'\to s$ and $t'\to t$, and the boundedness of $f(x)$.

Introduce process $w^*_{x,s}(t)=\tilde{w}(t)-\tilde{w}(s)+x$ and let $\tilde{\tau}_{x,s}$ be the first time $w^*_{x,s}(t)$ attains the boundary of (0, 1). Set $\tilde{w}_{x,s}(t)=w^*_{x,s}(t)$ for $t<\tilde{\tau}_{x,s}$ *and* $w^*_{x,s}(t)=w^*_{x,s}(\tilde{\tau}_{x,s})$ for $t\geqq\tilde{\tau}_{x,s}$. The results of § 12 imply that the measure in the function space corresponding to $\xi^*_{x,s}(t)$ will be absolutely continuous w.r.t. that corresponding to $w^*_{x,s}(t)$, and the density of the first w.r.t. the second measure will be

$$\rho=\exp\left\{\int_s^t\tilde{a}^*(u,w^*_{x,s}(u))\,d\tilde{w}(u)-\tfrac{1}{2}\int_s^t[\tilde{a}^*(u,w^*_{x,s}(u))]^2\,du\right\}.$$

In particular, if A is some Borel set on (0, 1) then

$$\mathbb{P}\{0<\xi^*_{x,s}(u)<1;\ s\leqq u\leqq t;\ \xi^*_{x,s}(t)\in A\}=\mathbb{M}\,\psi_A\,\rho,$$

where $\psi_A=1$ if $0<w^*_{x,s}(u)<1$ for $s\leqq u\leqq t$, $w^*_{x,s}(t)\in A$ and $\psi_A=0$ otherwise. The left side of the last equality is $\mathbb{P}\{\tilde{\xi}_{x,s}(t)\in A\}$ and the right side can be written in the form $\mathbb{M}\chi_A(\tilde{w}_{x,s}(t))\,\rho$. Thus for any continuous function $f(x)$ for which $f(0)=f(1)=0$ we have

$$\mathbb{M}f(\tilde{\xi}_{x,s}(t))=\mathbb{M}f(\tilde{w}_{x,s}(t))\exp\left\{\int_s^t\tilde{a}^*(u,w^*_{x,s}(u))\,d\tilde{w}(u)\right.$$
$$\left.-\tfrac{1}{2}\int_s^t[\tilde{a}^*(u,w^*_{x,s}(u))]^2\,du\right\}.\tag{5}$$

We proceed to the derivation of Kolmogorov's equation. Let $\tilde{a}^*(s,x)$ be sufficiently smooth and let $g(x)=0$ for $x\in[0,1]$, $g(x)>0$ for $x\notin[0,1]$. Furthermore, assume $g(x)$ has continuous, bounded derivatives up to and including second order. If $f(x)$ is defined for all x and has bounded continuous derivatives $f'(x)$ and $f''(x)$, then the function

$$u_n(s,x)=\mathbb{M} f(\xi^*_{x,s}(t))\exp\left\{-n\int_s^t g(\xi^*_{x,s}(u))\,du\right\} \tag{6}$$

in twice continuously differentiable in x and once in s and satisfies

$$-\frac{\partial u_n(s,x)}{\partial s}=\tilde{a}^*(s,x)\frac{\partial}{\partial x}u_n(s,x)+\frac{1}{2}\cdot\frac{\partial^2 u_n(s,x)}{\partial x^2}-n\,g(x)\,u_n(s,x) \tag{7}$$

with boundary condition $u_n(t,x)=f(x)$ (see §11, Remark 1). We also assume that $f(x)$ is nonnegative. Then $u_n(s,x)$ will be nonincreasing when $n\to\infty$ so that $\lim_{n\to\infty} u_n(s,x)=u_0(s,x)$ exists (here $u_0(t,x)=f(x)$).

Let $\varphi(x)$ be twice continuously differentiable with $\varphi(x)=0$ if $x\notin(0,1)$. Integrating (7) w.r.t. s from s_1 to t and then multiplying by $\varphi(x)$ and integrating w.r.t. x from $-\infty$ to $+\infty$, we find after noting that $\varphi(x)\,g(x)=0$ and using the formula for integration by parts that

$$-\int_{-\infty}^{\infty} f(x)\,\varphi(x)\,dx+\int_{-\infty}^{\infty} u_n(s_1,x)\,\varphi(x)\,dx$$
$$=-\int_{s_1}^{t}\int_{-\infty}^{\infty} u_n(s,x)\frac{\partial}{\partial x}(\tilde{a}^*(s,x)\,\varphi(x))\,dx+\frac{1}{2}\int_{s_1}^{t}\int_{-\infty}^{\infty} u_n(s,x)\frac{d^2}{dx^2}\varphi(x)\,dx.$$

Now let $n\to\infty$:

$$-\int_{-\infty}^{\infty} f(x)\,\varphi(x)\,dx+\int_{-\infty}^{\infty} u_0(s_1,x)\,\varphi(x)\,dx$$
$$=\int_{s_1}^{t}\int_{-\infty}^{\infty} u_0(s,x)\left[-\frac{\partial}{\partial x}(\tilde{a}^*(s,x)\,\varphi(x))+\frac{1}{2}\cdot\frac{d^2}{dx^2}\varphi(x)\right]dx.$$

Differentiating w.r.t. s_1, we obtain

$$-\frac{\partial}{\partial s_1}\int_0^1 u_0(s,x)\,\varphi(x)\,dx=\int_0^1 u_0(s,x)\left[-\frac{\partial}{\partial x}(\tilde{a}(s,x)\,\varphi(x))+\tfrac{1}{2}\varphi''(x)\right]dx. \tag{8}$$

Eq. (8) holds for all functions $\varphi(x)$, defined and twice continuously differentiable on [0, 1] which satisfy the conditions

$$\varphi(0)=\varphi(1)=\varphi'(0)=\varphi'(1)=\varphi''(0)=\varphi''(1)=0.$$

Relation (8) implies that $u_0(s,x)$ is the generalized solution of

$$-\frac{\partial}{\partial s}u_0(s,x)=\tilde{a}(s,x)\frac{\partial}{\partial x}u_0(s,x)+\frac{1}{2}\cdot\frac{\partial^2}{\partial x^2}u_0(s,x). \tag{9}$$

Let us determine the probabilistic significance of $u_0(s,x)$. We note that

$$\exp\left\{-n\int_s^t g(\xi^*_{x,s}(u))\,du\right\}\to 0,$$

provided that $\int_s^t g(\xi^*_{x,s}(u))\,du>0$, i.e., if there is a $u\in[s,t]$ for which $\xi^*_{x,s}(u)\notin[0,1]$ (in this case $g(\xi^*_{x,s}(u))>0$ for some u and because of the continuity in u of this function, $g(\xi^*_{x,s}(u))>0$ on some interval). Furthermore, if for $u\in[s,t]$ $\xi^*_{x,s}(u)\in[0,1]$, then $\int_s^t g(\xi^*_{x,s}(u))\,du=0$. Finally, as in the proof of Lemma 2

$$\mathbb{P}\{\sup_{s\leq u\leq t}\xi^*_{x,s}(u)=1\}=\mathbb{P}\{\inf_{s\leq u\leq t}\xi^*_{x,s}(u)=0\}=0.$$

Thus, if $f(0)=f(1)=0$, we obtain

$$\begin{aligned}\mathbb{P}&\left\{\lim_{n\to\infty}f(\xi^*_{x,s}(t))\exp\left\{-n\int_s^t g(\xi^*_{x,s}(u))\,du\right\}\neq f(\tilde{\xi}_{x,s}(t))\right\}\\&\leq\mathbb{P}\{\sup_{s\leq u\leq t}\xi^*_{x,s}(u)=1\}+\mathbb{P}\{\inf_{s\leq u\leq t}\xi^*_{x,s}(u)=0\}=0,\end{aligned}$$

so that, w.p. 1

$$\lim_{n\to\infty}f(\xi^*_{x,s}(t))\exp\left\{-n\int_s^t g(\xi^*_{x,s}(u))\,du\right\}=f(\tilde{\xi}_{x,s}(t)).$$

But then

$$\lim_{n\to\infty}\mathbb{M}f(\xi^*_{x,s}(t))\exp\left\{-n\int_s^t g(\xi^*_{x,s}(u))\,du\right\}=\mathbb{M}f(\tilde{\xi}_{x,s}(t))$$

and

$$u_0(s,x)=\mathbb{M}f(\tilde{\xi}_{x,s}(x)). \tag{10}$$

We now show that

$$\lim_{x\downarrow 0}u_0(s,x)=\lim_{x\uparrow 1}u_0(s,x)=0.$$

To this end we use (5) since $u_0(s,x)$ coincides with the left side of that formula. Since $\mathbb{M}\rho^2$ is uniformly bounded because $\tilde{a}^*(s,x)$ is bounded (see Lemma 2 § 12), we need only show that $f(\tilde{w}_{x,s}(t))\to 0$ in probability when either $x\downarrow 0$ or $x\uparrow 1$. But

$$\begin{aligned}\mathbb{P}\{f(\tilde{w}_{x,s}(t))\neq 0\}&\leq\mathbb{P}\{\sup_{s\leq u\leq t}[w(u)-w(s)]<1-x\}\\&=\frac{2}{\sqrt{2\pi(t-s)}}\int_0^{1-x}e^{-\frac{z^2}{2(t-s)}}\,dz\to 0\quad\text{when } x\uparrow 1.\end{aligned}$$

Analogously,

$$\mathbb{P}\{f(\tilde{w}_{x,s}(t))\neq 0\}\leqq \mathbb{P}\{\inf_{s\leqq u\leqq t}[w(u)-w(s)]>-x\}$$
$$=\mathbb{P}\{\sup_{s\leqq u\leqq t}[w(u)-w(s)]<x\}\to 0 \qquad \text{for } x\downarrow 0.$$

From what we have said above follows

Theorem 2. *Let $\tilde{a}(s,x)$ be a twice uniformly continuously differentiable function of x on $(0,1)$ and $f(x)$ twice continuously differentiable on $(0,1)$ with $f(0)=f(1)=0$. Then the function $u_0(s,x)=\mathbb{M}f(\tilde{\tilde{\xi}}_{x,s}(t))$ for $s<t$, $x\in(0,1)$ will be a continuous generalized solution of (9) satisfying the boundary conditions*

$$\lim_{s\uparrow t} u_0(s,x)=f(x); \qquad \lim_{x\uparrow 1} u_0(s,x)=\lim_{x\downarrow 0} u_0(s,x)=0.$$

Remark 1. If instead of the sequence of functions $u_n(s,x)$ we consider the sequence

$$V_n(s,x)=\mathbb{M}\exp\left\{-n\int_s^t g(\xi^*_{x,s}(u))\,du+\lambda\int_s^t \Phi(u,\xi^*_{x,s}(u))\,du\right\} f(\xi^*_{x,s}(t)),$$

where $g(x)$ and $f(x)$ are as in (6) and $\Phi(s,x)$ is twice continuously differentiable in x, then

$$\lim_{n\to\infty} V_n(s,x)=V_0(s,x)=\mathbb{M}\exp\left\{\lambda\int_s^t \Phi(u,\tilde{\tilde{\xi}}_{x,s}(u))\,du\right\} f(\tilde{\tilde{\xi}}_{x,s}(t)),$$

and $V_0(s,x)$ is a generalized solution of

$$-\frac{\partial V_0(s,x)}{\partial s}=\tilde{a}(s,x)\frac{\partial V_0(s,x)}{\partial x}+\frac{1}{2}\cdot\frac{\partial^2 V_0(s,x)}{\partial x^2}+\lambda\Phi(s,x)\,V_0(s,x),$$

satisfying the boundary conditions

$$\lim_{s\uparrow t} V_0(s,x)=f(x),$$
$$\lim_{x\downarrow 0} V_0(s,x)=\lim_{x\uparrow 1} V_0(s,x)=0.$$

Consider the probabilities

$$p_0(s,x,t)=\mathbb{P}\{\tilde{\tilde{\xi}}_{x,s}(t)=0\}, \qquad p_1(s,x,t)=\mathbb{P}\{\tilde{\tilde{\xi}}_{x,s}(t)=1\}.$$

Let $f_\varepsilon(x)=0$ if $x\notin(0,\varepsilon)$, $f_\varepsilon(x)>0$ if $x\in(0,\varepsilon)$ and assume $f_\varepsilon(x)$ is twice continuously differentiable. Then for each n, the function

$$\Phi_{\varepsilon,n}(s,x,t)=\mathbb{M}\exp\left\{-n\int_s^t f_\varepsilon(\tilde{\tilde{\xi}}_{x,s}(u))\,du\right\}\chi_{(0,1)}(\tilde{\tilde{\xi}}_{x,s}(t))$$

is a generalized solution of

$$-\frac{\partial}{\partial s}\Phi_{\varepsilon,n}(s,x,t)=\tilde{a}(s,x)\frac{\partial}{\partial x}\Phi_{\varepsilon,n}(s,x,t)+\frac{1}{2}\cdot\frac{\partial^2}{\partial x^2}\Phi_{\varepsilon,n}(s,x,t)$$
$$-nf_\varepsilon(x)\,\Phi_{\varepsilon,n}(s,x,t).$$

Thus in the interval $(\varepsilon, 1)\,\Phi_{\varepsilon,n}(s,x,t)$ is a generalized solution of

$$-\frac{\partial}{\partial s}\Phi_{\varepsilon,n}(s,x,t)=\tilde{a}(s,x)\frac{\partial}{\partial x}\Phi_{\varepsilon,n}(s,x,t)+\frac{1}{2}\cdot\frac{\partial^2}{\partial x^2}\Phi_{\varepsilon,n}(s,x,t). \quad (11)$$

The limit $\lim_{n\to\infty}\Phi_{\varepsilon,n}(s,x,t)=\Phi_\varepsilon(s,x,t)$ exists, so that letting $n\to\infty$ in (11) (multiply by some function and integrate as in (8)) we find that $\Phi_\varepsilon(s,x,t)$ is a generalized solution of

$$-\frac{\partial}{\partial s}\Phi_\varepsilon(s,x,t)=\tilde{a}(s,x)\frac{\partial}{\partial x}\Phi_\varepsilon(s,x,t)+\frac{1}{2}\cdot\frac{\partial^2}{\partial x^2}\Phi_\varepsilon(s,x,t),$$

where, as is easy to see,

$$\Phi_\varepsilon(s,x,t)=P\{\inf_{s\leqq u\leqq t}\tilde{\xi}_{x,s}(u)>\varepsilon\}.$$

Finally,

$$\lim_{\varepsilon\to 0}\Phi_\varepsilon(s,x,t)=P\{\inf_{s\leqq u\leqq t}\tilde{\xi}_{x,s}(u)>0\}=1-p_0(s,x,t),$$

exists so that $p_0(s,x,t)$ will be a generalized solution of

$$-\frac{\partial}{\partial s}p_0(s,x,t)=\tilde{a}(s,x)\frac{\partial}{\partial x}p_0(s,x,t)+\frac{1}{2}\cdot\frac{\partial^2}{\partial x^2}p_0(s,x,t)$$

on the interval (0, 1) (proceed again as in the case of (8)). It is also clear that $p_0(s,x,t)$ satisfies

$$\lim_{s\uparrow t}p_0(s,x,t)=0,\qquad \lim_{x\downarrow 0}p_0(s,x,t)=1,\qquad \lim_{x\uparrow 1}p_0(s,x,t)=0.$$

In the same way, one establishes that $p_1(s,x,t)$ is a generalized solution of

$$-\frac{\partial}{\partial s}p_1(s,x,t)=\tilde{a}(s,x)\frac{\partial}{\partial s}p_1(s,x,t)+\frac{1}{2}\cdot\frac{\partial^2}{\partial x^2}p_1(s,x,t)$$

with boundary conditions

$$\lim_{s\uparrow t}p_1(s,x,t)=0,\qquad \lim_{x\downarrow 0}p_1(s,x,t)=0,\qquad \lim_{x\uparrow 1}p_1(s,x,t)=1.$$

We note that

$$q(s,x,t)=p_0(s,x,t)+p_1(s,x,t)=\mathbb{P}\{\tau_{x,s}<t\}.$$

All the results above imply that $q(s, x, t)$ is a generalized solution of

$$-\frac{\partial}{\partial s} q(s, x, t)=\tilde{a}(s, x) \frac{\partial}{\partial x} q(s, x, t)+\frac{1}{2} \cdot \frac{\partial^2}{\partial x^2} q(s, x, t),$$

with boundary conditions

$$\lim_{s \uparrow t} q(s, x, t)=0, \quad \lim_{x \downarrow 0} q(s, x, t)=\lim_{x \uparrow 1} q(s, x, t)=1 .$$

We now turn to the case in which $\tilde{a}(s, x)$ independent of s: $\tilde{a}(s, x)=a(x)$. Then the finite-dimensional distributions of $\tilde{\xi}_{x, s}(s+t)$ coincide with those of $\tilde{\xi}_{x, 0}(t)$ (in particular, the distribution of $\tau_{x, s}-s$ is independent of s and coincides with that of $\tau_{x, 0}$), and

$$\mathbb{M} f(\tilde{\xi}_{x, s}(t))=\mathbb{M} f(\tilde{\xi}_{x, 0}(t-s)),$$

$$\mathbb{M} f(\tilde{\xi}_{x, s}(t)) \exp\left\{\int_s^t g(\tilde{\xi}_{x, s}(u))\, du\right\}=\mathbb{M} f(\tilde{\xi}_{x, 0}(t-s)) \exp\left\{\int_0^{t-s} g(\tilde{\xi}_{x, 0}(u))\, du\right\}.$$

From the preceeding results we can obtain

Theorem 3. *Let $\tilde{\xi}_x(t)$ satisfy $d\tilde{\xi}_x(t)=a(\tilde{\xi}_x(t))\, dt+dw(t)$ in the region G_1, with absorption on G_1's boundary, whereby $a(x)$ is twice continuously differentiable in $(0, 1)$. Then*

1) *the function*

$$u(t, x)=\mathbb{M} f(\tilde{\xi}_x(t)) \exp\left\{\int_s^t g(\tilde{\xi}_x(u))\, du\right\},$$

where $f(x)$ and $g(x)$ are twice continuously differentiable with $f(0)=f(1)=0$ is a generalized solution of

$$\frac{\partial u(t, x)}{\partial t}=L u(t, x)+g(x)\, u(t, x), \tag{12}$$

where $L u=a(x) \dfrac{\partial u}{\partial x}+\dfrac{1}{2} \cdot \dfrac{\partial^2 u}{\partial x^2}$ with boundary conditions

$$\lim_{t \downarrow 0} u(t, x)=f(x), \quad \lim_{x \downarrow 0} u(t, x)=\lim_{x \uparrow 1} u(t, x)=0;$$

2) *the functions*

$$p_0(t, x)=\mathbb{P}\{\tilde{\xi}_x(t)=0\}, \quad p_1(t, x)=\mathbb{P}\{\tilde{\xi}_x(t)=1\},$$

and

$$q(t, x)=\mathbb{P}\{\tau_{x, 0}<t\}$$

are generalized solutions of

$$\frac{\partial p_i(t, x)}{\partial t}=L p_i(t, x), \quad \frac{\partial q(t, x)}{\partial t}=L q(t, x),$$

with boundary conditions

$$\lim_{t\downarrow 0} p_i(t,x)=\lim_{t\downarrow 0} q(t,x)=0, \qquad \lim_{x\downarrow 0} p_1(t,x)=\lim_{x\uparrow 1} p_0(t,x)=0,$$

$$\lim_{x\downarrow 0} p_0(t,x)=\lim_{x\uparrow 1} p_1(t,x)=\lim_{x\downarrow 0} q(t,x)=\lim_{x\uparrow 1} q(t,x)=1.$$

In the homogeneous case we can solve these equations by means of the Laplace transform. Let L^* be the differential operator adjoint to L:

$$L^* V=-\frac{\partial}{\partial x}(a(x)\,V)+\frac{1}{2}\cdot\frac{\partial^2}{\partial x^2}V.$$

The fact that $u(t,x)$ is a generalized solution of (12) means that for any function $\varphi(x)$ which is twice continuously differentiable and satisfies $\varphi(0)=\varphi'(0)=0$, $\varphi(1)=\varphi'(1)=\varphi''(1)=0$,

$$\frac{d}{dt}\int_0^1 u(t,x)\,\varphi(x)\,dx=\int_0^1 u(t,x)\,L^*\varphi(x)\,dx+\int_0^1 g(x)\,u(t,x)\,\varphi(x)\,dx. \tag{13}$$

Since $u(t,x)\leqq A\,e^{Ct}$, where

$$C=\sup_{0\leqq x\leqq 1} g(x), \qquad \psi_\lambda(x)=\int_0^\infty u(t,x)\,e^{-\lambda t}\,dt$$

exists for $\lambda>C$. Multiplying (13) by $e^{-\lambda t}$ and integrating w.r.t. t from 0 to ∞ we obtain

$$\lambda\int_0^\infty e^{-\lambda t}\int_0^1 u(t,x)\,\varphi(x)\,dx\,dt-\int_0^1 f(x)\,\varphi(x)\,dx$$
$$=\int_0^{\prime}\int_0^1 e^{-\lambda t}u(t,x)\,[L^*\varphi(x)+g(x)\,\varphi(x)]\,dx,$$

or

$$\int_0^1 \psi_\lambda(x)\,[L^*\varphi(x)+g(x)\,\varphi(x)-\lambda\,\varphi(x)]\,dx+\int_0^1 f(x)\,\varphi(x)\,dx=0. \tag{14}$$

This relation holds for any function φ satisfying the conditions listed above. By proceeding to the limit we can easily convince ourselves that in addition to the continuity requirement for the second order derivatives it is sufficient that $\varphi(0)=\varphi(1)=0$.

Let $K_\lambda(x,y)$ be a function such that the solution of

$$L^*\varphi(x)+g(x)\,\varphi(x)-\lambda\,\varphi(x)=r(x) \tag{15}$$

with boundary conditions $\varphi(0)=\varphi(1)=0$ can be written in the form

$$\varphi(x)=\int_0^1 K_\lambda(x,y)\,r(y)\,dy$$

(K is called the Green's function for (15); under the indicated boundary conditions it has the form

$$K_\lambda(x,y)=\begin{cases}\dfrac{\varphi_1(x)\,\varphi_2(y)}{b(y)}, & x<y;\\[2ex] \dfrac{\varphi_2(x)\,\varphi_1(y)}{b(y)}, & x>y;\end{cases}$$

$$b(y)=\tfrac{1}{2}\left[\varphi_1(y)\,\varphi_2'(y)-\varphi_2(y)\,\varphi_1'(y)\right],$$

and $\varphi_i(x)$ fulfills $L^*\varphi_i(x)+g(x)\,\varphi_i(x)-\lambda\,\varphi_i(x)=0$ with $\varphi_1(0)=0$, $\varphi_1'(0)>0$, $\varphi_2(1)=0$ and $\varphi_2'(1)>0$).

Setting $\varphi(x)=\int_0^1 K_\lambda(x,y)\,r(y)\,dy$ in (14) we get

$$\begin{aligned}\int_0^1 \psi_\lambda(x)\,r(x)\,dx&=\int_0^1 f(x)\int_0^1 K_\lambda(x,y)\,r(y)\,dy\,dx\\&=\int_0^1 r(y)\left[\int_0^1 K_\lambda(x,y)\,f(x)\,dx\right]dy.\end{aligned}$$

Hence,

$$\psi_\lambda(x)=\int_0^1 K_\lambda(x,y)\,f(y)\,dy.$$

We have established that in the homogeneous case $\psi_\lambda(x)$ and $u(t,x)$ are defined in a unique way. It follows that if the boundary value problem (12) has a solution, then this solution coincides with $u(t,x)$. Analogous statements hold for the functions $p_0(t,x)$, $p_1(t,x)$ and $q(t,x)$. Since an application of the Laplace transform yields ordinary differential equations which always have a solution, we have proved

Theorem 4. *Suppose $\psi_\lambda(x)$, $\tilde p_0(\lambda,x)$, $\tilde p_1(\lambda,x)$ and $\tilde q(\lambda,x)$ are defined by means of*

$$\psi_\lambda(x)=\int_0^\infty e^{-\lambda t}\,u(t,x)\,dt,\qquad \tilde p_i(\lambda,x)=\int_0^\infty p_i(t,x)\,e^{-\lambda t}\,dt,$$

$$\tilde q(\lambda,x)=\int_0^\infty q(t,x)\,e^{-\lambda t}\,dt,$$

where $u(t,x)$, $p_i(t,x)$ and $q(t,x)$ have the same meaning as in Theorem 3. Then these functions are uniquely determined as solutions of

$$L\psi_\lambda(x)+g(x)\,\psi_\lambda(x)-\lambda\,\psi_\lambda(x)=-f(x),$$

$$L\tilde p_i(\lambda,x)-\lambda\,\tilde p_i(\lambda,x)=0,\qquad L\tilde q(\lambda,x)-\lambda\,\tilde q(\lambda,x)=0$$

with boundary conditions

$$\psi_\lambda(0)=\psi_\lambda(1)=0, \qquad \tilde{p}_0(\lambda,0)=\tilde{p}_1(\lambda,1)=\tilde{q}(\lambda,0)=\tilde{q}(\lambda,1)=\frac{1}{\lambda},$$
$$\tilde{p}_0(\lambda,1)=\tilde{p}_1(\lambda,1)=0.$$

§ 23. Instantaneous Reflection at the Boundary

We first consider a process on a semi-infinite interval which satisfies

$$d\xi(t)=a(t,\xi(t))\,dt+\sigma(t,\xi(t))\,dw(t), \tag{1}$$

where $a(t,x)$ and $\sigma(t,x)$ are defined for $t\in[0,T]$, $x>0$, are continuous in all their arguments and satisfy a Lipschitz condition in x: $|a(t,x)-a(t,y)|+|\sigma(t,x)-\sigma(t,y)|\leqq K|x-y|$ for $x>0$, $y>0$. Our goal is to construct a process $\xi(t)$ which is defined and continuous for $t\in[0,T]$, $\mathfrak{F}_t$-measurable for all $t\in[0,T]$ ($\mathfrak{F}_t$ is the σ-algebra generated by the variables $\xi(0)$ and $w(s)$ for $s\leqq t$; $\xi(0)$ is independent of $w(t)$), a solution of (1) for all t for which $\xi(t)>0$, and for which the Lebesgue measure of the set of t for which $\xi(t)=0$, is equal to zero.

Finding a process for which all the conditions are satisfied is equivalent to solving the following problem: find a pair of processes $\xi(t)$ and $\zeta(t)$ which satisfy: 1) $\xi(t)$ and $\zeta(t)$ are defined and continuous for $t\in[0,T]$ and are $\mathfrak{F}_t$-measurable for each t; 2) $\zeta(t)$ is nonincreasing and grows only at those points t for which $\xi(t)=0$; 3) for all $t\in[0,T]$, $\xi(t)\geqq 0$ and it satisfies

$$\xi(t)=\xi(0)+\int_0^t a(s,\xi(s))\,ds+\int_0^t \sigma(s,\xi(s))\,dw(s)+\zeta(t), \tag{2}$$

That $\zeta(t)$ grows only where $\xi(t)=0$ follows from the fact that $\xi(t)$ must satisfy (1). If Λ is the set of zeros of $\xi(t)$, then the ordinary and stochastic integrals over Λ will equal zero and $d\xi(t)\geqq 0$ for $t\in\Lambda$ so that $d\zeta(t)\geqq 0$ on Λ since $d\xi(t)=d\zeta(t)$ for $t\in\Lambda$.

Theorem 1. *The pair of random processes $\xi(t)$ and $\zeta(t)$ is uniquely determined by Conditions 1–3.*

Proof. Assume that $\xi_1(t), \zeta_1(t)$ and $\xi_2(t), \zeta_2(t)$ satisfy 1 and 2 and the equation

$$\xi_i(t)=\xi(0)+\int_0^t a(s,\xi_i(s))\,ds+\int_0^t \sigma(s,\xi_i(s))\,dw(s)+\zeta_i(t). \tag{3}$$

We will show that $\xi_1(t)=\xi_2(t)$ w.p. 1. Set $\Delta_{1,2}(t)=\xi_1(t)-\xi_2(t)$ if $\xi_1(t)-\xi_2(t)>0$, $\Delta_{1,2}(t)=0$ otherwise. The function $\Delta_{1,2}(t)$ is continuous w.p. 1. Let τ be the zero of $\Delta_{1,2}(s)$ farthest to the right on $[0, T]$. If $\tau<t$, then $\Delta_{1,2}(s)>0$ for $s\in(\tau, t]$. Hence $\xi_1(s)-\xi_2(s)>0$ for $s\in(\tau, t]$ and $\xi_1(\tau)=\xi_2(\tau)$. Since $\xi_1(s)>0$ for $s\in(\tau, t]$, $\zeta_1(s)$ is constant on $(\tau, t]$, i.e., $\zeta_1(t)-\zeta_1(\tau)=0$. Thus

$$\xi_1(t)-\xi_1(\tau)=\int_\tau^t a(s, \xi_1(s))\,ds+\int_\tau^t \sigma(s, \xi_1(s))\,dw(s),$$

$$\xi_2(t)-\xi_2(\tau)=\int_\tau^t a(s, \xi_2(s))\,ds+\int_\tau^t \sigma(s, \xi_2(s))\,dw(s)+\zeta_2(t)-\zeta_2(\tau).$$

Taking into account that $\xi_1(\tau)=\xi_2(\tau)$ and $\zeta_2(t)-\zeta_2(\tau)\geqq 0$ we can convince ourselves that under our assumptions $(\tau<t)$

$$\Delta_{1,2}(t)\leqq\int_\tau^t [a(s, \xi_1(s))-a(s, \xi_2(s))]\,ds+\int_\tau^t [\sigma(s, \xi_1(s))-\sigma(s, \xi_2(s))]\,dw(s).$$

For arbitrary t we have

$$\Delta_{1,2}(t)\leqq \sup_{0<t'<t}\left|\int_{t'}^t [a(s, \xi_1(s))-a(s, \xi_2(s))]\,ds + \int_{t'}^t [\sigma(s, \xi_1(s))-\sigma(s, \xi_2(s))]\,dw(s)\right|.$$

It is clear that an analogous estimate also holds for $\Delta_{2,1}(t)$, where $\Delta_{2,1}(t)=\xi_2(t)-\xi_1(t)$ for $\xi_2(t)-\xi_1(t)>0$ and $\Delta_{2,1}(t)=0$ for $\xi_2(t)-\xi_1(t)\leqq 0$. Since

$$|\xi_1(t)-\xi_2(t)|\leqq\int_0^t |a(s, \xi_1(s))-a(s, \xi_2(s))|\,ds + \sup_{0\leqq t'\leqq t}\left|\int_{t'}^t [\sigma(s, \xi_1(s))-\sigma(s, \xi_2(s))]\,dw(s)\right|.$$

Let $\chi_n(t)=1$ if $\sup_{0\leqq s\leqq t}|\xi_1(s)|\leqq n$, $\sup_{0\leqq s\leqq t}|\xi_2(s)|\leqq n$ and $\chi_n(t)=0$ otherwise. Using

$$\sup_{0\leqq t'\leqq t}\left|\int_{t'}^t g(s)\,dw(s)\right|\leqq 2\sup_{0\leqq t'\leqq t}\left|\int_0^{t'} g(s)\,dw(s)\right|$$

and Theorem 1 § 3, we obtain

$$\mathbb{M}\,|\xi_1(t)-\xi_2(t)|^2\,\chi_n(t)\leqq 2\mathbb{M}\left[\int_0^t \chi_n(s)\,|a(s, \xi_1(s))-a(s, \xi_2(s))|\,ds\right]^2 + 32\int_0^t \mathbb{M}\,\chi_n(s)[\sigma(s, \xi_1(s))-\sigma(s, \xi_2(s))]^2\,ds.$$

There is a K_1 for which

$$\mathbb{M}\,|\xi_1(t)-\xi_2(t)|^2\,\chi_n(t)\leqq K_1\int_0^t \mathbb{M}\,|\xi_1(s)-\xi_2(s)|^2\,\chi_n(s)\,ds$$

and from Lemma 1 §6 $\mathbb{M}|\xi_1(t)-\xi_2(t)|^2\chi_n(t)=0$ for all n. Hence, $\mathbb{P}\{\xi_1(t)=\xi_2(t)\}=1$. From the continuity w.p.1 of $\xi_1(t)$ and $\xi_2(t)$ it follows that $\mathbb{P}\{\sup_{0\leqq t\leqq T}|\xi_1(t)-\xi_2(t)|=0\}=1$. But then the integrals on the right side of (3) coincide for $i=1,2$. Therefore, $\zeta_1(t)$ and $\zeta_2(t)$ also coincide. □

Remark 1. It is necessary to direct attention to the fact that although two unknown functions appear in (2), the uniqueness of a pair of functions satisfying this equation has been proved. The function $\zeta(t)$ gives the total displacement of a particle due to reflection from the boundary. It is a non-trivial fact that such a function is unique, which means that for a process with instantaneous reflection, exit from the boundary can accur in only one way.

We now construct a process with instantaneous reflection from the boundary under the assumption that $\sigma(t,+0)>0$. We introduce a function $u(t,x)$ defined for $x\geqq 0$, $t\in[0,T]$, decreasing in x, possessing continuous bounded derivatives $u_t'(t,x)$, $u_x'(t,x)$, $u_{xx}''(t,x)$ and satisfying

$$u(t,0)=0,\qquad u_t'(t,0)+a(t,+0)\,u_x'(t,0)+\tfrac{1}{2}\sigma^2(t,+0)\,u_{xx}''(t,0)=0.$$

Set $\eta(t)=u(t,\xi(t))$. Then, if $\xi(t)$ is a process satisfying (1) with instantaneous reflection at the boundary, $\eta(t)$ will be a solution of

$$d\eta(t)=a_1(t,\eta(t))\,dt+\sigma_1(t,\eta(t))\,dw(t)\tag{4}$$

on the interval $(0,\infty)$ which is also a process with instantaneous reflection at the boundary. In (4)

$$\begin{aligned}a_1(t,x)&=u_t'(t,\varphi(t,x))+a(t,\varphi(t,x))\,u_x'(t,\varphi(t,x))\\&\quad+\tfrac{1}{2}\sigma^2(t,\varphi(t,x))\,u_{xx}''(t,\varphi(t,x));\\ \sigma_1(t,x)&=\sigma(t,\varphi(t,x))\,u_x'(t,\varphi(t,x)),\end{aligned}$$

where $\varphi(t,x)$ is such that $u(t,\varphi(t,x))=x$. Conversely, if $\eta(t)$ is a process with instantaneous boundary reflection satisfying (4), then the process $\xi(t)=\varphi(t,\eta(t))$ will be the one we are seeking. We note that we need only construct the solution of (2) for some arbitrary Wiener process $w(t)$. Thus, in the sequel it can be chosen in a special way.

Because of the assumptions made regarding $u(t,x)$, we have $a_1(t,+0)=0$. We extend the definition of the coefficients $a_1(t,x)$ and $\sigma_1(t,x)$ for $x\leqq 0$ by means of the relations

$$\begin{aligned}a_1^*(t,x)&=-a_1(t,-x)\quad(x<0),&\quad a_1^*(t,0)&=0,\\ \sigma_1^*(t,x)&=\sigma_1(t,-x)\quad(x<0),&\quad \sigma_1^*(t,0)&=\sigma_1(t,+0)\end{aligned}$$

(for $x>0$, a_1^* and σ_1^* coincide with a_1 and σ_1). Since $a_1(t,x)$ and $\sigma_1(t,x)$ satisfy a Lipschitz condition and $a_1^*(t,x)$ and $\sigma_1^*(t,x)$ are continuous for $x=0$, they will satisfy a Lipschitz condition for all $x\in(-\infty,\infty)$. Let $\eta^*(t)$ be a solution of

$$\eta^*(t)=\eta(0)+\int_0^t a_1^*(s,\eta^*(s))\,ds+\int_0^t \sigma_1^*(s,\eta^*(s))\,dw_1(s),$$

where $w_1(t)$ is some Wiener process and $\eta(0)$ is independent of $w_1(t)$. We will show that $\eta(t)=|\eta^*(t)|$ is a process with instantaneous reflection satisfying (4) with Wiener process $w(t)$ which can be expressed in terms of $w_1(t)$ by means of

$$w(t)=\int_0^t \operatorname{sign}\eta^*(s)\,dw_1(s). \tag{5}$$

For this purpose will need some lemmas.

Lemma 1. *The process* $w(t)$, *defined by* (5), *is a Wiener process independent of* $\eta^*(0)$.

Proof. Clearly, $w(t)$ is $\mathfrak{F}_t^{(1)}$-measurable, where $\mathfrak{F}_t^{(1)}$ is the minimum σ-algebra w.r.t. which $\eta^*(0)$ and $w_1(s)$ for $s\leqq t$ are measurable. The process $w(t)$ is continuous w.p. 1. Moreover,

$$\mathbb{M}\big(w(t+h)-w(t)/\mathfrak{F}_t^{(1)}\big)=0 \qquad (h>0),$$

$$\mathbb{M}\big([w(t+h)-w(t)]^2/\mathfrak{F}_t^{(1)}\big)=\mathbb{M}\left(\int_t^{t+h}[\operatorname{sign}\eta^*(s)]^2\,ds/\mathfrak{F}_t^{(1)}\right)$$
$$=h-\mathbb{M}\left(\int_t^{t+h}[1-\operatorname{sign}^2\eta^*(s)]\,ds/\mathfrak{F}_t^{(1)}\right),$$

and letting $\alpha=\inf\limits_{t\leqq s\leqq t+h}\sigma_1^*(s,0)$, we obtain

$$\mathbb{M}\left(\int_t^{t+h}[1-\operatorname{sign}^2\eta^*(s)]\,ds/\mathfrak{F}_t^{(1)}\right)$$
$$\leqq\frac{1}{\alpha^2}\mathbb{M}\left(\int_t^{t+h}[1-\operatorname{sign}^2\eta^*(s)]\,(\sigma_1^*(s,\eta^*(s)))^2\,ds/\mathfrak{F}_t^{(1)}\right)$$
$$\leqq\frac{1}{\alpha^2}\mathbb{M}\left(\int_t^{t+h}\frac{(\sigma_1^*(s,\eta^*(s)))^2}{1+n^2(\eta^*(s))^2}\,ds/\mathfrak{F}_t^{(1)}\right)$$
$$=\mathbb{M}\left(\frac{2}{\alpha^2 n}\int_{\eta^*(t)}^{\eta^*(t+\eta)}\arctan(nx)\,dx\right.$$
$$\left.-\frac{2}{\alpha^2 n}\int_t^{t+h}[\arctan n\eta^*(s)]\,a_1^*(s,\eta^*(s))\,ds/\mathfrak{F}_t^{(1)}\right)\to 0$$

for $n\to\infty$ $\left(\text{we used Itô's formula for the function } \frac{1}{n}\int\limits_0^x \arctan(nz)\,dz\right)$. Thus the fact that $w(t)$ is a Wiener process follows from Theorem 1 § 1 and the independence of $w(t)$ and $\eta^*(0)$ from Remark 1 § 1. □

Lemma 2. *Let $P^*(s,x,t,A)$ be the transition probability of $\eta^*(t)$. Then for any x and A*

$$P^*(s,x,t,A)=P^*(s,-x,t,-A),$$

where $-A$ is the set of y's for which $-y\in A$.

Proof. It suffices to show that the distributions of $\eta^*_{x,s}(t)$ and $-\eta^*_{-x,s}(t)$ coincide, where $\eta^*_{x,s}(t)$ satisfies

$$d\eta^*_{x,s}(t)=a^*_1\big(t,\eta^*_{x,s}(t)\big)\,dt+\sigma^*_1\big(t,\eta^*_{x,s}(t)\big)\,dw(t)$$

with initial condition $\eta^*_{x,s}(s)=x$. But

$$\begin{aligned}d\big(-\eta^*_{-x,s}(t)\big)&=-a^*_1\big(t,\eta^*_{-x,s}(t)\big)\,dt-\sigma^*_1\big(t,\eta^*_{-x,s}(t)\big)\,dw(t)\\&=a^*_1\big(t,-\eta^*_{-x,s}(t)\big)\,dt+\sigma^*_1\big(t,-\eta^*_{-x,s}(t)\big)\,d\big(-w(t)\big),\end{aligned}$$

since $a^*_1(t,x)$ is an odd and $\sigma^*_1(t,x)$ an even function of x. It remains to note that $-w(t)$ is also a Wiener process and $-\eta^*_{-x,s}(s)=x$. Hence, the equation for $-\eta^*_{-x,s}(t)$ differs only in the Wiener process which appears in it. The proof of the lemma follows.

Lemma 3. *Let $f(x)$ be a continuous even function which is monotone increasing for $x>0$. Then the process $\theta(t)=f\big(\eta^*(t)\big)$ will be a Markov process.*

Proof. It is sufficient to prove that

$$\mathbb{P}\{\theta(t+h)\in A/\mathfrak{F}^{(1)}_t\}=\mathbb{P}\{\theta(t+h)\in A/\theta(t)\}.$$

From the fact that $\eta^*(t)$ is a Markov process it follows that

$$\begin{aligned}\mathbb{P}\{\theta(t+h)\in A/\mathfrak{F}^{(1)}_t\}&=\mathbb{P}\{f\big(\eta^*(t+h)\big)\in A/\mathfrak{F}^{(1)}_t\}\\&=\mathbb{P}\{f\big(\eta^*(t+h)\big)\in A/\eta^*(t)\}=P^*\big(t,\eta^*(t),t+h,B\big),\end{aligned}$$

where B is the set of x's for which $f(x)\in A$. Since $f(x)$ is even, B is a set symmetric w.r.t. the origin. Thus,

$$P^*(t,-x,t+h,B)=P^*(t,x,t+h,B),$$

so that

$$P^*(t,x,t+h,B)=P^*(t,|x|,t+h,B).$$

But $|x|$ is uniquely defined by the value of $f(x)$ since $f(x)$ is even and increasing for $x>0$. Hence, denoting by $f^{-1}(x)$ those $y>0$ for which

$f(y)=x$ for $x>0$, we obtain

$$\mathbb{P}\{\theta(t+h)\in A/\mathfrak{F}_t^{(1)}\}=P^*(t,\eta^*(t),t+h,B)$$
$$=P^*(t,|\eta^*(t)|,t+h,B)=P^*(t,f^{-1}(\theta(t)),t+h,B).$$

The lemma is proved.

Set

$$f_\varepsilon(x)=\begin{cases}|x|, & |x|>\varepsilon,\\ \dfrac{1}{2\varepsilon}x^2+\dfrac{\varepsilon}{2}, & |x|\leqq\varepsilon;\end{cases}\qquad \theta_\varepsilon(t)=f_\varepsilon(\eta^*(t)).$$

Then $\theta_\varepsilon(t)$ satisfies

$$d\theta_\varepsilon(t)=\{f_\varepsilon'(\eta^*(t))\,a_1^*(t,\eta^*(t))+\tfrac{1}{2}f_\varepsilon''(\eta^*(t))[\sigma_1^*(t,\eta^*(t))]^2\}\,dt$$
$$+f_\varepsilon'(\eta^*(t))\,\sigma_1^*(t,\eta^*(t))\,dw_1(t),$$

from which

$$\theta_\varepsilon(t)-\theta_\varepsilon(0)=\int_0^t\psi_\varepsilon(|\eta^*(s)|)\,a_1^*(s,|\eta^*(s)|)\,ds$$
$$+\int_0^t\psi_\varepsilon(\eta^*(s))\,\sigma_1^*(s,|\eta^*(s)|)\,dw_1(s)+\frac{1}{2\varepsilon}\int_0^t\chi_\varepsilon(|\eta^*(s)|)[\sigma_1^*(s,\eta^*(s))]^2\,ds,$$

where

$$\psi_\varepsilon(x)=\begin{cases}-1, & x<-\varepsilon,\\ 1, & x>\varepsilon,\\ \dfrac{x}{\varepsilon}, & |x|\leqq\varepsilon,\end{cases}$$

so that $f_\varepsilon'(x)=\psi_\varepsilon(x)=\psi_\varepsilon(|x|)\operatorname{sign}x$, $\chi_\varepsilon(x)=1$ for $0\leqq x\leqq\varepsilon$, $\chi_\varepsilon(x)=0$ for $x>\varepsilon$. Letting $\varepsilon\to 0$ and taking into account the fact that $\psi_\varepsilon(x)\to\operatorname{sign}x$, $\theta_\varepsilon(t)\to|\eta^*(t)|$, we find that

$$\lim_{\varepsilon\to 0}\int_0^t\psi_\varepsilon(|\eta^*(s)|)\,a_1^*(s,|\eta^*(s)|)\,ds=\int_0^t a_1(s,|\eta^*(s)|)\,ds,$$

and

$$\lim_{\varepsilon\to 0}\int_0^t\psi_\varepsilon(\eta^*(s))\,\sigma_1^*(s,|\eta_1^*(s)|)\,dw_1(s)=\int_0^t\sigma_1(s,|\eta^*(s)|)\,dw(s),$$

exist in the sense of convergence in probability. This means that

$$\lim_{\varepsilon\to 0}\frac{1}{2\varepsilon}\int_0^t\chi_\varepsilon(|\eta^*(s)|)[\sigma_1^*(s,\eta^*(s))]^2\,ds=\zeta(t).$$

will also exist in this sense.

The process $\zeta(t)$ is constant on any interval (α, β) on which $|\eta^*(s)|>0$. Hence, the pair of processes $\eta(t)=|\eta^*(t)|$ and $\zeta(t)$ satisfy

$$\eta(t)=\eta(0)+\int_0^t a_1(s, \eta(s))\, ds+\int_0^t \sigma_1(s, \eta(s))\, dw(s)+\zeta(t)$$

along with Conditions 1 and 2 at the beginning of this section. This establishes the existence of a process with instantaneous reflection at the boundary.

Under certain smoothness conditions on the coefficients $a_1^*(t, x)$ and $\sigma_1^*(t, x)$ and the function $f(x)$, the function $u(s, x)=\mathbf{M} f(\eta_{x,s}^*(t))$, where $\eta_{x,s}^*(t)$ is defined in Lemma 2, also satisfies Kolmogorov's equation (see § 11)

$$-\frac{\partial}{\partial s} u(s, x)=a_1^*(s, x) \frac{\partial}{\partial x} u(s, x)+\frac{1}{2}[\sigma_1^*(s, x)]^2 \frac{\partial^2}{\partial x^2} u(s, x)$$

with initial condition $\lim\limits_{s \uparrow t} u(s, x)=f(x)$. Assume now that $f(x)$ is an even function and that $\eta_{x,s}(t)$ is a process satisfying for $t>s$ and $x>0$

$$\eta_{x,s}(t)=x+\int_s^t a_1(u, \eta_{x,s}(u))\, du+\int_s^t \sigma_1(u, \eta_{x,s}(u))\, dw(u)+\zeta(t)-\zeta(s)$$

along with Conditions 1 and 2. Then $u_1(s, x)=\mathbf{M} f(\eta_{x,s}(t))$ will satisfy for $x>0$ and $s<t$

$$-\frac{\partial}{\partial s} u_1(s, x)=a_1(s, x) \frac{\partial}{\partial s} u_1(s, x)+\frac{1}{2}[\sigma_1(s, x)]^2 \frac{\partial^2}{\partial x^2} u_1(s, x), \quad (6)$$

with the boundary conditions

$$\lim_{s \uparrow t} u_1(s, x)=f(x); \qquad \left.\frac{\partial u_1(s, x)}{\partial x}\right|_{x=0}=0.$$

The last condition follows from the fact that $u_1(s, x)=u(s, x)$ for $x>0$ and that $a(s, x)$ is even in x.

Analogous considerations lead to the equation

$$\begin{aligned}-\frac{\partial}{\partial s} v_1(s, x)=a_1(s, x) &\frac{\partial}{\partial x} v_1(s, x)\\ &+\frac{1}{2}[\sigma_1(s, x)]^2 \frac{\partial^2}{\partial x^2} v_1(s, x)+\lambda g(s, x) v_1(s, x)\end{aligned} \quad (7)$$

for the function

$$v_1(s,x)=\mathbb{M}\exp\left\{\lambda\int_s^t g\big(u,\eta_{x,s}(u)\big)\,du\right\} f\big(\eta_{x,s}(t)\big).$$

The function $v_1(s,x)$ also satisfies the boundary conditions

$$\lim_{s\uparrow t} v_1(s,x)=f(x), \qquad \left.\frac{\partial v_1(s,x)}{\partial x}\right|_{x=0}=0.$$

We now consider a process on the interval (0, 1) with instantaneous reflection at its endpoints. We will say that a process $\xi(t)$ satisfies (1) in the interval (0, 1) and suffers instantaneous reflection at its boundary if there exist $\zeta_0(t)$ and $\zeta_1(t)$ such that

1) $\xi(t)$, $\zeta_0(t)$ and $\zeta_1(t)$ are defined, $\mathfrak{F}_t$-measurable and continuous w.p.1 for $t\in[0,T]$ and $0\leqq\xi(t)\leqq 1$;

2) the function $\zeta_0(t)$ is nondecreasing and grows only at those t for which $\xi(t)=0$; the function $\zeta_1(t)$ is nonincreasing and decreases only at those t for which $\xi(t)=1$;

3) the functions $\xi(t)$, $\zeta_0(t)$ and $\zeta_1(t)$ are related by

$$\xi(t)=\xi(0)+\int_0^t a\big(s,\xi(s)\big)\,ds+\int_0^t \sigma\big(s,\xi(s)\big)\,dw(s)+\zeta_0(t)+\zeta_1(t). \tag{8}$$

The uniqueness of the process with reflection on the boundary of a finite interval can be deduced from Theorem 1. Indeed, denoting by $\tau_1,\tau_2,\dots,\tau_k,\dots$ the times of first attainment of the initial position after a previous visit to the boundary, we can partition the entire time axis into sections $[0,\tau_1]$, $[\tau_1,\tau_2]$, ..., on each of which the process coincides with a process with instantaneous reflection on an interval with a single boundary point.

For the construction of a process on a finite interval with instantaneous boundary reflection we assume that $a(t,x)$ vanishes at points of the boundary (this can always be arranged by considering in place of $\xi(t)$ the process $f(t,\xi(t))$ with suitable f). Now assume that $a(t,x)$ and $\sigma(t,x)$ are extended for all x with the help of the relations $a_1(t,x)=-a(t,-x)$, $\sigma_1(t,x)=\sigma(t,-x)$ for $-1\leqq x\leqq 0$, $a_1(t,x+2k)=a_1(t,x)$, $\sigma_1(t,x+2k)=\sigma_1(t,x)$ for $|x|\leqq 1$ and arbitrary integral k. Let $\xi_1(t)$ be a solution of

$$d\xi_1(t)=a_1\big(t,\xi_1(t)\big)\,dt+\sigma_1\big(t,\xi_1(t)\big)\,dw_1(t)$$

and assume that $s(x)=1$ for $2k<x<2k+1$ and $s(x)=-1$ for $2k-1<x<2k$ for arbitrary k. Set

$$l(x)=\int_0^x s(z)\,dz.$$

Then the process $\xi(t)=l(\xi_1(t))$ will satisfy Conditions 1 and 2 along with Eq. (8) if we take

$$w(t)=\int_0^t s(\xi_1(t))\,dw_1(t),$$

$$\zeta_0(t)=\lim_{\varepsilon\to 0}\frac{1}{2\varepsilon}\int_0^t \chi_\varepsilon(\xi(u))\,\sigma^2(u,\xi(u))\,du,$$

$$\zeta_1(t)=-\lim_{\varepsilon\to 0}\frac{1}{2\varepsilon}\int_0^t \chi_\varepsilon(1-\xi(u))\,\sigma^2(u,\xi(u))\,du$$

(these limits exist in the sense of convergence in probability).

The proofs that $w(t)$ is a Wiener process and $l(\xi_1(t))$ is a Markov process are analogous to the proofs of Lemmas 1–3. Eq. (8) is deduced with the help of a passage to the limit from Itô's formula for the function $f_\varepsilon(\xi_1(t))$, where $f_\varepsilon(x)$ is a periodic function with period two for which $f_\varepsilon(x)=|x|$ for $\varepsilon<|x|<1-\varepsilon$, $f_\varepsilon(x)=\dfrac{x^2}{2\varepsilon}+\dfrac{\varepsilon}{2}$ for $|x|\leqq\varepsilon$ and

$$f_\varepsilon(x)=1-\frac{(1-|x|)^2}{2\varepsilon}-\frac{\varepsilon}{2}$$

for $1-\varepsilon\leqq|x|\leqq 1$.

We remark that when $a_1(t,x)$, $\sigma_1(t,x)$, $g(s,x)$ and $f(x)$ are sufficiently smooth, then the function

$$V(s,x)=\mathbb{M}\exp\left\{\int_s^t g(u,\xi_{x,s}(u))\,du\right\}f(\xi_{x,s}(t)),$$

where $\xi_{x,s}(t)$ satisfies

$$\xi_{x,s}(t)=x+\int_s^t a(u,\xi_{x,s}(u))\,du+\int_s^t \sigma(u,\xi_{x,s}(u))\,dw(u)+\zeta_0(t)+\zeta_1(t),$$

for $t\geqq s$ and $\xi_{x,s}(t)$, $\zeta_0(t)$ and $\xi_1(t)$ fulfill Conditions 1 and 2, is a solution of

$$-\frac{\partial}{\partial s}V(s,x)=a(s,x)\frac{\partial}{\partial x}V(s,x)+\frac{1}{2}\sigma^2(s,x)\frac{\partial^2}{\partial x^2}V(s,x)+g(s,x)\,V(s,x)$$

with boundary conditions $\lim_{s\uparrow t}V(s,x)=f(x)$, $\left.\dfrac{\partial}{\partial x}V(s,x)\right|_{\substack{x=0\\x=1}}=0$.

We now investigate the asymptotic behavior of the distribution of $\xi(t)$ and of the mean $\dfrac{1}{T}\displaystyle\int_0^T f(\xi(t))\,dt$ for a process with instantaneous reflection at the boundary in the case where the underlying coefficients are time independent. We will need

Lemma 4. *Let $\xi(t)$, $\zeta_0(t)$ and $\zeta_1(t)$ satisfy Conditions 1 and 2 and relation (2). If $f(x)$ is a continuously differentiable function on $[0, 1]$, then*

$$f(\xi(t))-f(\xi(0))=\int_0^t [f'(\xi(s))\,a(s,\xi(s))+\tfrac{1}{2}f''(\xi(s))\,\sigma^2(s,\xi(s))]\,ds$$

$$+\int_0^t f'(\xi(s))\,\sigma(s,\xi(s))\,dw(s)+\int_0^t f'(\xi(s))\,d\zeta_0(s)$$

$$+\int_0^t f'(\xi(s))\,d\zeta_1(s).$$

Proof. Assume $0=t_0<t_1<\cdots<t_n=t$; $\Delta t_k=t_{k+1}-t_k$, then

$$f(\xi(t))-f(\xi(0))=\sum_{k=0}^{n-1}[f(\xi(t_{k+1}))-f(\xi(t_k))]$$

$$=\sum_{k=0}^{n-1}\left\{f'(\xi(t_k))\left[\int_t^{t_{k+1}} a(s,\xi(s))\,ds+\int_{t_k}^{t_{k+1}}\sigma(s,\xi(s))\,dw(s)\right]\right.$$

$$\left.+\tfrac{1}{2}f''(\xi(t_k)+\theta_k)\left[\int_{t_k}^{t_{k+1}}\sigma(s,\xi(s))\,dw(s)\right]^2\right\}$$

$$+\sum_{k=0}^{n-1} f'(\xi(t_k))\,[\zeta_0(t_{k+1})-\zeta_0(t_k)+\zeta_1(t_{k+1})-\zeta_1(t_k)]$$

$$+O\left(\sum_{k=0}^{n-1}\left(\Delta t_k\left|\int_{t_k}^{t_{k+1}}\sigma(s,\xi(s))\,dw(s)\right|+\Delta t_k^2\right.\right.$$

$$\left.\left.+(\zeta_0(t_{k+1})-\zeta_0(t_k))^2+(\zeta_1(t_{k+1})-\zeta_1(t_k))^2\right)\right);$$

$$|\theta_k|\leqq|\xi(t_{k+1})-\xi(t_k)|.$$

The first sum converges for $\max_k \Delta t_k\to 0$ to

$$\int_0^t [f'(\xi(s))\,a(s,\xi(s))+\tfrac{1}{2}f''(\xi(s))\,\sigma^2(s,\xi(s))]\,ds+\int_0^t f'(\xi(s))\,\sigma(s,\xi(s))\,dw(s)$$

which can be proved exactly as in the derivation of Itô's formula (§ 3). From the continuity of $f'(\xi(s))$ and the monotonicity of $\zeta_i(t)$ follows

$$\sum_{k=0}^{n-1} f'(\xi(t_k))\,[\zeta_i(t_{k+1})-\zeta_i(t_k)]\to\int_0^t f'(\xi(s))\,d\zeta_i(s)$$

when $\max_k \Delta t_k\to 0$. Furthermore,

$$\mathbb{M}\sum_{k=0}^{n-1}\left|\int_{t_k}^{t_{k+1}}\sigma(s,\xi(s))\,dw(s)\right|\Delta t_k\leqq C\sum_{k=0}^{n-1}(\Delta t_k)^{\frac{3}{2}}\leqq C\,t\sqrt{\max_k \Delta t_k},$$

$$\sum_{k=0}^{n-1}\Delta t_k^2\leqq t\max_k \Delta t_k,$$

and because of the monotonicity and continuity of $\zeta_i(t)$

$$\sum_{k=0}^{n-1}(\zeta_i(t_{k+1})-\zeta_i(t_k))^2 \leqq \max_k |\zeta_i(t_{k+1})-\zeta_i(t_k)| \cdot |\zeta_i(t)| \to 0$$

for $\max_k \Delta t_k \to 0$. This completes the proof.

Denote by $a(x)$ and $\sigma(x)$ the coefficients of the equation and let $\xi_x(t)$, $\zeta_0(t)$ and $\zeta_1(t)$ satisfy conditions 1 and 2 along with the equation

$$\xi_x(t)=x+\int_0^t a(\xi_x(s))\,ds+\int_0^t \sigma(\xi_x(s))\,dw(s)+\zeta_0(t)+\zeta_1(t).$$

Lemma 5. *Let $0<x<y<1$ and $V_0(x)$ be a solution of*

$$a(x)\,V_0'(x)+\tfrac{1}{2}\sigma^2(x)\,V_0''(x)=1,$$

for which $V_0'(0)=0$. Let $\tau_{x,y}$ be the first passage time to y of $\xi_x(t)$. Then

$$\mathbb{M}\tau_{x,y}=V_0(y)-V_0(x).$$

Proof. Applying Lemma 4 to $V_0(x)$, we have

$$\begin{aligned} V_0(\xi_x(t))-V_0(x)=&\int_0^t \left[a(\xi_x(s))\,V_0'(\xi_x(s))+\tfrac{1}{2}\sigma^2(\xi_x(s))\,V_0''(\xi_x(s))\right]ds\\ &+\int_0^t V_0'(\xi_x(s))\,\sigma(\xi_x(s))\,dw(s)\\ &+\int_0^t V_0'(\xi_x(s))\,d\zeta_0(s)+\int_0^t V_0'(\xi_x(s))\,d\zeta_1(s). \end{aligned}$$

Now put $t=\tau_{x,y}$. Since $\xi_x(\tau_{x,y})=y$, $\int_0^t V_0'(\xi_x(s))\,d\zeta_0(s)=0$ ($V_0'(\xi_x(s))=0$ at all points of growth of $\zeta_0(t)$) and $\zeta_1(s)=0$ for $s<\tau_{x,y}$, we have

$$V_0(y)-V_0(x)=\tau_{x,y}+\int_0^{\tau_{x,y}} V_0'(\xi_x(s))\,\sigma(\xi_x(s))\,dw(s). \tag{9}$$

From (9) it is easy to deduce that

$$\mathbb{M}\tau_{x,y}\leqq 2(V_0(y)-V_0(x))+\sup_z\,[V_0'(z)\,\sigma(z)]^2,$$

so that $\mathbb{M}\tau_{x,y}<\infty$ which means that

$$\mathbb{M}\int_0^{\tau_{x,y}} V_0'(\xi_x(s))\,\sigma(\xi_x(s))\,dw(s)=0, \tag{10}$$

and this completes the proof of the lemma.

Remark 2. It can be shown analogously that for $x>y$

$$\mathbb{M}\tau_{x,y}=V_1(y)-V_1(x),$$

where $V_1(x)$ satisfies

$$a(x)\,V_1'(x)+\tfrac{1}{2}\sigma^2(x)\,V_1''(x)=1, \quad \text{with } V_1'(1)=0.$$

We will now prove several ergodic theorems.

Theorem 2. *For any bounded measurable function $f(x)$ and any $x\in(0,1)$*

$$\lim_{t\to\infty}\mathbb{M}f(\xi_x(t))$$

exists and is independant of x.

Proof. Let $x_1<x_2$. From the uniqueness theorem it follows that for all t $\xi_{x_1}(t)\leqq\xi_{x_2}(t)$. If for some $t'>0$ $\xi_{x_1}(t')=\xi_{x_2}(t')$, then $\xi_{x_1}(t)=\xi_{x_2}(t)$ for all $t>t'$. Since $0\leqq\xi_{x_1}(t)\leqq\xi_{x_2}(t)\leqq 1$ and $\xi_{x_1}(\tau_{x_1,1})=1$, we also have $\xi_{x_2}(\tau_{x_1,1})=1$ since for $t>\tau_{x_1,1}$, $\xi_{x_1}(t)=\xi_{x_2}(t)$. Moreover, $\tau_{x_1,1}<\tau_{0,1}$, so that

$$\mathbb{P}\{\xi_{x_1}(t)\neq\xi_{x_2}(t)\}\leqq\mathbb{P}\{\tau_{0,1}>t\},$$

and hence also

$$\mathbb{P}\{f(\xi_{x_1}(t))\neq f(\xi_{x_2}(t))\}\leqq\mathbb{P}\{\tau_{0,1}>t\}.$$

Finally,

$$|\mathbb{M}f(\xi_x(t))-\mathbb{M}f(\xi_y(t))|\leqq\mathbb{M}\,|f(\xi_x(t))-f(\xi_y(t))|\leqq 2\,\|f\|\,\mathbb{P}\{\tau_{0,1}>t\},$$

where $\|f\|=\sup_x|f(x)|$. Therefore we can write

$$|\mathbb{M}f(\xi_x(t))-\mathbb{M}f(\xi_x(t+s))|=\left|\mathbb{M}f(\xi_x(t))-\int_0^1\mathbb{M}f(\xi_y(t))\,\mathbb{P}\{\xi(s)\in dy\}\right|$$

$$\leqq\int_0^1|\mathbb{M}f(\xi_x(t))-\mathbb{M}f(\xi_y(t))|\,\mathbb{P}\{\xi(s)\in dy\}\to 0$$

for $t\to\infty$ uniformly in s. This implies the existence of $\lim_{t\to\infty}\mathbb{M}f(\xi_x(t))$. That this limit is independent of x follows from the inequality

$$|\mathbb{M}f(\xi_x(t))-\mathbb{M}f(\xi_y(t))|\leqq 2\,\|f\|\,\mathbb{P}\{\tau_{0,1}>t\}. \quad \square$$

Corollary 1. *Set $F_t(x,y)=\mathbb{P}\{\xi_x(t)<y\}$. Then*

$$F(y)=\lim_{t\to\infty}F_t(x,y)$$

exists (in the sense of weak convergence of distributions).

We now find an explicit expression for $F(y)$.

Lemma 6. *If*

$$\Phi(x)=\exp\left\{\int_0^x \frac{2a(z)}{\sigma^2(z)}\,dz\right\},$$

then for any bounded measurable function $f(x)$ for which

$$\int_0^1 f(x)\frac{\Phi(x)}{\sigma^2(x)}\,dx=0$$

we have $\lim_{t\to\infty}\mathbb{M}f(\xi_x(t))=0$.

Proof. Set

$$G(x)=\int_0^x \frac{1}{\Phi(z)}\int_0^z \frac{2f(u)\Phi(u)}{\sigma^2(u)}\,du.$$

Then

$$G''(x)\frac{\sigma^2(x)}{2}+a(x)G'(x)=f(x) \quad\text{and}\quad G'(z)=\frac{1}{\Phi(z)}\int_0^z \frac{2f(u)\Phi(u)}{\sigma^2(u)}\,du,$$

so that $G'(0)=0$ and

$$G'(1)=\frac{2}{\Phi(1)}\int_0^1 f(u)\frac{\Phi(u)}{\sigma^2(u)}\,du=0.$$

Thus from Lemma 4

$$G(\xi_x(t))-G(x)=\int_0^t f(\xi_x(s))\,ds+\int_0^t G'(\xi_x(s))\,\sigma(\xi_x(s))\,dw(s)$$

and

$$\mathbb{M}G(\xi_x(t))-G(x)=\int_0^t \mathbb{M}f(\xi_x(s))\,ds,$$

so that

$$0=\lim_{t\to\infty}\frac{\mathbb{M}G(\xi_x(t))-G(x)}{t}=\lim_{t\to\infty}\frac{1}{t}\int_0^t \mathbb{M}f(\xi_x(s))\,ds=\lim_{t\to\infty}\mathbb{M}f(\xi_x(t)),$$

because the latter limit exists by Theorem 2. □

Corollary 2. *Let $f(x)$ be an arbitrary bounded measurable function. Then*

$$\lim_{t\to\infty}\mathbb{M}f(\xi_x(t))=c\int_0^1 \frac{f(y)\Phi(y)}{\sigma^2(y)}\,dy,$$

where

$$c=\left[\int_0^1 \frac{\Phi(y)}{\sigma^2(y)}\,dy\right]^{-1}.$$

This follows from the fact that

$$\bar{f}(x)=f(x)-c\int_0^1 \frac{f(y)\Phi(y)}{\sigma^2(y)}\,dy$$

fulfills the assumptions of Lemma 6. We have proved

Theorem 3. *If* $F_t(x, y) = \mathbb{P}\{\xi_x(t) < y\}$, *then for* $0 < y < 1$

$$\lim_{t\to\infty} F_t(x, y) = F(y),$$

where

$$F(y) = \left[\int_0^1 \frac{1}{\sigma^2(z)}\Phi(z)\,dz\right]^{-1} \int_0^y \frac{\Phi(z)}{\sigma^2(z)}\,dz. \tag{11}$$

We now prove an ergodic theorem for temporal means. Let $F(y)$ be defined at (11).

Theorem 4. *For any arbitrary measurable function* $f(y)$ *for which* $\int_0^1 |f(y)|\,dF(y) < \infty$,

$$\lim_{T\to\infty} \frac{1}{T}\int_0^T f(\xi(t))\,dt = \int_0^1 f(y)\,dF(y)$$

holds w.p. 1.

Proof. We assume initially that $f(y)$ is nonnegative and bounded. Let the sequence $0 < \tau_0 < \tau_1 < \cdots$ be defined as the times of first return to the point x after a preceding visit to the boundary, i.e., $\xi_x(\tau_k) = x$, $k = 0, 1, \ldots,$ between τ_k and τ_{k+1} the process attains the boundary and if τ_k^* is the first time the boundary is attained after time τ_k, then $\xi_x(t) \neq x$ for $t \in (\tau_k^*, \tau_{k+1})$. It is easy to see that the random variables

$$\int_0^{\tau_1} f(\xi_x(s))\,ds, \ldots, \quad \int_{\tau_k}^{\tau_{k+1}} f(\xi_x(s))\,ds, \ldots$$

are independent and identically distributed. From Lemma 5 we have

$$\mathbb{M}\int_0^{\tau_1} f(\xi_x(s))\,ds \leqq \|f\|(\mathbb{M}\tau_{x,0} + \mathbb{M}\tau_{0,x} + \mathbb{M}\tau_{x,1} + \mathbb{M}\tau_{1,x}) < \infty.$$

From the strong law of large numbers

$$\mathbb{P}\left\{\lim_{k\to\infty}\frac{1}{k}\sum_{i=0}^{k-1}\int_{\tau_i}^{\tau_{i+1}} f(\xi_x(s))\,ds = \mathbb{M}\int_0^{\tau_1} f(\xi_x(s))\,ds\right\} = 1.$$

Consequently,

$$\mathbb{P}\left\{\lim_{k\to\infty}\frac{1}{k-1}\int_0^{\tau_k} f(\xi_x(s))\,ds = \lim_{k\to\infty}\frac{1}{k+1}\int_0^{\tau_k} f(\xi_x(s))\,ds = \mathbb{M}\int_0^{\tau_1} f(\xi_x(s))\,ds\right\} = 1.$$

Taking $f(x) = 1$, we obtain

$$\mathbb{P}\left\{\lim_{k\to\infty}\frac{1}{k-1}\tau_k = \lim_{k\to\infty}\frac{1}{k+1}\tau_k = \mathbb{M}\tau_1\right\} = 1.$$

Let ν_T be an integer for which $\tau_{\nu_T} \leqq T < \tau_{\nu_T+1}$. Since τ_k is finite for all k, $\nu_T \to \infty$ for $T \to \infty$ and w.p. 1,

$$\lim_{T\to\infty} \frac{1}{\tau_{\nu_T+1}} \int_0^{\tau_{\nu_T}} f(\xi_x(s))\,ds = \lim_{T\to\infty} \frac{\frac{1}{\nu_T}\int_0^{\tau_{\nu_T}} f(\xi_x(s))\,ds}{\frac{1}{\nu_T}\tau_{\nu_T+1}} = \frac{\mathbb{M}\int_0^{\tau_1} f(\xi_x(s))\,ds}{\mathbb{M}\tau_1},$$

$$\lim_{T\to\infty} \frac{1}{\tau_{\nu_T}} \int_0^{\tau_{\nu_T+1}} f(\xi_x(s))\,ds = \frac{1}{\mathbb{M}\tau_1}\mathbb{M}\int_0^{\tau_1} f(\xi_x(s))\,ds.$$

Since

$$\frac{1}{\tau_{\nu_T+1}} \int_0^{\tau_{\nu_T}} f(\xi_x(s))\,ds \leqq \frac{1}{T}\int_0^T f(\xi_x(s))\,ds \leqq \frac{1}{\tau_{\nu_T}} \int_0^{\tau_{\nu_T+1}} f(\xi_x(s))\,ds$$

we have w.p. 1

$$\lim_{T\to\infty} \frac{1}{T}\int_0^T f(\xi_x(s))\,ds = \frac{1}{\mathbb{M}\tau_1}\mathbb{M}\int_0^{\tau_1} f(\xi_x(s))\,ds$$

for any bounded measurable nonnegative function f. Since

$$\left|\frac{1}{T}\int_0^T f(\xi_x(s))\,ds\right| \leqq \|f\|,$$

$$\lim_{T\to\infty} \mathbb{M}\frac{1}{T}\int_0^T f(\xi_x(s))\,ds = \mathbb{M}\lim_{T\to\infty}\frac{1}{T}\int_0^T f(\xi_x(s))\,ds,$$

but

$$\lim_{T\to\infty} \mathbb{M}\frac{1}{T}\int_0^T f(\xi_x(s))\,ds = \int_0^1 f(y)\,dF(y),$$

so that

$$\mathbb{P}\left\{\lim_{T\to\infty}\frac{1}{T}\int_0^T f(\xi_x(s))\,ds = \int_0^1 f(y)\,dF(y)\right\} = 1. \tag{12}$$

The theorem is proved for bounded, nonnegative measurable functions. The extension to all functions which fulfill the hypotheses of the theorem is carried through exactly as in Theorem 2 § 18. □

In a manner similar to that of Theorems 3 and 4 we can prove the following ergodic theorem for processes with instantaneous reflection at the boundary of $(0, \infty)$.

Theorem 5. *Let* $F_t(x, y) = \mathbb{P}\{\xi_x(t) < y\}$, *when* $\xi_x(t)$ *satisfies* (2) *with coefficients* $a(x)$ *and* $\sigma(x)$ *and initial condition* $\xi_x(0) = x$. *If*

$$\int_0^\infty \frac{1}{\sigma^2(z)}\Phi(z)\,dz < \infty,$$

then $\lim_{t\to\infty} F_t(x, y) = F(y)$, *where*

$$F(y)=\begin{cases} 0, & y<0;\\ \int_0^y \frac{1}{\sigma^2(z)}\Phi(z)\,dz\left(\int_0^\infty \frac{\Phi(z)}{\sigma^2(z)}\,dz\right)^{-1}, & y>0, \end{cases}$$

and for any measurable function $f(x)$ *for which* $\int_0^\infty f(x)\,dF(x)<\infty$ *we have w.p. 1*

$$\lim_{T\to\infty}\frac{1}{T}\int_0^T f(\xi_x(s))\,ds=\int_0^\infty f(y)\,dF(y).$$

§ 24. Delayed Reflection at the Boundary

Here we consider only homogeneous processes which suffer delayed reflection at the boundary, i.e., solutions of equations with time-independent coefficients on the interval $(0, \infty)$ with one boundary point. We will not prove uniqueness but will construct the solutions and study some of their properties; in particular, we will obtain equations which allow us to determine the transition probability for the solutions of the relevant stochastic equations.

Let the functions $a(x)$ and $\sigma(x)$ for $x>0$ be defined so as to satisfy the following properties: 1) There exists a K such that for $x>0$ and $y>0$ $|a(x)-a(y)|+|\sigma(x)-\sigma(y)|\leqq K|x-y|$; 2) $\sigma(x)>0$ for $x>0$, $\sigma(0)=0$; $a(0)>0$ (the functions $a(x)$ and $\sigma(x)$ can have jump discontinuities at zero).

Consider the solution of

$$d\xi(t)=a(\xi(t))\,dt+\sigma(\xi(t))\,dw(t), \tag{1}$$

satisfying $\xi(t)\geqq 0$. Such a solution on $(0, \infty)$ will be said to suffer delayed reflection at the boundary; $a(0)$ is the speed of exit from the latter.

In order to construct a solution of (1) we consider a process $\xi_1(t)$ satisfying

$$d\xi_1(t)=a(\xi_1(t))\,dt+\sigma(\xi_1(t))\,dw_1(t), \tag{2}$$

where $w_1(t)$ is some Wiener process on the interval $(0, \infty)$ which suffers instantaneous boundary reflection. Then (see § 23)

$$\xi_1(t)=\xi_1(0)+\int_0^t a(\xi_1(s))\,ds+\int_0^t \sigma(\xi_1(s))\,dw_1(s)+\zeta_1(t), \tag{3}$$

where $\zeta_1(t)$ is a nondecreasing continuous process whose points of growth are the zeros of the process $\xi_1(t)$. Define the family of random variables τ_t by

$$t=\tau_t+\frac{1}{a(0)}\zeta_1(\tau_t). \tag{4}$$

If $\mathfrak{F}_u^{(1)}$ denotes the minimum σ-algebra w.r.t. which the variables $\xi_1(0)$ and $w_1(s)$ for $s\leqq u$ are measurable ($\xi_1(0)$ is independent of the process $w_1(t)$), then τ_t will, for each $t>0$, be a Markov time w.r.t. the system of σ-algebras $\mathfrak{F}_u^{(1)}$. We consider a random time-substitution in (3) taking $\xi_1(\tau_t)=\eta(t)$. Let $\chi_+(x)=1$ for $x>0$, $\chi_+(x)=0$ for $x=0$. Then

$$\begin{aligned}\int_0^{\tau_t} a(\xi_1(s))\,ds&=\int_0^{\tau_t} a(\xi_1(s))\,\chi_+(\xi_1(s))\,ds+\frac{1}{a(0)}\int_0^{\tau_t} a(\xi_1(s))\,\chi_+(\xi_1(s))\,d\zeta_1(s)\\&=\int_0^t a(\xi_1(\tau_u))\,\chi_+(\xi_1(\tau_u))\left[d\tau_u+\frac{1}{a(0)}\,d\zeta_1(\tau_u)\right]\\&=\int_0^t a(\eta(u))\,\chi_+(\eta(u))\,du\end{aligned}$$

(we used here the fact that $\chi_+(\xi_1(s))\,d\zeta_1(s)=0$ and relation (4)). Moreover,

$$\begin{aligned}\zeta_1(\tau_t)&=\int_0^{\tau_t}[1-\chi_+(\xi_1(s))]\,d\zeta_1(s)\\&=\int_0^{\tau_t}[1-\chi_+(\xi_1(s))]\,d\zeta_1(s)+a(0)\int_0^{\tau_t}[1-\chi_+(\xi_1(s))]\,ds\\&=\int_0^t[1-\chi_+(\xi_1(\tau_s))]\,(d\zeta_1(\tau_s)+a(0)\,d\tau_s)\\&=\int_0^t a(0)\,[1-\chi_+(\xi_1(s))]\,ds=\int_0^t a(\eta(s))\,[1-\chi_+(\xi_1(s))]\,ds.\end{aligned} \tag{5}$$

From the chain of equalities (5) it follows in particular that

$$\int_0^t(1-\chi_+(\eta(u)))\,du=\frac{1}{a(0)}\zeta_1(\tau_t). \tag{6}$$

We now transform the stochastic integral

$$\int_0^{\tau_t}\sigma(\xi_1(s))\,dw_1(s)=\int_0^t\sigma(\xi_1(\tau_u))\,dw_1(\tau_u)=\int_0^t\sigma(\eta(u))\,dw_1(\tau_u).$$

Let $\bar{w}(t)$ be a Wiener process independent of $w_1(t)$ and $\xi_1(t)$. Set

$$w(t)=w_1(\tau_t)+\int_0^t(1-\chi_+(\eta(s)))\,d\bar{w}(s).$$

We will show that $w(t)$ is also a Wiener process. Denote by $\mathfrak{F}_t$ the minimal σ-algebra w.r.t. which $\xi_1(s)$, $w_1(s)$ for $s\leqq\tau_t$ and $\bar{w}(s)$ for $s\leqq t$

are measurable. Then for $h>0$ we have from Remark 2 §4

$$\mathbb{M}(w(t+h)-w(t)/\mathfrak{F}_t)=0,$$

$$\begin{aligned}\mathbb{M}([w(t+h)-w(t)]^2/\mathfrak{F}_t)&=\mathbb{M}([w_1(\tau_{t+h})-w_1(\tau_t)]^2/\mathfrak{F}_t)\\&\quad+\mathbb{M}\left(\int_t^{t+h}(1-\chi_+(\eta(s)))\,ds/\mathfrak{F}_t\right)\\&=\mathbb{M}\left[\tau_{t+h}-\tau_t+\frac{1}{a(0)}(\zeta_1(\tau_{t+h})-\zeta_1(\tau_t))/\mathfrak{F}_t\right]=h\end{aligned}$$

(here we used Eqs. (4) and (6)). Clearly, $w(t)$ is continuous w.p.1 so that from Theorem 1 §1 it is Wiener. Since

$$\int_0^t \sigma(\eta(u))(1-\chi_+(\eta(u)))\,d\bar{w}(u)=0,$$

because $\sigma(x)(1-\chi_+(x))=0$, we obtain

$$\int_0^t \sigma(\eta(u))\,dw_1(\tau_u)=\int_0^t \sigma(\eta(u))\,dw(u).$$

Thus, after the aforementioned time-substitution, we get

$$\begin{aligned}\eta(t)=\xi_1(\tau_t)&=\xi_1(0)+\int_0^t a(\eta(u))\,\chi_+(\eta(u))\,du\\&\quad+\int_0^t\sigma(\eta(u))\,dw(u)+\int_0^t a(\eta(u))(1-\chi_+(\eta(u)))\,du\\&=\eta(0)+\int_0^t a(\eta(u))\,du+\int_0^t\sigma(\eta(u))\,dw(u).\end{aligned}$$

The process $\eta(t)$ is nonnegative, continuous w.p.1 and satisfies (1). We have established the existence of a solution to (1).

Now let $\xi(t)$ be some solution of (1) satisfying the initial condition $\xi(t)|_{t=0}=\xi(0)$, continuous w.p.1 and measurable for all $t>0$ w.r.t. $\mathfrak{F}_t$, the σ-algebra generated by $\xi(0)$ and $w(s)$, $s\leqq t$. We make the time-substitution $\xi(\tau_t)=\xi_1(t)$ in (1), whereby τ_t is defined by

$$t=\int_0^{\tau_t}\chi_+(\xi(s))\,ds.$$

The function τ_t is continuous since $\lambda(u)=\int_0^u \chi_+(\xi(s))\,ds$ is continuous and strictly monotone ($\xi(s)$ cannot vanish on the entire interval). We transform each of the integrals on the right side of the integrated version of (1) after the above time-substitution as follows:

$$\int_0^{\tau_t} a(\xi(s))\,ds = \int_0^t a(\xi(\tau_s))\,d\tau_s$$

$$= \int_0^t a(\xi(\tau_s))\,\chi_+(\xi_1(s))\,ds + a(0)\int_0^t (1-\chi_+(\xi_1(s)))\,d\tau_s.$$

Since

$$t = \int_0^{\tau_t} \chi_+(\xi(s))\,ds = \int_0^t \chi_+(\xi_1(s))\,d\tau_s,$$

$dt = \chi_+(\xi_1(s))\,d\tau_s$ so that

$$\int_0^{\tau_t} a(\xi(s))\,ds = \int_0^t a(\xi_1(s))\,ds + \zeta_1(t),$$

where

$$\zeta_1(t) = a(0)\int_0^t [1-\chi_+(\xi_1(s))]\,d\tau_s.$$

Clearly, the points of growth of $\zeta_1(t)$ can only be zeros; $\xi_1(t)$ and $\zeta_1(t)$ are nondecreasing continuous functions. We now show that the time spent by $\xi_1(t)$ at the point zero is equal to zero, i.e.,

$$\int_0^t (1-\chi_+(\xi_1(s)))\,ds = t - \int_0^t \chi_+(\xi_1(s))\,ds = t - \int_0^t \chi_+(\xi_1(s))\,ds$$

$$= t - \int_0^t \chi_+(\xi_1(\tau_s))\,d\tau_s = t - \int_0^{\tau_t} \chi_+(\xi(s))\,ds = 0$$

(here we used $ds = \chi_+(\xi_1(s))\,d\tau_s$). Furthermore,

$$\int_0^{\tau_t} \sigma(\xi(s))\,dw(s) = \int_0^t \sigma(\xi_1(s))\,dw(\tau_s) = \int_0^t \sigma(\xi_1(s))\,\chi_+(\xi_1(s))\,dw(\tau_s).$$

We prove that

$$w_1(t) = \int_0^t \chi_+(\xi_1(s))\,dw(\tau_s) = \int_0^{\tau_t} \chi_+(\xi(s))\,dw(s)$$

is a Wiener process. This follows from Theorem 1 § 1 since $w_1(t)$ is continuous w.p.1 and for $\mathfrak{F}_t^{(1)} = \mathfrak{F}_{\tau_t}$

$$\mathbb{M}(w_1(t+h) - w_1(t)/\mathfrak{F}_t^{(1)}) = 0,$$

$$\mathbb{M}([w_1(t+h) - w_1(t)]^2/\mathfrak{F}_t^{(1)}) = \mathbb{M}\left(\int_{\tau_t}^{\tau_{t+h}} \chi_+(\xi(s))\,ds/\mathfrak{F}_{\tau_t}\right) = h.$$

This means that $\xi_1(t)$ satisfies (3) with $w_1(t)$ and $\zeta_1(t)$ as defined, i.e., $\xi_1(t)$ is a process with instantaneous reflection at the boundary.

Is it possible to again obtain $\xi(t)$ from $\xi_1(t)$ with the help of a random time-substitution? Clearly $\xi(t)=\xi_1(\varphi_t)$, where φ_t is a solution of $\tau_{\varphi_t}=t$. According to the definition of τ_s, this means that

$$\varphi_t=\int_0^t \chi_+(\xi(s))\,ds=t-\int_0^t(1-\chi_+(\xi(s)))\,ds=t-\int_0^{\varphi_t}[1-\chi_+(\xi(\tau_s))]\,d\tau_s$$

$$=t-\int_0^{\varphi_t}[1-\chi_+(\xi_1(s))]\,d\tau_s=t-\frac{1}{a(0)}\zeta_1(\varphi_t)$$

(we used the form of the process $\zeta_1(t)$). Hence, $\xi(t)=\xi_1(\varphi_t)$, where φ_t satisfies $t=\varphi_t+\frac{1}{a(0)}\zeta_1(\varphi_t)$. These considerations yield the proof of

Theorem 1. *The formulas* $\xi_1(t)=\xi(\tau_t)$, $\xi(t)=\xi_1(\varphi_t)$,

$$w_1(t)=\int_0^{\tau_t}\chi_+(\xi(s))\,dw(s),\qquad w(t)=w_1(\varphi_t)+\int_0^t[1-\chi_+(\xi(s))]\,d\bar{w}(s),$$

and

$$\zeta_1(t)=a(0)\int_0^t(1-\chi_+(\xi_1(s)))\,d\tau_s,$$

where $\bar{w}(t)$ *is independent of* $w_1(t)$ *and a Wiener process, and* τ_t *and* φ_t *are defined by*

$$t=\int_0^{\tau_t}\chi_+(\xi(s))\,ds,\qquad t=\varphi_s+\frac{1}{a(0)}\zeta_1(\varphi_t),$$

establish a one-to-one relationship between the solutions of (1) *and* (3).

Corollary 1. *The finite-dimensional distributions of the process* $\xi(t)$ *satisfying* (1) *are uniquely determined by the distribution of* $\xi(0)$ *and the coefficients* $a(x)$ *and* $\sigma(x)$.

Lemma 1. *If the function* $\Phi(t,x)$ *for* $t>0$ *and* $x>0$ *satisfies*

$$\frac{\partial}{\partial t}\Phi(t,x)=a(x)\frac{\partial}{\partial x}\Phi(t,x)+\frac{1}{2}\sigma^2(x)\frac{\partial^2}{\partial x^2}\Phi(t,x) \tag{7}$$

with boundary condition

$$\lim_{x\downarrow 0} a(0)\,\Phi'_x(t,x)=\Phi'_t(t,0), \tag{8}$$

and if the function $\sigma(x)\,\Phi'_x(t,x)$ *is bounded, then for* $t_1<t_2$

$$\mathbb{M}(\Phi(t_1,\xi(t_2))/\mathfrak{F}_{t_1})=\Phi(t_2,\xi(t_1)).$$

Proof. From Itô's formula

$$\begin{aligned} d\Phi(t_2-s,\xi(t_1+s)) = [&-\Phi'_t(t_2-s,\xi(t_1+s)) \\ &+\Phi'_x(t_2-s,\xi(t_1+s))\,a(\xi(t_1+s)) \\ &+\tfrac{1}{2}\sigma^2(\xi(t_1+s))\,\Phi''_{xx}(t_2-s,\xi(t_1+s))]\,ds \\ &+\Phi'_x(t_2-s,\xi(t_1+s))\,\sigma(\xi(t_1+s))\,dw(t_1+s). \end{aligned}$$

The expression in square brackets is zero because of (7) and (8), so that

$$\Phi(t_1,\xi(t_2))-\Phi(t_2,\xi(t_1)) = \int_0^{t_2-t_1} \Phi'_x(t_2-s,\xi(t_1+s))\,\sigma(\xi(t_1+s))\,dw(t_1+s).$$

The proof of the lemma follows from this relation and Remark 1 § 2.

Remark 1. In order to solve (7) with boundary condition (8) it is convenient to take Laplace transforms w.r.t. t. Let

$$u_\lambda(x) = \int_0^\infty e^{-\lambda t}\,\Phi(t,x)\,dt.$$

Then, applying the Laplace transform to (7), we obtain

$$\lambda u_\lambda(x) - \Phi(0,x) = a(x)\frac{d}{dx}u_\lambda(x) + \frac{\sigma^2(x)}{2}\cdot\frac{d^2}{dx^2}u_\lambda(x). \tag{9}$$

The boundary condition for $u_\lambda(x)$ is obtained by applying the Laplace transform to (8):

$$\lim_{x\downarrow 0} a(0)\frac{d}{dx}u_\lambda(x) = \lambda u_\lambda(0) - \Phi(0,0). \tag{10}$$

It turns out that we can find the Laplace transform for the mathematical expectation of a function of the trajectory of a process directly, without assuming the existence of a solution to (7) with boundary condition (8).

Lemma 2. *Let $\varphi(x)$ be a bounded, continuous function and $u_\lambda(x)$ a bounded solution of*

$$\lambda u_\lambda(x) - \varphi(x) = a(x)\frac{d}{dx}u_\lambda(x) + \frac{\sigma^2(x)}{2}\cdot\frac{d^2}{dx^2}u_\lambda(x) \tag{11}$$

for $x>0$ satisfying the boundary condition

$$a(0)\frac{d}{dx}u_\lambda(+0) - \lambda u_\lambda(0) = -\varphi(0) \tag{12}$$

and let $u'_\lambda(x)\,\sigma(x)$ be bounded. Then for all $t>0$

$$u_\lambda(\xi(t)) = e^{\lambda t}\,\mathbb{M}\left(\int_t^\infty e^{-\lambda s}\,\varphi(\xi(s))\,ds/\mathfrak{F}_t\right). \tag{13}$$

Proof. Applying Itô's formula to $u_\lambda(\xi(t))\,e^{-\lambda t}$, we obtain

$$d\left[u_\lambda(\xi(t))\,e^{-\lambda t}\right] = \left[-\lambda\,u_\lambda(\xi(t)) + \frac{d}{dx}\,u_\lambda(\xi(t))\,a(\xi(t))\right.$$

$$\left. + \frac{1}{2}\,\sigma^2(\xi(t))\,\frac{d^2}{dx^2}\,u_\lambda(\xi(t))\right] e^{-\lambda t} + u'_\lambda(\xi(t))\,\sigma(\xi(t))\,e^{-\lambda t}\,dw(t).$$

The expression in square brackets on the right side of the last equation is equal to $-\varphi(\xi(t))$ because of (11) and (12). Thus for $t<T$

$$e^{-\lambda T}\,u_\lambda(\xi(T)) - e^{-\lambda t}\,u_\lambda(\xi(t)) = -\int_t^T e^{-\lambda s}\,\varphi(\xi(s))\,ds$$

$$+\int_t^T \sigma(\xi(s))\,u'_\lambda(\xi(s))\,e^{-\lambda s}\,dw(s).$$

Letting $T\to\infty$ we obtain

$$u_\lambda(\xi(t))\,e^{-\lambda t} = \int_t^\infty e^{-\lambda s}\,\varphi(\xi(s))\,ds - \int_t^\infty \sigma(\xi(s))\,u'_\lambda(\xi(s))\,e^{-\lambda s}\,dw(s), \tag{14}$$

since $u_\lambda(\xi(T))\,e^{-\lambda T}\to 0$ because of the boundedness of $u_\lambda(x)$; the existence of the integral on the right side of (14) follows from the boundedness of $u'_\lambda(x)\,\sigma(x)$ and $\varphi(x)$. Taking expectations on both sides of (14) we obtain the proof of the lemma.

Corollary 2. *Assume that for $a(x)$ and $\sigma(x)$ there exists a class K of bounded continuous functions $\varphi(x)$ such that* 1) *for $\varphi\in K$ and all λ with* $\operatorname{Re}\lambda>0$ *there exists a bounded solution of* (11) *satisfying boundary condition* (12) *for which $u'_\lambda(x)\,\sigma(x)$ is bounded;* 2) *for any bounded continuous function $g(x)$ we can find a sequence of functions $\varphi_n(x)$ in K which converges everywhere to $g(x)$. Then any solution of* (1) *will be a homogeneous Markov process; if $P(t,x,dy)$ is the transition probability of this process, then for any continuous bounded function $g(x)$*

$$\int_0^\infty e^{-\lambda t}\int g(y)\,P(t,x,dy)\,dt = \lim_{n\to\infty} u_\lambda^{(n)}(x), \tag{15}$$

where $u_\lambda^{(n)}(x)$ satisfies (11) *with boundary condition* (12) *if we replace φ in them by $\varphi_n(x)$ and if $\varphi_n(x)\to g$ for $n\to\infty$.*

In fact, it follows from (13) that for all $\varphi\in K$ with $t'>t$, $\mathbb{M}(\varphi(\xi(t'))/\mathfrak{F}_t) = \psi(\xi(t))$, where ψ is some function (it is necessary to use the inversion formula for the Laplace transform). Hence,

$$\mathbb{M}(\varphi(\xi(t'))/\mathfrak{F}_t) = \mathbb{M}[\varphi(\xi(t'))/\xi(t)],$$

i.e., the process $\xi(t)$ is Markov. The homogeneity follows from (13) since

$$u_\lambda(\xi(t)) = \mathbb{M}\left(\int_0^\infty e^{-\lambda s}\varphi(\xi(t+s))\,ds/\mathfrak{F}_t\right) = \int_0^\infty e^{-\lambda s}\mathbb{M}[\varphi(\xi(t+s))/\xi(t)]\,ds$$

and

$$u_\lambda(\xi(0)) = \int_0^\infty e^{-\lambda s}\mathbb{M}[\varphi(\xi(s))/\xi(0)]\,ds,$$

so that the function $\Phi(t, t+s, x)$, for which

$$\Phi(t, t+s, \xi(t)) = \mathbb{M}[\varphi(\xi(t+s))/\xi(t)],$$

is independent of t. The formula

$$\int_0^\infty e^{-\lambda t}\int\varphi(y)\,P(t,x,dy)\,dt = u_\lambda(x) \tag{16}$$

follows from (13) for $\varphi\in K$ and (15) is obtained from (16) by means of a limit passage.

We turn now to ergodic theorems for processes with delayed reflection at the boundary.

Lemma 3. *Let* $\Phi(x) = \exp\left\{\int_0^x \frac{2a(u)}{\sigma^2(u)}\,du\right\}$ *and assume* $\int_0^\infty \frac{\Phi(z)}{\sigma^2(z)}\,dz < \infty$. *Then for any bounded continuous function $f(x)$ for which*

$$\frac{\sigma(x)}{\Phi(x)}\int_x^\infty f(z)\frac{\Phi(z)}{\sigma^2(z)}\,dz$$

is bounded and the integral $\int_0^\infty \frac{\Phi(z)f(z)}{\sigma^2(z)}\int_0^z \frac{dx}{\Phi(x)}\,dz$ *converges, we have*

$$\lim_{T\to\infty}\frac{1}{T}\int_0^T \mathbb{M}f(\xi(t))\,dt = \frac{f(0)+2a(0)\int_0^\infty f(z)\frac{\Phi(z)}{\sigma^2(z)}\,dz}{1+2a(0)\int_0^\infty \frac{\Phi(z)}{\sigma^2(z)}\,dz}. \tag{17}$$

Proof. First let f be such that

$$f(0)+2a(0)\int_0^\infty f(z)\frac{\Phi(z)}{\sigma^2(z)}\,dz = 0. \tag{18}$$

We will show that in this case

$$\lim_{T\to\infty}\frac{1}{T}\int_0^T \mathbb{M}f(\xi(t))\,dt = 0.$$

Let $V(x)$ satisfy

$$a(x)\,V'(x)+\tfrac{1}{2}\sigma^2(x)\,V''(x)=f(x)$$

for $x>0$ with boundary condition $a(0)\,V'(0)=f(0)$. Then

$$V'(x)=\left[\frac{f(0)}{a(0)}+\int_0^x f(z)\frac{2\Phi(z)}{\sigma^2(z)}\,dz\right](\Phi(x))^{-1}=-\frac{2}{\Phi(x)}\int_x^\infty f(z)\frac{\Phi(z)}{\sigma^2(z)}\,dz.$$

From the hypotheses follows the boundedness of $V(x)$ and $\sigma(x)\,V'(x)$. Applying Itô's formula to $V(\xi(t))$ we get

$$V(\xi(T))-V(\xi(0))=\int_0^T f(\xi(t))\,dt+\int_0^T \sigma(\xi(t))\,V'(\xi(t))\,dw(t),$$

thus

$$\frac{1}{T}\int_0^T \mathbb{M} f(\xi(t))\,dt=\frac{\mathbb{M}V(\xi(T))-\mathbb{M}V(\xi(0))}{T}\to 0$$

for $T\to\infty$.

Let f be an arbitrary function satisfying the hypotheses of the lemma. Set $\bar{f}(x)=f(x)-c$, where

$$c=\frac{f(0)+2a(0)\int_0^\infty f(z)\frac{\Phi(z)}{\sigma^2(z)}\,dz}{1+2a(0)\int_0^\infty \frac{\Phi(z)}{\sigma^2(z)}\,dz}.$$

Since

$$\frac{1}{T}\int_0^T \mathbb{M} f(\xi(t))\,dt=\frac{1}{T}\int_0^T \mathbb{M}\bar{f}(\xi(t))\,dt+c,$$

and

$$\lim_{T\to\infty}\frac{1}{T}\int_0^T \mathbb{M}\bar{f}(\xi(t))\,dt=0,$$

the lemma is proved.

Corollary 3. *Let $F_t(y)$ be the distribution function of $\xi(t)$. Then*

$$\lim_{T\to\infty}\frac{1}{T}\int_0^T F_t(y)\,dt=F(y),$$

where

$$F(y)=\begin{cases}0, & y\leqq 0;\\[6pt] \dfrac{1+2a(0)\int_0^y \frac{\Phi(z)}{\sigma^2(z)}\,dz}{1+2a(0)\int_0^\infty \frac{\Phi(z)}{\sigma^2(z)}\,dz}, & y>0.\end{cases}$$

Remark 2. $F(y)$ has a jump at zero equal to

$$\frac{1}{1+2a(0)\int_0^\infty \frac{\Phi(z)}{\sigma^2(z)}\,dz},$$

the probabilistic meaning of which is $\lim_{T\to\infty}\frac{1}{T}\int_0^T \mathbb{P}\{\xi(t)=0\}\,dt$.

Lemma 4. *Let $\xi(t)$ satisfy (1) and set $f(x)=\int_0^x \frac{dz}{\Phi(z)}$, where $\Phi(z)$ is the same as in Lemma 3. Then the process $\bar\xi(t)=f(\xi(t))$ satisfies*

$$d\bar\xi(t)=\chi_0(\bar\xi(t))\,a(0)\,dt+\bar\sigma(\bar\xi(t))\,dw(t), \tag{19}$$

where $\chi_0(x)=0$ for $x>0$; $\chi_0(x)=1$ for $x=0$ and $\bar\sigma(x)=\dfrac{\sigma(\varphi(x))}{\Phi(\varphi(x))}$. ($\varphi(x)$ is the inverse of $f(x)$.)

The proof of this lemma follows from Itô's formula.

Remark 3. If for any continuous bounded function $g(x)$ the limit $\lim_{T\to\infty}\frac{1}{T}\int_0^T g(\bar\xi(t))\,dt$ exists w.p. 1, then

$$\lim_{T\to\infty}\frac{1}{T}\int_0^T g(\xi(t))\,dt=\lim_{T\to\infty}\frac{1}{T}\int_0^T g(\varphi(\bar\xi(t)))\,dt,$$

will also exist w.p. 1.

Analogous statements also hold for the variables

$$\lim_{T\to\infty}\frac{1}{T}\int_0^T \mathbb{M}\,g(\xi(t))\,dt,\qquad \lim_{T\to\infty}\frac{1}{T}\int_0^T \mathbb{M}\,g(\bar\xi(t))\,dt,$$

and so in the sequel we will consider merely solutions of (19).

Corollary 4. *If $\bar\xi(t)$ satisfies (19) and $\int_0^\infty \frac{dz}{\bar\sigma^2(z)}<\infty$, then for any bounded continuous function $g(x)$ for which $\bar\sigma(x)\int_x^\infty \frac{g(z)}{\bar\sigma^2(z)}\,dz$ is bounded and the integral $\int_0^\infty \frac{z\,g(z)}{\bar\sigma^2(z)}\,dz$ converges we have*

$$\lim_{T\to\infty}\frac{1}{T}\int_0^T \mathbb{M}\,g(\bar\xi(t))\,dt=\frac{g(0)+2a(0)\int_0^\infty \frac{g(z)}{\bar\sigma^2(z)}\,dz}{1+2a(0)\int_0^\infty \frac{dz}{\bar\sigma^2(z)}}.$$

Lemma 5. *If* $\int_0^\infty \frac{dz}{\bar\sigma^2(z)} < \infty$, *then the limit*

$$\lim_{T\to\infty} \frac{1}{T} \int_0^T [1-\chi_+(\bar\xi(s))]\, ds, \tag{20}$$

exists with probability one.

Proof. We make the random time substitution $\bar\xi(\tau_t) = \bar\xi_1(t)$, where

$$t = \int_0^{\tau_t} \chi_+(\bar\xi(s))\, ds.$$

Then $\bar\xi_1(t)$ satisfies

$$\bar\xi_1(t) = \bar\xi_1(0) + \int_0^t \bar\sigma(\bar\xi_1(u))\, dw_1(u) + \bar\zeta_1(t),$$

where

$$w_1(t) = \int_0^{\tau_t} \chi_+(\bar\xi(s))\, dw(s) \quad \text{and} \quad \bar\zeta_1(t) = a(0) \int_0^{\tau_t} [1-\chi_+(\xi(s))]\, ds.$$

Hence, to prove the existence of the limit at (20) it is sufficient to show that the finite or infinite limit $\lim_{t\to\infty} \frac{\bar\zeta_1(t)}{t}$ exists since

$$\lim_{T\to\infty} \frac{1}{T} \int_0^T (1-\chi_+(\bar\xi(s)))\, ds = \lim_{T\to\infty} \frac{1}{\tau_T} \int_0^{\tau_T} (1-\chi_+(\bar\xi(s)))\, ds$$

$$= \frac{1}{a(0)} \lim_{T\to\infty} \frac{\bar\zeta_1(T)}{\tau_T} = \frac{1}{a(0)} \lim_{T\to\infty} \frac{\bar\zeta_1(T)}{T + \frac{1}{a(0)} \bar\zeta_1(T)} = \lim_{T\to\infty} \frac{\frac{\bar\zeta_1(T)}{T}}{a(0) + \frac{\bar\zeta_1(T)}{T}}.$$

Let the sequence $\nu_1, \nu_2, \ldots$ be the times some point $x>0$ is reached, in each case after a previous visit to the boundary, i.e., $\bar\xi_1(\nu_k) = x$, $\min_{\nu_k < t < \nu_{k+1}} \bar\xi_1(t) = 0$ and if $\nu_k < s' < \nu_{k+1}$ and $\bar\xi_1(s') = 0$, then $\bar\xi_1(u) < x$ for $u \in (s', \nu_{k+1})$. The ν_k are Markov times, $\nu_2 - \nu_1$, $\nu_3 - \nu_2, \ldots$, as well as $\bar\zeta_1(\nu_2) - \bar\zeta_1(\nu_1)$, $\bar\zeta_1(\nu_3) - \bar\zeta_1(\nu_2), \ldots$, are independent, identically distributed variables and $\mathbb{M}(\nu_2 - \nu_1) < \infty$, which follows from Lemma 1 § 18. Thus w.p. 1 there exists

$$\lim_{n\to\infty} \frac{\bar\zeta_1(\nu_n)}{\nu_n} = \lim_{n\to\infty} \frac{\bar\zeta_1(\nu_n)}{\nu_{n+1}} = \lim_{n\to\infty} \frac{\bar\zeta_1(\nu_{n+1})}{\nu_n} = \lim_{n\to\infty} \frac{\sum_{k=1}^n (\bar\zeta_1(\nu_{k+1}) - \bar\zeta_1(\nu_k))}{\sum_{k=1}^n (\nu_{k+1} - \nu_k)},$$

which is equal to $\dfrac{\mathbb{M}(\bar{\zeta}_1(v_2)-\bar{\zeta}_1(v_1))}{\mathbb{M}(v_2-v_1)}$. (This expression can take the value $+\infty$.) When $v_n \leqq T \leqq v_{n+1}$, the inequality

$$\frac{\bar{\zeta}_1(v_n)}{v_{n+1}} \leqq \frac{\bar{\zeta}_1(T)}{T} \leqq \frac{\bar{\zeta}_1(v_{n+1})}{v_n}$$

holds and this completes the proof of the lemma.

We can now establish an ergodic theorem for temporal means.

Theorem 2. *If the conditions of Lemma 5 are fulfilled, then for any measurable function* $g(x)$ *for which* $\int_0^\infty |g(z)| \frac{1}{\bar{\sigma}^2(z)} dz < \infty$, *we have w.p.1*

$$\lim_{T\to\infty} \frac{1}{T} \int_0^T g(\bar{\xi}(t))\,dt = \frac{g(0)+2a(0)\int_0^\infty g(z)\frac{1}{\bar{\sigma}^2(z)}}{1+2a(0)\int_0^\infty \frac{dz}{\bar{\sigma}^2(z)}}. \tag{21}$$

Proof. We first prove the existence of the limit on the left side of (21). Since

$$\frac{1}{T}\int_0^T g(\bar{\xi}(t))\,dt = g(0)\frac{1}{T}\int_0^T (1-\chi_+(\bar{\xi}(t)))\,dt + \frac{1}{T}\int_0^T g(\bar{\xi}(t))\,\chi_+(\bar{\xi}(t))\,dt$$

and the existence of the limit for the first integral follows from Lemma 5, it is sufficient to show that

$$\lim_{T\to\infty} \frac{1}{T}\int_0^T g(\bar{\xi}(t))\,\chi_+(\bar{\xi}(t))\,dt.$$

Using the notation of Lemma 5 we can write

$$\begin{aligned}\frac{1}{\tau_T}\int_0^{\tau_T} g(\bar{\xi}(t))\,\chi_+(\bar{\xi}(t))\,dt &= \frac{1}{\tau_T}\int_0^T g(\bar{\xi}(\tau_s))\,\chi_+(\bar{\xi}(\tau_s))\,d\tau_s \\ &= \frac{1}{\tau_T}\int_0^T g(\bar{\xi}_1(s))\,ds = \frac{T}{\tau_T}\cdot\frac{1}{T}\int_0^T g(\bar{\xi}_1(s))\,ds.\end{aligned}$$

The existence of

$$\lim_{T\to\infty}\frac{1}{T}\int_0^T g(\bar{\xi}_1(s))\,ds = \int_0^\infty g(z)\frac{dz}{\bar{\sigma}^2(z)}\left(\int_0^\infty \frac{dz}{\bar{\sigma}^2(z)}\right)^{-1}$$

follows from Theorem 5 § 23, and

$$\frac{T}{\tau_T} = 1 - \frac{1}{\tau_T}\int_0^{\tau_T} (1-\chi_+(\bar{\xi}(s)))\,ds,$$

so that $\lim_{T\to\infty} \frac{T}{\tau_T}$ exists because of Lemma 5. The existence of

$$\lim_{T\to\infty} \frac{1}{T}\int_0^T g(\bar{\xi}(t))\,dt$$

now follows from the fact that $\tau_T \to \infty$ for $T\to\infty$. For bounded functions fulfilling the conditions of Corollary 4 it is obvious that

$$\lim_{T\to\infty} \frac{1}{T}\int_0^T g(\bar{\xi}(s))\,ds = \lim_{T\to\infty} \frac{1}{T}\int_0^T \mathbf{M}\, g(\bar{\xi}(s))\,ds = \frac{g(0)+2a(0)\int_0^\infty g(z)\frac{dz}{\bar{\sigma}^2(z)}}{1+2a(0)\int_0^\infty \frac{dz}{\bar{\sigma}^2(z)}}.$$

Since we have established that

$$\lim_{T\to\infty} \frac{1}{T}\int_0^T g(\bar{\xi}(t))dt = g(0)\lim_{T\to\infty} \frac{1}{T}\int_0^T (1-\chi_+(\bar{\xi}(t)))dt + \frac{\int_0^\infty \frac{g(z)dz}{\bar{\sigma}^2(z)} \lim_{T\to\infty}\frac{\tau_T}{T}}{\int_0^\infty \frac{dz}{\bar{\sigma}^2(z)}}, \tag{22}$$

we have

$$\lim_{T\to\infty} \frac{1}{T}\int_0^T (1-\chi_+(\bar{\xi}(t)))\,dt = \left(1+2a(0)\int_0^\infty \frac{dz}{\bar{\sigma}^2(z)}\right)^{-1}, \tag{23}$$

and

$$\lim_{T\to\infty} \frac{T}{\tau_T} = 2a(0)\int_0^\infty \frac{dz}{\bar{\sigma}^2(z)}\left(1+2a(0)\int_0^\infty \frac{dz}{\bar{\sigma}^2(z)}\right)^{-1}. \tag{24}$$

The proof now follows from (24), (23) and (22).

§ 25. Processes with Jump Reflection at the Boundary

We consider a process $\xi(t)$ which satisfies in $(0, \infty)$ the equation

$$d\xi(t) = a(\xi(t))\,dt + \sigma(\xi(t))\,dw(t) \tag{1}$$

along with the following condition: the set of t's for which $\xi(t)=0$ is at most countable and if $\tau_1, \tau_2, \ldots$ are the points at which $\xi(\tau_k)=0$, then $\xi(\tau_k+0)=\eta_k$, $k=1, \ldots$, are independent, identically distributed random variables independent of $w(t)$. If $a(x)$ and $\sigma(x)$ satisfy a Lipschitz condition for $x>0$, one can construct the described solution in the following way. Let $\xi_{x,s}(t)$ satisfy

$$d\xi_{x,s}(t) = a(\xi_{x,s}(t))\,dt + \sigma(\xi_{x,s}(t))\,dw_s(t)$$

with initial condition $\xi_{x,s}(0)=x$, where $w_s(t)=w(s+t)-w(s)$, and $\eta_1, \eta_2, \ldots$ is a sequence of independent, identically distributed random variables independent of $w(t)$. Let $\tau_{x,s}$ be the time the process first reaches the boundary of the interval $(0, \infty)$. Set

$$\xi(t)=\xi_{\xi(0),0}(t) \qquad \text{for } 0 \leqq t \leqq \tau_{\xi(0),0}=\tau_1, \quad \text{and}$$
$$\xi(t)=\xi_{\eta_k,\tau_k}(t-\tau_k) \qquad \text{for } \tau_k < t \leqq \tau_{\eta_k,\tau_k}+\tau_k=\tau_{k+1}.$$

Since uniqueness up to exit from $(0, \infty)$ is sufficient for uniqueness of a solution of (1), the listed solution is unique. It is rather easy to convince oneself that this solution will be homogeneous Markov process and that its transition probability $P(t, x, A)$ coincides with $\mathbb{P}\{\xi_x(t) \in A\}$, where $\xi_x(t)$ is the indicated solution of (1) with initial condition $\xi(0)=x$. To determine $P(t, x, A)$ we consider

$$u_\lambda(x)=\mathbb{M}\int_0^\infty f(\xi_x(t))\, e^{-\lambda t}\, dt=\int_0^\infty e^{-\lambda t}\int f(y)\, P(t, x, dy)\, dt.$$

The following modified version of Itô's formula for processes of the type under consideration will be necessary.

Lemma 1. *If $\xi(t)$ satisfies (1) with jump reflection at the boundary and $f(t, x)$ is a continuous function for which $f_t'(t, x)$, $f_x'(t, x)$ and $f_{xx}''(t, x)$ are continuous, then*

$$\begin{aligned} &f(T, \xi(T))-f(0, \xi(0)) \\ &\quad=\int_0^T \left[f_t'(t, \xi(t))+a(\xi(t))\, f_x'(t, \xi(t))+\tfrac{1}{2}\sigma^2(\xi(t))\, f_{xx}''(t, \xi(t))\right] dt \qquad (2) \\ &\qquad+\int_0^T f_x'(t, \xi(t))\, \sigma(\xi(t))\, dw(t)+\sum_{\tau_k<T}\left[f(\tau_k, \eta_k)-f(\tau_k, 0)\right]. \end{aligned}$$

The proof of this lemma follows from the fact that Itô's formula is applicable on the intervals (τ_k, τ_{k+1}) so that for $\tau_k < T \leqq \tau_{k+1}$

$$\begin{aligned} f(T, \xi(T))=&f(\tau_k, \xi(\tau_k+0)) \\ &+\int_{\tau_k}^T \left[f_t'(t, \xi(t))+a(\xi(t))\, f_x'(t, \xi(t))+\tfrac{1}{2}\sigma^2(\xi(t))\, f_{xx}''(t, \xi(t))\right] dt \\ &+\int_{\tau_k}^T f_x'(t, \xi(t))\, \sigma(\xi(t))\, dw(t), \end{aligned}$$

and

$$f(\tau_k, \xi(\tau_k+0))-f(\tau_k, \xi(\tau_k))=f(\tau_k, \eta_k)-f(\tau_k, 0).$$

Lemma 2. *Let $f(x)$ be an arbitrary, bounded, twice continuously differentiable function for which the expression*

$$g(x) = -\lambda f(x) + a(x) f'(x) + \tfrac{1}{2}\sigma^2(x) f''(x)$$

is bounded. Then

$$-f(x) = \sum_k \mathbb{M}[f(\eta_k) - f(0)]\, e^{-\lambda \tau_k} + \int_0^\infty e^{-\lambda t} \mathbb{M} g(\xi_x(t))\, dt. \tag{3}$$

Proof. Applying formula (2) to the function $f(t,x) = e^{-\lambda t} f(x)$, we obtain

$$\begin{aligned} f(\xi_x(T))\, e^{-\lambda T} - f(x) &= \sum_{\tau_k < T} [f(\eta_k) - f(0)]\, e^{-\lambda \tau_k} \\ &+ \int_0^T e^{-\lambda t} [-\lambda f(\xi(t)) + a(\xi_x(t)) f'(\xi_x(t)) + \tfrac{1}{2}\sigma^2(\xi_x(t)) f''(\xi_x(t))]\, dt \\ &+ \int_0^T e^{-\lambda t} f'(\xi_x(t))\, \sigma(\xi_x(t))\, dw(t). \end{aligned}$$

Letting $T \to \infty$, we find

$$\begin{aligned} -f(x) = \sum_{k=1}^\infty [f(\eta_k) - f(0)]\, e^{-\lambda \tau_k} &+ \int_0^\infty e^{-\lambda t} g(\xi_x(t))\, dt \\ &+ \int_0^\infty e^{-\lambda t} f'(\xi_x(t))\, \sigma(\xi_x(t))\, dw(t). \end{aligned} \tag{4}$$

Furthermore,

$$\left| \int_0^\infty e^{-\lambda t} g(\xi_x(t))\, dt \right| \leqq \frac{1}{\lambda} \|g\| \qquad (\|g\| = \sup_x |g(x)|),$$

$$\left| \sum_{k=1}^\infty (f(\eta_k) - f(0))\, e^{-\lambda \tau_k} \right| \leqq 2 \|f\| \sum_{k=1}^\infty e^{-\lambda \tau_k},$$

$$\begin{aligned} \mathbb{M} \left(\sum_{k=1}^\infty e^{-\lambda \tau_k} \right)^2 &\leqq 2 \sum_{k \leqq j} \mathbb{M}\, e^{-\lambda \tau_k} e^{-\lambda \tau_j} = 2 \sum_{k \leqq j} \mathbb{M}\, e^{-2\lambda \tau_k} e^{-\lambda(\tau_j - \tau_k)} \\ &= 2 \sum_{k \leqq j} \mathbb{M}\, e^{-2\lambda \tau_k} \mathbb{M}\, e^{-\lambda(\tau_j - \tau_k)} = 2\mathbb{M}\, e^{-2\lambda \tau_1} \sum_{k \leqq j} \mathbb{M}\, e^{-2\lambda(\tau_k - \tau_1)} \mathbb{M}\, e^{-\lambda(\tau_j - \tau_k)} \\ &= \mathbb{M}\, e^{-2\lambda \tau_1} \sum_{k=1}^\infty [\psi(2\lambda)]^{k-1} \sum_{j=k}^\infty [\psi(\lambda)]^{j-k} = \frac{\mathbb{M}\, e^{-2\lambda \tau_1}}{[1 - \psi(2\lambda)](1 - \psi(\lambda))}, \end{aligned}$$

where $\psi(\lambda) = \mathbb{M}\, e^{-\lambda(\tau_2 - \tau_1)}$, since the variables $\tau_1, \tau_2 - \tau_1, \ldots, \tau_k - \tau_{k-1}$ are independent, and $\tau_2 - \tau_1, \ldots, \tau_k - \tau_{k-1}, \ldots$ are identically distributed.

Hence,

$$\mathbb{M}\left(\int_0^\infty e^{-\lambda t} f'(\xi_x(t))\,\sigma(\xi_x(t))\,dw(t)\right)^2 < \infty,$$

$$\mathbb{M}\int_0^\infty e^{-\lambda t} f'(\xi_x(t))\,\sigma(\xi_x(t))\,dw(t) = 0.$$

Taking expectations on both sides of (4), we obtain (3).

Remark 1. Using the notation of the Lemma we can write

$$\mathbb{M}\sum_k [f(\eta_k)-f(0)]\,e^{-\lambda\tau_k} = \mathbb{M}[f(\eta_1)-f(0)]\,\frac{\mathbb{M}\,e^{-\lambda\tau_{x,0}}}{1-\psi(\lambda)}. \tag{5}$$

Corollary 1. *Let $u_\lambda(x)$ for $x>0$ be a bounded solution of*

$$-\lambda u_\lambda(x) + a(x)\frac{d}{dx}u_\lambda(x) + \frac{1}{2}\sigma^2(x)\frac{d^2}{dx^2}u_\lambda(x) = g(x), \tag{6}$$

where $g(x)$ is a bounded, continuous function

$$u_\lambda(0) = \int_0^\infty u_\lambda(y)\,dF(y), \tag{7}$$

where $F(y)=\mathbb{P}\{\eta_k<y\}$. Then

$$u_\lambda(x) = \mathbb{M}\int_0^\infty e^{-\lambda t} g(\xi_x(t))\,dt.$$

Corollary 1 allows us to determine the transition probability of $\xi(t)$.

We now turn to an ergodic theorem for $\xi(t)$. Let $f(t)$ be a bounded, twice continuously differentiable function. Then

$$f(\xi(T)) - f(\xi(0)) = \sum_{\tau_k<T} [f(\eta_k)-f(0)]$$
$$+ \int_0^T \left[a(\xi(s))\,f'(\xi(s)) + \tfrac{1}{2}\sigma^2(\xi(s))\,f''(\xi(s))\right] ds + \int_0^T f'(\xi(s))\,\sigma(\xi(s))\,dw(s).$$

If $a(x)\,f'(x)+\frac{1}{2}\sigma^2(x)\,f''(x)=g(x)$, where $g(x)$ is bounded, then, since

$$\Big|\sum_{\tau_k\le T}[f(\eta_k)-f(0)]\Big| \le 2\,\|f\|\,e^{\lambda T}\sum_k e^{-\lambda\tau_k}$$

and the expression on the right side, as was shown in Lemma 2, has finite second moment, we have

$$\mathbb{M}\left(\int_0^T f'(\xi(t))\,\sigma(\xi(t))\,dw(t)\right)^2 < \infty$$

and

$$\mathbb{M}\int_0^T f'(\xi(t))\,\sigma(\xi(t))\,dw(t) = 0.$$

Thus,

$$\begin{aligned}\frac{\mathbb{M}[f(\xi(T))-f(\xi(0))]}{T}&=\frac{1}{T}\mathbb{M}\sum_{\tau_k<T}[f(\eta_k)-f(0)]+\frac{1}{T}\int_0^T \mathbb{M}\,g(\xi(t))\,dt\\&=\frac{1}{T}\mathbb{M}\,\nu_T\,\mathbb{M}[f(\eta_k)-f(0)]+\frac{1}{T}\int_0^T \mathbb{M}\,g(\xi(t))\,dt,\end{aligned}\tag{8}$$

where ν_T is the number of points τ_k which satisfy $\tau_k<T$.

Lemma 3. *Assume $\sigma(x)>0$ and set*

$$\Phi(x)=\exp\left\{\int_0^x \frac{2a(u)}{\sigma^2(u)}\,du\right\},\qquad B(x)=\frac{2\Phi(x)}{\sigma^2(x)}\int_0^x \frac{1-F(y)}{\Phi(y)}\,dy,$$

where $F(y)=\mathbb{P}\{\eta_k<y\}$. If $\int_0^\infty B(x)\,dx<\infty$, then for any bounded continuous function $g(x)$ for which

$$\int_0^\infty B(x)\,g(x)\,dx=0,\tag{9}$$

we have

$$\lim_{T\to\infty}\frac{1}{T}\int_0^T \mathbb{M}\,g(\xi(t))\,dt=0.\tag{10}$$

Proof. Let

$$u(x)=\int_0^x \frac{1}{\Phi(u)}\int_0^u \frac{2g(z)\Phi(z)}{\sigma^2(z)}\,dz+c_1\int_0^x \frac{du}{\Phi(u)}.$$

It is easy to verify that $a(x)\,u'(x)+\frac{1}{2}\sigma^2(x)\,u''(x)=g(x)$. Choose c_1 in such a way that

$$\int_0^\infty u(x)\,dF(x)=\int_0^\infty (1-F(x))\,du(x)=0.$$

For this purpose it is sufficient to take $c_1=\int_0^\infty \frac{2g(z)\Phi(z)}{\sigma^2(z)}\,dz$, since then

$$\begin{aligned}\int_0^\infty (1-F(x))\,du(x)&=\int_0^\infty \frac{1-F(x)}{\Phi(x)}\left[\int_0^x \frac{2g(z)\Phi(z)}{\sigma^2(z)}\,dz-\int_0^\infty \frac{2g(z)\Phi(z)}{\sigma^2(z)}\,dz\right]dx\\&=-\int_0^\infty \frac{1-F(x)}{\Phi(x)}\int_x^\infty \frac{2g(z)\Phi(z)}{\sigma^2(z)}\,dz\,dx=-\int_0^\infty \frac{2g(z)\Phi(z)}{\sigma^2(z)}\int_0^z \frac{1-F(x)}{\Phi(x)}\,dx\,dz\\&=-\int_0^\infty g(z)\,B(z)\,dz=0\end{aligned}$$

from (9). Therefore,

$$\mathbb{M}[u(\eta_k)-u(0)]=\int_0^\infty u(y)\,dF(y)=0$$

so that on the basis of (8)

$$\frac{1}{T}\int_0^T \mathbb{M}\, g(\xi(t))\,dt = \frac{\mathbb{M}\, u(\xi(T)) - \mathbb{M}\, u(\xi(0))}{T} \to 0$$

for $T \to \infty$ since $u(x)$ is bounded. □

Corollary 2. *Under the assumptions of Lemma 3 for any bounded continuous function* $g(x)$

$$\lim_{T\to\infty} \frac{1}{T}\int_0^T \mathbb{M}\, g(\xi(t))\,dt = \int_0^\infty B(x)\,g(x)\,dx \left(\int_0^\infty B(x)\,dx\right)^{-1}.$$

In order to show this it is sufficient to apply Lemma 3 to the function

$$\bar{g}(x) = g(x) - \int_0^\infty B(x)\,g(x)\,dx \left(\int_0^\infty B(x)\,dx\right)^{-1},$$

for which

$$\int_0^\infty \bar{g}(x)\,B(x)\,dx = 0.$$

We will merely treat the case $a(x)=0$ (we can always arrive at this case with a substitution of state variable). It is clear that the first passage time τ_x of $\xi_x(t)$ to zero coincides with the same time for the process $\tilde{\xi}_x(t)$ which satisfies $d\tilde{\xi}_x(t) = \tilde{a}(\tilde{\xi}_x(t))\,dt + \tilde{\sigma}(\tilde{\xi}_x(t))\,dw(t)$ with initial condition $\tilde{\xi}_x(0) = x$, where $\tilde{a}(x)$ and $\tilde{\sigma}(x)$ are defined for $x \in (-\infty, \infty)$, coincide with $a(x)$ and $\sigma(x)$ for $x > 0$ and fulfill the hypotheses of the existence and uniqueness theorem. If $\int_0^\infty \frac{dz}{\sigma^2(z)} < \infty$, then from Lemma 1 § 18

$$\mathbb{M}\,\tau_x = 2x\int_x^\infty \frac{dz}{\sigma^2(z)} + \int_0^x \frac{2z\,dz}{\sigma^2(z)},$$

so that

$$\mathbb{M}(\tau_{k+1} - \tau_k) = \int_0^\infty \left[2x\int_x^\infty \frac{dz}{\sigma^2(z)} + \int_0^x \frac{2z\,dz}{\sigma^2(z)}\right] dF(x)$$

$$= 2\int_0^\infty (1 - F(x)) \int_x^\infty \frac{dz}{\sigma^2(z)} = 2\int_0^\infty \frac{dz}{\sigma^2(z)} \int_0^z (1 - F(x))\,dx < \infty,$$

provided that $\mathbb{M}\,\eta_k = \int_0^\infty (1 - F(x))\,dx < \infty$. Consequently, employing the methodology we have already used several times for ergodic theorems, i.e., considering the identically distributed variables $\int_{\tau_k}^{\tau_{k+1}} f(\xi(s))\,ds$, we can establish

Theorem 1. *If* $\xi(t)$ *satisfies* $d\xi(t)=\sigma(\xi(s))\,dw(t)$ *in* $(0,\infty)$ *and* $\eta_1,\eta_2,\dots$ *are the sizes of the jumps of* $\xi(t)$ *from the boundary, whereby* $\mathbb{P}\{\eta_k<x\}=F(x)$ *and* $\mathbb{M}\,\eta_k<\infty$, *and if* $\sigma(x)$ *satisfies a Lipschitz condition for* $x>0$ *and* $\int_0^\infty \frac{dz}{\sigma^2(z)}<\infty$, *then for any measurable function* $g(x)$ *for which* $\int_0^\infty |f(x)|\frac{dx}{\sigma^2(x)}<\infty$, *the limit*

$$\lim_{T\to\infty}\frac{1}{T}\int_0^T f(\xi(t))\,dt=\frac{\int_0^\infty f(x)\frac{1}{\sigma^2(x)}\int_0^x(1-F(y))\,dy\,dx}{\int_0^\infty \frac{1}{\sigma^2(x)}\int_0^x(1-F(y))\,dy\,dx}$$

exists w.p.1.

Part II. Systems of Stochastic Differential Equations

Chapter 1. Vector Stochastic Differential Equations

In this chapter we will generalize in certain directions the concepts of stochastic integral and stochastic differential equation introduced in Part I: the solutions of stochastic differential equations will be random vector functions and we will be interested in equations whose solutions might be discontinuous and not necessarily Markov processes. In this connection we reformulate somewhat our approach to the definition of a stochastic differential equation.

Let us consider a differential equation in an m-dimensional vector space X:

$$\frac{dx}{dt}=A(t,x), \qquad x=x(t), \qquad t\geqq 0, \qquad x(0)=x_0, \tag{0.1}$$

which describes the motion of some system Σ in the phase space X. The function $A(t,x)$, $t\geqq 0$, $x\in X$ is a "local characteristic" of the motion which uniquely defines the trajectories of the system Σ in X provided that certain conditions are fulfilled. This motion is differentiable (it possesses a velocity). If the motion $x=x(t)=x(t,x_0)$, $x(0)=x_0$, $t\geqq 0$ (which need not possess a velocity and may even be discontinuous) is given with the help of the corresponding local characteristic, then one can a attempt to determine the motion with the aid of some function $A(t,x,h)$, $t,h\geqq 0$, $x\in X$ and the requirement that the difference $x(t+\Delta t)-x(t)-A(t,x(t),\Delta t)$ be, in a certain sense, smaller than Δt. The function $A(t,x,h)$ now plays the role of the local characteristic of the motion, and the previous requirement is naturally replaced by its "integrated" form

$$x(t)-x(0)=\int_0^t A(\tau,x(\tau),d\tau), \tag{0.2}$$

where the integral on the right is understood as the limit in some sense of integral sums of the form

$$\sum_{k=0}^{n-1} A(t_k,x(t_k),\Delta t_k), \qquad 0=t_0<t_1<\cdots<t_n=t, \qquad \Delta t_k=t_{k+1}-t_k \tag{0.3}$$

as $\max_k \Delta t_k \to 0$. The limit (0.3) can be considered as a variant of the definition of a line integral in the vector field $\{A(t, x, h), t \geqq 0, h \geqq 0, x \in X\}$ along the curve $x = x(\tau)$, $0 \leqq \tau \leqq t$.

The existence of discontinuities of the function $x(t)$ can be viewed as caused by the jumps of the function $A(\cdot, \cdot, h)$ at the point $h=0$ and the nondifferentiability or unboundedness of the variation of $x(t)$ as due to the "pathological" analytical nature of $A(\cdot, \cdot, h)$.

The corresponding theory for equation (0.2) for sufficiently general functions $A(t, x, h)$ has not yet been developed but it can be developed for a statistical approach to the problem, considering the space of "admissible functions $A(t, x, h)$" in which a certain (probability) measure is given, and then carrying through the necessary constructions in a statistical sense. In this chapter we will give such constructions in a rather general framework. In particular, we will not exclude the possibility of after-effect for the motions obtained. In the subsequent chapters we will return to motions without after-effect admitting discontinuous trajectories.

§ 1. Stochastic Line Integrals

We now define a stochastic line integral along a random curve and consider some of its properties.

Let X be a real, m-dimensional vector space; x, y, z elements of X; $(x|y)$ the scalar product in X; $|x|$ the norm of $x \in X$; $(\Omega, \mathfrak{S}, \mathbb{P})$ will denote some fixed probability space; Ω is a space of elementary events; $\mathfrak{S}$ a σ-algebra of subsets of Ω, and $\mathbb{P}$ a probability defined on $\mathfrak{S}$.

The family of random vectors in X

$$\{\alpha(t, x, h) = \alpha(t, x, h, \omega);\ t \in [0, T],\ h \in [0, h_0],\ x \in X,\ \omega \in \Omega\}$$

will be called a *random vector field in X*.

A stochastic line integral will be defined for random fields and random curves (integration paths), satisfying certain special probability-theoretic conditions. These conditions consist of the fact that the process of variation of the field $\alpha(t, x, h)$ after time t_0 $(t > t_0)$ depends only to a very small degree on the information about the state of the field at time t_0 $(t + h \leqq t_0)$; we will call such fields *fields with limited after-effect*. A field of the type $\alpha(t, x, h) = \alpha(t+h, x) - \alpha(t, x)$, where $\alpha(t, x)$ for constant x is a process with time-independent increments, will satisfy these conditions.

We consider the collection of all random vectors α in X for which

$$\mathbb{M}|\alpha|^2 < \infty,$$

and which becomes a Hilbert space if we take $\mathbb{M}(\alpha|\beta)$ as the scaler product of two elements α and β. Let $L_2(X)$ designate this space. Convergence in $L_2(X)$ is convergence in mean square and if a sequence $\alpha_n \in L_2(X)$ $(n=1, 2, \ldots)$ converges in mean square to α_0, i.e.,

$$\mathbb{M}|\alpha_n-\alpha_0|^2 \to 0 \quad \text{for } n\to\infty,$$

then we will write

$$\alpha_0 = \text{l.i.m.}\, \alpha_n.$$

Analogous notation also applies for convergence in $L_2(X)$ of an arbitrary partially-ordered set of random elements.

In this chapter we will always assume that the random field $\alpha(t, x, h)$ satisfies the following condition: for arbitrary $t\in[0, T]$ and $h\in[0, h_0]$ the random vector $\alpha(t, x, h)$ is w.p. 1 a Borel function of x and for fixed t, x and h

$$\mathbb{M}|\alpha(t, x, h)|^2 < \infty.$$

From this condition it follows, in particular, that if ξ is a $\mathfrak{S}$-measurable random vector, then $\alpha(t, \xi, h)$ will also be a $\mathfrak{S}$-measurable random vector.

Definition 1. The random field $\alpha(t, x, h)$ will be called a *field with limited after-effect* if there exists a monotone nondecreasing family of σ-algebras $\{\mathfrak{F}_t, 0\leqq t<T+h_0\}$, $\mathfrak{F}_t\subset\mathfrak{S}$ such that 1) the random vector $\alpha(t, x, h)$ is $\mathfrak{F}_{t+h}$-measurable; 2) there exists a monotone nondecreasing, left-continuous, bounded, nonrandom function $F(t)$ $(0\leqq t\leqq T+h_0,$ $F(0)=0)$ for which

$$|\mathbb{M}\{\alpha(t, x, h)/\mathfrak{F}_t\}| \leqq (1+|x|)\, F(t, t+h), \tag{1}$$

$$\mathbb{M}\{|\alpha(t, x, h)|^2/\mathfrak{F}_t\} \leqq (1+|x|^2)\, F(t, t+h), \tag{2}$$

where $F(t, t+h)=F(t+h)-F(t)$.

Definition 2. If for each $R>0$ there is a constant C_R (independent of t, x, h and chance) such that for all x, y with $|x|\leqq R$, $|y|\leqq R$

$$|\mathbb{M}\{\alpha(t, x, h)-\alpha(t, y, h)/\mathfrak{F}_t\}| \leqq C_R|x-y|\, F(t, t+h), \tag{3}$$

$$\mathbb{M}\{|\alpha(t, x, h)-\alpha(t, y, h)|^2/\mathfrak{F}_t\} \leqq C_R|x-y|\, F(t, t+h), \tag{4}$$

then we will say that the field $\alpha(t, h, x)$ satisfies a *local Lipschitz condition.* If we can put $C_R=C$, C independent of R, then we will say that $\alpha(t, x, h)$ satisfies a *uniform Lipschitz condition.*

We note that the second inequalities in Definitions 1 and 2 indicate that the mathematical expectation of the quadratic norm of the variables $\alpha(t, x, h)$ and $\alpha(t, x, h)-\alpha(t, y, h)$ is of the order $\Delta F=F(t, t+h)$. It therefore follows that the expectation of the norm of these variables is of order

$\sqrt{\Delta F}$. These requirements are insufficient for our purpose and the reason for the first inequalities is that the expectation of the variables $\alpha(t, x, h)$ and $\alpha(t, x, h) - \alpha(t, y, h)$ themselves (but not of their norms!) is of order ΔF. Furthermore, in Definitions 1 and 2 is contained the requirement that all the enumerated estimates hold in a certain sense uniformly w.r.t. the σ-algebras $\mathfrak{F}_t$, i.e., uniformly w.r.t. all possible information on the field $\alpha(\tau, z, h)$ for $\tau \geqq 0$, $h > 0$, $\tau + h < t$ and arbitrary $z \in X$.

It is clear that the previous conditions cannot be sufficient for the construction of a theory of integration. Indeed, what we require is that $\alpha(t, x, h)$ for small h (and in some generalized sense), should be "close" to the differential of some function, i.e., it should differ little from an additive function on $(t, t+h)$. We formulate this essential requirement with the aid of the following

Definition 3. The random field $\alpha(t, x, h)$ will be called *quasi-differential* if there exists a nonrandom, continuous, nondecreasing function $g(t)$ $(0 \leqq t \leqq T + h_0)$ such that $g(0) = 0$ and

$$|\mathbb{M}\{\alpha(t, x, h_1 + h_2) - [\alpha(t, x, h_1) + \alpha(t + h_1, x, h_2)]/\mathfrak{F}_{t+h_1}\}| \leqq g(h_1)(1 + |x|)\, F(t + h_1, t + h_1 + h_2), \tag{5}$$

$$\mathbb{M}\{|\alpha(t, x, h_1 + h_2) - [\alpha(t, x, h_1) + \alpha(t + h_1, x, h_2)]|^2/\mathfrak{F}_{t+h_1}\} \leqq g(h_1)(1 + |x|^2)\, F(t + h_1, t + h_1 + h_2). \tag{6}$$

We denote by means of $\Pi_0 = \Pi_0(F, g, C)$ the class of quasi-differential random fields with limited after-effect satisfying a uniform Lipschitz condition with given functions $F(t)$, $g(h)$ and constant C, and by means of $\Pi_1 = \Pi_1(F, g)$ the class of quasi-differential random fields with limited after-effect with fixed functions $F(t)$, $g(h)$ satisfying a local Lipschitz condition.

In order to present the content of Definitions 1–3 more clearly in concrete cases, we consider some very simple examples of random fields.

Example 1. Let

$$\alpha(t, x, h) = A(t, x)\, h + B(t, x)[\alpha(t + h) - \alpha(t)], \tag{7}$$

where $A(t, x)$ is an $\mathfrak{F}_t$-measurable random function with values in X, $B(t, x)$ is for fixed t and x $(t \geqq 0, x \in X)$ an $\mathfrak{F}_t$-measurable, random, bounded linear operator mapping the finite-dimensional vector space X_1 into X, $A(t, x)$ and $B(t, x)$ are for fixed t w.p.1 Borel functions of x and finally, $\alpha(t)$ $(t \geqq 0)$ is some random $\mathfrak{F}_t$-measurable process for each t, with values in X_1, with

$$\mathbb{M}\{\alpha(t + h) - \alpha(t)/\mathfrak{F}_t\} = 0,$$

$$\mathbb{M}\{|\alpha(t + h) - \alpha(t)|^2/\mathfrak{F}_t\} \leqq F(t + h) - F(t), \quad h > 0,$$

where $F(t)$ is a monotone nondecreasing, nonrandom function bounded on $[0, T]$. Without loss of generality we can assume that $F(t, t+h) \geqq h$. Then $\alpha(t, x, h)$ will be a field with limited after-effect if for some constant C w.p. 1

$$|A(t, x)| + |B(t, x)| \leqq C(1+|x|) \qquad (0 \leqq t \leqq T),$$

where $|B(t, x)|$ is the operator norm of the operator $B(t, x)$. If for arbitrary $R>0$ there is a constant $C_R>0$ such that for all x, y with $|x| \leqq R$, $|y| \leqq R$

$$|A(t, x) - A(t, y)| + |B(t, x) - B(t, y)| \leqq C_R |x-y|,$$

w.p. 1, then the field $\alpha(t, x, h)$ will satisfy a local Lipschitz condition. It is also easy to verify the conditions for quasi-differentiability. Since

$$\begin{aligned} &\alpha(t, x, h_1+h_2) - [\alpha(t, x, h_1) + \alpha(t+h_1, x, h_2)] \\ &\quad = [A(t, x) - A(t+h_1, x)]\, h_2 + [B(t, x) - B(t+h_1, x)] \\ &\quad\quad \times [\alpha(t+h_1+h_2) - \alpha(t+h_1)], \end{aligned}$$

the field will be quasi-differential if there exists a nonnegative function $g(h)$, $h>0$ such that $\lim_{h\to 0} g(h)=0$ and

$$\frac{|A(t, x) - A(t+h, x)| + |B(t, x) - B(t+h, x)|}{1+|x|} \leqq g(h) \tag{8}$$

for all t and $t+h$ in $[0, T]$.

Example 2. We obtain a further specialization of the field (7) by assuming that $A(t, x)$ and $B(t, x)$ are nonrandom and that $\alpha(t)$ is a homogeneous process with independent increments and finite second-order moments for which

$$\mathbb{M}[\alpha(t+h) - \alpha(t)] = 0.$$

Then automatically

$$\mathbb{M}|\alpha(t+h) - \alpha(t)|^2 = Lh.$$

Conditions (1)–(6) turn out to be satisfied with $F(t) = Lt$ if

a) the function

$$\frac{|A(t+h, x) - A(t, x)| + |B(t+h, x) - B(t, x)|}{1+|x|} \to 0$$

uniformly w.r.t. x, t ($x \in X$, $t \in [0, T]$ when $h \to 0$,

b) the functions $A(t, x)$ and $B(t, x)$ satisfy a Lipschitz condition

$$|A(t, x) - A(t, y)| + |B(t, x) - B(t, y)| \leqq C_R |x-y|$$

for arbitrary R and x, y with $|x| \leqq R$, $|y| \leqq R$, where C_R is independent of t.

Example 3. Let

$$\alpha(t, x, h) = \int_t^{t+h} \alpha(\tau, x)\, dG(\tau),$$

where $G(t)$ is a monotone nondecreasing function and $\alpha(t, x)$ is a random field whose sample functions are w.p. 1 bounded, Borel, $\mathfrak{F}_t$-measurable functions of t for any t ($t \in [0, T+h]$). Such a field is always quasi-differential (with $g(t) \equiv 0$) and has limited after-effect if there is a nonrandom constant L such that, w.p. 1

$$|\alpha(t, x)| \leqq L(1 + |x|).$$

It also satisfies a Lipschitz condition if w.p. 1

$$|\alpha(t, x) - \alpha(t, y)| \leqq C_R |x - y| \qquad \text{for } |x| \leqq R,\ |y| \leqq R,$$

where R is arbitrary and C_R depends neither on chance nor on t. Here we can take $LG(t)$ as the function $F(t)$ if $L \geqq 1$.

More general and complicated examples of fields with limited after-effect will be presented in the sequel.

We now turn to the definition of a stochastic line integral in the field $\alpha(t, x, h)$ along some random curve $\xi(t)$ ($0 \leqq t \leqq T$). We first present a simple definition.

Consider an arbitrary partition of the interval $[0, T]$ by the points $t_0, t_1, \ldots, t_n$, $0 = t_0 < \cdots < t_n = T$ and let δ be the modulus of this partition, $|\delta| = \max\limits_{0 \leqq k \leqq n-1} \Delta t_k$, $\Delta t_k = t_{k+1} - t_k$.

The sum

$$\sigma = \sum_{k=0}^{n-1} \alpha(t_k, \xi(t_k), \Delta t_k)$$

will be called an *integral sum.*

Definition 4. The limit in mean-square of an integral sum for $|\delta| \to 0$, if it exists, will be called a *stochastic line integral* in the field $\alpha(t, x, h)$ along the curve $\xi(t)$, $t \in [0, T]$:

$$\int_0^T \alpha(t, \xi(t), dt) = \underset{|\delta| \to 0}{\text{l.i.m.}} \sum_{k=0}^{n-1} \alpha(t_k, \xi(t_k), \Delta t_k).$$

We remark that the choice of the point $\xi(t_k)$ in the summand $\alpha(t_k, \xi(t_k), \Delta t_k)$ of the integral sums is essential, i.e., the existence of a limit of the sum

$$\sum_{k=0}^{n-1} \alpha(t_k, \xi(t_k'), \Delta t_k), \qquad t_k \leqq t_k' < t_{k+1},$$

and the value of this limit depends in general on the choice of the point t_k'.

In order to prove the existence of stochastic line integral and establish some of its properties we will need estimates of integral sums. The following lemmas will be required for this purpose.

Lemma 1. *If g_n, h_n and Δ_n $(n=0, 1, 2, \ldots)$ are sequences of nonnegative numbers and*

$$g_{n+1} \leqq (1+h_n)\, g_n + \Delta_n$$

or

$$g_{n+1} \leqq e^{h_n}(g_n + \Delta_n), \qquad n=1, 2, \ldots,$$

then

$$g_n \leqq e^{\sum_{k=0}^{n-1} h_k} \left[g_0 + \sum_{k=0}^{n-1} \Delta_k \right].$$

The proof is obtained immediately by induction if we use the inequality $1+h_n \leqq e^{h_n}$.

Lemma 2. *If $f(x, w)$ is a nonnegative $\mathfrak{B} \times \mathfrak{S}$-measurable function, where $\mathfrak{B}$ is a σ-algebra of Borel sets of X, $\mathfrak{S}$ a σ-algebra of sets of Ω and $\xi = \xi(w)$ is an $\mathfrak{F}$-measurable random vector in X, $\mathfrak{F} \subset \mathfrak{S}$, then*

$$\mathbb{M}\{f(\xi(\omega), \omega)/\mathfrak{F}\} = \mathbb{M}\{f(x, \omega)/\mathfrak{F}\}|_{x=\xi(\omega)}.$$

The proof follows by approximating $f(x, \omega)$ by means of functions of the form $\sum_{k=1}^{n} g_k(x)\, v_k(w)$.

In the sequel we will treat stochastic line integrals as functions of the upper limit of integration. In this connection we introduce integral sums with variable summation limit.

Set

$$\sigma(t) = \sigma^{(\delta)}(t) = \sum_{k=0}^{j(t)-1} \alpha(t_k, \xi(t_k), \Delta t_k) + \alpha(t_{j(t)}, \xi(t_{j(t)}), t - t_{j(t)}), \tag{9}$$

where $j(t)$ is the largest index k for which $t_k < t$.

For the values of integral sums $\sigma(t)$ for $t = t_r$ we introduce the simpler notation

$$\sigma_r = \sigma^{(\delta)}(t_r) = \sum_{k=0}^{r-1} \alpha(t_k, \xi(t_k), \Delta t_k).$$

Concerning the curve $\xi(t)$ along which we will consider the line integral, we assume that it is bounded in $L_2(X)$-norm. In the sequel this assumption will be replaced by a more general one.

Let H_1 denote the space of bounded (in $L_2(X)$) vector functions $\beta(t)$, $0 \leqq t \leqq T$ with values in X and norm

$$\|\beta(\cdot)\|_1 = \sup_{0 \leqq t \leqq T} \{\mathbb{M}\,|\beta(t)|^2\}^{\frac{1}{2}}$$

and such that $\beta(t) = \beta(t, w)$ is $\mathfrak{F}_t$-measurable for each t. H_0 will denote the subspace of H_1 consisting of random curves $\beta(t)$ continuous in $(L_2(X))$ everywhere on $[0, T]$ except possibly for a finite number of points in $[0, T]$ at which $F(t)$ is continuous.

Lemma 3. *If $\alpha(t, x, h)$ is a field with limited after-effect satisfying Conditions* (1) *and* (2), *and $\xi(t) \in H_1$, then*

$$\mathbb{M}\{|\sigma_r|^2/\mathfrak{F}_0\} \leqq 3 e^{F(t_r)} \left\{ F(t_r) + \sum_{k=0}^{r-1} \mathbb{M}\{|\xi(t_k)|^2/\mathfrak{F}_0\} F(t_k, t_{k+1}) \right\}. \tag{10}$$

Proof. Set

$$m_r = \mathbb{M}\{|\sigma_r|^2/\mathfrak{F}_0\}.$$

Since

$$\sigma_{r+1} = \sigma_r + \alpha(t_r, \xi(t_r), \Delta t_r),$$

we have

$$m_{r+1} = m_r + 2\mathbb{M}\{(\sigma_r | \alpha(t_r, \xi(t_r), \Delta t_r))/\mathfrak{F}_0\} + \mathbb{M}\{|\alpha(t_r, \xi(t_r), \Delta t_r)|^2/\mathfrak{F}_0\}.$$

Using known properties the expectation, Lemma 2 and (1), we obtain

$$\begin{aligned} &|\mathbb{M}\{(\sigma_r | \alpha(t_r, \xi(t_r), \Delta t_r))/\mathfrak{F}_0\}| \\ &\quad = |\mathbb{M}[\mathbb{M}\{(\sigma_r | \alpha(t_r, \xi(t_r), \Delta t_r))/\mathfrak{F}_{t_r}\}/\mathfrak{F}_0]| \\ &\quad = |\mathbb{M}[(\sigma_r | \mathbb{M}\{\alpha(t_r, \xi(t_r), \Delta t_r)/\mathfrak{F}_{t_r}\})/\mathfrak{F}_0]| \\ &\quad \leqq \mathbb{M}[|\sigma_r|(1 + |\xi(t_r)|) F(t_r, t_{r+1})/\mathfrak{F}_0]. \end{aligned}$$

Since $2ab \leqq a^2 + b^2$, we can show that the previous expression does not exceed

$$\left[\frac{m_r}{2} + \mathbb{M}\{1 + |\xi(t_r)|^2/\mathfrak{F}_0\}\right] F(t_r, t_{r+1}).$$

Using (2) we find

$$\mathbb{M}\{|\alpha(t_r, \xi(t_r), \Delta t_r)|^2/\mathfrak{F}_0\} \leqq \mathbb{M}\{1 + |\xi(t_r)|^2/\mathfrak{F}_0\} F(t_r, t_{r+1}),$$

whence

$$m_{r+1} \leqq (1 + F(t_r, t_{r+1}))\, m_r + 3\mathbb{M}\{1 + |\xi(t_r)|^2/\mathfrak{F}_0\} F(t_r, t_{r+1}). \tag{11}$$

From Lemma 1 and the last relation we obtain (10).

Lemma 4. *Let $\alpha(t,x,h)\in\Pi_0(F,g,C)$, and let η be an $\mathfrak{F}_0$-measurable vector in X with $\mathbb{M}|\eta|^2<\infty$. Then*

$$\mathbb{M}\{|\sigma_n-\alpha(0,\xi(0),t_n)+\eta|^2/\mathfrak{F}_0\}\leqq\exp\{(C+g(t_n))F(t_n)\}\times\Big[|\eta|^2+4g(t_n)(1+|\xi(0)|^2)F(t_n)+3\sum_{r=0}^{n-1}\mathbb{M}\{|\xi(t_r)-\xi(0)|^2/\mathfrak{F}_0\}F(t_r,t_{r+1})\Big]. \tag{12}$$

Proof. Set

$$\beta_r=\sigma_r-\alpha(0,\xi(0),t_r)+\eta,\qquad v_r=\mathbb{M}\{|\beta_r|^2/\mathfrak{F}_0\},$$
$$\gamma_{r+1}=\alpha(0,\xi(0),t_{r+1})-\alpha(0,\xi(0),t_r)-\alpha(t_r,\xi(0),\Delta t_r).$$

Since

$$\beta_{r+1}=\beta_r-\gamma_{r+1}+\alpha(t_r,\xi(t_r),\Delta t_r)-\alpha(t_r,\xi(0),\Delta t_r),$$

we have

$$v_{r+1}=v_r+\mathbb{M}\{|\gamma_{r+1}|^2/\mathfrak{F}_0\}+\mathbb{M}\{|\alpha(t_r,\xi(t_r),\Delta t_r)-\alpha(t_r,\xi(0),\Delta t_r)|^2/\mathfrak{F}_0\}$$
$$-2\mathbb{M}\{(\beta_r|\gamma_{r+1})+(\beta_r|\alpha(t_r,\xi(t_r),\Delta t_r)-\alpha(t_r,\xi(0),\Delta t_r)) \tag{13}$$
$$-(\gamma_{r+1}|\alpha(t_r,\xi(t_r),\Delta t_r)-\alpha(t_r,\xi(0),\Delta t_r))/\mathfrak{F}_0\}.$$

From (5) and (6)

$$|\mathbb{M}\{\gamma_{r+1}/\mathfrak{F}_{t_r}\}|\leqq g(t_r)(1+|\xi(0)|)F(t_r,t_{r+1}),$$
$$\mathbb{M}\{|\gamma_{r+1}|^2/\mathfrak{F}_{t_r}\}\leqq g(t_r)(1+|\xi(0)|^2)F(t_r,t_{r+1}),$$

thus,

$$|\mathbb{M}\{(\beta_r|\gamma_{r+1})/\mathfrak{F}_0\}|=|\mathbb{M}\{(\beta_r|\mathbb{M}[\gamma_{r+1}/\mathfrak{F}_{t_r}])/\mathfrak{F}_0\}|$$
$$\leqq\frac{g(t_r)}{2}(v_r+(1+|\xi(0)|^2)F(t_r,t_{r+1}).$$

From (3) and (4)

$$\mathbb{M}\{|\alpha(t_r,\xi(t_r),\Delta t_r)-\alpha(t_r,\xi(0),\Delta t_r)|^2/\mathfrak{F}_0\}$$
$$\leqq CM\{|\xi(t_r)-\xi(0)|^2/\mathfrak{F}_0\}F(t_r,t_{r+1}),$$
$$|\mathbb{M}\{(\beta_r|\alpha(t_r,\xi(t_r),\Delta t_r)-\alpha(t_r,\xi(0),\Delta t_r))/\mathfrak{F}_0\}|$$
$$\leqq CM\{|\beta_r||\xi(t_r)-\xi(0)|/\mathfrak{F}_0\}F(t_r,t_{r+1}).$$

Moreover,

$$|\mathbb{M}\{(\gamma_{r+1}|\alpha(t_r,\xi(t_r),\Delta t_r)-\alpha(t_r,\xi(0),\Delta t_r))/\mathfrak{F}_0\}|$$
$$\leqq\tfrac{1}{2}[\mathbb{M}\{|\gamma_{r+1}|^2/\mathfrak{F}_0\}+\mathbb{M}\{|\alpha(t_r,\xi(t_r),\Delta t_r)-\alpha(t_r,\xi(0),\Delta t_r)|^2/\mathfrak{F}_0\}]$$
$$\leqq\frac{g(t_r)}{2}(1+|\xi(0)|^2)F(t_r,t_{r+1})+\frac{C}{2}\mathbb{M}\{|\xi(t_r)-\xi(0)|^2/\mathfrak{F}_0\}F(t_r,t_{r+1}).$$

Replacing the summands on the right side of (13) by those obtained for their estimates we arrive at

$$v_{r+1} \leqq [1+(C+g(t_r))F(t_r, t_{r+1})]\, v_r + 4g(t_r)(1+|\xi(0)|^2)F(t_{r_1}, t_{r+1})$$
$$+3C\mathbb{M}\{|\xi(t_r)-\xi(0)|^2/\mathfrak{F}_0\}\, F(t_r, t_{r+1}),$$

so that using Lemma 1 we get

$$v \leqq \exp\left\{\sum_{r=0}^{n-1}(C+g(t_r))F(t_r, t_{r+1})\right\}$$
$$\times\left[|\eta|^2+\sum_{r=0}^{n-1}[4(1+|\xi(0)|^2)g(t_r)+3\mathbb{M}\{|\xi(t_r)-\xi(0)|^2/\mathfrak{F}_0\}]\,F(t_r, t_{r+1})\right]. \square$$

The estimates obtained also allow us to estimate the first moment of β_r without difficulty. If we set

$$v'_r = \mathbb{M}\{\beta_r/\mathfrak{F}_0\},$$

then

$$v'_{r+1} = v'_r - \mathbb{M}\{\gamma_{r+1}/\mathfrak{F}_0\} + \mathbb{M}\{\alpha(t_r, \xi(t_r), \Delta t_r) - \alpha(t_r, \xi(0), \Delta t_r)/\mathfrak{F}_0\}.$$

This implies that

$$|v'_{r+1}| \leqq |v'_r| + [g(t_r)(1+|\xi(0)|) + C\mathbb{M}\{|\xi(t_r)-\xi(0)|/\mathfrak{F}_0\}]\, F(t_r, t_{r+1}),$$

and so

$$|\mathbb{M}\{\sigma_n - \alpha(0, \xi(0), t_n)/\mathfrak{F}_0\}| \leqq g(t_n)(1+|\xi(0)|)F(t_n)$$
$$+C\sum_{k=0}^{n-1}\mathbb{M}\{|\xi(t_k)-\xi(0)|/\mathfrak{F}_0\}\, F(t_r, t_{r+1}). \tag{14}$$

We turn our attention now to the difference between two integral sums constructed for different partitions of $[0, T]$. It suffices to limit ourselves to the case where one of these is a sub-partition of the other. Hence, let

$$\delta = \{0 = t_0, t_1, t_2, \ldots, t_n = T\},$$
$$\delta' = \{0 = t_{00}, t_{01}, \ldots, t_{0p_1} = t_1 = t_{10}, t_{11}, \ldots, t_{n-1, p_n} = t_n = T\},$$

and $\sigma(t) = \sigma^{(\delta)}(t)$, $\sigma'(t) = \sigma^{(\delta')}(t)$ be the corresponding integral sums with variable summation limits

$$\sigma_{rj} = \sigma(t_{rj}), \qquad \sigma'_{rj} = \sigma'(t_{rj}).$$

We estimate the difference $\sigma'(T) - \sigma(T)$. To this end we note that from the equality

$$\sigma'_{rj} - \sigma_{rj} = \sigma'_{r0} - \sigma_{r0} + \tilde{\sigma}_{rj},$$

where

$$\tilde{\sigma}_{rj} = \sum_{k=0}^{j-1} \alpha(t_{rk}, \xi(t_{rk}), \Delta t_{rk}) - \alpha(t_{r0}, \xi(t_{r0}), t_{rj} - t_{r0}),$$

and Lemma 4 it follows that

$$\begin{aligned} \mathbb{M}\{|\sigma'_{rj} - \sigma_{rj}|^2/\mathfrak{F}_{t_r 0}\} \leqq \exp\{[C + g(t_{rj} - t_{r0})]\, F[t_{r0}, t_{rj})\} \\ \times \Big[|\sigma'_{r0} - \sigma_{r0}|^2 + 4(1 + |\xi(t_{r0})|^2)\, F[t_{r0}, t_{rj})\, g(t_{rj} - t_{r0}) \\ + 3 \sum_{k=0}^{j-1} \mathbb{M}\{|\xi(t_{rk}) - \xi(t_{r0})|^2/\mathfrak{F}_{t_r 0}\}\, F[t_{rk}, t_{r,k+1})\Big]. \end{aligned} \tag{15}$$

Set

$$d_r = \mathbb{M}\{|\sigma'_{r0} - \sigma_{r0}|^2/\mathfrak{F}_0\}.$$

From (15)

$$d_{r+1} \leqq \exp(F'_r)\, d_r + z_r,$$

where

$$\begin{aligned} F'_r &= [C + g(t_{r+1} - t_r)]\, F[t_r, t_{r+1}), \\ z'_r &= \exp(F'_r)(4 g(t_{r+1} - t_r)\, F[t_r, t_{r+1}))\, \mathbb{M}\{(1 + |\xi(t_r)|^2)/\mathfrak{F}_0\} \\ &\quad + 3 \sum_{k=0}^{p_{r+1}-1} \mathbb{M}\{|\xi(t_{rk}) - \xi(t_{r0})|^2/\mathfrak{F}_0\}\, F[t_{rk}, t_{r,k+1}). \end{aligned}$$

Applying Lemma 1 again, we arrive at

Lemma 5.

$$\begin{aligned} \mathbb{M}\{|\sigma'_n - \sigma_n|^2/\mathfrak{F}_0\} \leqq \exp\{(C + g(|\delta|))\, F(T)\} \\ \times \Big[4 g(|\delta|) \sum_{k=0}^{n-1} \mathbb{M}\{(1 + |\xi(t_r)|^2)/\mathfrak{F}_0\}\, F(t_r, t_{r+1}) \\ + 3 \sum_{r=0}^{n-1} \sum_{k=0}^{p_{r+1}-1} \mathbb{M}\{|\xi(t_{rk}) - \xi(t_{r0})|^2/\mathfrak{F}_0\}\, F(t_{rk}, t_{r,k+1})\Big]. \end{aligned}$$

Theorem 1. *If $\alpha(t, x, h) \in \Pi_0(F, g, C)$ for some $F(t)$, $g(h)$ and C, then the stochastic line integral exists along an arbitrary random curve $\xi(t) \in H_0$. Here, almost surely*

$$\left|\mathbb{M}\left\{\int_0^T \alpha(\tau, \xi(\tau), d\tau)/\mathfrak{F}_0\right\}\right| \leqq F(T) + \int_0^T \mathbb{M}\{|\xi(\tau)|/\mathfrak{F}_0\}\, dF(\tau), \tag{16}$$

$$\mathbb{M}\left\{\left|\int_0^T \alpha(\tau, \xi(\tau), d\tau)\right|^2 \Big/ \mathfrak{F}_0\right\} \leqq 3 e^{F(T)} \left\{F(T) + \int_0^T \mathbb{M}\{|\xi(\tau)|^2/\mathfrak{F}_0\}\, dF(\tau)\right\}. \tag{17}$$

Proof. From Lemma 5 it is not difficult to conclude that if $\xi(t)$, as a function with values in $L_2(X)$, is continuous except possibly for a finite number of points at which $F(t)$ is continuous, then an arbitrary sequence

σ_n with $|\delta|=|\delta_n|\to 0$ is fundamental in $L_2(X)$. Hence the existence of the line integral follows. Inequality (16) is easy to obtain by considering the sum σ_n and going to the limit.

Without changing the assumptions on the field $\alpha(t,x,h)$ we generalize the definition of the stochastic line integral to a wider class of curves $\xi(t)$.

Let $H=H(F)$ denote the class of curves $\xi(t)=\xi(t,\omega)$ which satisfy

1) $\xi(t)=\xi(t,\omega)\in L_2(X)$ and $\xi(t,\omega)$ is $\mathfrak{F}_t$-measurable $(0\leqq t\leqq T)$;

2) $\mathbb{M}|\xi(t)|^2$ is a Borel function of t and

$$\|\xi(\cdot)\|_2^2=\int_0^T \mathbb{M}|\xi(t)|^2 F(dt)<\infty.$$

The proof of the next lemma is easy.

Lemma 6. *If H_0 is naturally imbedded in H, then H_0 is everywhere dense in H.*

We show now that if

$$\|\xi_n(\cdot)-\xi(\cdot)\|_2\to 0, \qquad n\to\infty,$$

where $\xi_n(\cdot)\in H_0$ $(n=1,2,\dots)$, then a sequence of stochastic line integrals in the field $\alpha(t,x,h)$ along the curves $\xi_n(\cdot)$ converges in $L_2(X)$ to a definite limit. This circumstance justifies

Definition 5. The stochastic line integral

$$\int_0^T \alpha(t,\xi(t),dt)$$

along $\xi(\cdot)\in H$ is the mean square limit of a sequence of integrals along the curves $\xi_n(\cdot)$, where $\xi_n(\cdot)\in H_0$ and

$$\int_0^T \mathbb{M}|\xi(t)-\xi_n(t)|^2 F(dt)\to 0, \qquad n\to\infty.$$

The proof that a sequence of stochastic integrals along $\xi_n(\cdot)$ is fundamental in $L_2(X)$ if $\xi_n(\cdot)$ is a fundamental sequence of elements of H_0 follows from

Lemma 7. *Assume that $\xi(t)$ and $\eta(t)\in L_2(X)$ $(0\leqq t\leqq T)$, and that the random field $\alpha(t,x,h)$ satisfies a uniform Lipschitz condition, and let*

$$\sigma_r'=\alpha+\sum_{k=0}^{r-1}\alpha(t_k,\xi(t_k),\Delta t_k),$$

$$\sigma_r''=\beta+\sum_{k=0}^{r-1}\alpha(t_k,\eta(t_k),\Delta t_k),$$

where α and β are $\mathfrak{F}_0$-measurable random vectors in $L_2(X)$. Then

$$\mathbb{M}\{|\sigma_n''-\sigma_n'|^2/\mathfrak{F}_0\} \leqq e^{CF(T)}\left(|\beta-\alpha|^2+2C\sum_{k=0}^{n-1}\mathbb{M}\{|\eta(t_k)-\xi(t_k)|^2/\mathfrak{F}_0\}\,F(t_k,t_{k+1})\right). \tag{18}$$

Proof. We have

$$\sigma_{r+1}''-\sigma_{r+1}'=\sigma_r''-\sigma_r'+\alpha(t_r,\eta(t_r),\Delta t_r)-\alpha(t_r,\xi(t_r),\Delta t_r).$$

Let $d_r=\mathbb{M}\{|\sigma_r''-\sigma_r'|^2/\mathfrak{F}_0\}$. Then

$$\begin{aligned}d_{r+1}=d_r&+\mathbb{M}[\mathbb{M}\{|\alpha(t_r,\eta(t_r),\Delta t_r)-\alpha(t_r,\xi(t_r),\Delta t_r)|^2/\mathfrak{F}_{t_r}\}/\mathfrak{F}_0]\\&+2\mathbb{M}[\mathbb{M}\{(\sigma_r''-\sigma_r'|\alpha(t_r,\eta(t_r),\Delta t_r)-\alpha(t_r,\xi(t_r),\Delta t_r))/\mathfrak{F}_{t_r}\}/\mathfrak{F}_0].\end{aligned}$$

Using inequalities (3) and (4) and assuming $C_R=C$, is easy to establish that

$$d_{r+1}\leqq d_r(1+CF(t_r,t_{r+1}))+2C\mathbb{M}\{|\eta(t_r)-\xi(t_r)|^2/\mathfrak{F}_0\}\,F(t_r,t_{r+1}).$$

An application of Lemma 1 yields (18).

Corollary 1. *If $\xi(\cdot)\in H_0$, $\eta(\cdot)\in H_0$, the random field $\alpha(t,x,h)$ fulfills the conditions of Theorem 1 and*

$$\xi_1=\alpha+\int_0^T\alpha(t,\xi(t),dt),\qquad \eta_1=\beta+\int_0^T\alpha(t,\eta(t),dt),$$

then

$$\mathbb{M}\{|\eta_1-\xi_1|^2/\mathfrak{F}_0\}\leqq e^{CF(T)}\left(|\beta-\alpha|^2+2C\int_0^T\mathbb{M}\{|\eta(t)-\xi(t)|^2/\mathfrak{F}_0\}\,dF(t)\right). \tag{19}$$

Theorem 2. *If the field $\alpha(t,x,h)$ fulfills the assumptions of Theorem 1, then a stochastic line integral in the sense of Definition 5 exists along an arbitrary curve $\xi(\cdot)\in H$ and (16), (17) and (19) hold for arbitrary $\xi(\cdot)$ and $\eta(\cdot)\in H$.*

Corollary 2. *If the assumptions of Theorem 1 hold and $\xi(\cdot)\in H$, then*

$$\left|\mathbb{M}\left\{\int_0^T\alpha(t,\xi(t),dt)-\alpha(0,\xi(0),T)/\mathfrak{F}_0\right\}\right| \leqq g(T)(1+|\xi(0)|)\,F(T)+C\int_0^T\mathbb{M}\{|\xi(t)-\xi(0)|/\mathfrak{F}_0\}\,dF(t), \tag{20}$$

$$\begin{aligned}\mathbb{M}\left\{\left|\int_0^T\alpha(t,\xi(t),dt)-\alpha(0,\xi(0),T)\right|^2\Big/\mathfrak{F}_0\right\}&\leqq\exp\{(C+g(T))\,F(T)\}\\&\times\left[4g(T)(1+|\xi(0)|^2)\,F(T)+3\int_0^T\mathbb{M}\{|\xi(t)-\xi(0)|^2/\mathfrak{F}_0\}\,dF(t)\right].\end{aligned} \tag{21}$$

Corollary 3. *Under the previous assumptions*

$$\mathbb{M}\left\{\left|\int_0^T \alpha(t,\xi(t),dt)-\sigma_n\right|^2 \Big/ \mathfrak{F}_0\right\} \leqq \exp\{(C+g(|\delta|))F(T)\}$$
$$\times\left[4g(|\delta|)\left(\sum_{r=0}^{n-1}\mathbb{M}\{|\xi(t_r)|^2/\mathfrak{F}_0\}F(t_r,t_{r+1})+F(T)\right)\right. \quad (22)$$
$$\left.+3\sum_{r=0}^{n-1}\int_{t_r}^{t_{r+1}}\mathbb{M}\{|\xi(t)-\xi(t_r)|^2/\mathfrak{F}_0\}\,dF(t)\right].$$

The claims of Corollaries 2 and 3 follow from Eqn. (20), (21) and Lemma 5 with the help of a double limit passage: the first from the integral sums σ_n to the stochastic integral for $\xi(\cdot)\in H_0$, and the second passage in H from the sequence $\xi_n(\cdot)\in H_0$ to an arbitrary function $\xi(\cdot)\in H$.

Lemma 8. *Let $\alpha(t,x,h)\in\Pi_0(F,g,C)$ and $\eta(t)$, $\xi(t)\in H$. Then for any $A>0$ and $N>0$*

$$\begin{aligned}\mathbb{P}&\left\{\left|\int_0^T \alpha(t,\xi(t),dt)-\int_0^T \alpha(t,\eta(t),dt)\right|>A/\mathfrak{F}_0\right\}\\ &\leqq \frac{2C\,e^{CF(T)}N}{A^2}+\mathbb{P}\left\{\int_0^T|\xi(t)-\eta(t)|^2\,dF(t)>N/\mathfrak{F}_0\right\}.\end{aligned} \quad (23)$$

Proof. Assume

$$\xi_N(t)=\xi(t) \quad\text{and}\quad \eta_N(t)=\eta(t), \qquad \text{if } \int_0^t|\xi(t)-\eta(t)|^2\,dF(t)\leqq N,$$

and

$$\xi_N(t)=\eta_N(t)=0, \qquad \text{if } \int_0^t|\xi(t)-\eta(t)|^2\,dF(t)>N.$$

Clearly, $\xi_N(t)$ and $\eta_N(t)$ belong to H. The proof follows from the inequalities

$$\mathbb{P}\left\{\left|\int_0^T \alpha(t,\xi(t),dt)-\int_0^T \alpha(t,\eta(t),dt)\right|>A/\mathfrak{F}_0\right\}$$
$$\leqq \mathbb{P}\left\{\left|\int_0^T \alpha(t,\xi_N(t),dt)-\int_0^T \alpha(t,\eta_N(t),dt)\right|>A/\mathfrak{F}_0\right\}$$
$$+\mathbb{P}\left\{\left|\int_0^T \alpha(t,\xi(t),dt)-\int_0^T \alpha(t,\xi_N(t),dt)\right.\right.$$
$$\left.\left.+\int_0^T \alpha(t,\eta_N(t),dt)-\int_0^T \alpha(t,\eta(t),dt)\right|>0/\mathfrak{F}_0\right\}$$

$$\leqq \frac{\mathbb{M}\left\{\left|\int_0^T \alpha(t, \xi_N(t), dt) - \int_0^T \alpha(t, \eta_N(t), dt)\right|^2 \Big/ \mathfrak{F}_0\right\}}{A^2}$$

$$+\mathbb{P}\left\{\int_0^T |\xi(t)-\eta(t)|^2\, dF(t) > N/\mathfrak{F}_0\right\}$$

and the remark that from (19) and the definition of the variables $\xi_N(t)$ and $\eta_N(t)$

$$\mathbb{M}\left\{\left|\int_0^T \alpha(t, \xi_N(t), dt) - \int_0^T \alpha(t, \eta_N(t), dt)\right|^2 \Big/ \mathfrak{F}_0\right\}$$

$$\leqq 2\, C e^{CF(T)}\, \mathbb{M}\left\{\int_0^T |\eta_N(t)-\xi_N(t)|^2\, dF(t)/\mathfrak{F}_0\right\} \leqq 2\, C e^{CF(T)} N.$$

The preceding lemma allows a generalization of the stochastic line integral to a wider class of integration curves.

Let $H_2 = H_2(F)$ denote the class of measurable random process $\beta(t)$, $0 \leqq t \leqq T$ with values in X, $\mathfrak{F}_t$-measurable for each t, for which, w.p. 1

$$\int_0^T |\beta(t)|^2\, dF(t) < \infty.$$

Consider the sequence $\xi_n(t)$, $n=1, 2, \ldots, \xi_n(\cdot) \in H$ and assume that

$$\mathbb{P}\text{-}\lim_{n\to\infty} \int_0^T |\xi(t)-\xi_n(t)|^2\, dF(t) = 0. \tag{24}$$

From (23) taking $A=\delta$, $2\, C e^{CF(t)N} = A^2 \varepsilon$, where ε and δ are arbitrary given positive numbers, we obtain

$$\mathbb{P}\left\{\left|\int_0^T \alpha(t, \xi_n(t), dt) - \int_0^T \alpha(t, \xi_{n+m}(t), dt) > \delta\right.\right\}$$

$$\leqq \varepsilon + \mathbb{P}\left\{\int_0^T |\xi_n(t)-\xi_{n+m}(t)|^2\, dF(t) > N\right\},$$

from which it follows that

$$\mathbb{P}\text{-}\lim \int_0^T \alpha(t, \xi_n(t), dt), \qquad n\to\infty.$$

exists.

Definition 6. If $\xi_0(t) \in H_2(F)$, $\xi_n(t) \in H(F)$ and (24) holds, then we define

$$\int_0^T \alpha(t, \xi_0(t), dt) = \mathbb{P}\text{-}\lim \int_0^T \alpha(t, \xi_n(t), dt).$$

Since an arbitrary process $\xi_0(\cdot)\in H_2(F)$ can be correspondingly approximated by a sequence $\xi_n(\cdot)\in H(F)$, the stochastic line integral can be defined (in the same way) for arbitrary curves $\xi(\cdot)\in H_2$ and arbitrary fields $\alpha(t, x, h)\in \Pi_0(F, g, C)$.

Another type of generalization of the stochastic line integral which is connected with a desire to free ourselves from the uniform Lipschitz condition is obtained as follows. Assume

$$\mathbb{P}\{\sup_{0\leqq t\leqq T}|\xi(t)|=\infty\}=0. \tag{25}$$

and set

$$\xi_N(t)=\xi(t), \qquad \text{if } \sup_{0\leqq \tau\leqq t}|\xi(\tau)|\leqq N$$

and

$$\xi_N(t)=0, \qquad \text{if } \sup_{0\leqq \tau\leqq t}|\xi(\tau)|>N.$$

Then $\xi_N(t)$ is for each t $\mathfrak{F}_t$-measurable and $\xi_N(\cdot)\in H$. On the other hand, if $\alpha(t, x, h)$ is a quasi-differential field with limited after-effect satisfying a uniform Lipschitz condition, then

$$\int_0^T \alpha(t, \xi_N(t), dt)=\int_0^T \alpha_N(t, \xi_N(t), dt), \tag{26}$$

where the integral on the right is taken w.r.t an arbitrary quasi-differential field with limited after-effect satisfying a uniform Lipschitz condition and coinciding with $\alpha(t, x, h)$ for $|x|\leqq N$. The integral on the left exists in the previously defined sense (Definitions 4 and 5). Let Λ_N denote the sets $\{\sup_{0\leqq t\leqq T}|\xi(t)|\leqq N\}$ in Ω. Clearly, the integrals in (26) coincide for $N>N_0$ on Λ_N. Since $\Lambda_{N'}\supset\Lambda_N$ for $N'>N$ and $\mathbb{P}\left(\bigcup_1^\infty \Lambda_N\right)=1$, then for $N\to\infty$ the integrals in (26) converge in probability to some limit. These arguments can be combined with the preceding ones relating to functions from H_2.

Definition 7. Let $\xi(\cdot)\in H_2$,

$$\mathbb{P}\text{-}\lim_{n\to\infty}\int_0^T |\xi(t)-\xi_n(t)|^2\, dF(t)=0 \quad \text{and} \quad \sup_{0\leqq t\leqq T}|\xi_n(t)|\leqq n.$$

Then we set

$$\int_0^T \alpha(t, \xi(t), dt)=\mathbb{P}\text{-}\lim_{n\to\infty}\int_0^T \alpha_n(t, \xi_n(t), dt), \tag{27}$$

where $\alpha_n(t, x, h)$ is a quasi-differentiable field with limited after-effect satisfying a local Lipschitz condition and

$$\alpha_n(t, x, h)=\alpha(t, x, h) \qquad \text{for } |x|\leqq n,\ x\in X,\ 0\leqq h\leqq h_0.$$

Theorem 3. *The stochastic line integral* (27) *exists for an arbitrary quasi-differential field with limited after-effect satisfying a local Lipschitz condition.*

Remark 1. If $\xi(\cdot)\in H$, then the integral (27) has finite second-order moments and satisfies (16) and (17) w.p. 1.

We turn to the following question. Suppose that the random fields $\alpha(t,x,h)$ and $\beta(t,x,h)$ satisfy the hypotheses of Theorem 3. Under what conditions do the stochastic line integral determined by $\alpha(t,x,h)$ and $\beta(t,x,h)$ coincide (w.p. 1) along an arbitrary curve $\xi(\cdot)\in H_2$?

A partial answer can be obtained from

Lemma 9. *Assume that the random fields $\alpha(t,x,h)$ and $\beta(t,x,h)$ fulfill the conditions of Theorem 3 and that*

$$|\mathbb{M}\{\alpha(t,x,h)-\beta(t,x,h)/\mathfrak{F}_t\}|\leqq F_1(t,h)(1+|x|), \tag{28}$$

$$\mathbb{M}\{|\alpha(t,x,h)-\beta(t,x,h)|^2/\mathfrak{F}_t\}\leqq F_1(t,h)(1+|x|^2), \tag{29}$$

where $F_1(t,h)$ is a nonrandom function and

$$V_{(0,T)}(F_1)=\overline{\lim_{|\delta|\to 0}}\sum_{k=0}^{n-1}F_1(t_k,\Delta t_k)<\infty.$$

Then for an arbitrary curve $\xi(t)\in H_1$

$$\mathbb{M}\left\{\left|\int_0^T\alpha(\tau,\xi(\tau),d\tau)-\int_0^T\beta(\tau,\xi(\tau),d\tau)\right|^2\Big/\mathfrak{F}_0\right\}$$
$$\leqq V_{(0,T)}(F_1)\,e^{V_{(0,T)}(F_1)}\big(1+\sup_{0\leqq t\leqq T}\mathbb{M}\{|\xi(t)|^2/\mathfrak{F}_0\}\big). \tag{30}$$

Proof. We can limit ourselves to functions $\xi(\cdot)$ from H_0. Assume σ_r and $\tilde\sigma_r$ are integral sums constructed for such a $\xi(\cdot)$ and for fields $\alpha(t,h,x)$ and $\beta(t,x,h)$, resp. Then

$$\tilde\sigma_{r+1}-\sigma_{r+1}=\tilde\sigma_r-\sigma_r+\alpha(t_r,\xi(t_r),\Delta t_r)-\beta(t_r,\xi(t_r),\Delta t_r).$$

Using (28) and (29) and the usual procedure, we get

$$\mathbb{M}\{|\tilde\sigma_{r+1}-\sigma_{r+1}|^2/\mathfrak{F}_0\}\leqq\mathbb{M}\{|\bar\sigma_r-\sigma_r|^2/\mathfrak{F}_0\}\,(1+F_1(t_r,\Delta t_r))$$
$$+\mathbb{M}\{(1+|\xi(t)|^2)/\mathfrak{F}_0\}\,F_1(t_r,\Delta t_r),$$

from which

$$\mathbb{M}\{|\tilde\sigma_n-\sigma_n|^2/\mathfrak{F}_0\}\leqq\exp\left\{\sum_{r=0}^{n-1}F_1(t_r,\Delta t_r)\right\}\sum_{r=0}^{n-1}\mathbb{M}\{(1+|\xi(t_r)|^2)/\mathfrak{F}_0\}\,F_1(t_r,\Delta t_r),$$

which finishes the proof.

Corollary 4. *If*

$$V_{(0,T)}(F_1)=0,$$

then for arbitrary $\xi(\cdot)\in H_2$

$$\int_0^T \alpha(t,\xi(t),dt)=\int_0^T \beta(t,\xi(t),dt).$$

Corollary 5. *If* $\beta(t,x,h)=\beta_n(t,x,h)$ *for any* $n=1,2,\ldots$ *satisfies the conditions of Lemma* 9 *and*

$$V_{(0,T)}(F_1)=V_{(0,T)}(F_1^n)\to 0, \qquad n\to\infty,$$

then for arbitrary $\xi(t)\in H_1$

$$\underset{n\to\infty}{\text{l.i.m.}}\int_0^T \beta_n(t,\xi(t),dt)=\int_0^T \alpha(t,\xi(t),dt).$$

§ 2. Stochastic Line Integrals as Function of the Upper Limit

Assume the field $\alpha(t,x,h)$ and the random process $\xi(t)$ fulfill the hypotheses of Theorem 3 § 1 ($t\in[0,T]$). Consider the process

$$\eta(t)=\int_0^t \alpha(\tau,\xi(\tau),d\tau). \tag{1}$$

For each $t\in[0,T]$ $\eta(t)$ is defined w.p. 1. In all of the sequel we will assume that $\eta(t)$ is a separable process. In a manner analogous to the proof of Lemma 1 § 3, Part 1 we can obtain the following result.

Lemma 1. *Let* $\xi_1,\xi_2,\ldots,\xi_n$ *be a sequence of real random variables,* $\mathfrak{F}_k$ $(k=0,1,\ldots,n)$ *nondecreasing* σ*-algebras,* ξ_k $\mathfrak{F}_k$*-measurable and*

$$|\mathbb{M}\{\xi_k/\mathfrak{F}_{k-1}\}|\leqq\psi_k.$$

Set

$$\eta_k=\sum_{i=1}^k \xi_i, \qquad \zeta=\max\{|\eta_1|,|\eta_2|,\ldots,|\eta_n|\}.$$

Then

$$\mathbb{M}\zeta^\alpha\leqq\left(\frac{\alpha}{\alpha-1}\right)^\alpha\mathbb{M}(\rho_n+|\eta_n|)^\alpha, \tag{2}$$

where $\alpha>1$ *and*

$$\rho_n=\sum_{k=1}^n \psi_k.$$

We apply this inequality to the sum

$$\sigma_k=\sum_{r=1}^{k}\alpha\big(t_{r-1},\xi(t_{r-1}),\Delta t_{r-1}\big),$$

corresponding to the partition δ of $[0,T]$, set $\alpha=2$ in (2) with

$$\xi_k=\alpha\big(t_{k-1},\xi(t_{k-1}),\Delta t_{k-1}\big)$$

and get

$$\mathbb{M}\max_{1\leqq k\leqq n}|(\sigma_k|x)|^2\leqq 8(\mathbb{M}\rho_n^2+\mathbb{M}|\sigma_n|^2)|x|^2, \tag{3}$$

where x is an arbitrary vector, $x\in X$ and

$$\rho_n=\sum_{k=0}^{n-1}(1+|\xi(t_k)|)F(t_k,t_{k+1}).$$

Since

$$\mathbb{M}\rho_n^2\leqq 2F(T)\left[F(T)+\sum_{k=0}^{n-1}\mathbb{M}|\xi(t_k)|^2F(t_k,t_{k+1})\right],$$

$$\mathbb{M}|\sigma_n|^2\leqq 3\,e^{F(T)}\left[F(T)+\sum_{k=0}^{n-1}\mathbb{M}|\xi(t_k)|^2F(t_k,t_{k+1})\right],$$

it follows from (3) that

$$\mathbb{M}\zeta^2\leqq 40\,s\,e^{F(T)}\left[F(T)+\sum_{k=0}^{n-1}\mathbb{M}|\xi(t_k)|^2F(t_k,t_{k+1})\right], \tag{4}$$

where s is the dimension of the space X and

$$\zeta=\max\{|\sigma_k|^2,\quad k=1,2,\dots,n\}.$$

Lemma 2. *Let $\alpha(t,x,h)$ be a quasi-differential field with limited after-effect satisfying a local Lipschitz condition, $\xi(\cdot)\in H$ and $\eta(t)$ be the separable process at* (1). *Then*

$$\mathbb{M}\{\sup_{1\leqq t\leqq T}|\eta(t)|^2\}\leqq C_1\left[F(T)+\int_0^T\mathbb{M}|\xi(t)|^2\,dF(t)\right], \tag{5}$$

where $C_1=80\,s\,e^{F(T)}$.

The transition from (4) to (5) is justified by the same arguments as in Lemma 2 § 3, Part 1.

Remark 1. If we assume that the field $\alpha(t,x,h)$ satisfies a uniform Lipschitz condition with constant C and $\xi_i\in H$ $(i=1,2)$, then the same arguments show that

$$\mathbb{M}\left\{\sup_{1\leqq t\leqq T}\left|\int_0^t \alpha(\tau,\xi_1(\tau),d\tau)-\int_0^t \alpha(\tau,\xi_2(\tau),d\tau)\right|^2\right\} \leqq C_1\, C\int_0^T M\,|\xi_1(t)-\xi_2(t)|^2\, dF(t). \tag{6}$$

The preceding remark together with the last lemma allow us to sharpen the result of Lemma 8 § 1 as follows.

Lemma 3. *If* $\alpha(t,x,h)$ *is a quasi-differential field with limited after-effect satisfying a uniform Lipschitz condition with constant* C *and* $\xi_i(\cdot)\in H$ $(i=1,2)$, *then for arbitrary* $A>0$, $N>0$ *we have*

$$\mathbb{P}\left\{\sup_{1\leqq t\leqq T}\left|\int_0^t \alpha(\tau,\xi_1(\tau),d\tau)-\int_0^t \alpha(\tau,\xi_2(\tau),d\tau)\right|\geqq A\right\} \leqq \frac{C_1 CN}{A^2}+\mathbb{P}\left\{\int_0^T|\xi_1(t)-\xi_2(t)|^2\, dF(t)>N\right\}. \tag{7}$$

The proof runs analogously to that of Lemma 8 § 1.

Assume $\alpha(t,x,h)$ satisfies a uniform Lipschitz condition and $\xi(t)\in H_0$. In this case the stochastic line integral is the limit (in the sense of mean-square convergence) of the corresponding integral sums. As in § 1, we denote the integral sum by $\sigma(t)$ and shall estimate the value of $\sup_{0\leqq t\leqq T}|\eta(t)-\sigma(t)|$. We note that (14) § 1 can be reduced with the help of a limit passage to the following

$$\left|\mathbb{M}\left\{\int_0^t \alpha(\tau,\xi(\tau),d\tau)-\alpha(0,\xi(0),t)/\mathfrak{F}_0\right\}\right| \leqq g(t)(1+|\xi(0)|)F(t)+C\int_0^t \mathbb{M}\{|\xi(\tau)-\xi(0)|/\mathfrak{F}_0\}\, dF(\tau). \tag{8}$$

This inequality will be of immediate use. Let the integral sum $\sigma(t)$ be constructed for given partition $\delta=\{0,t_1,\dots,t_m=T\}$. Consider the sequence Λ_n of sets of finite numbers of points in $[0,T]$ for which a) $\Lambda_n\subset\Lambda_{n+1}$; b) $\bigcup_{n=1}^{\infty}\Lambda_n=\Lambda$, where Λ is a separability set for the process $\eta(t)$; c) Λ_n contains the points appearing in δ.

We write points from Λ_n in the form t_{jk} $(j=0,1,\dots,m-1,\ k=0,1,\dots,r_j)$, whereby $t_{jr_j}=t_{j+1,0}=t_{j+1}$. Let

$$\eta_{jk}=\eta(t_{jk})-\sigma(t_{jk}),\qquad \zeta_{jk}=\eta_{jk}-\eta_{jk-1}.$$

We can then write

$$\zeta_{jk}=\left(\int_{t_{jk-1}}^{t_{jk}}\alpha(\tau,\xi(\tau),d\tau)-\alpha(t_{jk-1},\xi(t_{jk-1}),\Delta t_{jk-1})\right)$$
$$+(\alpha(t_{jk-1},\xi(t_{jk-1}),\Delta t_{jk-1})-\alpha(t_{jk-1},\xi(t_{j0}),\Delta t_{jk-1}))$$
$$-(\alpha(t_{j0},\xi(t_{j0}),t_{jk}-t_{j0})-\alpha(t_{j0},\xi(t_{j0}),t_{jk-1}-t_0)$$
$$-\alpha(t_{jk-1},\xi(t_{j0}),\Delta t_{jk-1})),$$

where $\Delta t_{jk-1}=t_{jk}-t_{jk-1}$. Using (8) and (3), (5) § 1, we get

$$|\mathbb{M}\{\zeta_{jk}/\mathfrak{F}_{jk-1}\}|\leqq\psi_{jk},$$

where

$$\psi_{jk}=[g(\Delta t_{jk-1})(1+|\xi(t_{jk-1})|)+g(t_{jk}-t_{j0})(1+|\xi(t_{j0})|)]$$
$$\times F(t_{jk-1},t_{jk})+C[|\xi(t_{jk-1})-\xi(t_{j0})|\,F(t_{jk-1},t_{jk})]$$
$$+\int_{t_{jk-1}}^{t_{jk}}\mathbb{M}\{|\xi(\tau)-\xi(t_{jk-1})|/\mathfrak{F}_{t_{jk-1}}\}\,dF(\tau).$$

From the preceding assumptions we have w.p. 1

$$\sup_{0\leqq t\leqq T}|\eta(t)-\sigma(t)|=\lim_{n\to\infty}\cdot\sup_{t_{jk}\in\Lambda_n}|\eta_{jk}|,$$

so that

$$\mathbb{M}\{\sup_{0\leqq t\leqq T}|\eta(t)-\sigma(t)|^2\}=\lim_{n\to\infty}\mathbb{M}\{\sup_{t_{jk}\in\Lambda_n}|\eta_{jk}|^2\}.$$

From Lemma 1

$$\mathbb{M}\{\sup_{t_{jk}\in\Lambda_n}|\eta_{jk}|^2\}\leqq 8s\{\mathbb{M}\rho_n^2+\mathbb{M}|\eta(T)-\sigma(T)|^2\},$$

where $\rho_n=\sum_{j,k}\psi_{jk}$. From the preceding expression for ψ_{jk} we find that for some constant C' depending only on C and $F(T)$

$$\lim\mathbb{M}\rho_n^2\leqq C'\left(g(|\delta|)\sum_{j=0}^{m-1}\mathbb{M}(1+|\xi(t_j)|^2)\,F(t_j,t_{j+1})\right.$$
$$\left.+\sum_{j=0}^{m-1}\int_{t_j}^{t_{j+1}}\mathbb{M}|\xi(\tau)-\xi(t_j)|^2\,dF(\tau)\right).$$

In an analogous way we can find an estimate for $\mathbb{M}|\eta(T)-\sigma(T)|^2$ (see (22) § 1). We have proved

Lemma 4. *If $\alpha(t,x,h)\in\Pi_0(F,g,C)$ and $\xi(t)\in H_0$, then there exists a constant C'' depending only on C and $F(T)$ such that*

$$\mathbb{M}\{\sup|\eta(t)-\sigma(t)|^2\}\leq C''\Bigg[g(|\delta|)\sum_{j=0}^{m-1}\mathbb{M}(1+|\xi(t_j)|^2)F(t_j,t_{j+1}) + \sum_{j=0}^{m-1}\int_{t_j}^{t_{j+1}}\mathbb{M}|\xi(\tau)-\xi(t_j)|^2\,dF(\tau)\Bigg]. \tag{9}$$

Theorem 1. *Assume that $\xi(\cdot)\in H_2$, that the field $\alpha(t,x,h)$ is quasi-differentiable without after-effect and that it satisfies a local Lipschitz condition. If for fixed t and x $\alpha(t,x,h)$ has no discontinuities of the second kind, then the separable stochastic integral*

$$I(t)=\int_0^t\alpha(\tau;\xi(\tau),d\tau)$$

will also be free of discontinuities of the second kind (w.p.1). If $\alpha(t,x,h)$ is a continuous function of h w.p.1, then $I(t)$ is continuous w.p.1.

The proof is similar to that of Theorem 1 § 3 Part I. We first assume that $\alpha(t,x,h)$ satisfies a uniform Lipschitz condition and $\xi(\cdot)\in H_0$. In agreement with (9) we construct a sequence of integral sums σ_n for which

$$\mathbb{M}\{\sup_{0\leq t\leq T}|\eta(t)-\sigma_n(t)|^2\}\leq\frac{1}{2^{n+2}}.$$

Then

$$\mathbb{P}\left\{\sup_{0\leq t\leq T}|\sigma_{n+1}(t)-\sigma_n(t)|>\frac{1}{2^{n/4}}\right\}\leq 2^{-\frac{n}{2}}.$$

Hence, the series

$$\sigma_1(t)+(\sigma_2(t)-\sigma_1(t))+\cdots+(\sigma_{n+1}(t)-\sigma_n(t))+\cdots$$

converges uniformly since from the convergence of

$$\sum_{n=2}^{\infty}\mathbb{P}\left\{\sup_{0\leq t\leq T}|\sigma_{n+1}(t)-\sigma_n(t)|>\frac{1}{2^{n/4}}\right\}$$

and The Borel-Cantelli lemma it follows that, w.p.1, only a finite number of the events

$$\sup_{0\leq t\leq T}|\sigma_{n+1}(t)-\sigma_n(t)|>\frac{1}{2^{n\,4}}.$$

can occur. This proves the theorem in the special case considered. If $\xi(t)\in H$ and $\alpha(t,x,h)\in\Pi_1$, then the claim of the theorem is carried through analogously. It is merely necessary to approximate $\xi(t)$ by random curves $\xi_n(t)\in H_0$ in such a way that the integrals

$$\eta_n(t)=\int_0^t \alpha(t, \xi_n(t), dt)$$

converge uniformly w.r.t. t with probability one. That this can be done follows from (6) and arguments analogous to those above. Finally, the theorem's validity in the general case follows from the fact that for $\sup_{0\leqq t\leqq T}|\xi(t)|\leqq N$

$$\int_0^t \alpha(\tau, \xi(\tau), d\tau)=\int_0^t \alpha_N(\tau, \xi_N(\tau), d\tau)$$

for all $t\in[0, T]$, where $\alpha_N(t, x, h)\in \Pi_0$, $\xi_N(\cdot)\in H$, i.e., $\eta(t)$ coincides w.p.1 with a function having no discontinuities of the second kind (or, resp., a continuous one).

Remark 2. If $F(t)$ is continuous, $\alpha(t, x, h)\in \Pi_1$ and $\alpha(t, x, h)$ (for fixed t and x) is a separable process, then $\alpha(t, x, h)$, as a function of h has no discontinuities of the second kind.

In fact, from (2) and (6) §1 it follows that

$$\mathbb{M}\{|\alpha(t, x, h+\Delta)-\alpha(t, x, h)|^2/\mathfrak{F}_{t+h}\}\leqq K[F(t+h+\Delta)-F(t+h)],$$

where K depends neither on chance nor on t, h or Δ. The statement now follows from a known theorem [27].

Corollary. *If $F(t)$ is continuous and $\alpha(t,x,h)$ and $\eta(t)$ are separable processes w.r.t. the arguments h and t, resp., then $\eta(t)$ will have no discontinuities of the second kind.*

§ 3. Stochastic Differential Equations

We will now define a stochastic differential equation in a vector space, will prove existence and uniqueness theorems for its solutions and establish their simplest properties. The definition of a stochastic differential equation is based on the notion of a stochastic line integral introduced in the previous paragraph. However, we now assume that the function $F(t)$ oppearing in inequalities (1)–(6) of the last paragraph is equal to Lt, where L is some constant. The basic notation introduced in §1 will be maintained in the sequel.

In particular, $\mathfrak{F}_t(0\leqq t\leqq T)$ will denote some fixed, nondecreasing family of σ-algebras which are sub-algebras of $\mathfrak{S}$, $\{\Omega, \mathfrak{S}, \mathbb{P}\}$ the basic probability space, X a finite-dimensional vector space. $\Pi_0(L, C, g)$ will be the class of random fields $\alpha(t, x, h)$ satisfying the conditions of limited after-effect with fixed constant L

$$|\mathbb{M}\{\alpha(t,x,h)/\mathfrak{F}_t\}| \leqq L(1+|x|)h, \tag{1}$$

$$\mathbb{M}\{|\alpha(t,x,h)|^2/\mathfrak{F}_t\} \leqq L(1+|x|^2)h, \tag{2}$$

the uniform Lipschitz condition

$$|\mathbb{M}\{\alpha(t,x,h)-\alpha(t,y,h)/\mathfrak{F}_t\}| \leqq C|x-y|h, \tag{3}$$

$$\mathbb{M}\{|\alpha(t,x,h)-\alpha(t,y,h)|^2/\mathfrak{F}_t\} \leqq C|x-y|^2 h \tag{4}$$

and the following conditions for quasi-differentiability:

$$\begin{aligned}&|\mathbb{M}\{\alpha(t,x,h_1+h_2)-\alpha(t,x,h_1)-\alpha(t+h_1,x,h_2)/\mathfrak{F}_{t+h_1}\}|\\ &\quad\leqq Lh_2(1+|x|)\,g(h_1),\end{aligned} \tag{5}$$

$$\begin{aligned}&\mathbb{M}\{|\alpha(t,x,h_1+h_2)-\alpha(t,x,h_1)-\alpha(t+h_1,x,h_2)|^2/\mathfrak{F}_{t+h_1}\}\\ &\quad\leqq Lh_2(1+|x|^2)\,g(h_1),\end{aligned} \tag{6}$$

where $g(h)$ is a fixed, nonnegative, nondecreasing function, $h\to 0$ and $\lim_{h\to 0} g(h)=0$.

If Conditions (3) and (4) are weakened, requiring that for each $R>0$ there exists a constant $C=C_R$ for which (3) and (4) hold for all x and y for which $|x|\leqq R$, $|y|\leqq R$, then we obtain a wider class of random fields which we designate by $\Pi_1=\Pi_1(L,g)$.

We denote by H_C the space of random processes (random curves) $\xi(t)=\xi(t,\omega)$, $t\in[0,T]$, $\omega\in\Omega$ with values in X, measurable for each ω, $\mathfrak{F}_t$-measurable for each $t\in[0,T]$, and for which $\mathbb{M}|\xi(t+h)-\xi(t)|^2\to 0$ for $h\to 0$, t, $t+h\in[0,T]$.

In H_C we introduce the norm

$$\|\xi(\cdot)\|_C=\max_{0\leqq t\leqq T}\{\mathbb{M}|\xi(t)|^2\}^{\frac{1}{2}}. \tag{7}$$

Along an arbitrary curve $\xi(\cdot)\in H_C$ the stochastic integral

$$\xi_1(t)=x+\int_0^t \alpha(\tau,\xi(\tau),d\tau), \tag{8}$$

exists, where x is $\mathfrak{F}_0$-measurable random vector, $x\in L_2(X)$, $\xi_1(t)$ is $\mathfrak{F}_t$-measurable and

$$\mathbb{M}|\xi_1(t+h)-\xi_1(t)|^2 \leqq C_1\left\{h+\int_t^{t+h}\mathbb{M}|\xi(t)|^2\,dt\right\}, \tag{9}$$

where C_1 is some constant independent of $\xi(t)$. Hence the correspondence $\xi_1(\cdot)=I(\xi(\cdot))$, defined by formula (8) maps H_C into itself.

Definition 1. A solution of the stochastic differential equation

$$d\xi=\alpha(t,\xi,dt), \qquad \xi(0)=x, \tag{10}$$

will be a random process $\xi(t)$ $(t\in[0,T], \xi(\cdot)\in H_C)$, defined for each t w.p.1 and such that

$$\xi(t)=x+\int_0^t \alpha(\tau,\xi(\tau),d\tau).$$

The quantity $\xi(0)=x$ will be called the initial condition for (10).

Theorem 1. *If the random field* $\alpha(t,x,h)\in\Pi_0$, *then the stochastic differential equation* (10) *has a unique solution in* H_C *for arbitrary initial value* x.

Proof. We have already noted that the integration operation (8) maps H_C into H_C. Let

$$\xi_n(\cdot)=I(\xi_{n-1}(\cdot)), \qquad \eta_n(\cdot)=I(\eta_{n-1}(\cdot)), \qquad n=1,2,\ldots,$$

and $\xi_0(t)$, $\eta_0(t)$ be arbitrary processes in H_C.

From (19) § 1

$$\begin{aligned}\mathbb{M}|\xi_n(t)-\eta_n(t)|^2 &\leqq C_2\int_0^t \mathbb{M}|\xi_{n-1}(\tau)-\eta_{n-1}(\tau)|^2\,d\tau\\ &\leqq \frac{C_2^n t^{n-1}}{(n-1)!}\int_0^t \mathbb{M}|\xi_0(\tau)-\eta_0(\tau)|^2\,d\tau,\end{aligned}$$

where C_2 is some constant independent of t. Thus some power of the operator I is a contraction and I itself is continuous. Consequently, the operator I has a unique fixed point – the solution of (10).

Remark 1. Under the conditions of the theorem the solution of a stochastic differential equation can be obtained by means of the method of successive approximations starting from an arbitrary initial-approximation $\xi_0(t)\in H_C$. In this regard, the mean square convergence w.p.1 of sequences of approximations still holds if the mathematical expectation is replaced by the conditional expectation w.r.t. $\mathfrak{F}_0$.

Remark 2. If the random fields $\alpha(t,x,h)$ and $\beta(t,x,h)$ fulfill the conditions of Lemma 9 § 1 and $V_{0,T}(F_1)=0$, then on the interval $[0,T]$ the solutions of the equations

$$d\xi=\alpha(t,\xi,dt), \qquad d\eta=\beta(t,\eta,dt), \qquad \eta(0)=\xi(0)$$

coincide w.p.1 for each $t\in[0,T]$.

Remark 3. Let the field $\alpha(t,x,h)$ satisfy the hypotheses of Theorem 1 and assume that random process $\varphi(t)$ $(0\leqq t\leqq T)$ is $\mathfrak{F}_t$-measurable,

$\varphi(t) \in L_2(X)$ and is continuous in $L_2(X)$. The equation

$$\xi(t) = \varphi(t) + \int_0^t \alpha(\tau, \xi(\tau), d\tau) \tag{11}$$

has a unique solution in H_C.

The proof of this claim differs only slightly from that of Theorem 1.

Definition 2. An Euler approximate solution of (10) will be a random process $\xi_\delta(t)$ satisfying the recursion relation

$$\xi_\delta(0) = x, \qquad \xi_\delta(t) = \xi_\delta(t_k) + \alpha(t_k, \xi_\delta(t_k), t - t_k), \qquad t_k \leqq t \leqq t_{k+1},$$
$$k = 1, 2, \ldots, n-1,$$

where $\delta = \{t_0, t_1, \ldots, t_n)$ is some partition of $[0, T]$.

We will show that an Euler approximation gives a good approximation to the solution of (10) for all $\alpha(t, x, h) \in \Pi_0(L, g, C)$.

Lemma 1. *If* $\xi(t)$ *satisfies* (10) *and* $\alpha(t, x, h)$ (1)–(6), *then*

$$\mathbb{M}\{|\xi(t) - x|^2 / \mathfrak{F}_0\} \leqq (e^{C_1 t} - 1)(1 + |x|^2),$$

where C_1 *is a constant depending only on* C, L *and* $T (0 \leqq t \leqq T)$.

Proof. Using Theorem 1 and Corollary 1 § 1, we get

$$\begin{aligned} \mathbb{M}\{|\xi_n(t) - x|^2 / \mathfrak{F}_0\} &\leqq 2\mathbb{M}\left\{\left|\int_0^t \alpha(\tau, \xi_{n-1}(\tau), d\tau) - \alpha(\tau, x, d\tau)\right|^2 \Big/ \mathfrak{F}_0\right\} \\ &\quad + 2\mathbb{M}\left\{\left|\int_0^t \alpha(\tau, x, d\tau)\right|^2 \Big/ \mathfrak{F}_0\right\} \\ &\leqq C_1 \int_0^t \mathbb{M}\{|\xi_{n-1}(\tau) - x|^2 / \mathfrak{F}_0\}\, d\tau + C_1(1 + |x|^2)\, t. \end{aligned}$$

Here C_1 is some constant depending only on C, L and T, and $\xi_n(t)$ is the n-th succesive approximation to the solution of (10) where we took as initial approximation $\xi_0(t) = x$. Iterating the preceding inequality we find

$$\mathbb{M}\{|\xi_n(t) - x|^2 / \mathfrak{F}_0\} \leqq (1 + |x|^2) \sum_{k=1}^n \frac{C_1^k t^k}{k!}.$$

Letting $n \to \infty$ and taking into account the remark following Theorem 1 we obtain the required result.

Lemma 2. *For arbitrary* $\varepsilon > 0$ *we can find an* $\varepsilon_0 = \varepsilon_0(\varepsilon)$ *such that for* $|\delta| < \varepsilon_0$

$$\mathbb{M}\{\sup_{0 \leqq t \leqq T} |\xi(t) - \xi_\delta(t)|^2 / \mathfrak{F}_0\} \leqq \varepsilon(1 + |x|^2),$$

where ε_0 *depends only on* $|\delta|$, $g(|\delta|)$, T *and the constants L and C in Conditions* (1)–(6).

Proof. Since

$$\mathbb{M}\{|\xi(t+h)-\xi(t)|^2/\mathfrak{F}_0\} \leqq 2\mathbb{M}\left\{\left|\int_t^{t+h}\alpha(\tau,\xi(\tau),d\tau)-\int_t^{t+h}\alpha(\tau,x,d\tau)\right|^2\Big/\mathfrak{F}_0\right\}$$
$$+2\mathbb{M}\left\{\left|\int_t^{t+h}\alpha(\tau,x,d\tau)\right|^2\Big/\mathfrak{F}_0\right\},$$

using (17) and (19) § 1 we find

$$\mathbb{M}\{|\xi(t+h)-\xi(t)|^2/\mathfrak{F}_0\} \leqq C_1(1+|x|^2)h+\int_t^{t+h}\mathbb{M}\{|\xi(\tau)-x|^2/\mathfrak{F}_0\}\,d\tau,$$

so that

$$\mathbb{M}\{|\xi(t+h)-\xi(t)|^2/\mathfrak{F}_0\} \leqq K(1+|x|^2)(e^{C_1 h}-1).$$

It now remains only to apply Lemma 4 § 2, from which

$$\mathbb{M}\{\sup|\xi(t)-\xi_\delta(t)|^2/\mathfrak{F}_0\} \leqq C_2\big(g(|\delta|)+\delta\big)(1+|x|^2), \tag{12}$$

where C_2 is a constant depending, like C_1, only on C, L and T.

Corollary 1. *If* $\xi(t)$ *satisfies* (10) *and the hypotheses of Theorem 1 are fulfilled, then* $\xi(t)=\underset{|\delta|\to 0}{\text{l.i.m.}}\ \xi_\delta(t)$.

Remark 4. Lemma 1 implies that if $\xi(t)$ satisfies (10) and $\alpha(t,x,h)\in\Pi_0$, then

$$\mathbb{M}|\xi(t+h)-\xi(t)|^2 \leqq C'h,$$

where C' is some constant. In particular, $\xi(t)$ is a stochastically continuous process. Since $\xi(t)$ has no discontinuities of the second kind, the process $\xi(t)$ is stochastically equivalent to a random process whose sample functions are right-continuous w.p.1 at each point of $[0, T]$. An analogous conclusion applies to the stochastic process $\alpha(t,x,h)$ considered as a function of h. Therefore, in what follows we will view the sample functions of the processes $\alpha(\cdot,\cdot,h)$ and $\xi(t)$ as right-continuous. We can hereby regard $\xi(t)$ as the limit w.p.1 of a uniformly convergent sequence of integral sums on $[0, T]$.

Remark 5. If $\alpha(\cdot,\cdot,h)$ is w.p.1 continuous as a function of h, then a separable solution of (10) is also continuous w.p.1.

If G is an open set in X, then the set of all t for which $\xi(t)\notin G$ is closed. Let

$$\tau=\min\{t;\ \xi(t)\notin G\}.$$

The random variable τ is measurable since $\tau=\inf\{t;\ t\in\Lambda,\ \xi(t)\notin G\}$ where Λ is the countable set of points of separability of $\xi(t)$. We will call τ the first exit time from the region G.

Corollary 2. *Assume $\xi_i(t)$ $(i=1,2)$ are separable solutions of the equations $d\xi_i(t)=\alpha_i(t,\xi_i(t),dt)$, $i=1,2$, $\xi_1(0)=\xi_2(0)$, that the field $\alpha(t,x,h)$ satisfies the conditions of Theorem 1 and that for some open set G, $G\subset X$,*

$$\alpha_1(t,x,h)=\alpha_2(t,x,h), \qquad x\in G,\ t\in[0,T],\ h\in[0,h_0].$$

Let τ_i denote the first exit time $\xi_i(t)$ from G_0, where G_0 together with its closure lies in G. Then $\tau_1=\tau_2$ and $\xi_1(t)=\xi_2(t)$ for all $t<\tau$ w.p.1.

Indeed, for sufficiently small $|\delta|$, the Euler polygonal segments $\xi_i^{(\delta)}(t)$ lie w.p.1 in G for $t<\tau_1$. But then $\xi_1^{(\delta)}(t)=\xi_2^{(\delta)}(t)$ for $t\subset\tau_1$ so that $\xi_1(t)=\xi_2(t)$ for $t<\tau$ (see the previous remark). Hence, $\tau_2\leqq\tau_1$. Exchanging the roles of the processes $\xi_1(t)$ and $\xi_2(t)$ in this argument, we obtain $\tau_1=\tau_2=\tau$ and $\xi_1(t)=\xi_2(t)$ for $t<\tau$ w.p.1. This property expresses the local dependence of the solutions of stochastic differential equations on the field.

We now extend the conditions for applicability of Theorem 1, replacing the uniform Lipschitz condition by a local one. We will need

Lemma 3. *If $\alpha(t,x,h)\in\Pi_0(L,g,C)$ and*

$$\mathbb{M}\sup_{0\leqq t\leqq T}|\varphi(t)|^2<\infty,$$

then there exists a constant C_2 depending only on L and T such that for the solution $\xi(t)$ of (11) we have

$$\mathbb{M}\{\sup_{0\leqq t\leqq T}|\xi(t)|^2/\mathfrak{F}_0\}\leqq C_2[1+\mathbb{M}(\sup_{0\leqq t\leqq T}|\varphi(t)|^2)].$$

Proof. Let

$$z(t)=\mathbb{M}\{\sup_{0\leqq\tau\leqq t}|\xi(\tau)|^2/\mathfrak{F}_0\}, \qquad \gamma=\mathbb{M}(\sup_{0\leqq t\leqq T}|\varphi(t)|^2)$$

from Lemma 2 § 2

$$z(t)\leqq 2\gamma+2LC_1\left(T+\int_0^t\mathbb{M}\{|\xi(\tau)|^2/\mathfrak{F}_0\}\,d\tau\right)\leqq 2\gamma+C_1'+C_2'\int_0^t z(\tau)\,d\tau,$$

where C_1' and C_2' are constants depending only on L and T. From the last inequality

$$z(t)\leqq(2\gamma+C_1')e^{C_1' t}$$

which was to be proved.

Theorem 2. *If $\alpha(t,x,h)\in\Pi_1$, $\mathbb{P}\{\sup_{0\leqq t\leqq T}|\varphi(t)|=\infty\}=0$ and $\varphi(t)$ is right-continuous, then the stochastic differential equation (11) has a unique w.p.1 right-continuous solution which is $\mathfrak{F}_t$-measurable for each $t\in[0,T]$. If $\varphi(t)$ and $\alpha(\cdot,\cdot,h)$ are w.p.1 continuous in t and h, resp., then $\xi(t)$ is also continuous w.p.1.*

We prove the existence of a solution. First we approximate the field $\alpha(t, x, h)$ by the fields $\alpha_N(t, x, h) \in \Pi_0$. Set

$$\alpha_N(t, x, h) = g_N(x)\, \alpha(t, x, h),$$

where $g_N(x)=1$ if $|x| \leqq N$, $g_N(x)=N+1-|x|$ if $N \leqq |x| < N+1$ and $g_N(x)=0$ for $|x| > N+1$. Clearly, $\alpha_N(t, x, h) \in \Pi_0(L, g, C_N)$ so that

$$\xi_N(t) = \varphi_N(t) + \int_0^t \alpha_N(\tau, \xi_N(\tau), d\tau), \quad \text{where } \varphi_N(t) = \varphi(t)\, g_N(\varphi(t))$$

has a right-continuous solution. Let τ_N be the first exit time of $\xi_N(t)$ from the sphere $|x| \leqq N$. Then $\xi_{N+1}(t) = \xi_N(t)$, if $t \leqq \tau_N$. From Lemma 3

$$\mathbb{P}\{\tau_N \leqq T\} = \mathbb{P}\{\sup_{0 \leqq t \leqq T} |\xi_N(t)| > N\} \to 0 \quad \text{for } N \to \infty.$$

Thus, w.p.1 the limit $\lim \xi_N(t) = \xi(t)$ $(0 \leqq t \leqq T)$ exists and w.p.1 there is a number $N_0 = N(\omega)$ such that

$$\xi(t) = \xi_N(t)\, (0 \leqq t \leqq T), \quad N > N_0.$$

Consequently, $\xi(t)$ is w.p.1 right-continuous and it also follows that

$$\mathbb{P}\text{-}\lim_{N \to \infty} \int_0^t \alpha_N(\tau, \xi_N(\tau), d\tau) = \int_0^t \alpha(\tau, \xi(\tau)\, d\tau) \tag{13}$$

exists by definition. Moreover, w.p.1 the integral on the right side of (13) coincides with that on the left for all $t \in [0, T]$ and all $N \geqq N_0 = N_0(\omega)$. Thus, $\xi(t)$ satisfies (11). It is clear that $\xi(t)$ is $\mathfrak{F}_t$-measurable for each t.

The proof of the uniqueness runs analogously to that of Theorem 3 §6 Part I. If $\xi_i(t)$ are two right-continuous solutions of (11), $\mathfrak{F}_t$-measurable for each t, and if $\psi_N(t)=1$ when

$$\sup_{0 \leqq \tau \leqq t} |\xi_i(\tau)| \leqq N \quad (i=1, 2),$$

and $\psi_N(t)=0$ when $\max_i \sup_{0 \leqq \tau \leqq t} |\xi_i(\tau)| > N$, then

$$\begin{aligned}
&\mathbb{M}\, \psi_N(t)\, |\xi_1(t) - \xi_2(t)|^2 \\
&\quad = \mathbb{M}\, \psi_N(t) \left| \int_0^t \alpha_{N+1}(\tau, \xi_1(\tau), d\tau) - \int_0^t \alpha_{N+1}(\tau, \xi_2(\tau), d\tau) \right|^2 \\
&\quad \leqq C'_{N+1}\, \mathbb{M}\, \psi_N(t) \int_0^t |\xi_1(\tau) - \xi_2(\tau)|^2\, d\tau \\
&\quad \leqq C'_{N+1}\, \mathbb{M} \int_0^t \psi_N(\tau)\, |\xi_1(\tau) - \xi_2(\tau)|^2\, d\tau,
\end{aligned}$$

where C'_{N+1} is some constant. From the preceding inequality it follows that $\psi_N(t)\, |\xi_1(t) - \xi_2(t)|^2 = 0$ for each N and t w.p.1, from which the uniqueness follows. The remaining claim of the theorem on the continuity of $\xi(t)$ follows from the fact that w.p.1 $\xi(t)$ coincides with some $\xi_N(t)$ and $\xi_N(t)$ is continuous as a consequence of Theorem 1 § 2.

Remark 6. If $\alpha(t, x, h) \in \Pi_1(L, g)$ and

$$\mathbb{M} \sup_{0 \leqq t \leqq T} |\varphi(t)|^2 < \infty,$$

then the solution $\xi(t)$ of (11) belongs to L_2 for each t and

$$\mathbb{M}\left(\sup_{0 \leqq t \leqq T} |\xi(t)|^2\right) \leqq C_2\left[1 + \mathbb{M}\left(\sup_{0 \leqq t \leqq T} |\varphi(t)|^2\right)\right], \tag{14}$$

where the constant C_2 is as in Lemma 3.

Indeed, since $\xi(t)$ begins at some $N_0 = N_0(\omega)$ to coincide with $\xi_N(t)$, $N \geqq N_0$, we have $\sup_{0 \leqq t \leqq T} |\xi(t)| \leqq \varliminf_{N \to \infty} \sup_{0 \leqq t \leqq T} |\xi_N(t)|$. Now we need only use Fatou's lemma and Lemma 3 to obtain (14).

Chapter 2. Stochastic Differential Equations without After-effect

§ 4. Preliminary Remarks

The random motions studied in the one-dimensional case (Part I) possessed the following properties: the main part of the displacement of a point moving during $(t, t+h)$ consisted of "regular" displacement equal to $A(t, x)\, h$ of the same character as one defined by an ordinary differential equation; on this was superimposed a random component, characterizing the fluctuations of the point w.r.t. the regular motion, of the form $\sigma(t, x)[w(t+h)-w(t)]$, where $w(t)$ was a Wiener process.

In this chapter we will treat motions in several dimensions whose local characteristic is complicated by an additional factor – a discontinuous process with independent increments and arbitrary set of jumps. We will denote this additional component by $\gamma(t, x, h)$ and begin with some preliminary remarks concerning it.

As before, we consider processes with finite second-order moments. As is well-known, an arbitrary homogeneous vector process with independent increments, finite second-order moments, mean zero and no continuous component can be represented as

$$\gamma(t, x, h)=\int u\, \tilde{\nu}_{tx}(h, du), \tag{1}$$

where u is an n-dimensional vector, the integration is to be carried out over all of n-dimensional space, $\nu_{tx}(h, A)$ is Poisson random measure and

$$\tilde{\nu}_{tx}(h, A)=\nu_{tx}(h, A)-h\,\Pi_{tx}(A), \tag{2}$$

$$h\,\Pi_{tx}(A)=\mathbb{M}\, \nu_{tx}(h, A),$$

with

$$\int |u|^2\, \Pi_{tx}(dx)<\infty\,. \tag{3}$$

The variables t and x play the role here of parameters of the measures under consideration and since the dependence of measures on a parameter causes a certain inconvenience, it is more expedient to proceed to other representations containing certain point-functions depending on

t and x, i.e., to representations of the form

$$\gamma(t, x, h) = \int f(t, x, u)\, \tilde{\nu}(h, du) \tag{4}$$

(with fixed "standard" Poisson measure $\nu(h, A)$). In this transition to (4) no generality is lost. In order to see this, we find an expression for the characteristic function of the distribution of the variable at (4). The integral there is to be understood as the limit in mean square of the sum $(n \to \infty)$

$$\sum_k f_k^{(n)}\, \tilde{\nu}(h, A_k^{(n)}), \tag{5}$$

where $f^{(n)}(u)$ is a piecewise-continuous function

$$f^{(n)}(u) = f_k^{(n)} \quad \text{for } u \in A_k^{(n)}, \qquad \bigcup_k A_k^{(n)} = X,$$

$$\int |f(u) - f^{(n)}(u)|^2\, \Pi(du) \to 0.$$

The dependence of $f^{(n)}(u)$ on t and x will no longer be explicitly written. Taking into account that $\nu(t, A)$ has a Poisson distribution and that on the non-intersecting sets A_k, $k = 1, 2, \ldots, n$ the random variables

$$\nu(h, A_1),\ \nu(h, A_2), \ldots, \nu(h, A_n)$$

are independent, we obtain for the characteristic function $g_n(\lambda)$ of the random vector (5) the following expression:

$$\begin{aligned}
g_n(\lambda) &= \mathbb{M} \exp\{i(\lambda | \sum_k f_k^{(n)}\, \tilde{\nu}(h, A_k^{(n)}))\} \\
&= \prod_k \mathbb{M} \exp\{i(\lambda | f_k^{(n)})\, \tilde{\nu}(h, A_k^{(n)})\} \\
&= \prod_k \exp\{(e^{i(\lambda | f_k^{(n)})} - 1 - i(\lambda | f_k^{(n)}))\, h\, \Pi(A_k^{(n)})\} \\
&= \exp\{h \int [e^{i(\lambda | f^{(n)}(u))} - 1 - i(\lambda | f^{(n)}(u))]\, \Pi(du)\}.
\end{aligned}$$

Letting $n \to \infty$ we obtain for the characteristic function $g(\lambda)$ of (4)

$$\ln g(\lambda) = h \int [e^{i(\lambda | f(u))} - 1 - i(\lambda | f(u))]\, \Pi(du), \tag{6}$$

where $f(u) = f(t, x, u)$ is an arbitrary function satisfying

$$\int |f(u)|^2\, \Pi(du) < \infty. \tag{7}$$

The substitution $v = f(u)$ transforms (6) into

$$\ln g(\lambda) = h \int [e^{i(\lambda | v)} - 1 - i(\lambda | v)]\, \Pi f^{-1}(dv), \tag{8}$$

and (7) is then

$$\int |v|^2\, \Pi f^{-1}(dv) < \infty. \tag{9}$$

On the other hand, for any measure $\Pi'(A)$ on the σ-algebra of sets of the n-dimensional space X there exists a measurable mapping $f(x)$ of X into X for which

$$\Pi f^{-1}(A)=\Pi'(A),$$

where $\Pi(A)$ is the Lebesgue measure of A. Hence, formula (6) together with condition (7) includes arbitrary distributions corresponding to the processes (1), i.e., to arbitrary homogeneous processes with independent increments, finite second-order moments and no continuous component.

In this chapter we will thus consider motions in an s-dimensional space X described by a stochastic differential equation

$$d\xi=\alpha(t,\xi,dt),$$

where

$$\alpha(t,x,h)=a(t,x)\,h+B(t,x)[w(t+h)-w(t)) + \int f(t,x,u)\,\tilde{\nu}(h,du),$$

$a(t,x)$, $f(t,x,u)$ are nonrandom n-dimensional functions of the arguments $t\in[0,T]$, $x\in X$, and $u\in X$, $B(t,x)$ is a nonrandom operator function mapping (for fixed t and x) X linearly into X, $w(t)$ is an s-dimensional vector Wiener process, i.e., a vector process whose components in an arbitrary orthonormal basis are independent homogeneous Wiener processes and

$$\tilde{\nu}(t,A)=\nu(t,A)-t\,\Pi(A),$$

where $\nu(t,A)$ is the Poisson measure,

$$t\,\Pi(A)=\mathbb{M}\,\nu(t,A).$$

§ 5. Some Special Types of Stochastic Integrals

We shall need the definitions of several stochastic integrals in special cases. The definitions and properties of the integrals are analogous to those we considered previously. First we consider integrals of the form

$$\int_0^T (\varphi(t)|d\psi(t)), \tag{1}$$

where $\varphi(t)$ and $\psi(t)$ are certain random processes. Such integrals are a very special case of a stochastic line integral corresponding to the random field $\alpha(t,x,h)=(x|\psi(t+h)-\psi(t))$. However, we will consider here assumptions which are somewhat different from those made earlier: we will assume that the process $\psi(t)$ is a D-martingale.

Definition 1. A random vector process $\psi(t)$, $t\in[0,T]$ will be called a *martingale*, if $\mathbb{M}|\psi(t)|<\infty$ and there is given a family of σ-algebras

$\mathfrak{F}_t (t \in [0, T], \mathfrak{F}_t \subset \mathfrak{S})$ for which $\psi(t)$ is $\mathfrak{F}_t$-measurable for each t and

$$\mathbb{M}\{\psi(s)/\mathfrak{F}_t\} = \psi(t) \quad \text{for } s > t.$$

A martingal $\psi(t)$ will be called a *D-martingale* if $\mathbb{M}|\psi(t)|^2 < \infty$ and there exists a nonnegative random matrix $\rho(t)$, $\mathfrak{F}_t$-measurable for each t w.p. 1 such that

$$\mathbb{M}\{(\psi(s)-\psi(t))(\psi(s)-\psi(t))^*/\mathfrak{F}_t\} = \mathbb{M}\left\{\int_t^s \rho(\tau)\, d\tau/\mathfrak{F}_t\right\}, \quad 0 \leqq t \leqq s \leqq T.$$

Vectors will again be viewed as column vectors and the symbol "*" will denote the transpose of a vector (a row vector). We mention the following special case of a D-martingale.

If

$$\mathbb{M}\{[\psi_k(s) - \psi_k(t)]^2/\mathfrak{F}_t\} = F_k(s) - F_k(t),$$

where $\psi_k(t)$ are the components of the vector $\psi(t)$ and $F_k(t)$ are non-random functions, then the field $\alpha(t, x, h) = (x \mid \psi(t+h) - \psi(t))$ satisfies all the assumtions of Chapter I, i.e., it is a field with limited after-effect, is quasi-differential $(g(h)=0)$ and satisfies a uniform Lipschitz condition. Consequently, all of the results on line integrals of the previous chapter apply to the integral (1).

In the more general case of a D-martingale the construction of the stochastic integral and the derivation of its properties are carried out as before. It is sufficient to limit ourselves to the scalar case, i.e., to assume that $\Phi(t)$ and $\psi(t)$ are random processes assuming numerical values since (1) can be considered as the sum of integrals of scalar processes.

First we assume that $\varphi(t)$ is piece-wise constant, i.e.,

$$\varphi(t) = \alpha_k \quad \text{for } t \in [t_k^*, t_{k+1}^*], \; k = 0, 1, \ldots, n-1,$$

where $0 = t_0^* < t_1^* < \cdots < t_n^*$, and α_k is an $\mathfrak{F}_{t_k^*}$-measurable random variable. We then have

$$\int_0^T \varphi(t)\, d\psi(t) = \sum_{k=0}^{n-1} \alpha_k [\psi(t_{k+1}^*) - \psi(t_k^*)]. \tag{2}$$

As in § 2, Part I we verify that (2) has the following properties:

1) if α_1 and α_2 are constant random variables, then

$$\int_0^T [\alpha_1 \varphi_1(t) + \alpha_2 \varphi_2(t)]\, d\psi(t) = \alpha_1 \int_0^T \varphi_1(t)\, d\psi(t) + \alpha_2 \int_0^T \varphi_2(t)\, d\psi(t); \tag{3}$$

2) if $\chi_{[\alpha, \beta)}(t)$ is the indicator of $[\alpha, \beta)$, then

$$\int_0^T \chi_{[\alpha, \beta)}(t)\, d\psi(t) = \psi(\beta) - \psi(\alpha);$$

3) if $\mathbb{M}\int_0^T \varphi^2(t)\,\rho(t)\,dt<\infty$, then

$$\mathbb{M}\left\{\int_0^T \varphi(t)\,d\psi(t)/\mathfrak{F}_0\right\}=0,$$

$$\mathbb{M}\left\{\left|\int_0^T \varphi(t)\,d\psi(t)\right|^2\Big/\mathfrak{F}_0\right\}=\mathbb{M}\left\{\int_0^T \varphi^2(t)\,\rho(t)\,dt/\mathfrak{F}_0\right\}; \tag{4}$$

4) for arbitrary C and N we have

$$\mathbb{P}\left\{\left|\int_0^T \varphi(t)\,d\psi(t)\right|>C\right\}\leqq\mathbb{P}\left\{\int_0^T \varphi^2(t)\,\rho(t)\,dt>N\right\}+\frac{N}{C^2}. \tag{5}$$

Let $H(\rho)$ and $H_2(\rho)$ denote the classes of random processes $\varphi(t)$ defined on $[0, T]$, $\mathfrak{F}_t$-measurable for each t and satisfying w.p. 1

$$\mathbb{M}\int_0^T \varphi^2(t)\,\rho(t)\,dt<\infty, \qquad \int_0^T \varphi^2(t)\,\rho(t)\,dt<\infty$$

respectively. Properties 1–4 allow a generalization of the definition of the stochastic integral to the classes $H(\rho)$ and $H_2(\rho)$. If $\varphi(t)\in H(\rho)$ and $\varphi_n(t)$ is an arbitrary sequence of step processes in $H(\rho)$ for which

$$\mathbb{M}\int_0^T [\varphi(t)-\varphi_n(t)]^2\,\rho(t)\,dt\to 0,$$

then we assume

$$\int_0^T \varphi(t)\,d\psi(t)=\text{l.i.m.}\int_0^T \varphi_n(t)\,d\psi(t). \tag{6}$$

The existence this limit independent of the choice of the approximating sequence $\varphi_n(t)$ and the validity of Properties 1–4 for the integral (6) are proved like the analogous claims in § 2, Part I for integrals w.r.t. a Wiener process. If $\varphi(t)\in H_2(\rho)$, then we choose a sequence $\varphi_n(t)\in H(\rho)$ such that

$$\int_0^T [\varphi(t)-\varphi_n(t)]^2\,\rho(t)\,dt\to 0$$

in probability and assume

$$\int_0^T \varphi(t)\,d\psi(t)=\mathbb{P}\text{-lim}\int_0^T \varphi_n(t)\,d\psi(t)$$

(see § 2, Part I). The corresponding limit exists independent of the choice of the sequence $\varphi_n(t)$ approximating $\varphi(t)$ and possesses Properties 1 and 4.

For the sequel we also note that for arbitrary $\varphi_i(t) \in H(\rho)$ $(i=1,2)$

$$\mathbb{M}\left\{\int_0^T \varphi_1(t)\, d\psi(t) \cdot \int_0^T \varphi_2(t)\, d\psi(t)/\mathfrak{F}_0\right\} = \mathbb{M}\left\{\int_0^T \varphi_1(t)\, \varphi_2(t)\, \rho(t)\, dt/\mathfrak{F}_0\right\}. \quad (7)$$

We now consider the properties of a stochastic integral as a function of its upper limit. Set

$$\eta(t) = \int_0^t \varphi(t)\, d\psi(t). \quad (8)$$

We remark that the proofs of Theorems 1 and 2 § 3, Part I depend only on the fact that the Wiener process $w(t)$ is a continuous D-martingale. In this connection we have the following theorems.

Theorem 1. *Let $\varphi(t) \in H(\rho)$ and let $\eta(t)$ be a separable stochastic integral. Then*

$$\mathbb{P}\left\{\sup_{0 \leqq t \leqq T}\left|\int_0^t \varphi(s)\, d\psi(s)\right| > a\right\} \leqq \frac{1}{a^2}\int_0^T \mathbb{M}\varphi^2(s)\, \rho(s)\, ds, \quad (9)$$

$$\mathbb{M}\left\{\sup_{0 \leqq t \leqq T}\left|\int_0^t \varphi(s)\, d\psi(s)\right|^2\right\} \leqq 4\mathbb{M}\left[\int_0^T \varphi(s)\, d\psi(s)\right]^2 = 4\int_0^T \mathbb{M}\varphi^2(s)\rho(s)\, ds. \quad (10)$$

Theorem 2. *Let $\varphi(t) \in H_2(\rho)$ and let $\eta(t)$ be a separable process. If the process $\psi(t)$ has w.p. 1 no discontinuities of the second kind (is continuous), then $\eta(t)$ also has no discontinuities of the second kind w.p. 1 (is continuous). In this regard*

$$\mathbb{P}\left\{\sup_{0 \leqq t \leqq T}\left|\int_0^t \varphi(s)\, d\psi(s)\right| > C\right\} \leqq \frac{N}{C^2} + \mathbb{P}\left\{\int_0^T \varphi^2(s)\, \rho(s)\, ds > N\right\}.$$

Theorem 3. *Assume the separable martingale $\psi(t)$ has no discontinuities of the second kind. If, moreover,*

$$\sum_{k=1}^n \mathbb{P}\{|\psi(t_{nk}) - \psi(t_{nk-1})| > a\} \to 0 \quad \text{for } \max_k |t_{nk} - t_{nk-1}| \to 0,$$

where $0 = t_{n0} < t_{n1} < t_{n2} < \cdots < t_{nn} = T$, then $\psi(t)$ is continuous w.p. 1.

We note that if $\varphi(t) \in H(\rho)$ then $\eta(t)$ is a D-martingale.

We turn now to the definition of a stochastic integral in which we will distinguish the space and time variables, which play different roles in our constructions.

Let X be s-dimensional Euclidean space, $\mathfrak{B}_\varepsilon$ a σ-algebra of sets of X lying in the set $\left\{x, \varepsilon \leqq |x| \leqq \frac{1}{\varepsilon}\right\}$, $\varepsilon > 0$ and $\mathfrak{B}_0$ the sum of all $\mathfrak{B}_\varepsilon$ with $\varepsilon > 0$. Assume that each $A \in \mathfrak{B}_0$ is related to a random variable $\nu(A)$ having a Poisson distribution with parameter $\Pi(A)$ $(0 \leqq \Pi(A) < \infty)$, whereby

a) if $A_1, A_2, \ldots, A_n$ are pair-wise disjunct, $A_k \in \mathfrak{B}_0$, then the random variables $\nu(A_1)$, $\nu(A_2)$ are mutually independent.

b) if $A_n \in \mathfrak{B}_\varepsilon$ ($\varepsilon > 0$, $n = 1, 2, \ldots$) then w.p. 1

$$\nu\left(\bigcup_{n=1}^{\infty} A_n\right) = \sum_{n=1}^{\infty} \nu(A_n).$$

The random set function $\nu(A)$ will be called a Poisson measure on $\mathfrak{B}_0$. From the definition it follows that

$$\Pi\left(\bigcup_{n=1}^{\infty} A_n\right) = \sum_{n=1}^{\infty} \Pi(A_n), \qquad A_n \in \mathfrak{B}_\varepsilon.$$

Hence without loss of generality we can assume that $\Pi(A)$ is a measure on the σ-algebra of Borel sets of X. It is finite on an arbitrary bounded Borel set of X whose closure does not contain the point zero. We can assume that $\Pi\{0\} = 0$, where $\{0\}$ is a one-point set, but, generally speaking, it is allowed that

$$\Pi\{x; |x| < \varepsilon\} = \infty, \qquad \Pi\{x; |x| > \varepsilon\} = \infty.$$

We introduce the random measure $\tilde{\nu}(A)$ on $\mathfrak{B}_0$,

$$\tilde{\nu}(A) = \nu(A) - \Pi(A),$$

and define the stochastic integral $\int \alpha(x)\, \tilde{\nu}(dx)$ in a way analogous to that constructed for the Wiener process (Part I, Chapter I, § 2). Assume that some σ-algebra of events $\mathfrak{A}$ is fixed in a basic probability space in such a way that $\nu(A)$ does not depend on $\mathfrak{A}$. Consider a random function $\alpha(x)$, $x \in X$ which is $\mathfrak{A}$-measurable for arbitrary x. If $\alpha(x)$ is a step function, i.e., $\alpha(x) = \alpha_k$ when $x \in A_k$, where the sets A_k are pairwise disjoint, ($k = 1, 2, \ldots, n$) and $\alpha(x) = 0$ for $x \notin \bigcup_{k=1}^{N} A_k$ with $A_k \in \mathfrak{B}_\varepsilon$, then we take

$$\int \alpha(x)\, \tilde{\nu}(dx) = \sum_{k=1}^{n} \alpha_k\, \tilde{\nu}(A_k).$$

If $\alpha(x)$ is also such that

$$\int \mathbb{M}\, \alpha^2(x)\, \Pi(dx) < \infty, \tag{11}$$

then we set

$$\int \alpha(x)\, \tilde{\nu}(dx) = \text{l.i.m.} \int \alpha_n(x)\, \tilde{\nu}(dx), \tag{12}$$

where $\alpha_n(x)$ is an arbitrary sequence of simple functions for which

$$\int_X \mathbb{M}\, |\alpha(x) - \alpha_n(x)|^2\, \Pi(dx) \to 0.$$

The existence of the limit in (12) above follows from the easy to verify equality

$$\mathbb{M}\,|\int \alpha_1(x)\,\tilde{\nu}(dx) - \int \alpha_2(x)\,\tilde{\nu}(dx)|^2 = \int \mathbb{M}\,|\alpha_1(x) - \alpha_2(x)|^2\,\Pi(dx) \tag{13}$$

for the simple functions $\alpha_i(x)$ $(i=1, 2)$.

From (12) it follows that (13) also holds for arbitrary $\alpha_i(x)$ ($\mathfrak{A}$-measurable for arbitrary x) satisfying Condition 1. The stochastic integral introduced clearly possesses the following properties:

1) $\int [C_1\,\alpha_1(x) + C_2\,\alpha_2(x)]\,\tilde{\nu}(dx) = C_1 \int \alpha_1(x)\,\tilde{\nu}(dx) + C_2 \int \alpha_2(x)\,\tilde{\nu}(dx)$;

2) if $\chi_A(x)$ is the indicator of the set $A \in \mathfrak{B}_0$, then

$$\int \chi_A(x)\,\tilde{\nu}(dx) = \tilde{\nu}(A);$$

3) $\mathbb{M}\{\int \alpha(x)\,\tilde{\nu}(dx)/\mathfrak{A}\} = 0$, $\mathbb{M}\{|\int \alpha(x)\,\tilde{\nu}(dx)|^2/\mathfrak{A}\} = \int \mathbb{M}\,|\alpha(x)|^2\,\Pi(dx)$;

4) for arbitrary $N>0$ and $C>0$

$$\mathbb{P}\{|\int \alpha(x)\,\tilde{\nu}(dx)| > C\} \leqq \frac{N}{C^2} + \mathbb{P}\{\int \alpha^2(x)\,\Pi(dx) > N\}. \tag{14}$$

The proof of (14) can be obtained initially for step functions $\alpha(x)$ and then with the help of a limit passage for arbitrary $\alpha(x)$ satisfying (11) (in the same way as the corresponding equality of §2 Chapter I, Part I). Inequality (14) allows an extension of the definition of the integral to functions $\alpha(x)$ for which

$$\int \alpha^2(x)\,\Pi(dx) \tag{15}$$

is finite w.p.1. In fact, let $\alpha_n(x)$ be a sequence of $\mathfrak{A}$-measurable random variables with

$$\int \mathbb{M}\,|\alpha_n(x)|^2\,\Pi(dx) < \infty$$

and

$$\mathbb{P}\text{-}\lim \int |\alpha(x) - \alpha_n(x)|^2\,\Pi(dx) = 0.$$

Then from (14)

$$\begin{aligned}&\mathbb{P}\{|\int [\alpha_{n+m}(x) - \alpha_n(x)]\,\tilde{\nu}(dx)| > \delta\}\\ &\quad\leqq \varepsilon + \mathbb{P}\{|\int [\alpha_{n+m}(x) - \alpha_n(x)]^2\,\Pi(dx)| > \varepsilon\,\delta^2\},\end{aligned}$$

i.e., the sequence of integrals

$$\int \alpha_n(x)\,\tilde{\nu}(dx)$$

is fundamental in the sense of convergence in probability. Define

$$\int \alpha(x)\,\tilde{\nu}(dx) = \mathbb{P}\text{-}\lim \int \alpha_n(x)\,\tilde{\nu}(dx).$$

It is not difficult to show that the value of this integral does not depend on the choice of the approximating sequence $\alpha_n(x)$ and in this connection the extended definitions of the Properties 1 and 4 of the integral still hold.

In addition to the integral w.r.t. the measure $\tilde{\nu}(\cdot)$ we will also treat integrals defined directly w.r.t. $\nu(\cdot)$. To assure the existence of such an integral, outside of the existence w.p. 1 of (15) it is also necessary to require the finiteness (w.p. 1) of

$$\int |\alpha(x)|\, \Pi(dx).$$

Then, by definition

$$\int \alpha(x)\, \nu(dx) = \int \alpha(x)\, \tilde{\nu}(dx) + \int \alpha(x)\, \Pi(dx).$$

We now consider the Poisson measure in the space $[0, T] \times X$ whose values on a set of the form $\Delta \times A$ will be denoted by $\nu(\Delta, A)$. Assume that the random measure ν is homogeneous w.r.t. translation on $[0, T]$, i.e., that

$$\mathbb{M}\, \nu(\Delta, A) = |\Delta|\, \Pi(A),$$

where $|\Delta|$ is the Lebesgue measure of Δ. Set

$$\tilde{\nu}(\Delta, A) = \nu(\Delta, A) - |\Delta|\, \Pi(A).$$

As in Chapter 1 we assume that the fixed, nondecreasing family of σ-algebras $\mathfrak{F}_t$ $(0 \leqq t \leqq T)$ belong to $\mathfrak{S}$ where

a) for arbitrary $A \in \mathfrak{B}_0$ the random variables $\nu([0, t], A)$ are $\mathfrak{F}_t$-measurable;

b) the family of random variables $\{\nu((t, t+h), A), A \in \mathfrak{B}_0,\ h > 0\}$ is independent of the σ-algebra $\mathfrak{F}_t$.

We now define the integral

$$\int_0^T \int \varphi(t, y)\, \tilde{\nu}(dt, dy) \tag{16}$$

for a certain class of functions $\varphi(t, y)$ $(t \in [0, T], y \in X)$, $\mathfrak{F}_t$-measurable for arbitrary t and y. We assume initially that $\varphi(t, y) = \varphi_k(y)$ if $t \in [t_{k-1}, t_k)$, $k = 1, 2, \ldots, n$, $0 = t_0 < t_1 < \cdots < t_n = T$, $\varphi_k(y)$ is $\mathfrak{F}_{t_{k-1}}$-measurable and

$$\int \mathbb{M}\, \varphi_k^2(y)\, \Pi(dy) < \infty.$$

Then we set

$$\int_0^T \int \varphi(t, y)\, \tilde{\nu}(dt, dy) = \sum_{k=1}^{n} \int \varphi(t_{k-1}, y)\, \tilde{\nu}((t_{k-1}, t_k), dy).$$

The integrals on the right side of this expression make sense in virtue of the definitions just given. It is not difficult to verify that the introduced integral possesses the following properties:

1) $$\int_0^T \int [C_1 \varphi_1(t, y) + C_2 \varphi_2(t, y)]\, \tilde{\nu}(dt, dy) = C_1 \int_0^T \int \varphi_1(t, y)\, \tilde{\nu}(dt, dy) + C_2 \int_0^T \int \varphi_2(t, y)\, \tilde{\nu}(dt, dy); \tag{17}$$

2) if $\chi_{\Delta\times A}(t, y)$ is the indicator of $\Delta\times A$, then

$$\int_0^T\int \chi_{\Delta\times A}(t, y)\,\tilde{v}(dt, dy)=\tilde{v}(\Delta, A); \tag{18}$$

3) $$\mathbb{M}\left\{\int_0^T\int \varphi(t, y)\,\tilde{v}(dt, dy)/\mathfrak{F}_0\right\}=0, \tag{19}$$

$$\mathbb{M}\left\{\left|\int_0^T\int \varphi(t, y)\,\tilde{v}(dt, dy)\right|^2\Big/\mathfrak{F}_0\right\}=\int_0^T\int \mathbb{M}\{\varphi^2(t, y)/\mathfrak{F}_0\}\,\Pi(dy)\,dt; \tag{20}$$

4) for arbitrary $N>0$, $C>0$

$$\begin{aligned}&\mathbb{P}\left\{\left|\int_0^T\int \varphi(t, y)\,\tilde{v}(dt, dy)\right|>C\right\}\\ &\quad\leqq\frac{N}{C^2}+\mathbb{P}\left\{\int_0^T\int \varphi^2(t, y)\,\Pi(dy)\,dt>N\right\}.\end{aligned} \tag{21}$$

The integral defined at (16) is easily generalized to a wider class of functions. Set

$$\int_0^T\int \varphi(t, y)\,\tilde{v}(dt, dy)=\underset{n\to\infty}{\text{l.i.m.}}\int_0^T\int \varphi_n(t, y)\,\tilde{v}(dt, dy), \tag{22}$$

where $\varphi(t, y)$ is an arbitrary random function, $\mathfrak{F}_t$-measurable with $t\in[0, T]$, $y\in X$, such that

$$\int_0^T\int \mathbb{M}\,\varphi^2(t, y)\,\Pi(dy)\,dt<\infty,$$

and $\varphi_n(t, y)$ is an arbitrary sequence of $\mathfrak{F}_t$-measurable functions for which

$$\int_0^T\int \mathbb{M}\,|\varphi(t, y)-\varphi_n(t, y)|^2\,\Pi(dy)\,dt\to 0 \qquad \text{when } n\to\infty.$$

The limit on the right side of (22) exists and is independent of the choice of the sequence $\varphi_n(t, y)$ and Relations 1–4 are also satisfied for the integral on the left side of (22). We will say that $\varphi(t, y)\in H(\Pi)$ if for arbitrary $t\in[0, T]$ and $y\in X$ the random variable $\varphi(t, y)$ is measurable as a function of the three arguments ω, t and y, is $\mathfrak{F}_t$-measurable and

$$\int_0^T\int \mathbb{M}\,\varphi^2(t, y)\,\Pi(dy)\,dt<\infty,$$

and that $\varphi(t, y)\in H_2(\Pi)$ if it satisfies the previous conditions except for the last which is replaced by

$$\int_0^T\int \varphi^2(t, y)\,\Pi(dy)\,dt<\infty$$

(w.p.1).

The inequality (21) allows generalization of the definition of the integral (16) to functions $\varphi(t, y)$ in $H_2(\Pi)$. We obtain, namely

$$\int_0^T\int \varphi(t, y)\,\tilde{\nu}(dt, dy) = \mathbb{P}\text{-}\lim \int_0^T\int \varphi_n(t, y)\,\tilde{\nu}(dt, dy),$$

where $\varphi_n(t, y) \in H(\Pi)$ and

$$\int_0^T\int |\varphi(t, y) - \varphi_n(t, y)|^2\,\Pi(dy)\,dt \to 0 \qquad \text{for } n \to \infty$$

in probability. (The existence of the corresponding limit is established as in the analogous case treated somewhat earlier.)

We now turn to the properties of the integral

$$\Phi(t) = \int_0^t\int \varphi(\tau, y)\,\tilde{\nu}(d\tau, dy),$$

where $\varphi(t, y) \in H$ as function of t. It follows from the previous formulas that

$$\mathbb{M}\{\Phi(t) - \Phi(s)/\mathfrak{F}_s\} = 0, \qquad 0 \leqq s < t \leqq T, \tag{23}$$

$$\mathbb{M}\{|\Phi(t) - \Phi(s)|^2/\mathfrak{F}_s\} = \int_s^t\int \mathbb{M}\{\varphi^2(\tau, y)/\mathfrak{F}_s\}\,\Pi(dy)\,d\tau, \tag{24}$$

i.e., the process $\Phi(t)$ is a D-martingale. Since $\Phi(t)$ is defined only w.p.1 for each t, we can extend the definition of $\Phi(t)$ in such a way that it represents a separable process. Then $\Phi(t)$ is bounded w.p.1 and has no discontinuities of the second kind.

Theorem 1′. *If $\varphi(t, y) \in H$, then*

$$\mathbb{P}\{\sup_{0 \leqq t \leqq T} |\Phi(t)| > a\} \leqq \frac{1}{a^2}\int_0^T\int \mathbb{M}\,\varphi^2(t, y)\,\Pi(dy)\,d\tau,$$

$$\mathbb{M}\{\sup_{0 \leqq t \leqq T} |\Phi(t)|^2\} \leqq 4\,\mathbb{M}\left[\int_0^T\int \varphi(\tau, y)\,\tilde{\nu}(d\tau, dy)\right]^2 = 4\int_0^T\int \mathbb{M}\,\varphi^2(\tau, y)\,\Pi(dy)\,d\tau.$$

Lemma 1. *If $\varphi(t, y) \in H_2(\Pi)$ and N and a are arbitrary, then*

$$\mathbb{P}\{\sup_{0 \leqq t \leqq T} |\Phi(t)| > a\} \leqq \frac{N}{a^2} + \mathbb{P}\left\{\int_0^T\int \varphi^2(t, y)\,\Pi(dy)\,dt > N\right\}.$$

The proof coincides with that of Theorem 2 § 3, Part I.

Lemma 2. *Let $A\in\mathfrak{B}_0$, $B\in\mathfrak{B}_0$ and let $\delta=\{0=t_0<t_1<\cdots<t_n=T\}$ be an arbitrary partition of $[0, T]$, $\Delta_k=[t_k, t_{k+1}]$. Then*

$$\underset{|\delta|\to 0}{\text{l.i.m.}} \sum_{k=0}^{n-1} \tilde{v}(\Delta_k, A)\, \tilde{v}(\Delta_k, B)=v([0, T], A\cap B). \tag{25}$$

Set

$$x=\sum_{k=0}^{n-1} x_k, \qquad x_k=\tilde{v}(\Delta_k, A)\, \tilde{v}(\Delta_k, B)-v(\Delta_k, A\cap B).$$

First, we have

$$\mathbb{M}\, x_k=\mathbb{M}\, \tilde{v}^2(\Delta_k, A\cap B)-\mathbb{M}\, v(\Delta_k, A\cap B)=0,$$

so that $\mathbb{M}\, x=0$. We will show that $\operatorname{Var} x\to 0$ for $|\delta|\to 0$. We have

$$\begin{aligned}\operatorname{Var} x_k &=\operatorname{Var}\big[\big(\tilde{v}(\Delta_k, A\setminus B)+\tilde{v}(\Delta_k, A\cap B)\big)\big(\tilde{v}(\Delta_k, B\setminus A)+\tilde{v}(\Delta_k, A\cap B)\big)\\ &\qquad -v(\Delta_k, A\cap B)\big]\\ &=\operatorname{Var}\big[\tilde{v}\big(\Delta_k, (A\setminus B)\cup(B\setminus A)\big)\, \tilde{v}(\Delta_k, A\cap B)\\ &\qquad +\tilde{v}(\Delta_k, A\setminus B)\, \tilde{v}(\Delta_k, B\setminus A)+\tilde{v}^2(\Delta_k, A\cap B)-v(\Delta_k, A\cap B)\big]\\ &\leqq 3\,\big\{\mathbb{M}\big[\tilde{v}\big(\Delta_k, (A\setminus B)\cup(B\setminus A)\big)\, \tilde{v}(\Delta_k, A\cap B)\big]^2\\ &\qquad +\mathbb{M}\big[\tilde{v}(\Delta_k, A\setminus B)\, \tilde{v}(\Delta_k, B\setminus A)\big]^2\\ &\qquad +\mathbb{M}\big[\tilde{v}^2(\Delta_k, A\cap B)-v(\Delta_k, A\cap B)\big]^2\big\}.\end{aligned}$$

For the proof it is sufficient to establish that for arbitrary disjoint C and D from $\mathfrak{B}_0$ we have for $|\delta|\to 0$

$$I_1=\sum_{k=0}^{n-1} \mathbb{M}\big[\tilde{v}(\Delta_k, C)\, \tilde{v}(\Delta_k, D)\big]^2\to 0,$$

$$I_2=\sum_{k=0}^{n-1} \mathbb{M}\big[\tilde{v}^2(\Delta_k, C)-v(\Delta_k, C)\big]^2\to 0.$$

The first of these relations is verified as follows

$$I_1=\sum_{k=0}^{n-1} \mathbb{M}\, \tilde{v}^2(\Delta_k, C)\, \mathbb{M}\, \tilde{v}^2(\Delta_k, D)=\sum_{k=0}^{n-1} |\Delta_k|^2\, \Pi(C)\, \Pi(D)\to 0$$

($|\Delta_k|=t_{k+1}-t_k$). Moreover,

$$\begin{aligned}&\mathbb{M}\big[\tilde{v}^2(\Delta_k, C)-v(\Delta_k, C)\big]^2\\ &\quad=\mathbb{M}\big[\tilde{v}^4(\Delta_k, C)-2\tilde{v}^3(\Delta_k, C)-2|\Delta_k|\, \Pi(C)\, \tilde{v}^2(\Delta_k, C)+v^2(\Delta_k, C)\big],\end{aligned}$$

whereby

$$\mathbb{M}\,\tilde{v}^4(\Delta_k, C) = |\Delta_k|\,\Pi(C)\big(1+3|\Delta_k|\,\Pi(C)\big),$$
$$\mathbb{M}\,\tilde{v}^3(\Delta_k, C) = |\Delta_k|\,\Pi(C),$$
$$\mathbb{M}\,v^2(\Delta_k, C) = |\Delta_k|\,\Pi(C)\big(1+|\Delta_k|\,\Pi(C)\big),$$

so that

$$I_2 = \sum_{k=0}^{n-1} 2|\Delta_k|^2\,\Pi^2(C) \to 0 \quad \text{for } |\delta| \to 0.$$

Lemma 3. *If* $\varphi_i(t, x)$ *is a step function* $(i=1, 2)$, *then*

$$\begin{aligned}
\mathbb{P}\text{-}\lim_{|\delta|\to 0} \sum_{k=0}^{n-1} \int_{t_k}^{t_{k+1}} \int \varphi_1(t, y)\,\tilde{v}(dt, dy) \int_{t_k}^{t_{k+1}} \int \varphi_2(t, y)\,\tilde{v}(dt, dy) \\
= \int_0^T \int \varphi_1(t, y)\,\varphi_2(t, y)\,v(dt, dy).
\end{aligned} \tag{26}$$

Since $\varphi_i(t, y)$ $(i=1, 2)$ is a step function, it is $\mathfrak{F}_t$-measurable and there exists a finite number of half-open pairwise disjoint intervals

$$\Delta_k = [t_k^*, t_{k+1}^*) \quad (k=0, 1, \ldots, m-1), \qquad \bigcup_{k=1}^{m} \Delta_k = [0, T]$$

and a finite number of sets A_r $(r=1, 2, \ldots, p)$, $A_r \in \mathfrak{B}_0$ such that $\varphi_i(t, x)$ is constant on $\Delta_k \times A_r$ and equal to zero outside of $\bigcup_{r=1}^{p} A_r$. In other words,

$$\varphi_i(t, y) = \sum_{k, r} \alpha_{k, r}^{(i)}\,\chi_{\Delta_k}(t)\,\chi_{A_r}(y),$$

where the $\alpha_{kr}^{(i)}$ are $\mathfrak{F}_{t_k}$-measurable random variables. Applying (25) to the set $A=A_r$ and the interval $[t_s^*, t_{s+1}^*]$ instead of $[0, T]$, multiplying by $\alpha_{sr}^{(i)}$ and summing over r, we obtain

$$\begin{aligned}
\mathbb{P}\text{-}\lim_{|\delta|\to 0} \sum \int_{t_k}^{t_{k+1}} \int \varphi_1(t, y)\,\tilde{v}(dt, dy)\,\tilde{v}\big((t_k, t_{k+1}), B\big) \\
= \int_{t_k^*}^{t_{k+1}^*} \int \varphi_1(t, x)\,\chi_B(x)\,v(dt, dx),
\end{aligned} \tag{27}$$

where δ is partition of $[t_s^*, t_{s+1}^*]$. Setting $B=A_r$, multiplying by $\alpha_{sr}^{(2)}$, summing over all r and then summing the entire equation thus obtained over s, we obtain (26).

Set

$$\begin{aligned}
\Phi_i(t) &= \int_0^t \int \varphi_i(\tau, x)\,\tilde{v}(d\tau, dx), \qquad i=1, 2, \\
\Phi(t) &= \Phi_1(t)\,\Phi_2(t).
\end{aligned} \tag{28}$$

Consider an arbitrary partition $\delta=\{t_0, t_1, \ldots, t_n\}$ of $[0, T]$. We have

$$\Phi(T)=\sum_{i=0}^{n-1}[\Phi(t_{i+1})-\Phi(t_i)]=\sum_{i=0}^{n-1}\Delta\Phi(t_i)$$

and

$$\begin{aligned}\Delta\Phi(t_i)&=\Phi(t_{i+1})-\Phi(t_i)\\&=\Phi_1(t_i)\,\Delta\Phi_2(t_i)+\Phi_2(t_i)\,\Delta\Phi_1(t_i)+\Delta\Phi_1(t_i)\,\Delta\Phi_2(t_i).\end{aligned}\tag{29}$$

We assume that $\varphi_i(t,x)\in H$ and that they are step functions. Then it is not difficult to establish that

$$\mathbb{P}\text{-}\lim_{|\delta|\to 0}\sum\Phi_1(t_i)\,\Delta\Phi_2(t_i)=\int_0^T\Phi_1(t)\,d\Phi_2(t).$$

The integral on the right is taken in the sense of an integral w.r.t. a martingale introduced at the beginning of this section. To prove the last relation it is sufficient to show that

$$\int_0^T(\Phi_1^{(\delta)}(t)-\Phi_1(t))^2\,\rho_2(t)\,dt\to 0 \qquad \text{for } |\delta|\to 0$$

in the sense of convergence in probability, where $\Phi_1^{(\delta)}(t)=\Phi_1(t_i)$ if $t_i\leqq t\leqq t_{i+1}$, $\rho_2(t)=\int\varphi_2^2(t,y)\,\Pi(dy)$. In turn, for the step function $\varphi_2(t,y)$ this follows from the fact that

$$\mathbb{P}\text{-}\lim\int_0^T(\Phi_1^{(\delta)}(t)-\Phi_1(t))^2\,dt=0.$$

But

$$\begin{aligned}\mathbb{M}\int_0^T(\Phi_1^{(\delta)}(t)-\Phi_1(t))^2\,dt&=\sum_{i=0}^{n-1}\int_{t_i}^{t_{i+1}}\int_{t_i}^{t}\int\mathbb{M}\,\varphi_1^2(\tau,x)\,\Pi(dx)\,dt\,d\tau\\&\leqq|\delta|\int_0^T\int\mathbb{M}\,\varphi_1^2(t,x)\,\Pi(dx)\,dt\to 0\end{aligned}$$

for $|\delta|\to 0$ which proves the claim. Letting $|\delta|\to 0$ in (29) and taking Lemma 3 into account, we get

$$\begin{aligned}\Phi_1(T)\,\Phi_2(T)&=\int_0^T\Phi_1(t)\,d\Phi_2(t)+\int_0^T\Phi_2(t)\,d\Phi_1(t)\\&\quad+\int_0^T\int\varphi_1(t,x)\,\varphi_2(t,x)\,\nu(dt,dx).\end{aligned}\tag{30}$$

Let $H_n(\Pi)$ denote the class of functions $\varphi(t,x)$ $(0\leqq t\leqq T,\ x\in X)$ possessing the following properties:

a) for arbitrary $t\in[0,T]$ and $x\in X$ $\varphi(t,x)$ is $\mathfrak{F}_t$-measurable and is for fixed ω a Borel function of (t,x);

b) w.p.1

$$\int_0^T \int |\varphi(t, x)|^n \Pi(dx)\, dt < \infty .$$

Using Hölder's inequality it is easy to show that if $\varphi \in H_s(\Pi) \cap H_r(\Pi)$, where $s<r$, then $\varphi \in H_l(\Pi)$, where $s<l<r$.

Let us generalize (30) to functions $\varphi_i(t, x) \in H_{2,4}(\Pi)$, where

$$H_{2,4}(\Pi) = H_4(\Pi) \cap H_2(\Pi).$$

To this end we introduce an arbitrary sequence of step functions

$$\varphi_i^{(n)}(t, x) \in H(\Pi)$$

such that

$$\text{P-lim} \int_0^T \int [\varphi_i^{(n)}(t, x) - \varphi_i(t, x)]^{2k} \Pi(dx)\, dt = 0 \qquad (k=1, 2;\ i=1, 2). \quad (31)$$

The integrals of the form (28) corresponding to the functions $\varphi_i^{(n)}(t, x)$ will be denoted by $\Phi_i^{(n)}(t)$. Then $\Phi_i^{(n)}(t)$ converges in probability to $\Phi_i(t)$ by definition. Moreover, (Theorem 2)

$$\text{P}\left\{\left|\int_0^T \Phi_1^{(n)}(t)\, d\Phi_2(t) - \int_0^T \Phi_1(t)\, d\Phi_2(t)\right| > \varepsilon\right\}$$
$$\leqq \frac{N}{\varepsilon^2} + \text{P}\left\{\int_0^T \int (\Phi_1^{(n)}(t) - \Phi_1(t))^2 \varphi_2^2(t, y)\, \Pi(dy)\, dt > N\right\}.$$

In turn (Lemma 1),

$$\text{P}\{\sup_{0 \leqq t \leqq T} |\Phi_1^{(n)}(t) - \Phi_1(t)| > C\}$$
$$\leqq \frac{N}{C^2} + \text{P}\left\{\int_0^T \int (\varphi_1^{(n)}(t, y) - \varphi_1(t, y))^2\, \Pi(dy)\, dt > N\right\},$$

which together with the previous inequality yields

$$\text{P-}\lim_{n\to\infty} \int_0^T \Phi_1^{(n)}(t)\, d\Phi_2(t) = \int_0^T \Phi_1(t)\, d\Phi_2(t)$$

for arbitrary $\varphi_2 \in H(\Pi)$ and any sequence $\varphi_1^{(n)} \in H(\Pi)$ satisfying (31). Under the same conditions one can show that

$$\text{P-}\lim_{n\to\infty} \int_0^T \Phi_2(t)\, d\Phi_1^{(n)}(t) = \int_0^T \Phi_2(t)\, d\Phi_1(t).$$

Up to now we have used condition (31) only for $k=1$. Using it now for $k=2$ along with the definition of the stochastic integral w.r.t. the measure

$\nu(dt, dx)$ it is not difficult to verify that

$$\begin{aligned}
&\mathbb{P}\text{-}\lim_{n\to\infty} \int_0^T\int \varphi_1^{(n)}(t,x)\,\varphi_2(t,x)\,\nu(dt,dx)\\
&\quad=\int_0^T\int \varphi_1(t,x)\,\varphi_2(t,x)\,\nu(dt,dx)\\
&\quad=\int_0^T\int \varphi_1(t,x)\,\varphi_2(t,x)\,\tilde\nu(dt,dx)+\int_0^T\int \varphi_1(t,x)\,\varphi_2(t,x)\,\Pi(dx)\,dt.
\end{aligned}$$

Theorem 4. *For the stochastic integrals* (28) ($\varphi_i\in H_2(\Pi)\cap H_4(\Pi)$, $i=1,2$) *the following formula for integration by parts holds:*

$$d(\Phi_1(t)\,\Phi_2(t))=\Phi_1(t)\,d\Phi_2(t)+\Phi_2(t)\,d\Phi_1(t)+\int \varphi_1(t,x)\,\varphi_2(t,x)\,\nu(dt,dx). \quad (32)$$

This is merely another way of writing (30).

Corollary 1. *If* $\varphi_i(t,x)\in H_1(\Pi)\cap H_4(\Pi)$, $i=1,2$, *then*

$$\begin{aligned}
\Psi_1(T)\,\Psi_2(T)&=\int_0^T \Psi_1(t)\,d\Psi_2(t)+\int_0^T \Psi_2(t)\,d\Psi_1(t)\\
&\quad+\int_0^T \varphi_1(t,x)\,\varphi_2(t,x)\,\nu(dt,dx),
\end{aligned} \quad (33)$$

where

$$\Psi_i(t)=\int_0^t\int \varphi_i(\tau,x)\,\nu(d\tau,dx).$$

Indeed, the passage from (32) to (33) can be carried through by using the usual formula for integration by parts. We recall that from $\varphi_i\in H_1(\Pi)\cap H_4(\Pi)$ it follows that $\varphi_i\in H_2(\Pi)$.

We mention some special cases of (30). First, for arbitrary A and $B\in\mathfrak{B}_0$ ($\tilde\nu(t,A)=\tilde\nu([0,t],A)$)

$$\tilde\nu(t,A)\,\tilde\nu(t,B)=\int_0^t \tilde\nu(t,A)\,\tilde\nu(dt,B)+\int_0^t \tilde\nu(t,B)\,\tilde\nu(dt,A)+\nu(t,A\cap B)$$

and

$$\tilde\nu^2(t,A)=2\int_0^t \tilde\nu(t,A)\,\tilde\nu(dt,A)+\nu(t,A).$$

Such formulas also hold if the measure $\tilde\nu(t,A)$ is replaced by $\nu(t,A)$.

Lemma 4. *If* $A\in\mathfrak{B}_0$, *then*

$$d\nu^m(t,A)=[(\nu(t,A)+1)^m-\nu^m(t,A)]\,\nu(dt,A). \quad (34)$$

The proof can be carried out by induction. From (32), assuming that (34) holds for m, we obtain

$$\begin{aligned} d\nu^{m+1}(t, A) &= \nu(t, A)\, d\nu^m(t, A) + \nu^m(t, A)\, \nu(dt, A) \\ &\quad + \left[(\nu(t, A)+1)^m - \nu^m(t, A)\right] \nu(dt, A) \\ &= \left[(\nu(t, A)+1)^{m+1} - \nu^{m+1}(t, A)\right] \nu(dt, A). \end{aligned}$$

Corollary 2. *If $P(x)$ is a polynomial, then*

$$dP(\nu(t, A)) = \left[P(\nu(t, A)+1) - P(\nu(t, A))\right] \nu(dt, A).$$

Analogous formulas also hold for the stochastic integrals

$$\Psi(t) = \int_0^t \int \varphi(\tau, x)\, \nu(d\tau, dx), \tag{35}$$

where $\varphi(t, x)$ satisfies conditions of Corollary 1, i.e.,

$$d\Psi^2(t) = 2\Psi(t)\, d\Psi(t) + \int \varphi^2(t, x)\, \nu(dt, dx) \tag{36}$$

and if $\varphi \in H_1(\Pi) \cap H_{2m}(\Pi)$,

$$d\Psi^m(t) = \int \left([\Psi(t) + \varphi(t, x)]^m - \Psi^m(t)\right) \nu(dt, dx). \tag{37}$$

Formula (34) is a special case of (37) $(\varphi(t, x) = \chi_A(x))$. The proof of (37) goes as in Lemma 4. We thus obtain a formula the stochastic differential of a polynomial:

$$dP(\Psi(t)) = \int \left(P[\Psi(t) + \varphi(t, x)] - P(\Psi(t))\right) \nu(dt, dx), \tag{38}$$

where $\varphi \in H_1(\Pi) \cap H_{2m}(\Pi)$ and m is the degree of the polynomial $P(x)$. Formula (37) can be used to estimate the moments of stochastic integrals w.r.t. the Poisson measure. Let $\varphi(t, x) \in H_1(\Pi) \cap H_{2m}(\Pi)$ and set

$$a_m(t) = \sup_{0 \leqq \tau \leqq t} M|\Psi(\tau)|^m, \qquad b_m(t, x) = M|\varphi(t, x)|^m.$$

From (37)

$$a_m(t) \leqq \sum_{k=0}^{m-1} C_m^k \int_0^t \int M|\Psi(\tau)|^k |\varphi(\tau, x)|^{m-k}\, \Pi(dx)\, d\tau.$$

From Hölder's inequality

$$M|\Psi(t)|^k |\varphi(t, x)|^{m-k} \leqq a_m^{\frac{k}{m}}(t)\, b_m^{1-\frac{k}{m}}(t, x).$$

So long as $a_m(t) \leqq 1$ we have

$$a_m(t) \leqq \int_0^t \int \left[(b_m^{\frac{1}{m}}(t, x) + 1)^m - 1\right] \Pi(dx)\, dt;$$

if, however,

$$a_m(t) \geqq 1,$$

then

$$a_m(t) \leqq a_m^{\frac{m-1}{m}}(t) \int_0^t \int [(b_m^{\frac{1}{m}}(t,x)+1)^m - 1]\, \Pi(dx)\, dt,$$

whence

$$a_m(t) \leqq \left[\int_0^t \int [(b_m^{\frac{1}{m}}(t,x)+1)^m - 1]\, \Pi(dx)\, dt\right]^m.$$

Thus,

$$a_m(T) \leqq \max\left\{\left[\int_0^T \int [(b_m^{\frac{1}{m}}(t,x)+1)^m - 1]\, \Pi(dx)\, dt\right]^j, j=1, m\right\}. \tag{39}$$

We now generalize (38) to a wider class of functions. Let

$$H_{m,n}(\Pi) = H_m(\Pi) \cap H_n(\Pi),$$

and assume initially that $\varphi \in H_{1,2}(\Pi)$, $\sup_{t,x} |\varphi(t,x)| < \infty$ w.p. 1 and that $f(x)$ is continuous along with its derivative. Then w.p. 1 there exists a constant $\gamma = \gamma(\omega)$ for which

$$|f[\Psi(t) + \varphi(t,x)] - f[\Psi(t)]| \leqq \gamma(\omega)\, |\varphi(t,x)|$$

and consequently, the integral

$$\int_0^T \int [f(\Psi(t) + \varphi(t,x)) - f(\Psi(t))]\, \nu(dt, dx)$$

exists. We construct a sequence of polynomials $P_n(x)$ converging uniformly along with first order derivatives on any compact to $f(x)$ and its derivative, resp. Then

$$\mathbb{P}\text{-}\lim \int_0^T \int [P_n(\Psi(t) + \varphi(t,x)) - P_n(\Psi(t))]\, \nu(dt, dx)$$
$$= \int_0^T \int [f(\Psi(t) + \varphi(t,x)) - f(\Psi(t))]\, \nu(dt, dx).$$

Hence,

$$d f(\Psi(t)) = \int [f(\Psi(t) + \varphi(t,x)) - f(\Psi(t))]\, \nu(dt, dx). \tag{40}$$

Now let $\varphi(t,x)$ be an arbitrary function from $H_{1,2}(\Pi)$. Denote by D_φ the class of nonrandom continuously differentiable functions in X for which

$$f(\Psi(t) + \varphi(t,x)) - f(\Psi(t)) \in H_{1,2}(\Pi).$$

Set $\varphi_N(t,x)=0$ if $|\varphi(t,x)|\geqq N$ and let $\Psi(t)$ be the stochastic integral (35) in which $\varphi(t,x)=\varphi_N(t,x)$ is set. Then $\mathbb{P}$-lim $\Psi_N(t)=\Psi(t)$. Also,

$$\mathbb{P}\left\{\int_0^T\int|f(\Psi(t)+\varphi(t,x))-f(\Psi(t))-f(\Psi_N(t)+\varphi_N(t,x))\right.$$
$$\left.+f(\Psi_N(t))|^k\,\Pi(dx)\,dt>\varepsilon\right\}$$
$$=\mathbb{P}\Big\{\int_{B_N}\int|f(\Psi(t)+\varphi(t,x))-f(\Psi(t))|^k\,\Pi(dx)\,dt>\varepsilon\Big\},$$

where B_N is the (random) set $\{t,x;\,|\varphi(t,x)|\geqq N\}$. Its $\Pi\times l$-measure (l is Lebesgue measure on $[0,T]$) tends to zero w.p. 1. Hence,

$$\int_{B_N}\int|f(\Psi(t)+\varphi(t,x))-f(\Psi(t))|^k\,\Pi(dx)\,dt\to 0,\qquad k=1,2,$$

w.p. 1 and certainly then in probability. Thus,

$$\mathbb{P}\text{-}\lim_{N\to\infty}\int_0^T\int[f(\Psi_N(t)+\varphi_N(t,x))-f(\Psi_N(t))]\,\nu(dt,dx)$$
$$=\int_0^T\int[f(\Psi(t)+\varphi(t,x))-f(\Psi(t))]\,\nu(dt,dx).$$

We have proved

Lemma 5. *If* $\varphi\in H_{1,2}(\Pi)$, $f(x)$ *is continuously differentiable and* $\varphi(x)\in D_\varphi$, *then* $f(\Psi(t))$ *satisfies* (40).

§ 6. The Generalized Itô Formula for Stochastic Differentials

Let $\mathfrak{F}_t$ $(0\leqq t\leqq T)$ be a fixed family of σ-algebras, $\alpha(t)$ a vector process, $\gamma(t,u)$ a vector field with values in X, $0\leqq t\leqq T$, $u\in X$ and $\beta(t)$ a linear map of X into X, whereby $\alpha(t)$, $\beta(t)$ and $\gamma(t,u)$ are $\mathfrak{F}_t$-measurable for any $t\in[0,T]$, $\sqrt{\alpha(t)}$ and $\beta(t)$ belong to H_2 and $\gamma(t,u)\in H_2(\Pi)$. Here, H denotes a space of random vector processes $\mathfrak{F}_t$-measurable for each $t\in[0,T]$ w.p. 1, t-measurable, and such that

$$\int_0^T|\alpha(t)|^2\,dt<\infty$$

w.p. 1. If we also have

$$\mathbb{M}\int_0^T|\alpha(t)|^2\,dt<\infty,$$

then we will say $\alpha(t)\in H_2$. These assumptions will be assumed fulfilled throughout the entire section. If

$$\zeta(t)=\zeta(0)+\int_0^t \alpha(\tau)\,d\tau+\int_0^t \beta(\tau)\,dw+\int_0^t\int \gamma(\tau,u)\,\tilde{\nu}(d\tau,du),$$

where $w(t)$ is an s-dimensional Wiener process (i.e., a continuous process with values in X whose components in an arbitrary orthonormal basis of X are homogeneous, continuous, mutually independent Wiener processes), $\nu(t,A)$ is a Poisson measure with parameter $t\Pi(A)$, $\tilde{\nu}(t,A)=\nu(t,A)-\Pi(A)\,t$, and $w(t)$ and $\nu(t,A)$ are $\mathfrak{F}_t$-measurable for arbitrary $A\in\mathfrak{B}_0$, $t\in[0,T]$, then we will say that the process $\zeta(t)$ has on $[0,T]$ the stochastic differential

$$d\zeta=\alpha(t)\,dt+\beta(t)\,dw+\int\gamma(t,u)\,\tilde{\nu}(dt,du)=\alpha(t,dt). \tag{1}$$

In this section we will generalize Itô's formula, established in § 3, Part I for one-dimensional stochastic differentials with $\gamma(t,u)=0$, to processes with stochastic differential (1).

We will need a series of auxiliary results which are immediate extensions of arguments of the preceding section.

Lemma 1. *If*

$$\Phi(t)=\int_0^t\int\varphi(\tau,u)\,\tilde{\nu}(d\tau,du),\qquad \varphi(t,u)\in H_2(\Pi) \tag{2}$$

and $\alpha(t)$ is w.p.1 a differentiable function whose derivative has no discontinuities of the second kind and $\varphi(t,u)$ is a scalar function, then

$$d(\alpha(t)\,\Phi(t))=\alpha'(t)\,\Phi(t)\,dt+\alpha(t)\,d\Phi(t). \tag{3}$$

Proof. From

$$\int_0^t t\,d\Phi(t)=\mathbb{P}\text{-}\lim_{n\to\infty}\sum_{k=0}^{n-1}t_k[\Phi(t_{k+1})-\Phi(t_k)],$$

and

$$\int_0^t \Phi(t)\,dt=\mathbb{P}\text{-}\lim_{n\to\infty}\sum_{k=0}^{n-1}\Phi(t_{k+1})(t_{k+1}-t_k)$$

it follows that

$$\Phi(t)\,t=\int_0^t\Phi(t)\,dt+\int_0^t t\,d\Phi(t),$$

so that

$$d(t\,\Phi(t))=\Phi(t)\,dt+t\,d\Phi(t).$$

Hence for an arbitrary continuous piecewise linear function $a(t)$

$$d\big(a(t)\,\Phi(t)\big)=\Phi(t)\,a'(t)\,dt+a(t)\,d\Phi(t).$$

Approximating an arbitrary w.p. 1 differentiable function $\alpha(t)$ whose derivative has no discontinuities of the second kind with the help of a sequence of piecewise linear functions $\alpha_n(t)$ converging uniformly along with their derivatives to $\alpha(t)$ and $\alpha'(t)$, we obtain (3) in the general case.

Corollary 1. *Let $\Phi_i^*(t)=\alpha_i(t)+\Phi_i(t)$, $d\Phi_i(t)=\int\varphi_i\,\tilde{v}(dt,du)$ and $\varphi_i\in H_{1,2}$, $i=1,2$, where the $\alpha_i(t)$ satisfy the conditions of Lemma 1. Then,*

$$\begin{aligned}d\big(\Phi_1^*(t)\,\Phi_2^*(t)\big)&=\Phi_1^*(t)\,d\Phi_2^*(t)+\Phi_2^*(t)\,d\Phi_1^*(t)\\&\quad+\int\varphi_1(t,u)\,\varphi_2(t,u)\,v(dt,du).\end{aligned}\tag{4}$$

Formula (4) is easy to verify using merely Theorem 4 § 5 and the lemma just proved.

We generalize the previously introduced class D_φ to functions $g(t,z)$ of two real variables t and z $\big(t\in[0,T],\ z\in(-\infty,\infty)\big)$. We will say that $g(t,z)\in D_\varphi$ if $g(t,z)$ is continuously differentiable in t and z and $g\big(t,\Psi(t)+\varphi(t,x)\big)-g\big(t,\Psi(t)\big)\in H_{1,2}(\Pi)$, where $\Psi(t)$ is the stochastic integral defined by (35) § 5.

Lemma 2. *If $\varphi(t,z)\in H_{1,2}(\Pi)$ and $g(t,z)\in D_\varphi$, then*

$$\begin{aligned}dg\big(t,\Psi(t)\big)&=g_t'\big(t,\Psi(t)\big)\,dt\\&\quad+\int\big[g\big(t,\Psi(t)+\varphi(t,x)\big)-g\big(t,\Psi(t)\big)\big]\,v(dt,dx).\end{aligned}\tag{5}$$

Proof. Assume $g(t,x)=h(t)\,g(x)$. Corresponding to (40) § 5 we have

$$dg\big(\Psi(t)\big)=\int\big[g\big(\Psi(t)+\varphi(t,x)\big)-g\big(\Psi(t)\big)\big]\,v(dt,dx).$$

Using Lemma 1:

$$\begin{aligned}dg\big(t,\Psi(t)\big)&=d\big[h(t)\,g\big(\Psi(t)\big)\big]=h'(t)\,g\big(\Psi(t)\big)\,dt\\&\quad+\int h(t)\big[g\big(\Psi(t)+\varphi(t,x)\big)-g\big(\Psi(t)\big)\big]\,v(dt,dx),\end{aligned}$$

i.e., formula (5) for functions of the form $g(t,x)=h(t)\,g(x)$. It now remains to note that the class of functions $g(t,x)$ for which (5) holds is linear and closed in the class of functions for which $g(t,x)$, $g_t'(t,x)$ and $g_x'(t,x)$ converge uniformly on each finite interval.

Remark 1. Formula (5) also holds for random functions $g(t,x)$ if $g(t,x)$ satisfies the assumptions of Lemma 2 w.p. 1 and is $\mathfrak{F}_t$-measurable for arbitrary $t\in[0,T]$.

We proceed to the multi-dimensional case. Let $\gamma(t,u)$ be an m-dimensional vector field ($t\in[0,T]$, $u\in X$, $\gamma(t,u)\in X$). We will say that $\gamma(t,u)\in H_s(\Pi)$ ($\gamma(t,u)\in H_{s,p}(\Pi)$) if its components belong to $H_s(\Pi)$ (resp. $H_{s,p}(\Pi)$). Stochastic integrals determined by (2) or by (35) §4 in which the scalar function $\varphi(t,u)$ is replaced by the vector function $\gamma(t,u)$ will be denoted by $\Phi_\gamma(t)$ and $\Psi_\gamma(t)$, resp. From Lemma 1 and the formula for integration by parts (32) §5 it follows that

$$d(\Phi^*_{\gamma_1}(t)|\Phi^*_{\gamma_2}(t))=(\Phi^*_{\gamma_1}(t)|d\Phi^*_{\gamma_2}(t))+(\Phi^*_{\gamma_2}(t)|d\Phi^*_{\gamma_1}(t)) + \int(\gamma_1(t,u)|\gamma_2(t,u))\,\nu(dt,du), \tag{6}$$

where

$$\Phi^*_{\gamma_i}(t)=\Phi_{\gamma_i}(t)+\alpha_i(t) \qquad (i=1,2),\ \gamma_i\in H_{1,2}(\Pi),$$

and $\alpha_i(t)$ is w.p. 1 a piecewise-differentiable vector function.

Lemma 3. *Let $\gamma\in H_{1,2}(\Pi)$ and $g(t,x)\in D_\gamma$. Then*

$$dg(t,\Psi_\gamma(t))=g'_t(t,\Psi_\gamma(t))\,dt + \int[g(t,\Psi_\gamma(t)+\gamma(t,u))-g(t,\Psi_\gamma(t))]\,\nu(dt,du). \tag{7}$$

Assume first that in addition to the conditions of the lemma, $\gamma(t,x)$ is bounded w.p. 1 and that $g(t,x)$ is an arbitrary continuously differentiable function of the form

$$g(t,x_1,x_2,\dots,x_m)=h(t)\,g_1(x_1)\,g_2(x_2)\dots g_m(x_m).$$

For such functions (7) is established by induction on m with recourse to (40) §4, Lemma 1 and Corollary 1 of this section. For example, when $m=2$

$$d(h(t)\,g_1(\Psi_1(t))\,g_2(\Psi_2(t)))=h'(t)\,g_1(\Psi_1(t))\,g_2(\Psi_2(t))\,dt + h(t)\,d(g_1(\Psi_1(t))\,g_2(\Psi_2(t))).$$

From Corollary 1 and (40) §5

$$\begin{aligned}d(g_1(\Psi_1(t))\,g_2(\Psi_2(t)))&=g_1(\Psi_1(t))\int[g_2(\Psi_2(t)+\gamma_2(t,u))-g_2(\Psi_2(t))]\,\nu(dt,du)\\&\quad+g_2(\Psi_2(t))\int[g_1(\Psi_1(t)+\gamma_1(t,u))-g_1(\Psi_1(t))]\,\nu(dt,du)\\&\quad+\int[g_1(\Psi_1(t)+\gamma_1(t,u))-g_1(\Psi_1(t))]\\&\quad\times[g_2(\Psi_2(t)+\gamma_2(t,u))-g_2(\Psi_2(t))]\,\nu(dt,du)\\&=\int[g_1(\Psi_1(t)+\gamma_1(t,u))\,g_2(\Psi_2(t)+\gamma_2(t,u))\\&\quad-g_1(\Psi_1(t))\,g_2(\Psi_2(t))]\,\nu(dt,du),\end{aligned}$$

whence (7) follows for $m=2$. The class of functions $g(t,x_1,\dots,x_m)$ for which (7) holds is in the case considered ($\gamma(t,x)$ bounded w.p. 1) obviously linear and closed w.r.t. limit passage, where $g(t,x)$ and its first derivative converge uniformly on each compact in $[0,T]\times X$ and (7) holds for arbitrary continuously differentiable functions $g(t,x)$. The transition to the general case $g\in D_\gamma$, $\gamma\in H_{1,2}(\Pi)$ is justified as in the proof of Lemma 5 §5.

Let $\tilde{D}_\gamma$ denote the class of continuously differentiable functions $g(t,x)$, $x\in X$ for which

$$g(t,\Phi_\gamma(t)+\gamma(t,u))-g(t,\Phi_\gamma(t))\in H_2(\Pi),$$
$$g(t,\Phi_\gamma(t)+\gamma(t,u))-g(t,\Phi_\gamma(t))-(\nabla g(t,\Phi_\gamma)|\gamma(t,u))\in H_1(\Pi),$$

where $\nabla g(t,x)$ is the vector with components $\dfrac{\partial g}{\partial x_i}$ $(i=1,2,\ldots,m)$.

Lemma 4. *If* $\gamma(t,u)\in H_2(\Pi)$, *and* $g(t,x)\in\tilde{D}_\gamma$, *then*

$$\begin{aligned}dg(t,\Phi(t))=&\left[g_t'(t,\Phi_\gamma(t))+L_\pi(g)\right]dt\\&+\int\left[g(t,\Phi_\gamma(t)+\gamma(t,u))-g(t,\Phi_\gamma(t))\right]\tilde{\nu}(dt,du),\end{aligned}\tag{8}$$

where L_π *is the integro-differential operator*

$$L_\pi(g)=\int\left[g(t,\Phi_\gamma(t)+\gamma(t,u))-g(t,\Phi_\gamma(t))-(\nabla g(t,\Phi_\gamma(t))|\gamma(t,u))\right]\Pi(du).\tag{9}$$

To prove this we apply (7) and the previous limit passage.

Formulas (7) and (9) are analogs of Itô's formula for stochastic differentials when the integration is carried out w.r.t. the Poisson measure.

We proceed to the general case of (1)

Lemma 5. *Let* $w(t)$ *be a one-dimensional Wiener process and assume* $w(t)$ *and* $\nu(t,A)$ $(A\in\mathfrak{B}_0,\ t\in[0,T])$ *are independent. Then*

$$d(w(t)\,\nu(t,A))=\nu(t,A)\,dw+w(t)\,\nu(dt,A).$$

Since for an arbitrary partition $\delta=\{t_0,t_1,\ldots,t_n\}$ of $[0,T]$ we have

$$w(t)\,\nu(t,A)=\sum_{k=0}^{n-1}\left[\Delta w(t_k)\,\Delta\nu(t_k,A)+\nu(t_k,A)\,\Delta w(t_k)+w(t_k)\,\Delta\nu(t_k,A)\right],$$

where $\Delta w(t_k)=w(t_{k+1})-w(t_k)$ and $\Delta\nu(t_k,A)=\nu(t_{k+1},A)-\nu(t_k,A)$, it is sufficient to show that

$$\text{l.i.m.}\sum_{k=0}^{n-1}\Delta w(t_k)\,\Delta\nu(t_k,A)=0.$$

This follows from

$$\mathbb{M}\sum_{k=0}^{n-1} \Delta w(t_k)\,\Delta v(t_k, A)=0,$$

$$\begin{aligned}\operatorname{Var}\left[\sum_{k=0}^{n-1} \Delta w(t_k)\,\Delta v(t_k, A)\right] &= \sum_{k=0}^{n-1} \operatorname{Var}[\Delta w(t_k)\,\Delta v(t_k, A)]\\ &= \sum_{k=0}^{n-1} \mathbb{M}[\Delta w(t_k)]^2\,\mathbb{M}[\Delta v(t_k, A)]^2\\ &= \sum_{k=0}^{n-1} \Delta t_k^2\,\Pi(A)\to 0 \quad \text{for } |\delta|\to 0.\end{aligned}$$

Lemma 6. *If $w_1(t)$ and $w_2(t)$ are independent Wiener process, then*

$$d(w_1(t)\,w_2(t))=w_1(t)\,dw_2(t)+w_2(t)\,dw_1(t).$$

The proof goes as in the previous lemma.

Lemma 7. *If*

$$d\xi_i=\alpha_i(t)\,dt+\sum_{j=1}^{n}\beta_{ij}(t)\,dw_j(t)+\int\gamma_i(t,u)\,\tilde{v}(dt,du), \qquad i=1,2,$$

α_i, β_{ij}, γ_i are one-dimensional random functions (satisfying the conditions at the beginning of this paragraph), $\gamma_i\in H_{2,4}(\Pi)$ and $w_j(t)$ $(j=1,\dots,n)$ are independent Wiener processes, then

$$d(\xi_1(t)\,\xi_2(t))=\xi_1(t)\,d\xi_2+\xi_2(t)\,d\xi_1+\sum_{j=1}^{n}\beta_{1j}\beta_{2j}\,dt+\int\gamma_1\gamma_2\,v(dt,du). \tag{10}$$

As in the proof of Lemma 1 it is easy to show that the correctness of (10) in the general case follows from the fact that it is valid for functions $\alpha_i(t)$ and $\beta_{ij}(t)$ independent of time and for functions $\gamma_i(t,u)=\gamma\,\chi_A(u)$, $A\in\mathfrak{B}_0$. Since the class of functions for which (10) holds is linear and (10) is linear in ξ_1 and ξ_2, taking into account its correctness when either $\beta_{1j}=\beta_{2j}=0$ or $\gamma_1-\gamma_2=0$, and Lemma 6, it suffices to show that

$$d(\zeta'\zeta'')=\zeta'\,d\zeta''+\zeta''\,d\zeta',$$

if

$$\zeta'=a\,t+\beta\,w(t), \qquad \zeta''=c\,v(t,A).$$

But the equality

$$d(t\,v(t,A))=v(t,A)\,dt+t\,v(dt,A)$$

follows from Lemma 1, and

$$d(w(t)\,v(t,A))=v(t,A)\,dw(t)+w(t)\,v(dt,A)$$

from Lemma 5. □

Lemma 8. *If $g(x)$ is a twice continuously differentiable function of x $(x \in (-\infty, \infty))$, $g(x) \in D_\gamma$,*

$$\xi(t) = \xi_0 + \int_0^t \alpha(\tau)\,d\tau + \int_0^t \sum_{j=1}^n \beta_j(\tau)\,dw_j(\tau) + \Psi(t),$$

$$\Psi(t) = \int_0^t \int \gamma(\tau, u)\,\nu(d\tau, du),$$

and the processes $\nu(t, A)$ and $w_j(t)$ are independent, then

$$\begin{aligned} dg(\xi(t)) = {} & \left[g'(\xi(t))\,\alpha(t) + \tfrac{1}{2} g''(\xi(t)) \sum_{j=1}^n \beta_j^2(t)\right] dt \\ & + g'(\xi(t)) \sum_{j=1}^n \beta_j(t)\,dw_j(t) + \int \left[g(\xi(t) + \gamma(t, u)) - g(\xi(t))\right] \nu(dt, du). \end{aligned} \tag{11}$$

Proof. As in the proof of Lemma 5 (Chapter 3, Part I) and Lemma 5 § 5, it is sufficient to prove (11) for $g(x) = x^m$. For such a function (10) takes the form

$$\begin{aligned} d\xi^m(t) = {} & \left[m\,\xi^{m-1}(t)\,\alpha(t) + \tfrac{1}{2} m(m-1)\,\xi^{m-2}(t) \sum_{j=1}^n \beta_j^2(t)\right] dt \\ & + m\,\xi^{m-1}(t) \sum_{j=1}^n \beta_j(t)\,dw_j(t) + \int \left[(\xi(t) + \gamma(t, u))^m - \xi^m(t)\right] \nu(dt, du). \end{aligned} \tag{12}$$

The proof of (12) can be obtained by induction. Assume it holds for m. Using the last lemma we write

$$\begin{aligned} d[\xi^{m+1}(t)] = {} & \xi^m(t)\,d\xi(t) + \xi(t)\,d(\xi^m(t)) + \sum_{j=1}^n \beta_j^2\, m\,\xi^{m-1}(t)\,dt \\ & + \int \left[(\xi(t) + \gamma(t, u))^m - \xi^m(t)\right] \gamma(t, u)\,\nu(dt, du), \end{aligned}$$

substituting here the expressions for $d\xi(t)$ and $d\xi^m(t)$ from (12) we obtain on the right side an expression which coincides with the right side of formula (12) if we replace m by $m+1$ in the latter. This proves the lemma.

Theorem 1. *Let*

$$d\xi = \alpha(t)\,dt + \beta(t)\,dw(t) + \int \gamma(t, u)\,\nu(dt, du),$$

where

$$\alpha(t) = \{\alpha_1(t), \ldots, \alpha_m(t)\}, \qquad \beta(t) = \{\beta_{jk}(t), j = 1, \ldots, m;\ k = 1, \ldots, n\},$$

$$\gamma(t, u) = \{\gamma_1(t, u), \ldots, \gamma_m(t, u)\}, \qquad \sqrt{|\alpha_i(t)|} \in H_2, \qquad \beta_{ik}(t) \in H_2,$$

$\gamma(t, u) \in H_{1,2}(\Pi)$, $w(t)$ is an n-dimensional Wiener process ($w_j(t)$ and $\nu(t, A)$ independent), $g(t, x) = g(t, x_1, x_2, \ldots, x_m)$ is a twice continuously differen-

tiable function of the x_j and $g(t,x)\in D_\gamma$. Then

$$dg(t,\xi(t))=\Big[g_t'(t,\xi)+\sum_{k=1}^{m} g_{x_k}'(t,\xi)\alpha_k(t)+\tfrac{1}{2}\sum_{j,k=1}^{m} g_{x_k x_j}''(t,\xi)\sum_{i=1}^{m}\beta_{ki}(t)\beta_{ji}(t)\Big]dt$$
$$+\sum_{i=1}^{n}\sum_{k=1}^{m} g_{x_k}'(t,\xi)\beta_{ki}(t)\,dw_i(t)+\int[g(t,\xi+\gamma)-g(t,\xi)]\,\nu(dt,dx). \tag{13}$$

The proof proceeds by induction on the number of independent arguments of the functions $g(t,x)$. We first assume that $g(t,x)$ is independent of t: $(g(t,x)=g(x))$. If $g(x)$ is a function of a single variable and we consider $g(\xi_k(t))$, where $\xi_k(t)$ is an arbitrary component of the vector $\xi(t)$, then (13) is established by Lemma 8. Assume (13) has been proved for functions of s components of the vector $\xi(t)$. We will show that it still holds for functions of $s+1$ variables. With this in mind we first consider a function $g_{s+1}(x_1,\dots,x_s,x_{s+1})$ of the following special form:

$$g_{s+1}(x_1,\dots,x_s,x_{s+1})=g(x_1,\dots,x_s)\,f(x_{s+1}).$$

Then from Lemma 7 and the induction assumption it follows that

$$dg_{s+1}(\xi(t))=f(\xi_{s+1}(t))\,dg(\xi_1,\dots,\xi_s)+g(\xi(t))\,df(\xi_{s+1}(t))$$
$$+\Big(\sum_{i=1}^{n}\sum_{k=1}^{s} g_{x_k}'(\xi)\beta_{ki}(t)f'(\xi_{s+1})\beta_{s+1,i}(t)\Big)dt$$
$$+\int[g(\xi+\gamma)-g(\xi)]\,[f(\xi_{s+1}+\gamma_{s+1})-f(\xi_{s+1})]\,\nu(dt,dx).$$

Substituting here the expression for $dg(\xi_1,\dots,\xi_s)$ from (13) and for $df(\xi_{s+1})$ from (11) we can convince ourselves that (13) also holds for functions of $s+1$ variables in our special case. But then it will also hold for linear combinations of functions of the form g_{s+1}, and with the help of a limit passage we find that it also holds for arbitrary twice differentiable functions $g(x_1,\dots,x_{s+1})$ satisfying the hypotheses of the theorem. Hence, (13) is proved for autonomous functions. To effect the passage to time-dependent functions we first verify (13) for functions of the form $g(t,x)=f(t)\,g(x)$ with the help of Lemma 7 and then use arguments analogous to those just carried out. The theorem is proved.

Formula (13) is the generalization of Itô's formula we spoke about in the first paragraph. We write it out in the one-dimensional case. If $\alpha(t)$, $\beta(t)$, $\gamma(t,u)$ and $w(t)$ are one-dimensional and

$$d\xi=\alpha\,dt+\beta(t)\,dw(t)+\int\gamma(t,u)\,\nu(dt,du),$$

then

$$dg(t,\xi(t))=[g_t'(t,\xi(t))+g_x'(t,\xi(t))\alpha(t)+\tfrac{1}{2}g_{xx}''(t,\xi(t))\beta^2(t)]\,dt$$
$$+g_x'(t,\xi(t))\beta(t)\,dw(t) \tag{14}$$
$$+\int[g(t,\xi(t)+\gamma(t,u))-g(t,\xi(t))]\,\nu(dt,du).$$

As an example of the application of (14) we consider the solution of the linear, homogeneous differential equation

$$d\xi=\xi\left[\alpha(t)\,dt+\beta(t)\,dw(t)+\int\gamma(t,u)\,\nu(dt,du)\right]. \tag{15}$$

Introducing a new undetermined function $\eta(t)=\ln|\xi(t)|$ and using (14) we obtain for $\eta(t)$ the equation

$$d\eta=\left(\alpha(t)-\tfrac{1}{2}\beta^2(t)\right)dt+\beta(t)\,dw(t)-\int\ln|1+\gamma(t,u)|\,\nu(dt,du),$$

so that

$$\eta=\eta_0+\int_0^t\left(\alpha(\tau)-\tfrac{1}{2}\beta^2(\tau)\right)d\tau+\int_0^t\beta(\tau)\,dw(\tau)$$
$$+\int_0^t\int\ln|1+\gamma(t,u)|\,\nu(dt,du)$$

and

$$\xi(t)=\xi_0\exp\left\{\int_0^t\left(\alpha(\tau)-\tfrac{1}{2}\beta^2(\tau)\right)d\tau+\int_0^t\beta(\tau)\,dw(\tau)\right.$$
$$\left.+\int_0^t\int\ln|1+\gamma(t,u)|\,\nu(dt,du)\right\}.$$

Now consider the inhomogeneous linear equation

$$d\xi=(\alpha_0+\alpha_1\xi)\,dt+(\beta_0+\beta_1\xi)\,dw+\int(\gamma_0+\gamma_1\xi)\,\nu(dt,du) \tag{16}$$

and assume that $\gamma_1(t,u)>-1$. We put

$$\xi_0(t)=\exp\left\{-\int_0^t\left(\alpha_1(\tau)-\tfrac{1}{2}\beta_1^2(\tau)\right)d\tau-\int_0^t\beta_1(\tau)\,dw(\tau)\right.$$
$$\left.-\int_0^t\int\ln(1+\gamma_1(t,u))\,\nu(dt,du)\right\}.$$

With the help of (14) it is easy to verify that $\xi_0(t)$ satisfies

$$d\xi_0=\xi_0\left[(\beta_1^2-\alpha_1)\,dt-\beta_1\,dw-\int\frac{\gamma_1}{1+\gamma_1}\,\nu(dt,du)\right].$$

Introduce the process $\eta(t)$:

$$\eta(t)=\xi_0(t)\,\xi(t).$$

Using (10), we obtain

$$d\eta=\xi_0(\alpha_0-\beta_0\beta_1)\,dt+\beta_0\xi_0\,dw+\int\frac{\gamma_0\xi_0}{1+\gamma_1}\,\nu(dt,du),$$

whence

$$\eta(t)=\xi(0)+\int_0^t \xi_0(\tau)(\alpha_0(\tau)-\beta_0(\tau)\beta_1(\tau))\,d\tau+\int_0^t \beta_0(\tau)\xi_0(\tau)\,dw(\tau)$$
$$+\int_0^t\int \frac{\gamma_0(\tau)\xi_0(\tau)}{1+\gamma_1(\tau)}\,\nu(d\tau,du).$$

Thus for the solution of (16) we obtain

$$\xi(t)=\frac{\eta(t)}{\xi_0(t)}=\eta(t)\exp\left\{\int_0^t(\alpha_1(\tau)-\tfrac{1}{2}\beta_1^2(\tau))\,d\tau+\int_0^t\beta_1(\tau)\,dw\right.$$
$$\left.+\int_0^t\int\ln(1+\gamma(t,u))\,\nu(dt,du)\right\}. \tag{17}$$

We now present another, somewhat more general version of Theorem 1 and the generalized Itô formula. Let $\eta(t)$ possess the stochastic differential

$$d\eta=\alpha(t)\,dt+\beta(t)\,dw(t)+\int\gamma(t,u)\,\tilde{\nu}(dt,du). \tag{18}$$

Theorem 2. *If $\alpha(t)$ and $\beta(t)$ satisfy the hypotheses of Theorem 1,*

$$\gamma(t,u)\in H_2(\Pi),\qquad g(t,x)$$

is a twice continuously differentiable function of x and $g(t,x)\in\tilde{D}_\gamma$, then

$$dg(t,\eta(t))=L(g)\,dt+(\nabla g(t,\eta(t))\,|\,\beta\,dw)$$
$$+\int[g(t,\eta(t)+\gamma(t,u))-g(t,\eta(t))]\,\tilde{\nu}(dt,du). \tag{19}$$

where L is the integro-differential operator

$$L(g)=L_0(g)+L_\pi(g)+L_N(g),$$

with

$$L_0(g)=g_t'(t,\xi(t))+(\nabla g(t,\xi(t))\,|\,\alpha(t)). \tag{20}$$

The operator $L_\pi(g)$ is connected with a stochastic integral w.r.t. the Poisson measure and was introduced earlier (see (9)):

$$L_\pi(g)=\int[g(t,\eta(t)+\gamma(t,u))-g(t,\eta(t))-(\nabla g(t,\eta(t))\,|\,\gamma(t,u))]\,\Pi(du). \tag{21}$$

The operator $L_N(g)$ is related to the stochastic integral w.r.t. Wiener measure:

$$L_N(g)=\tfrac{1}{2}\operatorname{sp}(\nabla^2(g(t,\eta(t))\beta(t)\beta^*(t)))$$
$$=\tfrac{1}{2}\sum_{j,k=1}^{m}\sum_{i=1}^{n}g''_{x_jx_k}(t,\eta(t))\,\beta_{ki}(t)\,\beta_{ji}(t). \tag{22}$$

We recall that $\nabla g(t,x)$ denotes the vector whose components equal $g'_{x_k}(t)$, $\nabla^2 g(t,x)$ is the matrix with elements $g''_{x_jx_k}(t,x)$, $\operatorname{sp}A$ is the trace of A and $\beta^*(t)$ is the transpose of $\beta(t)$.

§ 7. Stochastic Differential Equations without After-effect

We consider the stochastic differential equation

$$d\xi(t)=a(t,\xi(t))\,dt+b(t,\xi(t))\,dw(t)+\int f(t,\xi(t),u)\,\tilde{\nu}(dt,du), \tag{1}$$

where $a(t,x)$, $b(t,x)$ and $f(t,x,u)$ are non-random, $a(t,x)$ and $f(t,x,u)$ are vector functions with values in X, $t\in[0,T]$, $x\in X$; $u\in X$, $b(t,x)$ is a linear operator function mapping X into X, $w(t)$ is an m-dimensional Wiener process,

$$\tilde{\nu}(t,A)=\nu(t,A)-t\,\Pi(A),$$

$\nu(t,A)$ is a Poisson measure in X, $\mathbb{M}\,\nu(t,A)=t\,\Pi(A)$ and the process $w(t)$ and measure $\nu(t,A)$ are independent of each other. Eq. (1) is a special case of (10 §2. We must assume in this connection that

$$\alpha(t,x,h)=a(t,x)\,h+\beta(t,x,h),$$

$$\beta(t,x,h)=b(t,x)\,[w(t+h)-w(t)]+\int f(t,x,u)\,\tilde{\nu}([t,t+h),du).$$

We denote the components of the field $\beta(t,x,h)$ in some fixed basis by means of $\beta_j(t,x,h)$ $(j=1,2,\dots,m)$. Then

$$\beta_j(t,x,h)=\bar{\beta}_j(t,x,h)+\gamma_j(t,x,h),$$

where

$$\bar{\beta}_j(t,x,h)=\sum_{k=1}^{m} b_{jk}(t,x)\,[w_k(t+h)-w_k(t)],$$

$$\gamma_j(t,x,h)=\int f_j(t,x,u)\,\tilde{\nu}([t,t+h),du),$$

$$\mathbb{M}\{\beta_j(t,x,h)/\mathfrak{F}_t\}=0;\qquad \mathbb{M}\{\gamma_j(t,x,h)\,\gamma_k(t,x,h)/\mathfrak{F}_t\}=h\int f_j\,f_k\,\Pi(du),$$

whence

$$\mathbb{M}\{|\beta(t,x,h)|^2/\mathfrak{F}_t\}=h\left[\int|f(t,x,u)|^2\,\Pi(du)+|b(t,x)|^2\right],$$

$$\mathbb{M}\{|\beta(t,x,h)-\beta(t,y,h)|^2/\mathfrak{F}_t\}$$
$$=h\left[\int|f(t,x,u)-f(t,y,u)|^2\,\Pi(du)+|b(t,x)-b(t,y)|^2\right],$$

where

$$|b(t,x)|^2=\sum_{j,k=1}^{m}|b_{jk}(i,x)|^2.$$

Hence, the field $\alpha(t,x,h)$ will be one with limited after-effect if there is an L for which

$$|a(t,x)|^2+|b(t,x)|^2+\int|f(t,x,u)|^2\,\Pi(du)\leqq L(1+|x|^2), \tag{2}$$

and will satisfy a local Lipschitz condition if for arbitrary $R>0$ there is a constant C_R such that when $|x|<R$, $|y|<R$

$$\begin{aligned}&|a(t,x)-a(t,y)|^2+|b(t,x)-b(t,y)|^2\\&\quad+\int|f(t,x,u)-f(t,y,u)|^2\,\Pi(du)\leqq C_R\,|x-y|^2.\end{aligned} \tag{3}$$

It is easy to verify that $\alpha(t, x, h)$ will be quasi-differential if there is a constant—we will again denote it by L—and a function $g(h)$, $g(h) \geqq 0$, $g(h) \downarrow 0$ for $h \to 0$, for which

$$|a(t+h, x)-a(t, x)|^2+|b(t+h, x)-b(t, x)|^2 \\ +\int |f(t+h, x, u)-f(t, x, u)|^2 \Pi(du) \leqq L(1+|x|^2) g(h). \tag{4}$$

If (2), (3) and (4) hold, then from Theorem 2 §3 we see that (1) has a unique solution which is right continuous w.p. 1. If $f(t, x, u) \equiv 0$, then the solution $\xi(t)$ is continuous w.p. 1. In what follows, when we refer to the existence and uniqueness theorem for (1), we will have this theorem in mind and the hypotheses of said theorem will be taken as (2), (3) and (4).

We estimate the moments of the solution of (1). To this end we use the generalized Itô formula ((19) §6), assuming that $g(t, x)=|x|^{2m}$. Since

$$\nabla g(x)=2m\,|x|^{2m-2}\, x,$$
$$\nabla^2 g(x)=4m(m-1)\,|x|^{2m-4}(x \times x)+2m\,|x|^{2m-2} E,$$

where E is the identity matrix and $x \times x$ is the matrix with elements $x_{jk}=x_j x_k$, we have in the case under consideration

$$\alpha_1(t)=L_N(g)=2m(m-1)\,|\xi(t)|^{2m-4}\,|b^*(t, \xi(t))\,\xi(t)|^2 \\ +m\,|\xi(t)|^{2m-2} \operatorname{sp}\left[b(t, \xi(t))\, b^*(t, \xi(t))\right]$$

and

$$\alpha_2(t)=L_\pi(g)=\int \{|\xi(t)+f(t, \xi(t), u)|^{2m}-|\xi(t)|^{2m} \\ -2m\,|\xi(t)|^{2m-2}(\xi(t) \,|\, f(t, \xi(t), u))\}\, \Pi(du).$$

From (19) §6

$$|\xi(t)|^{2m}=|x|^{2m}+\int_0^t \left[2m\,|\xi(\tau)|^{2m-2}(\xi(\tau) \,|\, a(\tau, \xi(\tau)))+\alpha_1(\tau)+\alpha_2(\tau)\right] d\tau \\ +\int_0^t 2m\,|\xi(\tau)|^{2m-2}(\xi(\tau) \,|\, b(\tau, \xi(\tau))\, dw(\tau)) \\ +\int_0^t \int \left[|\xi(\tau)+f(\tau, \xi(\tau),)|^{2m}-|\xi(\tau)|^{2m}\right] \tilde{v}(d\tau, du). \tag{5}$$

Now assume that in addition to (2) we also have

$$\int |f(t, x, u)|^p\, \Pi(du) \leqq L(1+|x|^p), \qquad p=2, 3, \ldots, 2m. \tag{6}$$

It is easy to show that there then exists a constant L' depending only on L and m for which

$$|L_N(g)|+|L_\pi(g)| \leqq L'(1+|\xi(t)|^{2m}).$$

Consider now the functions $a_N(t, x)$, $b_N(t, x)$ and $f_N(t, x, u)$, coinciding with $a(t, x)$, $b(t, x)$ and $f(t, x, u)$ for $|x| \leqq N$, equal to zero for $|x| \geqq N+1$, and satisfying the assumptions of the existence and uniqueness theorem (see the proof of Theorem 2 § 3) and Eqs. (2) and (6) with constant independent of N. Let $\xi_N(t)$ be the solution of the stochastic equation with coefficients $a_N(t, x)$, $b_N(t, x)$ and $f_N(t, x, u)$. Since the factors in front of $dw(\tau)$ and $\tilde{v}(d\tau, du)$ in the generalized Itô formula are bounded, the stochastic integrals on the right side of (5) possess finite moments. Taking expectations on both sides of (5) we easily obtain

$$\mathbb{M}|\xi_N(t)|^{2m} \leqq |x|^{2m} + L \int_0^t (1 + \mathbb{M}|\xi_N(\tau)|^{2m})\, d\tau,$$

whence $\mathbb{M}|\xi_N(t)|^{2m} \leqq L_1(1+|x|^{2m})$, where L_1 is independent of N. Since $\xi(t) = \mathbb{P}\text{-lim}\, \xi_N(t)$, the moments of $\xi(t)$ up to and including $2m$-th order are finite. Taking expectations in (5) we obtain the estimate

$$\mathbb{M}|\xi(t)|^{2p} \leqq L_{1,p}(1+|x|^{2p}), \qquad p = 1, 2, \ldots, m, \tag{7}$$

where $L_{1,p}$ only depends on L_1, T_1 and p.

Theorem 1. *If* (2), (3), (4), *and* (6) *hold, then the solution of* (1) *will possess finite moments up to and including* $2m$*-th order and* (7) *will be valid*

Remark 1. If the hypotheses of Theorem 1 are fulfilled, then analogous arguments show that there exists a constant L_{2p} such that

$$\mathbb{M}|\xi(t) - x|^{2p} \leqq L_{2p}(1+|x|^{2p})\, t, \qquad p = 1, 2, \ldots, m. \tag{8}$$

Remark 2. If the term contaning the Poisson measure is missing in the stochastic equation, then its solution will possess moments of all orders and

$$\mathbb{M}|\xi(t) - x|^{2p} \leqq L_{2p}(1+|x|^{2p})\, t^p. \tag{9}$$

The proof of (9) is analogous to that of (9) § 6 in Chapter II, Part I.

§ 8. Stochastic Differential Equations Depending on a Parameter. Differentiability w.r.t. the Initial Data

Questions concerning the continuous dependence of the solutions of stochastic equations on a parameter and their differentiability w.r.t. a parameter and w.r.t. the initial data were treated for the one-dimensional case in § 7, 8, Part I. A large part of the statements and proofs can be carried over without difficulty to more general equations. In this section we will present the corresponding theorems-sometimes with slight differences in formulation and short proofs when this seems necessary.

We first present several theorems related to the general stochastic equations considered in § 3.

Theorem 1. *Let* $\xi_n(t)$ $(n=0,1,2,\ldots)$ *satisfy*

$$\xi_n(t)=\varphi_n(t)+\int_0^t \alpha_n(\tau,\xi_n(\tau),d\tau), \qquad t\geqq 0,$$

and assume that the field $\alpha_n(t,x,h)\in\Pi_0(L,C,g)$, C, L *and* g *independent of* n. *Moreover, assume*

$$\lim_{n\to\infty}\sup_{0\leqq t\leqq T}\mathbb{M}|\varphi_n(t)-\varphi_0(t)|^2=0,$$

$$|\mathbb{M}\{\alpha_0(t,x,h)-\alpha_n(t,x,h)/\mathfrak{F}_t\}|\leqq F_1^{(n)}(t,h)(1+|x|),$$

$$\mathbb{M}\{|\alpha_0(t,x,h)-\alpha_n(t,x,h)|^2/\mathfrak{F}_t\}\leqq F_1^{(n)}(t,h)(1+|x|^2)$$

and

$$V_{(0,T)}(F_1^{(n)})=\lim_{|\delta|\to 0}\sum_{k=0}^{n-1}F_1^{(n)}(t_k,\Delta t_k)\to 0 \qquad \textit{for } n\to\infty.$$

Then

$$\sup_{0\leqq t\leqq T}\mathbb{M}|\xi_n(t)-\xi_0(t)|^2\to 0 \qquad \textit{for } n\to\infty.$$

If

$$\mathbb{M}\sup_{0\leqq t\leqq T}|\varphi_n(t)-\varphi_0(t)|^2\to 0,$$

then also

$$\mathbb{M}\sup_{0\leqq t\leqq T}|\xi_n(t)-\xi_0(t)|^2\to 0.$$

The proof of this theorem is analogous to that of Theorem 2 § 7, Part I and uses the boundedness of $\mathbb{M}|\xi_0(t)|^2$ (see Lemma 1 § 3), Lemma 9 § 1 and Corollary 5 of the same section.

With the help of Theorem 1 we can easily obtain a theorem on the differentiability of solutions of stochastic differential equations w.r.t. a parameter (see Theorem 4 § 7, Part I).

Theorem 2. *Assume given a family of random fields* $\alpha_\mu(t,x,h)$, *depending on a scalar parameter* μ $(\mu\geqq 0)$, $\alpha_\mu(t,x,h)\in\Pi_0(L,C,g)$ $(L, C, g$ *independent of* $\mu)$ *and possessing the properties*

a) the fields $\alpha_\mu(t,x,h)$ *is differentiable w.r.t.* μ *at the point* $\mu=0$ *in the following sense: there exists a random field* $\alpha'(t,x,h)$, $\alpha'(t,x,h)\in\Pi(L,g)$ *such that*

$$1)\quad \left|\mathbb{M}\left\{\frac{\alpha_\mu(t,x,h)-\alpha_0(t,x,h)}{\mu}-\alpha'(t,x,h)/\mathfrak{F}_t\right\}\right|\leqq(1+|x|)F_1(t,h,\mu)h;$$

$$2)\quad \mathbb{M}\left\{\left|\frac{\alpha_\mu(t,x,h)-\alpha_0(t,x,h)}{\mu}-\alpha'(t,x,h)\right|^2\Big/\mathfrak{F}_t\right\}\leqq(1+|x|^2)F_1(t,h,\mu)h$$

and

$$\lim_{\mu\to 0}\overline{\lim_{|\delta|\to 0}}\sum_{k=0}^{n-1} F_1(t_k, \Delta t_k, \mu)\,\Delta t_k = 0;$$

b) $\alpha_\mu(t, x, h)$ is differentiable in x for $\mu\in[0,\mu_0]$ in the following sense: there exists an random operator field $\nabla\alpha_\mu(t, x, h)$ such that

1) $$\left|\mathbb{M}\left\{\frac{\alpha_\mu(t, x+sz, h)-\alpha_\mu(t, x, h)}{s}-(\nabla\alpha_\mu(t, x, h)|z)/\mathfrak{F}_t\right\}\right| \leqq (1+|z|^2)\,F_2(t, h, s)\,h;$$

2) $$\mathbb{M}\left\{\left|\frac{\alpha_\mu(t, x+sz, h)-\alpha_\mu(t, x, h)}{s}-(\nabla\alpha_\mu(t, x, h)|z)\right|^2\Big/\mathfrak{F}_t\right\} \leqq (1+|z|)\,F_2(t, h, s)\,h,$$

whereby

$$\lim_{s\to 0}\overline{\lim_{|\delta|\to 0}}\sum_{k=0}^{n-1} F_2(t_k, \Delta t_k, s)\,\Delta t_k = 0,$$

c) $\nabla\alpha_\mu(t, x, h)$ possesses the following properties of continuity w.r.t. μ:

1) $$|\mathbb{M}\{\nabla\alpha_\mu(t, x, h)-\nabla\alpha_0(t, x, h)/\mathfrak{F}_t\}|\leqq F_3(t, h, \mu);$$

2) $$\mathbb{M}\{|\nabla\alpha_\mu(t, x, h)-\nabla\alpha_0(t, x, h)|^2/\mathfrak{F}_t\}\leqq F_3(t, h, \mu),$$

whereby

$$\lim_{\mu\to 0}\overline{\lim_{|\delta|\to 0}}\sum_{k=0}^{n-1} F_3(t_k, \Delta t_k, \mu) = 0;$$

d) the random field

$$\beta(t, x, h) = \alpha'(t, \xi_0(t), h) + (\nabla\alpha_0(t, \xi_0(t), h)|x),$$

where $\xi_0(t)$ satisfies

$$d\xi = \alpha_0(t, \xi(t), dt),$$

fulfills the conditions of the existence theorem, in particular

1) $$|\mathbb{M}\{\nabla\alpha_0(t, \xi_0(t), h)/\mathfrak{F}_t\}|\leqq Ch;$$

2) $$\mathbb{M}\{|\nabla\alpha_0(t, \xi_0(t), h)|^2/\mathfrak{F}_t\}\leqq Ch.$$

e) The derivative w.r.t. μ of the initial value $\xi_\mu(0)$ exists and

$$\mathbb{M}\left|\frac{\xi_\mu(0)-\xi_0(0)}{\mu}-\xi'(0)\right|^2\to 0 \quad \textit{for } \mu\to 0.$$

Then

$$\sup_{0\leqq t\leqq T}\mathbb{M}\left|\frac{\xi_\mu(t)-\xi_0(t)}{\mu}-\zeta(t)\right|^2\to 0 \quad \textit{for } \mu\to 0$$

where $\zeta(t)$ satisfies

$$d\zeta(t)=\alpha'(t,\xi_0(t),dt)+(\nabla\alpha_0(t,\xi_0(t),dt)|\zeta(t)).$$

Proof. Set

$$\zeta_\mu(t)=\frac{\xi_\mu(t)-\xi_0(t)}{\mu}.$$

This satisfies

$$d\zeta_\mu(t)=\beta(t,\zeta_\mu(t),dt)+\tilde\beta_\mu(t,\zeta_\mu(t),dt),$$

where $\beta(t,x,h)$ is the random field in condition d) and

$$\begin{aligned}\tilde\beta(t,x,h)=&\left[\frac{\alpha_\mu(t,\xi_0(t)+\mu x,h)-\alpha_\mu(t,\xi_0(t),h)}{\mu}-(\nabla\alpha_\mu(t,\xi_0,h)|x)\right]\\&+(\nabla\alpha_\mu(t,\xi_0(t),h)-\nabla\alpha_0(t,\xi_0(t),h)|x)\\&+\left[\frac{\alpha_\mu(t,\xi_0(t),h)-\alpha_0(t,\xi_0(t),h)}{\mu}-\alpha'(t,\xi_0(t),h)\right].\end{aligned}$$

It is easy to verify that the hypotheses of Theorem 1 are fulfilled (with the obvious modification of that theorem consisting of passage from the discrete parameter n ($n\to\infty$) to the continuous one μ ($\mu\to 0$)). The claim of our theorem then follows immediatly from Theorem 1.

Let us consider in more detail equations of the form

$$\xi(t)=\varphi(t)+\int_0^t\alpha(\tau,\xi(\tau))\,d\tau+\int_0^t\beta(\tau,\xi(\tau))\,dw(\tau)+\int_0^t\int\gamma(\tau,\xi(\tau),u)\,\tilde\nu(d\tau,du),$$

where $\varphi(t)$, $\alpha(t,x)$, $\beta(t,x)$ and $\gamma(t,x,u)$ are, generally speaking, random and $\mathfrak{F}_t$-measurable for each t, $w(t)$ *and* $\nu(t,A)$ are, resp., a Wiener process and Poisson measure with parameter $\Pi(A)t$ with $w(t)-w(s)$ and $\nu(t,A)-\nu(s,A)$ ($t\geqq s$) independent of $\mathfrak{F}_s$ and of each other: $\varphi(t)$, $\alpha(t,x)$ and $\gamma(t,x,u)$ are n-dimensional random vectors with values in X and $\tilde\nu(t,A)=\nu(t,A)-t\,\Pi(A)$.

We assume that the following conditions are fulfilled – they guarantee the applicability of the existence and uniqueness theorem: there exists a constant C such that w.p. 1

$$|\alpha(t,x)|^2+|\beta(t,x)|^2+\int|\gamma(t,x,u)|^2\,\Pi(du)\leqq C(1+|x|^2), \tag{1}$$

$$|\alpha(t,x)-\alpha(t,y)|^2+|\beta(t,x)-\beta(t,y)|^2+\int|\gamma(t,x,u)-\gamma(t,y,u)|^2\,\Pi(du)\leqq C|x-y|^2 \tag{2}$$

and, moreover,

$$\mathbb{P}\{\sup_{0\leqq t\leqq T}|\varphi(t)|=\infty\}=0. \tag{3}$$

Theorem 3. *Assume*

$$\begin{aligned}\xi_p(t)=\varphi_p(t)+\int_0^t \alpha_p(\tau,\xi_p(\tau))\,d\tau+\int_0^t \beta_p(\tau,\xi_p(\tau))\,dw(\tau)\\ +\int_0^t\int \gamma_p(\tau,\xi_p(\tau),u)\,\tilde{\nu}(d\tau,du),\end{aligned} \tag{4}$$

whereby $\alpha_p(t,x)$, $\beta_p(t,x)$ *and* $\gamma_p(t,x,u)$ *satisfy* (1) *and* (2) *with the same constant* C, $p=0,1,2,\ldots$. *If for arbitrary* $N>0$, $t\in[0,T]$, $\varepsilon>0$

$$\begin{aligned}\lim_{n\to\infty}\mathbb{P}\{\sup_{|x|\leqq N}(|\alpha_p(t,x)-\alpha_0(t,x)|+|\beta_p(t,x)-\beta_0(t,x)|\\ +\int|\gamma_p(t,x,u)-\gamma_0(t,x,u)|^2\,\Pi(du))>\varepsilon\}=0\end{aligned}$$

and

$$\lim_{n\to\infty}\sup_{0\leqq t\leqq T}\mathbb{M}|\varphi_p(t)-\varphi_0(t)|^2=0,$$

then

$$\lim_{n\to\infty}\sup_{0\leqq t\leqq T}\mathbb{M}|\xi_p(t)-\xi_0(t)|^2=0.$$

If, in addition

$$\lim_{p\to\infty}\mathbb{M}\sup_{0\leqq t\leqq T}|\varphi_p(t)-\varphi_0(t)|^2=0,$$

then also

$$\lim_{p\to\infty}\mathbb{M}\sup_{0\leqq t\leqq T}|\xi_p(t)-\xi_0(t)|^2=0.$$

The proof goes as in Theorem 2 §7 Chapter 1, Part I with the use of the properties of an integral w.r.t. the Poisson measure.

Theorem 4. *Let* $\xi_p(t)$, $0\leqq p\leqq a$ *satisfy*

$$\xi_p(t)=\xi_p(0)+\int_0^t \alpha_p(\tau,\xi_p(\tau),d\tau), \tag{5}$$

$$\begin{aligned}\alpha_p(t,x,h)=a_p(t,x)\,h+b_p(t,x)[w(t+h)-w(t)]\\ +\int f_p(t,x,u)\,\tilde{\nu}([t,t+h),du),\end{aligned} \tag{6}$$

whereby $a_p(t,x)$, $b_p(t,x)$, $f_p(t,x,u)$ *are non-random and satisfy for each* p *the conditions for existence and uniqueness of the solution of a stochastic equation and assume that for* $p\to 0$

$$a_p(t,x)\to a_0(t,x),\qquad b_p(t,x)\to b_0(t,x),\qquad f_p(t,x,u)\to f_0(t,x,u)$$

uniformly in x *in any bounded region* $|x|\leqq R$ *and that there exists a* C *independent of* p *such that*

$$|a_p(t,x)|^2+|b_p(t,x)|^2+\int|f_p(t,x,u)|^2\,\Pi(du)\leqq C(1+|x|^2),$$

and that $\xi_p(0)$ *converges in probability to* $\xi_0(0)$. *Then*

$$\mathbb{P}\text{-}\lim \sup_{0 \leqq t \leqq T} |\xi_p(t) - \xi_0(t)| = 0$$

(see the proof of Theorem 3 § 7, Part I).

We now turn to the question of differentiability of solutions of (4) w.r.t. a parameter. Let the parameter p vary in some interval $[a, b]$. By the derivative of a solution w.r.t. p we will understand the derivative in mean square

$$\frac{\partial}{\partial p} \xi_p(t) = \underset{\Delta p \to 0}{\text{l.i.m.}} \frac{\xi_{p+\Delta p}(t) - \xi_p(t)}{\Delta p}.$$

Theorem 5. *Assume that the following conditions are satisfied:*

1) $\varphi_p(t) \in H_1$, $\frac{\partial}{\partial p} \varphi_p(t)$ *exists and*

$$\lim_{\Delta p \to 0} \left\| \frac{\partial}{\partial p} \varphi_p(t) - \frac{1}{\Delta p} [\varphi_{p+\Delta p}(t) - \varphi_p(t)] \right\|_1 = 0;$$

2) $\frac{\partial}{\partial p} a_p(t, x)$, $\frac{\partial}{\partial p} b_p(t, x)$, $\frac{\partial}{\partial p} f_p(t, x, u)$ $\left(\frac{\partial}{\partial p} f_p(t, x, u) \in H_2(\Pi)\right)$

exist, whereby

$$\lim_{\Delta p \to 0} \mathbb{M} \int_0^T \left\{ \left| \frac{1}{\Delta p} [a_{p+\Delta p}(t, \xi_p(t)) - a_p(t, \xi_p(t))] - \frac{\partial}{\partial p} a_p(t, \xi_p(t)) \right|^2 \right.$$
$$\left. + \left| \frac{1}{\Delta p} [b_{p+\Delta p}(t, \xi_p(t)) - b_p(t, \xi_p(t))] - \frac{\partial}{\partial p} b_p(t, \xi_p(t)) \right|^2 \right\} dt = 0,$$

$$\lim_{\Delta p \to 0} \mathbb{M} \left| \int_0^T \left\{ \frac{1}{\Delta p} [\Phi_{p+\Delta p}(t, \xi_p(t), dt) - \Phi_p(t, \xi_p(t), dt)] \right. \right.$$
$$\left. \left. - \Phi'_p(t, \xi_p(t), dt) \right\} \right|^2 = 0,$$

with

$$\Phi_p(t, x, h) = \int f_p(t, x, u) \, \tilde{v}([t, t+h), du),$$

and

$$\Phi'_p(t, x, h) = \int \frac{\partial}{\partial p} f_p(t, x, u) \, \tilde{v}([t, t+h), du);$$

3) *the continuous partial derivatives*

$$\frac{\partial a_p(t, x)}{\partial x_j}, \quad \frac{\partial b_p(t, x)}{\partial x_j}, \quad \frac{\partial f_p(t, x, u)}{\partial x_j}, \quad j = 1, 2, \ldots, n,$$

exist and for all $t, x, 0 \leqq t \leqq T$, $x \in X$, *w.p.1*

$$\sum_{j=1}^{n} \left(\left| \frac{\partial}{\partial x_j} a_p(t, x) \right| + \left| \frac{\partial}{\partial x_j} b_p(t, x) \right| + \int \left| \frac{\partial}{\partial x_j} f_p(t, x, u) \right|^2 \Pi(du) \right) \leqq C.$$

Then $\xi_p(t)$ *is differentiable w.r.t.* p, $\xi'_p(t) = \frac{\partial}{\partial p} \xi(t)$ *satisfies*

$$\xi'_p(t) = \frac{\partial}{\partial p} \varphi_p(t) + \int_0^t \frac{\partial}{\partial p} \alpha_p(\tau, \xi_p(\tau), d\tau) + \int_0^t (\nabla \alpha_p(\tau, \xi_p(\tau), d\tau) \mid \xi'_p(\tau)), \quad (7)$$

where $\frac{\partial}{\partial p} \alpha_p(t, x, h)$ *denotes the random field*

$$\begin{aligned} \frac{\partial}{\partial p} \alpha_p(t, x, h) = {} & \frac{\partial}{\partial p} a_p(t, x) h + \frac{\partial}{\partial p} b_p(t, x) [w(t+h) - w(t)] \\ & + \int \frac{\partial}{\partial p} f_p(t, x, u) \tilde{\nu}([t, t+h), du), \end{aligned} \quad (8)$$

and if $y = (y_1, y_2, \ldots, y_n)$, *then*

$$\begin{aligned} (\nabla \alpha_p(t, x, h) \mid y) = \sum_{k=1}^{n} \Bigg[& \frac{\partial a_p(t, x)}{\partial x_k} h + \frac{\partial b_p(t, x)}{\partial x_k} [w(t+h) - w(t)] \\ & + \int \frac{\partial f_p(t, x, u)}{\partial x_k} \tilde{\nu}([t, t+h), du) \Bigg] y_k. \end{aligned} \quad (9)$$

The proof is analogous to that of Theorem 4 § 7, Part I.

We now dwell on the question of dependence of solutions on initial conditions. As in Part I, we will consider solutions at the initial time $t = s$ as functions of s and the initial value $z \in X$.

To this end we introduce the process $\xi_{z,s}(t)$ satisfying

$$\xi_{z,s}(t) = z + \int_s^t \alpha(\tau, \xi_{z,s}(\tau), d\tau), \quad (10)$$

where

$$\begin{aligned} \alpha(t, x, h) = {} & a(t, x) h + b(t, x) [w(t+h) - w(t)] \\ & + \int f(t, x, u) \tilde{\nu}([t, t+h), du), \end{aligned} \quad (11)$$

and $a(t, x)$, $b(t, x)$ and $f(t, x, u)$ satisfy the conditions for existence and uniqueness of the solution of (10).

Assume that

$$\frac{\partial a(t, x)}{\partial x_k}, \quad \frac{\partial b(t, x)}{\partial x_k}, \quad \int \left| \frac{\partial f(t, x, u)}{\partial x_k} \right|^2 \Pi(du), \qquad k = 1, 2, \ldots, n, \quad (12)$$

exist, are continuous and bounded ($t\in[0,T]$, $x\in X$) and that

$$\int\left|\frac{\partial f(t,x,u)}{\partial x_k}-\frac{\partial f(t',x',u)}{\partial x_k}\right|^2 \Pi(du)\to 0 \quad \text{for } (t',x')\to(t,x).$$

Theorem 5 implies that the $\frac{\partial}{\partial z_k}\xi_{z,s}(t)$ exist and satisfy

$$\frac{\partial}{\partial z_k}\xi_{z,s}(t)=\delta_k+\int_s^t\left(\nabla\alpha(\tau,\xi_{z,s}(\tau),d\tau)\middle|\frac{\partial}{\partial z_k}\xi_{z,s}(\tau)\right), \tag{13}$$

where δ_k is a vector whose k-th component equals one and the rest zero, and the field $(\nabla\alpha(t,\xi_{z,s}(t),h)|x)$ satisfies (9).

If the corresponding hypotheses are fulfilled we can apply Theorem 5 to (13) and establish the existence of derivatives of higher order.

Thus, for $\frac{\partial^2}{\partial z_k\partial z_j}\xi_{z,s}(t)$ we obtain

$$\begin{aligned}\frac{\partial^2}{\partial z_k\partial z_j}\xi_{z,s}(t)=&\int_s^t\left(\nabla^2\alpha(\tau,\xi_{z,s}(\tau),d\tau)\middle|\frac{\partial\xi}{\partial z_k}\middle|\frac{\partial\xi}{\partial z_j}\right)\\&+\int_s^t\left(\nabla\alpha(\tau,\xi_{z,s}(\tau),d\tau)\middle|\frac{\partial^2\xi_{z,s}(\tau)}{\partial z_k\partial z_j}\right),\end{aligned} \tag{14}$$

where $(\nabla^2\alpha(\tau,z,h)|x|y)$ denotes the random field

$$\begin{aligned}(\nabla^2\alpha(t,z,h)|x|y)=&\left(\sum_{p,q=1}^n\frac{\partial^2 a(t,z)}{\partial z_p\partial z_q}x_p y_q\right)h\\&+\left(\sum_{p,q=1}^n\frac{\partial^2 b(t,z)}{\partial z_p\partial z_q}x_p y_q\right)[w(t+h)-w(t)]\\&+\int\left(\sum_{p,q=1}^n\frac{\partial^2 f(t,z,u)}{\partial z_p\partial z_q}x_p y_q\right)\tilde{\nu}([t,t+h),du).\end{aligned}$$

We will not carry out the verification of Theorem 5's hypotheses but will prove the existence of the derivatives $\frac{\partial^2\xi_{z,s}(t)}{\partial z_k\partial z_j}$ and the validity of (14) directly. To this end, we will understand the derivative as the limit of the corresponding quotient in the sense of convergence in probability. Set

$$\xi_{\Delta z_j}(t)=\frac{1}{\Delta z_j}\left(\frac{\partial}{\partial z_k}\xi_{z+\Delta z_j,s}(t)-\frac{\partial}{\partial z_k}\xi_{z,s}(t)\right).$$

For $\xi_{\Delta z_j}(t)$ we have the equation

$$\xi_{\Delta z_j}(t)=\varphi_{\Delta z_j}(t)+\int_s^t(\nabla\alpha(\tau,\xi_{z,s}(\tau),d\tau)|\xi_{\Delta z_j}(\tau)), \tag{15}$$

where

$$\varphi_{\Delta z_j}(t)=\int_s^t \frac{1}{\Delta z_j}\left(\nabla\alpha\left(\tau, \xi_{z+\Delta z_j, s}(\tau), d\tau\right)\right.$$
$$\left.-\nabla\alpha\left(\tau, \xi_{z, s}(\tau), d\tau\right)\middle|\frac{\partial}{\partial z_k}\xi_{z+\Delta z_j, s}(\tau)\right). \tag{16}$$

Consider

$$\zeta(t)=\varphi(t)+\int_s^t \left(\nabla\alpha(\tau, \xi_{z, s}(\tau), d\tau)\,|\,\zeta(\tau)\right), \tag{17}$$

where $\varphi(t)$ is $\mathfrak{F}_t$-measurable for each $t\in[0, T]$. This has a solution for any $\varphi(t)$ for which

$$\mathbb{P}\{\sup_{0\leqq t\leqq T}|\varphi(t)|=\infty\}=0.$$

Lemma 1. *If in* (17), $\varphi(t)=\varphi_h(t)$; $h\geqq 0$ *converges in probability to* $\varphi_0(t)$ *for* $h\to 0$ *and*

$$\lim_{N\to\infty}\sup_{0\leqq h\leqq h_0}\mathbb{P}\{\sup_{0\leqq t\leqq T}|\varphi_h(t)|>N\}=0,$$

then the corresponding solution $\zeta_h(t)$ *of* (17) *converges for each* t *in probability to* $\zeta_0(t)$.

We note that if $\mathbb{M}|\varphi(t)|^2\leqq K$, then it follows from (17) that there exists a constant C' such that

$$\mathbb{M}|\zeta(t)|^2\leqq C'\left[\mathbb{M}|\varphi(t)|^2+\int_s^t \mathbb{M}|\zeta(\tau)|^2\,d\tau\right],$$

whence, in turn (C'' is independent of φ),

$$\mathbb{M}|\zeta(t)|^2\leqq C''\left[\mathbb{M}|\varphi(t)|^2+\int_s^t \mathbb{M}|\varphi(\tau)|^2\,d\tau\right].$$

The preceding inequality shows that if $\mathbb{M}|\varphi(t)-\varphi_n(t)|^2\to 0$, then

$$\mathbb{M}|\zeta_n(t)-\zeta_0(t)|^2\to 0.$$

In the general case we set $\varphi_n^N(t)=\varphi_n(t)$ if $\sup_{0\leqq\tau\leqq t}|\varphi_n(\tau)|\leqq N$ and $\varphi_n^N(t)=0$ otherwise. Then

$$\begin{aligned}
&\mathbb{P}\{|\zeta_n(t)-\zeta_0(t)|>\varepsilon\}\\
&\quad\leqq\mathbb{P}\{|\zeta_n^N(t)-\zeta_0^N(t)|>\varepsilon\}+\mathbb{P}\{\zeta_n^N(t)\neq\zeta_n(t)\}+\mathbb{P}\{\zeta_0^N(t)\neq\zeta_0(t)\}\\
&\quad=\mathbb{P}\{|\zeta_n^N(t)-\zeta_0(t)|>\varepsilon\}+\mathbb{P}\{\sup_{0\leqq t\leqq T}|\varphi_n(t)|>N\}\\
&\qquad+\mathbb{P}\{\sup_{0\leqq t\leqq T}|\varphi_0(t)|>N\},
\end{aligned}$$

where $\zeta_n^N(t)$ satisfies (17) for $\varphi(t)=\varphi_n^N(t)$. The latter inequality proves the lemma.

Lemma 2. *If Condition 3 of Theorem 5 holds for* (10), *then there is a constant C' for which*

$$\mathbb{M} \sup_{0 \leq t \leq T} |\xi_{z,s}(t)|^2 \leq C'(1+|z|^2).$$

Indeed, from the properties of the stochastic integral it follows that

$$\mathbb{M} \sup_{0 \leq t \leq T} \left| \int_s^t b(\tau, \xi(\tau))\, dw(\tau) \right|^2 \leq \int_s^T 4\mathbb{M}\, |b(\tau, \xi(\tau))|^2\, d\tau$$

$$\leq 4C \int_s^T (1+\mathbb{M}\,|\xi(\tau)|^2)\, d\tau,$$

$$\mathbb{M} \sup_{0 \leq t \leq T} \left| \int_s^t \int f(\tau, \xi(\tau), u)\, \tilde{v}(d\tau, du) \right|^2 \leq \mathbb{M} \int_s^T \int |f(\tau, \xi(\tau), u)|^2\, \Pi(du)\, d\tau$$

$$\leq 4C \int_s^T (1+\mathbb{M}\,|\xi(\tau)|^2)\, d\tau.$$

It is now easy to show from (10) that there is a C_1 independent of t, s and z such that for the variable $\eta(t) = \sup_{0 \leq \tau \leq t} |\xi_{z,s}(t)|^2$ we have

$$\mathbb{M}\, \eta(t) \leq C_1 \left(|z|^2 + t + \int_s^t \mathbb{M}\, \eta(\tau)\, d\tau \right),$$

which implies the proof of the lemma.

Lemma 3. *If the variables at* (12) *are bounded, then there is a constant C' for which*

$$\mathbb{M} \left\{ \sup_{0 \leq t \leq T} \left| \frac{\partial}{\partial z_k} \xi_{z,s}(t) \right|^2 + \sup_{0 \leq t \leq T} \left| \frac{\xi_{z+\Delta z_k, s}(t) - \xi_{z,s}(t)}{\Delta z_k} \right|^2 \right\} \leq C'.$$

The proof is similar to the preceding one.

Corollary. *If the variables*

$$\frac{\partial^2 a(t,x)}{\partial x_k \partial x_j}, \quad \frac{\partial^2 b(t,x)}{\partial x_k \partial x_j}, \quad \int \left| \frac{\partial^2 f(t,x,u)}{\partial x_k \partial x_j} \right|^2 \Pi(du), \qquad k, j = 1, 2, \ldots, m, \tag{18}$$

exist and are continuous in all arguments, then $\varphi_{\Delta z_j}(t)$ converges for $\Delta z_j \to 0$ in probability to $\varphi_0(t)$,

$$\varphi_0(t) = \int_s^t \left(\nabla^2 \alpha(\tau, \xi_{z,s}(\tau), d\tau) \left| \frac{\partial}{\partial z_k} \xi_{z,s}(\tau) \right| \frac{\partial}{\partial z_j} \xi_{z,s}(\tau) \right)$$

and

$$\lim_{N \to \infty} \sup_{|\Delta z_j| \leq h} \mathbb{P} \{ \sup_{s \leq t \leq T} |\varphi_{\Delta z_j}(t)| > N \} = 0.$$

From the above results follows

Theorem 6. *If the variables at* (12) *are bounded and those at* (18) *continuous, then*

$$\frac{\partial}{\partial z_k}\xi_{s,z}(t) \quad \textit{and} \quad \frac{\partial^2}{\partial z_k \partial z_j}\xi_{z,s}(t)$$

exist (in the sense of convergence in probability) and satisfy (13) *and* (14), *resp.*

Remark 1. One proves analogously that if the variables at (12) are bounded, continuous partial derivatives w.r.t. x of $a(t,x)$ and $b(t,x)$ up to and including m-th order exist, and the derivatives of $f(t,x,u)$ exist, are square integrable in u w.r.t. the measure Π and continuous in t and x in mean square (w.r.t. the measure Π), then $\xi_{z,s}(t)$ possesses derivatives (in probability) w.r.t. z_k up to m-th order inclusively.

Remark 2. Given the hypotheses of Theorem 6 we have

$$\mathbf{M}|\xi_{z,s}(t)-\xi_{z_1,s_1}(t)|^2 \leqq C_1\big(|z-z_1|^2+(s-s_1)(1+|z|^2)\big).$$

If the assumptions of Remark 1 are fulfilled we can show that the

$$\frac{\partial^p}{\partial z_k^p}\xi_{z,s}(t) \qquad (p=1,2,\ldots,m)$$

are continuous in z and s (in probability).

In some cases, the results obtained above are in inadequate for our purposes. By sharpening the assumptions on the coefficients of (10), we can obtain the existence of higher-order moments of the processes

$$\frac{\partial}{\partial z_k}\xi_{z,s}(t)$$

and also the existence of second-order partial derivatives in mean square w.r.t. the initial data.

Assume that conditions (1) and (2) are satisfied for (10) and that

$$\int |f(t,x,u)|^k\,\Pi(du) \leqq C(1+|x|^k), \qquad k=3,4. \tag{20}$$

From Theorem 1 § 7, the solution of (10) has finite moments of 4-th order.

Consider

$$d\zeta = \nabla\alpha(t,\zeta(t),dt), \qquad \zeta(0)=\delta, \tag{21}$$

where

$$\nabla\alpha(t,x,h) = (\nabla a(t,\xi_{z,s}(t))\,|\,x)\,h + (\nabla b(t,\xi_{z,s}(t))\,[w(t+h)-w(t)]\,|\,x)$$
$$+\int (\nabla f(t,\xi_{z,s}(t),u)\,|\,x)\,\tilde\nu([t,t+h),du).$$

Assuming that

$$\int |\nabla f(t,x,u)|^k\,\Pi(du) \leqq C, \qquad k=2,3,4, \tag{22}$$

where the constant C does not depend on x, and using the generalized Itô formula we find as in the proof of Theorem 1 § 7 that the solution of (21) has bounded 4-th order moments satisfying

$$\begin{aligned}\mathbb{M}|\zeta(t)|^4 = |\delta|^4 + \mathbb{M}\int_0^t \{4|\zeta(\tau)|^2 (\zeta(\tau)|\nabla a(\tau, \xi_{z,s}(\tau))|\zeta(\tau))\\ +2|\zeta(\tau)|^2 \operatorname{sp}(\nabla b(\tau, \xi_{z,s}(\tau))|\zeta(\tau))(\nabla b(\tau, \xi_{z,s}(\tau))|\zeta(\tau)) \quad (23)\\ +\int [|\zeta(\tau)+(\nabla f(\tau, \xi_{z,s}(\tau), u)|\zeta(\tau))|^4 - |\zeta(\tau)|^4\\ -4|\zeta(\tau)|^2 (\zeta(\tau)|\nabla f(\tau, \xi_{z,s}(\tau), u)|\zeta(\tau))]\, \Pi(du)\}\, d\tau.\end{aligned}$$

It is easy to show that the factor in square brackets behind $\Pi(du)$ is majorized by

$$K|\zeta(\tau)|^4 \sum_{k=2}^{4} |\nabla f(\tau, \xi_{z,s}(\tau), u)|^k,$$

where K is an absolute constant. Hence, taking into account the uniform boundedness of the variables

$$\nabla a(t,x), \quad \nabla b(t,x), \quad \int |\nabla f(t,x,u)|^k\, \Pi(du), \qquad k=2,3,4,$$

we get

$$\mathbb{M}|\zeta(t)|^4 \leqq d + C_1 \int_s^t \mathbb{M}|\zeta(\tau)|^4\, d\tau,$$

where d and C_1 are independent of t and z. Thus, $\mathbb{M}|\zeta(t)|^4$ is uniformly bounded (w.r.t. $z \in X$, $t \in [s, T]$). Under the same assumptions we also have

$$\mathbb{M}|\zeta(t)-\zeta(s)|^4 \leqq C_2(t-s).$$

We now consider the solution of (21) as a function of the parameter z_k and apply Theorem 5 to it. Conditions 1 and 3 of that theorem are fulfilled in an obvious way. To verify Condition 2 we look at one of the variables appearing there

$$\begin{aligned}I_1 = \mathbb{M}\int_s^T \Big| \frac{1}{\Delta z_k} [(\nabla a(t, \xi_{z+\Delta z, s}(t))|\zeta(t)) - (\nabla a(t, \xi_{z,s}(t))|\zeta(t))]\\ - \frac{\partial}{\partial z_k}(\nabla a(t, \xi_{z,s}(t))|\zeta(t))\Big|^2 dt,\end{aligned}$$

where Δz is a vector whose k-th component equals Δz_k with the remaining ones zero, and $\zeta(t)$ satisfies (21).

It is easy to show that if $\nabla^2 \alpha(t,x)$ exists and is bounded in x, then the derivative $\frac{\partial}{\partial z_k}(\nabla^2 \alpha(t, \xi_{z,s}(t))|\zeta(t))$ exists in the mean-square convergence sense and

$$\frac{\partial}{\partial z_k}(\nabla a(t,\xi_{z,s}(t))|\zeta(t)) = \left(\nabla^2 a(t,\xi_{z,s}(t))\left|\frac{\partial}{\partial z_k}\xi_{z,s}(t)\right|\zeta(t)\right).$$

In addition, we have

$$I_1 \leqq 2\,\mathbb{M}\int_s^T \left\{|\nabla^2 a(t,\tilde{\xi})|\,|\zeta(t)|^2 \left|\frac{\xi_{z+\Delta z,s}(t)-\xi_{z,s}(t)}{\Delta z_k} - \frac{\partial}{\partial z_k}\xi_{z,s}(t)\right|^2 \right.$$
$$\left. + |\nabla^2 a(t,\tilde{\xi}) - \nabla^2 a(t,\xi_{z,s}(t))|^2 \frac{|\zeta(t)|^4 + \left|\frac{\partial}{\partial z_k}\xi_{z,s}(t)\right|^4}{2}\right\} dt,$$

where $\tilde{\xi} = \xi_{z,s}(t) + \Theta[\xi_{z+\Delta z,s}(t) - \xi_{z,s}(t)]$, $0 \leqq \Theta \leqq 1$. If the boundedness assumption on $\nabla^2 a(t,x)$ holds, then the second term of the integrand above is majorized by

$$K\left(|\zeta(t)|^4 + \left|\frac{\partial}{\partial z_k}\xi_{z,s}(t)\right|^4\right),$$

which is independent of Δz and has expectation integrable w.r.t. t. Moreover, this summand converges in probability to zero. Hence, the part of I_1 which corresponds to this summand tends to zero for $\Delta z_k \to 0$. Using the generalized Itô formula and (20) we can show that

$$\mathbb{M}\left|\frac{\xi_{z+\Delta z,s}(t)-\xi_{z,s}(t)}{\Delta z_k} - \frac{\partial}{\partial z_k}\xi_{z,s}(t)\right|^4 \to 0 \quad \text{for } \Delta z_k \to 0.$$

Hence, $I_1 \to 0$. We estimate the remaining variables appearing in Condition 2 of Theorem 5 similarly and obtain

Theorem 7. *If* (10) *satisfies Conditions* (1) *and* (2) *and*

$$|\nabla^2 a(t,x)|^2 + |\nabla^2 b(t,x)|^2 + \int |\nabla^2 f(t,x,u)|^2 \,\Pi(du) \leqq C, \tag{24}$$

$$\int (|f(t,x,u)|^k + |\nabla f(t,x,u)|^k)\,\Pi(du) \leqq C, \quad k=2,3,4, \tag{25}$$

then the solution of (10) *has m.s. derivatives in* $\frac{\partial^2}{\partial z_k \partial z_j}\xi_{z,s}(t)$ *which satisfy* (14).

Remark 3. If in (10) $f(t,x,u)\equiv 0$, then its solution will possess moments of arbitrary order. Arguments analogous to those used in Theorem 7 show that the conclusions of the theorem remain in force if (24) is replaced by the following condition:

for some $m>0$

$$|\nabla^2 a(t,x)| + |\nabla^2 b(t,x)| \leqq C(1+|x|^m).$$

A similar remark applies to the general case $(f(t,x,u)\not\equiv 0)$ if we impose conditions on $f(t,x,u)$ which assure the existence of sufficiently high-order moments of the solution of (10).

§ 9. Solutions of Stochastic Differential Equations as Markov Processes

Theorem 1. *If $\xi(t)$ satisfies a stochastic equation without after-effect for which the assumptions for existence and uniqueness of solution are fulfilled, then $\xi(t)$ is a Markov process and its transition probability can be defined by*

$$P(t,x,s,A)=\mathbb{P}\{\xi_{x,t}(s)\in A\},$$

where $\xi_{x,t}(s)$ satisfies

$$\xi_{xt}(s)=x+\int_t^s \alpha(\tau,\xi_{xt}(\tau),d\tau),\qquad t<s.$$

The proof of this theorem is similar to that of Theorem 1 § 10, Part I.

Remark 1. If $f(x)$ $(x\in X)$ is continuous and bounded, then

$$\varphi_s(x,t)=\mathbb{M}f(\xi_{xt}(s))=\int P(t,x,s,dy)f(y)$$

is a continuous function of (x,t) for arbitrary s $(t<s)$. Furthermore, if $f(x)$ is continuous, $|f(x)|\leqq C(1+|x|)^\rho$ and $\mathbb{M}|\xi_{xt}(s)|^p$ is uniformly bounded in an arbitrary finite region of variation of x with $p>\rho$, then $\varphi_s(x,t)$ will also be continuous in (x,t).

Indeed, if $f(x)$ is continuous, then $f(\xi_{xt}(s))$ is continuous in probability as a function of x. The formulated condition guarantees the possibility of passing to the limit under the expectation sign. Now apply Remark 2 § 8. The condition $p>\rho$ can be replaced by the following: the assumptions of Theorem 1 § 7 hold – these guarantee the existence of the moments of $\xi_{x,t}(s)$ up to and including the $2m$-th –, $f(x)$ has continuous partial derivatives up to and including the $2m$-th and the $2m$-th partials are bounded. In fact, here $|D^k f(x)|\leqq C_k(1+|x|^{2m-k})$, where D^k denotes some k-th-order partial derivative and C_k is a constant. Thus, using Taylor's formula

$$\begin{aligned}&|\mathbb{M}f(\xi_{x+\Delta x,t}(s))-\mathbb{M}f(\xi_{xt}(s))|\\&\leqq C'\sum_{k=1}^{2m}\mathbb{M}(1+|\xi_{xt}(s)|^{2m-k})\,|\xi_{x+\Delta x,t}(s)-\xi_{xt}(s)|^k\\&\leqq C'\sum_{k=1}^{2m}\left(1+[\mathbb{M}|\xi_{xt}(s)|^{2m}]^{\frac{2m-k}{2m}}\right)[\mathbb{M}|\xi_{x+\Delta x,t}(s)-\xi_{xt}(s)|^{2m}]^{\frac{k}{2m}},\end{aligned}$$

from which the claim follows.

Remark 2. Under the assumptions of the previous remark

$$\lim_{\substack{s,t\to t_0\\ t<t_0<s}} \mathbb{M} f(\xi_{xt}(s)) = f(x).$$

Definition 1. By the *infinitesimal operator* (or *generator*) of a family of transition probabilities $P(t, x, s, A)$ of a Markov process we mean the operator A_s $(0 \leqq s \leqq T)$, defined on functions $f(x)$ $(x \in X)$ by

$$\begin{aligned}(A_s f)(x) &= \lim_{\substack{t,t'\to s\\ t<s<t'}} \frac{\mathbb{M} f(\xi_{tx}(t')) - f(x)}{t'-t}\\ &= \lim_{\substack{t,t'\to s\\ t<s<t'}} \frac{\int P(t, x, t', du) f(u) - f(x)}{t'-t}. \end{aligned} \tag{1}$$

The domain of definition $D(A)$ of the infinitesimal operator consists of all continuous functions $f(x)$ for which the limit at (1) exists for each $s \in [0, T]$, $x \in X$.

Let us calculate the infinitesimal operator of the process defined by

$$d\xi = a(t, \xi)\, dt + b(t, \xi)\, dw(t) + \int c(t, \xi, u)\, \tilde{\nu}(dt, du), \qquad \xi(0) = x, \tag{2}$$

which satisfies Conditions (2) and (3) § 7. Set

$$\eta(t) = \xi(t) - \xi_1(t),$$

$$\xi_1(t) = x + \int_0^t [a(\tau, x)\, d\tau + b(\tau, x)\, dw + \int c(\tau, x, u)\, \tilde{\nu}(d\tau, du)].$$

Lemma 1. *If the coefficients of* (2) *satisfy* (2) *and* (3) § 7 *with constant* C_R *independent of* R, *then there is a* $C > 0$ *such that*

$$|\mathbb{M}\{(\xi(t) - \xi_1(t))/\mathfrak{F}_0\}| \leqq C(1 + |x|)\, t^{\frac{3}{2}}, \tag{3}$$

$$\mathbb{M}\{|\xi(t) - \xi_1(t)|^2/\mathfrak{F}_0\} \leqq C(1 + |x|^2)\, t^2. \tag{3'}$$

Indeed,

$$\mathbb{M}\{(\xi(t) - \xi_1(t))/\mathfrak{F}_0\} = \int_0^t \mathbb{M}\{[a(\tau, \xi(\tau)) - a(\tau, x)]/\mathfrak{F}_0\}\, d\tau,$$

whence, using Condition (3) § 7 and Lemma 1 § 3, we get

$$\begin{aligned}|\mathbb{M}\{(\xi(t) - \xi_1(t))/\mathfrak{F}_0\}| &\leqq C \int_0^t \mathbb{M}\{|\xi(\tau) - x|/\mathfrak{F}_0\}\, d\tau\\ &\leqq C' \int_0^t (1 + |x|)\, \tau^{\frac{1}{2}}\, d\tau = C''(1 + |x|)\, t^{\frac{3}{2}},\end{aligned}$$

which verifies (3). Analogously

$$\begin{aligned}
&\mathbb{M}\{|\xi(t)-\xi_1(t)|^2/\mathfrak{F}_0\} \\
&\quad\leqq C\left\{\int_0^t \mathbb{M}\{|\xi(\tau)-\xi_1(\tau)|^2/\mathfrak{F}_0\}\,d\tau+\int_0^t \mathbb{M}\{|b(\tau,\xi(\tau))-b(\tau,x)|^2/\mathfrak{F}_0\,d\tau\right. \\
&\qquad\left.+\int_0^t\int \mathbb{M}\{|c(\tau,\xi(\tau),u)-c(\tau,x,u)|^2/\mathfrak{F}_0\}\,\Pi(du)\right\}d\tau \\
&\quad\leqq C'\int_0^t \mathbb{M}\{|\xi(\tau)-x|^2/\mathfrak{F}_0\}\,d\tau\leqq C''(1+|x|^2)\,t^2.
\end{aligned}$$

The lemma is proved.

We turn to the calculation of the infinitesimal operator of the transition probabilities. Let $f(x)$ $(x\in X)$ be an arbitrary twice continuously differentiable function with bounded partial derivatives of first and second orders. Consider

$$z(t,t')=\frac{1}{t'-t}\,\mathbb{M}\{f(\xi_{tx}(t'))-f(x+\alpha(t,x,t'-t))\},$$

where

$$\alpha(t,x,h)=\int_t^{t+h}\{a(s,x)\,ds+b(s,x)\,dw(s)+\int c(s,x,u)\,\tilde{v}(ds,du)\}.$$

Since

$$\begin{aligned}
&f(\xi_{t,x}(t'))-f(x+\alpha(t,x,t'-t)) \\
&\quad=(\nabla f(x+\alpha(t,x,t'-t))\,|\,\eta(t'))+\tfrac{1}{2}(\nabla^2\tilde{f}\,\eta(t')|\,\eta(t')) \\
&\quad=(\nabla f(x)|\,\eta(t'))+(\nabla f(x+\alpha(t,x,t'-t))-\nabla f(x)|\,\eta(t')) \\
&\qquad+\tfrac{1}{2}(\nabla^2\tilde{f}\,\eta(t')|\,\eta(t')),
\end{aligned}$$

where $\nabla^2\tilde{f}$ is the value of $\nabla^2 f(y)$ at the point y lying in the interval connecting $\xi_{x,t}(t')$ and $x+\alpha(t,x,t'-t)$, considering the boundedness of $\nabla f(x)$ we get

$$\begin{aligned}
(t'-t)\,|z(t',t)|&\leqq C\,|\mathbb{M}\{\eta(t')/\mathfrak{F}_t\}| \\
&\quad+C[\mathbb{M}\{|\nabla f(x+\alpha(t,x,t'-t))-\nabla f(x)|^2/\mathfrak{F}_t\}\,\mathbb{M}\{|\eta(t')|^2/\mathfrak{F}_t\}]^{\frac{1}{2}} \\
&\quad+C\,\mathbb{M}\{|\eta(t')|^2/\mathfrak{F}_t\}.
\end{aligned}$$

From Lemma 1 and the boundedness of $\nabla^2 f(x)$ follows

$$\begin{aligned}
(t'-t)\,|z(t,t')|&\leqq C'(1+|x|)(t'-t)^{\frac{3}{2}} \\
&\quad+C'(1+|x|)(t'-t)\,\mathbb{M}\{|\alpha(t,x,t'-t)|^2/\mathfrak{F}_t\}^{\frac{1}{2}} \qquad (4)\\
&\leqq C''(1+|x|)(t'-t)^{\frac{3}{2}}.
\end{aligned}$$

Hence,

$$|z(t,t')| \leqq C''(1+|x|)(t'-t)^{\frac{1}{2}}$$

and $z(t,t') \to 0$ for $t'-t \to 0$.

Moreover, using the generalized Itô formula we obtain

$$\begin{aligned}
&\mathbb{M}\{f(x+\alpha(t,x,h))\}-f(x)\\
&\quad=\mathbb{M}\Big\{\int_t^{t+h}\big[(\nabla f(x+\alpha(t,x,s-t))\,|\,a(s,x))\\
&\qquad+\tfrac{1}{2}\operatorname{sp}\{\nabla^2 f(x+\alpha(t,x,s-t))\,b(s,x)\,b^*(s,x)\}\\
&\qquad+\int\{f(x+\alpha(t,x,s-t)+c(s,x,u))-f(x+a(t,x,s-t))\\
&\qquad-(\nabla f(x+\alpha(t,x,s-t))\,|\,c(s,x,u))\}\,\Pi(du)\big]\,ds\Big\}.
\end{aligned}$$

Using the boundedness and continuity of $\nabla f(x)$, $\nabla^2 f(x)$ in x, of $a(s,x)$, $b(s,x)$ and $c(s,x,u)$ as functions of s and the convergence of $\alpha(t,x,h)$ to zero in probability for $h\to 0$, it is easy to see that

$$\begin{aligned}
&\lim_{\substack{t'-t\to 0\\ t\leqq s\leqq t'}}\frac{1}{t'-t}\big[\mathbb{M}\{f(x+\alpha(t,x,t'))\}-f(x)\big]\\
&\quad=(\nabla f(x)\,|\,a(s,x))+\tfrac{1}{2}\operatorname{sp}[\nabla^2 f(x)\,b(s,x)\,b^*(s,x)]\\
&\qquad+\int[f(x+c(s,x,u))-f(x)-(\nabla f\,|\,c(s,x,u))]\,\Pi(du).
\end{aligned} \tag{5}$$

Finally, since

$$\begin{aligned}
&\lim_{\substack{t'-t\to 0\\ t<s<t'}}\frac{1}{t'-t}\{\mathbb{M}f(\xi_{xt}(t'))-f(x)\}\\
&\quad=\lim_{\substack{t'-t\to 0\\ t<s<t'}}\frac{1}{t'-t}\{\mathbb{M}f(x+a(t,x,t'-t))-f(x)\}+\lim_{\substack{t'-t\to 0\\ t<s<t'}}z(t,t'),
\end{aligned}$$

we have from (4) and (5) the following

Theorem 2. *If $f(x)$ is a twice continuously differentiable function with bounded partial derivatives of first and second orders and $\xi_{x,t}(s)$ is a solution of* (1) § 7 *satisfying* (2), (3) *and* (4) § 7, *then $f(x)\in D(A)$ and*

$$\begin{aligned}
(A_s f)(x)&=(\nabla f(x)|a(s,x))+\tfrac{1}{2}\operatorname{sp}[\nabla^2 f(x)\,b(s,x)\,b^*(s,x)]\\
&\quad+\int[f(x+c(s,x,u))-f(x)-(\nabla f(x)|c(s,x,u))]\,\Pi(du).
\end{aligned} \tag{6}$$

If the stochastic equation is of continuous type $(c(t,x,u)\equiv 0)$, then the boundedness condition on the first and second partials of $f(x)$ in Theorem 2 can be replaced by the requirement that $f(x)$ and its first and second partials satisfy an inequality of the type

$$|f(x)| \leqq C(1+|x|^m), \qquad m>0. \tag{7}$$

We first remark that the existence of arbitrarily high-order moments of the solution of a stochastic differential equation implies the existence of $\mathbb{M} f(\xi_{tx}(s))$, $\mathbb{M} \nabla f(\xi_{tx}(s))$ and $\mathbb{M} \nabla^2 f(\xi_{tx}(s))$. From (19) § 6 we have

$$\begin{aligned}\mathbb{M}|\eta(t)|^4 = & \int_0^t \mathbb{M}\{4|\eta(s)|^2 (\eta(s)|a(s,\xi(s))-a(s,x) \\ & +4|[b^*(s,\xi(s))-b^*(s,x)]\,\eta|^2 \\ & +4|\eta(s)|^2 \operatorname{sp}[b(t,\xi(s))-b(t,x)][b^*(t,\xi(s))-b^*(t,x)]\}\,ds \\ & \leqq C\int_0^t \mathbb{M}\{|\eta(s)|^3\,|\xi(s)-x|+|\eta(s)|^2\,|\xi(s)-x|^2\}\,ds.\end{aligned}$$

Applying Hölder's inequality we get

$$\begin{aligned}\mathbb{M}|\eta(t)|^4 \leqq & C\left(\int_0^t \mathbb{M}|\eta(s)|^4\,ds\right)^{\frac{3}{4}}\left(\int_0^t \mathbb{M}|\xi(s)-x|^4\,ds\right)^{\frac{1}{4}} \\ & + C\left(\int_0^t \mathbb{M}|\eta(s)|^4\,ds \int_0^t \mathbb{M}|\xi(s)-x|^4\,ds\right)^{\frac{1}{2}}.\end{aligned}$$

Set

$$z(t)=\int_0^t \mathbb{M}|\eta(s)|^4\,ds,$$

since $z(t)$ is bounded $(0 \leqq t \leqq T)$ there exists a constant C_1 such that $z^{\frac{3}{4}}(t) \leqq C_1\, z^{\frac{1}{2}}(t)$. Using (8) § 7 we can write the previous inequality as

$$z'(t) \leqq C'\, z^{\frac{1}{2}}\, t^{\frac{3}{4}},$$

where C' is some constant (depending on the initial value x). Hence, $z(t)=o(t^{\frac{7}{2}})$ and

$$\mathbb{M}|\eta(t)|^4 = o(t^{\frac{11}{4}}) = o(t^2). \tag{8}$$

This result proves that by considering the variable $z(t,t')$ we can get

$$\begin{aligned}& \frac{1}{t'-t}\,|\mathbb{M}\{(\nabla^2 f\,|\eta(t')|\,\eta(t'))/\mathfrak{F}_t\}| \\ & \quad \leqq \frac{1}{t'-t}\,\mathbb{M}\{|\nabla^2 f|^2/\mathfrak{F}_t\}^{\frac{1}{2}}\,\mathbb{M}\{|\eta(t')|^4/\mathfrak{F}_t\}^{\frac{1}{2}} \to 0.\end{aligned}$$

Moreover, the assumptions made imply the existence of a constant C such that $|\nabla f(x+y)-\nabla f(x)| \leqq C(|y|+|y|^m)$ (C depends, in general, on x).

Hence

$$\begin{aligned}&\mathbb{M}\{|(\nabla f(x+\alpha(t,x,t'-t))-\nabla f(x)\,|\eta(t')|/\mathfrak{F}_t\}\\&\leqq C\big[\mathbb{M}\{|\alpha(t,x,t'-t)|^2+|\alpha(t,x,t'-t)|^{2m}/\mathfrak{F}_t\}\\&\quad\times\mathbb{M}\{|\eta(t')|^2/\mathfrak{F}_t\}\big]^{\frac{1}{2}}=o(t'-t),\end{aligned}$$

so that the relation $z(t,t')\to 0$ for $t'-t\to 0$ also holds in the case under consideration. Its now easy to see that (5) likewise remains valid $(c(t,x,u)\equiv 0)$. We have thus obtained

Theorem 3. *If $f(x)$ is a twice continuously differentiable and along with its partial derivatives of first and second orders satisfies* (7) *for some $m>0$, and $\xi(t)$ satisfies*

$$d\xi=a(t,\xi)\,dt+b(t,\xi)\,dw(t),$$

whose coefficients fulfill (2), (3) *and* (4) § 7, *then $f(x)\in D(A_s)$ and*

$$(A_s f)(x)=\sum_{j=1}^{n} a_j(s,x)\frac{\partial f(x)}{\partial x_j}+\frac{1}{2}\sum_{j,k=1}^{n}\sigma_{jk}(s,x)\frac{\partial^2 f(x)}{\partial x_j\,\partial x_n}, \tag{9}$$

$$\sigma_{jk}(s,x)=\sum_{r=1}^{n} b_{jr}(s,x)\,b_{kr}(s,x). \tag{10}$$

Remark 3. If $f(t,x)$ along with its first and second order partials w.r.t. $x_1,\dots,x_n$ are continuous in (t,x) $(0\leqq t\leqq T,\ x\in X)$ and satisfy the assumptions of Theorem 2 or Theorem 3 with constant independent of t, then

$$\lim_{\substack{t'-t\to 0\\ t\leqq s\leqq t'}}\frac{\mathbb{M}f(t',\xi_{tx}(t))-f(t',x)}{t'-t}=A_s f(s,x).$$

We now proceed to the derivation of Kolmogorov's backward equation for Markov processes generated by stochastic equations. We will show that

$$u(t,x)=\mathbb{M}f(\xi_{xt}(s)) \tag{11}$$

satisfies for $t\in(0,s)$, $x\in X$ a certain integro-differential equation which, when $c(t,x,u)\equiv 0$, assumes the form of an n-dimensional second order partial differential equation of parabolic type.

Lemma 2. *If the hypotheses of Theorem 2 or Theorem 3 are fulfilled, then $u(t,x)$ possesses first-order partial derivatives w.r.t. $x_1,\dots,x_n$ which are continuous in (t,x) and*

$$\frac{\partial u(t,x)}{\partial x_j}=\mathbb{M}\left(\nabla f(\xi_{xt}(s))\,\middle|\,\frac{\partial}{\partial x_j}\xi_{xt}(s)\right). \tag{12}$$

We have, in fact

$$\begin{aligned}\frac{u(t,x+\Delta x)-u(t,x)}{\Delta x_j} &- \mathbb{M}\left(\nabla f(\xi_{xt}(s))\,\middle|\,\frac{\partial}{\partial x_j}\xi_{xt}(s)\right)\\ &= \mathbb{M}\left\{\left(\nabla f(\xi_{xt}(s)+\theta\,\Delta\xi)-\nabla f(\xi_{xt}(s))\,\middle|\,\frac{\Delta\xi}{\Delta x_j}\right)\right.\\ &\quad\left.+\left(\nabla f(\xi_{xt}(s))\,\middle|\,\frac{\Delta\xi}{\Delta x_j}-\frac{\partial}{\partial x_j}\xi_{xt}(s)\right)\right\},\end{aligned}$$

where Δx denotes a vector whose j-th component equals Δx_j with remaining components zero, and θ is a number lying in $(0, 1)$.

When $\nabla f(x)$ and $\nabla^2 f(x)$ are bounded, the right side of the preceding equation tends to zero when $\Delta x_j \to 0$ since it does not exceed

$$\begin{aligned}&\left\{\mathbb{M}\,|\nabla f(\xi_{xt}(s)+\theta\,\Delta\xi)-\nabla f(\xi_{xt}(s))|^2\,\mathbb{M}\left|\frac{\Delta\xi_j}{\Delta x_j}\right|^2\right\}^{\frac{1}{2}}\\ &\quad+\left\{\mathbb{M}\,|\nabla f(\xi_{xt}(s))|^2\,\mathbb{M}\left|\frac{\Delta\xi}{\Delta x_j}-\frac{\partial}{\partial x_j}\xi_{xt}(s)\right|^2\right\}^{\frac{1}{2}},\end{aligned}$$

which tends to zero when $\Delta x_j \to 0$. Under the assumptions of Theorem 3 the same conclusion follows from the existence of the arbitrary order moments of $\xi_{xt}(s)$, from the fact that $\mathbb{M}\,|\Delta\xi|^m \to 0$ for arbitrary $m>0$ and the fact that $\nabla f(x)$ and $\nabla^2 f(x)$ grow no faster at infinity than some power of $|x|$. Making use of the continuity in mean square of the variable $\xi_{x,t}(s)$ w.r.t. (x, t) it is easy to show that the derivatives (12) are continuous in (x, t).

Remark 4. Under the conditions of Theorem 2 the partial derivatives of $u(t, x)$ are bounded w.r.t. x and under those of Theorem 3, $u(t, x)$ grows no faster than some power of $|x|$ when $|x| \to \infty$.

Lemma 3. *If the hypotheses of Theorem 2 hold and* (24) *and* (25) §8 *are fulfilled, then* $u(t,x)$ *possesses second order partials w.r.t.* $x_1, \dots, x_n$ *and*

$$\begin{aligned}\frac{\partial^2 u}{\partial x_k\,\partial x_j} &= \mathbb{M}\left(\nabla^2 f(\xi_{xt}(s))\,\middle|\,\frac{\partial}{\partial x_k}\xi_{xt}(s)\,\middle|\,\frac{\partial}{\partial x_j}\xi_{xt}(s)\right)\\ &\quad+\mathbb{M}\left(\nabla f(\xi_{xt}(s))\,\middle|\,\frac{\partial^2}{\partial x_k\,\partial x_j}\xi_{xt}(s)\right).\end{aligned}\tag{13}$$

This result also holds if the conditions of Theorem 3 and Remark 3 §8 obtain.

Proof. We have

$$\frac{1}{\Delta x_k}\left(\frac{\partial u(t,x+\Delta x)}{\partial x_j}-\frac{\partial u(t,x)}{\partial x_j}\right)$$
$$-\mathbb{M}\left(\nabla^2 f(\xi_{xt}(s))\left|\frac{\partial}{\partial x_k}\xi_{xt}(s)\right|\frac{\partial}{\partial x_j}\xi_{xt}(s)\right)$$
$$-\mathbb{M}\left(\nabla f(\xi_{xt}(s))\left|\frac{\partial^2}{\partial x_k\,\partial x_j}\xi_{xt}(s)\right.\right)=z_1+z_2+z_3+z_4,$$

where

$$z_1=\mathbb{M}\left(\nabla f(\xi_{xt}(s))\left|\frac{1}{\Delta x_k}\left[\frac{\partial}{\partial x_j}\xi_{x+\Delta x,t}(s)-\frac{\partial}{\partial x_j}\xi_{xt}(s)\right]-\frac{\partial^2}{\partial x_k\,\partial x_j}\xi_{xt}(s)\right.\right),$$

$$z_2=\mathbb{M}\left(\nabla^2 f(\tilde{\xi})-\nabla^2 f(\xi_{xt}(s))\left|\frac{\Delta\xi}{\Delta x_k}\right|\frac{\partial}{\partial x_j}\xi_{x+\Delta x,t}(s)\right),$$

$$z_3=\mathbb{M}\left(\nabla^2 f(\xi_{xt}(s))\left|\frac{\Delta\xi}{\Delta x_k}-\frac{\partial}{\partial x_k}\xi_{xt}(s)\right|\frac{\partial}{\partial x_j}\xi_{x+\Delta x,t}(s)\right),$$

and

$$z_4=\mathbb{M}\left(\nabla^2 f(\xi_{xt}(s))\left|\frac{\partial}{\partial x_k}\xi_{xt}(s)\right|\frac{\partial}{\partial x_j}\xi_{x+\Delta x,t}(s)-\frac{\partial}{\partial x_j}\xi_{xt}(s)\right).$$

Here $\tilde{\xi}$ is some point in the segment connecting $\xi_{x,t}(s)$ and $\xi_{x+\Delta x,t}$ and Δx is a vector whose only non-zero component is the k-th, which equals Δx_k. Let the first group of assumptions of the lemma hold. The variable $z_1\to 0$ for $\Delta x_k\to 0$ because $\nabla f(x)$ is bounded and the $\frac{\partial^2}{\partial x_k\,\partial x_j}\xi_{x,t}(s)$ exist in the mean-square convergence sense. For z_2 we have the estimate

$$|z_2|\leq\frac{1}{\sqrt{2}}\left\{\mathbb{M}|\nabla^2 f(\tilde{\xi})-\nabla^2 f(\xi_{xt}(s))|^2\right.$$
$$\left.\times\left[\mathbb{M}\left|\frac{\Delta\xi}{\Delta x_k}\right|^4+\mathbb{M}\left|\frac{\partial}{\partial x_j}\xi_{x+\Delta x,t}(s)\right|^4\right]\right\}^{\frac{1}{2}}.$$

The expression in square brackets is uniformly bounded as was proved at the end of the last section and the first factor tends to zero because the expression immediately following the expectation sign is bounded and tends to zero in probability. Hence, $z_2\to 0$. It is also easy to establish that $z_3\to 0$ and $z_4\to 0$. This proves the first part of the lemma. The second part is proved analogously. Instead of the boundedness of ∇f and $\nabla^2 f$, one uses the boundedness of the moments of arbitrary order of $\xi_{x,t}(s)$.

Remark 5. Under the hypotheses of Lemmas 2 and 3 the partial derivatives

$$\frac{\partial u}{\partial x_j}(t,x), \quad \frac{\partial^2 u}{\partial x_k \partial x_j}, \quad 0 \leqq t \leqq T, \; x \in X,$$

are continuous in (t, x).

This claim is proved with the help of arguments similar to those used in the proof of Lemma 3, formulas (12) and (13) and the mean-square continuity w.r.t. (t, x) of the derivatives

$$\frac{\partial}{\partial x_j}\xi_{xt}(s) \quad \text{and} \quad \frac{\partial}{\partial x_k \partial x_j}\xi_{xt}(s).$$

Assume the conditions of Lemma 3 hold and $t' < t'' < s$, s fixed. We have

$$\begin{aligned} u(t',x) &= \mathbb{M} f(\xi_{xt}(s)) = \mathbb{M} f(\xi_{\xi_{xt'}(t''),t''}(s)) \\ &= \mathbb{M}\{\mathbb{M}[f(\xi_{\xi_{xt'}(t''),t''}(s))/\mathfrak{F}_{t''}]\} = \mathbb{M}\, u(t'', \xi_{xt'}(t'')). \end{aligned}$$

Consequently,

$$\frac{u(t',x)-u(t'',x)}{t''-t'} = \frac{\mathbb{M}\, u(t'',\xi_{xt'}(t''))-u(t'',x)}{t''-t'}.$$

From Remarks 4 and 5 it follows that depending on which group of assumptions in Lemma 3 hold we can use either Theorem 2 or 3. We hence obtain

Theorem 4. *Let* (2) *satisfy the conditions of the existence and uniqueness theorem and*

$$|\nabla^2 a(t,x)|^2 + |\nabla^2 b(t,x)|^2 + \int |\nabla^2 c(t,x,u)|^2 \Pi(du) \leqq C, \tag{14}$$

$$\int (|c(t,x,u)|^k + |\nabla c(t,x,u)|^k)\, \Pi(du) \leqq C, \quad k=3,4, \tag{15}$$

and assume that $f(x)$ *is twice continuously differentiable, and that its first and second order partials are bounded. Then*

$$u(t,x) = \mathbb{M} f(\xi_{xt}(s)), \quad t<s,$$

is twice continuously differentiable in x, *differentiable in* t *and satisfies*

$$\frac{\partial u}{\partial t} + A_t u = 0, \tag{16}$$

where

$$\begin{aligned} A_t u = (u(t,x)|\nabla u) + \tfrac{1}{2}\operatorname{sp}[b(t,x)\, b^*(t,x)\, \nabla^2 u] \\ + \int [u(t, x+c(t,x,u)) - u(t,x) - (c(t,x,u)|\nabla u(t,x))]\, \Pi(du), \end{aligned} \tag{17}$$

and the boundary condition is

$$\lim_{t \uparrow s} u(t,x) = f(x). \tag{18}$$

Theorem 5. *Assume*

$$d\xi = a(t, \xi)\, dt + b(t, \xi)\, dw(t)$$

satisfies the existence and uniqueness conditions, that $a(t, x)$ and $b(t, x)$ are twice continuously differentiable in x, that their first order partials in x are bounded, that the second order partials of the coefficients grow no faster than some power of x for $x \to \infty$,

$$|\nabla^2 a(t, x)| + |\nabla^2 b(t, x)| \leqq c(1 + |x|^m), \qquad m > 0,$$

that $f(x)$ is twice continuously differentiable and that its first and second order partials also grow for $x \to \infty$ no faster then some power of x. Then the function $u(t, x)$ is twice continuously differentiable in x, differentiable in t and satisfies the 2nd order parabolic equation

$$\frac{\partial u}{\partial t} + \sum_{j=1}^{n} a_j(t, x) \frac{\partial u}{\partial x_j} + \frac{1}{2} \sum_{j,k=1}^{n} \sigma_{jk}(t, x) \frac{\partial^2 u}{\partial x_j \partial x_k} = 0, \qquad t < s, \tag{19}$$

with boundary condition (18).

Theorem 6. *If $a(t, x)$, $b(t, x)$ and $c(t, x, u)$ $(a(t, x)$ and $b(t, x))$ and $f(x)$ satisfy the conditions of Theorem 4 (Theorem 5), then the integro-differential equation* (16) *(partial differential equation* (19)*) has a solution for $x \in X$, $t \in [0, s]$ satisfying the boundary condition* (18).

Up to this point we have considered equations characterizing the dependence of $\mathbb{M} f(\xi_{x,t}(s))$ on the variables (t, x). What can be said about the behavior of $\mathbb{M} f(\xi_{x,t}(s))$ considered as a function of s?

In order to investigate this, we use the equality

$$\mathbb{M} f(\xi_{xt}(s'')) - \mathbb{M} f(\xi_{xt}(s')) = \mathbb{M}\{\mathbb{M}[f(\xi_{ys'}(s'')) - f(y)]_{y = \xi_{xt}(s')}\},$$

where $s'' > s' > t$. Dividing this equation by $s'' - s'$ and letting s'' and s' tend to s, we obtain as in the proof of Theorem 2

$$\frac{\partial}{\partial s} \mathbb{M} f(\xi_{xt}(s)) = \mathbb{M}(A_s f)(\xi_{xt}(s)). \tag{20}$$

Lemma 4. *If the hypotheses of Theorem 4 or Theorem 5 hold, then* (20) *is valid.*

Assume the transition probability generated by a stochastic equation possesses the density $p(t, x, s, y)$,

$$P(t, x, s, A) = \mathbb{P}(\xi_{xt}(s) \in A) = \int_A p(t, x, s, y)\, dy,$$

where A is any Borel set in X and the integral is taken w.r.t. Lebesgue measure in X.

Corollary 1. *If the transition probability has a density which is continuously differentiable in t and twice continuously differentiable in x and the stochastic equation satisfies the conditions of Theorem 4 or Theorem 5, then the density satisfies*

$$\frac{\partial p(t,x,s,y)}{\partial t}+(A_t\,p(t,\cdot,s,y))(x)=0.$$

Writing $A_t\,p(t,\cdot,s,y)$ will denote that the operator A_t acts on the function $p(t,x,s,y)$ in which (t,s,y) are considered as parameters. The proof of the stated claim follows from Theorems 4 and 5, the representation

$$u(t,x)=\int f(y)\,p(t,x,s,y)\,dy$$

and the possibility of differentiating this expression once w.r.t. t and twice w.r.t. x under the integral sign provided that $f(y)\equiv 0$ outside some compact.

We now consider how the equations are modified in a special case (which is still sufficiently broad and of practical importance). Assume that

$$\int \Pi(du)<\infty. \tag{21}$$

This assumption means that we will consider only stochastic equations in which the jump intensity during an infinitesimal period of time is finite. Assumption (21) seems to be applicable in all practically important cases. The expression

$$\int\left[f(x+c(x,u))-f(x)-(\nabla f(x)\,|\,c(t,x,u))\right]\Pi(du)$$

in the case under consideration can be written as

$$-\pi f(x)-(\nabla f(x)\,|\,z(t,x))+\int f(x+c(t,x,u))\,\Pi(du),$$

where

$$\pi=\int\Pi(du),\qquad z(t,x)=\int c(t,x,u)\,\Pi(du).$$

Consequently, the backward Kolmogorov equation for

$$v(t,x)=\mathbf{M}\,f(\xi_{xt}(T))$$

assumes the form

$$\begin{aligned}\frac{\partial v(t,x)}{\partial t}&-\pi v(t,x)+(a(t,x)-z(t,x)\,|\,\nabla v(t,x))\\&+\tfrac{1}{2}\operatorname{sp}(\nabla^2 v(t,x)\,b(t,x)\,b^*(t,x))+\int v(t,x+c(t,x,u))\,\Pi(du)=0.\end{aligned} \tag{22}$$

If the transition probability density exists satisfying the conditions of Corollary 1, then it is a solution of

$$\frac{\partial p}{\partial t} - \pi p + \sum_{k=1}^{n} \tilde{a}_k(t,x) \frac{\partial p}{\partial x_k} + \frac{1}{2} \sum_{k,j=1}^{n} \sigma_{kj}(t,x) \frac{\partial^2 p}{\partial x_k \partial x_j} + \int p(t, x + c(t,x,u), s, y) \Pi(du) = 0, \tag{23}$$

where

$$p = p(t,x,s,y), \qquad \tilde{a}_k(t,x) = a_k(t,x) - z_k(t,x).$$

Corollary 2. *Assume $\Pi(X) < \infty$, that the transition probability possesses a density and that the derivatives*

$$\frac{\partial p}{\partial s}, \quad \frac{\partial(\tilde{a}_k p)}{\partial y_k}, \quad \frac{\partial^2(\sigma_{kj} p)}{\partial y_k \partial y_j}, \qquad k, j = 1, 2, \ldots, n,$$

exist and are continuous, where $\tilde{a}_k = \tilde{a}_k(s,y)$, $\sigma_{kj} = \sigma_{kj}(s,y)$ and $p = p(t,x,s,y)$. We assume in addition that for any s and u the mapping

$$z = y + c(s,y,u) \tag{23}$$

maps X one-to-one onto itself and that the inverse mapping

$$y = z - c^*(s,z,u) \tag{24}$$

is differentiable. Then, $p(t,x,s,y)$, considered as a function of s and y, satisfies

$$\frac{\partial p}{\partial s} + (A_s^* p)(y) = \frac{\partial p}{\partial s} + \sum_{k=1}^{n} \frac{\partial(\tilde{a}_k p)}{\partial y_k} - \frac{1}{2} \sum_{k,j=1}^{n} \frac{\partial^2(\sigma_{kj} p)}{\partial y_k \partial y_j} - \int p(t,x,s,y - c^*(t,y,u)) D(y) \Pi(du) + \pi p = 0, \tag{25}$$

where $D(y)$ is the Jacobian of the transformation (24).

To prove this, it is necessary to use formula (20) which can be written as

$$\frac{\partial}{\partial s} \int p(t,x,s,y) f(y)\, dy = \int p(t,x,s,y) \Big[-\pi f(y) + \sum_{k=1}^{n} \tilde{a}_k(s,y) \frac{\partial f}{\partial y_k} + \frac{1}{2} \sum_{k,j=1}^{n} \sigma_{kj}(s,y) \frac{\partial^2 f(y)}{\partial y_k \partial y_j} + \int f(y + c(t,y,u)) \Pi(du) \Big] dy.$$

Integrating by parts on the right side of this formula and performing the change of variables (23) we arrive at

$$\int f(y) \left(\frac{\partial p}{\partial s} + (A_s^* p)(y) \right) dy = 0,$$

valid for an arbitrary twice continuously differentiable function $f(y)$ differing from zero in some bounded region. From this follows (25).

Analogous formulas can be obtained without the restriction

$$\Pi(X) < \infty.$$

§ 10. The Distribution of Functionals of the Solutions of Stochastic Differential Equations

Let $\xi_{x,t}(s)$ be the solution of a stochastic differential equation without after-effect satisfying the hypotheses of the existence and uniqueness theorem. Set

$$v_\lambda(t, x) = \mathbb{M} f(\xi_{xt}(T)) \exp\left\{i\left(\int_t^T g(s, \xi_{xt}(s))\, ds \,|\, \lambda\right)\right\}, \tag{1}$$

where $f(x)$ is a continuous scalar and $g(t, x)$ a continuous vector function, whereby $g(t, x)$ takes values in X and λ is a vector with values in X. If we set $f(x) = 1$, $t = 0$, then $v_\lambda(0, x)$ is the characteristic function of the distribution of the vector random variable

$$\int_0^T g(s, \xi_{x0}(s))\, ds.$$

If, however, $f(x) = \exp\{i(x \,|\, \mu)\}$, then $v_\lambda(0, x) = v_{\lambda,\mu}(0, x)$ coincides with the characteristic function of the joint distribution of the random vectors

$$\xi_{x0}(T), \qquad \int_0^T g(s, \xi_{x0}(s))\, ds.$$

Moreover, if $f(x)$ is such that for some $p > 1$

$$M|f(\xi_{xt}(T))|^p < C_x \qquad (0 \leqq t \leqq T),$$

where C_x is a constant depending only on x, then $v_\lambda(t, x)$ satisfies

$$\lim_{t \uparrow T} v_\lambda(t, x) = f(x), \tag{2}$$

which will be referred to in the sequel as the boundary condition. The validity of (2) follows from the fact $\xi_{x,t}(T)$ tends for $t \to T$ to x in probability and that we can go to the limit on the right side of (1) under the expectation sign.

Assume that the conditions guaranteeing the existence of 4-th order moments of the solution of a stochastic equation obtain (§ 7, Theorem 1) and that the functions $f(x)$ and $g(x)$ are twice continuously differentiable with bounded derivatives of first and second orders. It is then easy to see that $v_\lambda(t, x)$ is twice continuously differentiable in x, differentiable in t and

$$\frac{\partial v_\lambda(t, x)}{\partial t}+L v_\lambda(t, x)+i(\lambda \mid g(t, x))\, v_\lambda(t, x)=0. \tag{3}$$

If the stochastic equation is of continuous type $(c(t, x, u) \equiv 0)$, then the assumptions on the remaining coefficients of the equation, as also those on the functions $f(x)$ and $g(t, x)$ can be weakened. With respect to the latter, it is sufficient to require that they and their derivatives of first and second orders grow for $|x| \rightarrow \infty$ no faster than some power of $|x|$ (uniformly in t). Of course, other formulations of analogous theorems are possible, in which it is not assumed that the discontinuous part of the equation is equal to zero but rather that existence conditions for sufficiently high-order moments of the equation's solution hold.

As in §9, we remark here that these results also represent existence theorems for the boundary-value problem (2) for the integro-differential equation (3) (this is an equation of parabolic type if the discontinuous part is zero).

We introduce the generalized solution of (3) for stochastic equations of continuous type $(c(t, x, u) \equiv 0)$. Let $C_{\mathrm{fin}}^{(r)}$ be the class of twice continuously differentiable functions vanishing outside some compact in X.

If $\psi(x) \in C_{\mathrm{fin}}^2$, then with the help of simple transformations it is easy to verify that

$$\int_X \psi(x)\, L[v]\, d x=\int_X v(t, x)\, L^*[\psi]\, d x, \tag{4}$$

where

$$L^*[\psi]=-\sum_{k=1}^n \frac{\partial}{\partial x_k}\left(a_k(t, x)\, \psi(x)\right)+\frac{1}{2} \sum_{k, r=1}^n \frac{\partial^2\left(\sigma_{k r}(t, x)\, \psi(x)\right)}{\partial x_k\, \partial x_r}. \tag{5}$$

For this formula to hold it is sufficient that $a_k(t, x)$ be continuously differentiable and that $b_{kj}(t, x)$ be twice continuously differentiable.

Hence, if $v(t, x)$ is twice continuously differentiable in x and $\psi(x) \in C_{\mathrm{fin}}^{(2)}$, we have

$$\frac{\partial}{\partial t} \int_X v(t, x)\, \psi(x)\, d x+\int_X v(t, x)\left[L^*[\psi]+i(\lambda \mid g(t, x))\, \psi\right] d x=0. \tag{6}$$

A function $v(t, x)$ satisfying (6) for arbitrary $\psi(x) \in C_{\mathrm{fin}}^{(2)}$ will be called a *generalized solution* of (3).

As in the proof of the claims in Remark 4 § 11, Part I we obtain

Theorem 1. *If a stochastic equation satisfies the conditions of the existence and uniqueness theorem, $\nabla a(t, x)$ and $\nabla b(t, x)$ are bounded,*

$b(t,x)$, $f(x)$ and $g(t,x)$ are twice continuously differentiable and for some $m>0$

$$|f(x)|+|\nabla f|+|\nabla^2 f|+|g|+|\nabla g|+|\nabla^2 g|\leqq C(1+|x|^m),$$

then $v(t,x)$, defined at (1) is a generalized solution of (3).

We turn now to the more general problem of determining the distributions of stochastic integrals from the solutions of stochastic differential equations. Let

$$\begin{aligned}\eta_{yt}(s)=y&+\int_t^s g(\tau,\xi_{xt}(\tau))\,d\tau+\int_t^s h(\tau,\xi_{xt}(\tau))\,dw(\tau)\\&+\int_t^s\int q(\tau,\xi_{xt}(\tau),u)\,\tilde{\nu}(d\tau,du),\end{aligned}\tag{7}$$

where $g(t,x)$, $h(t,x)$ and $q(t,x,u)$ satisfy the same conditions as $a(t,x)$, $b(t,x)$ and $c(t,x,u)$, resp., in the existence and uniqueness theorem for stochastic equations.

Joining the equation for $\xi_{x,t}(s)$ to (7), we obtain a system of stochastic differential equations for the vector $(\xi_{x,t}(s),\eta_{y,t}(s))$ satisfying the assumptions of the existence and uniqueness theorem. We assume that $c(t,x,u)$ and $q(t,x,u)$ fulfill (6) §7 for $m=2$ so that the solution of the equation under consideration possesses finite 4th-order moments. Take $F(x,y)$, $x\in X$, $y\in X$ as a twice continuously differentiable function with bounded first and second order partials (or satisfying the more general assumptions under which the Kolmogorov equation used below is valid – see §9). Set

$$v(t,x,y)=\mathbb{M}F(\xi_{xt}(T),\eta_{yt}(T)),\qquad t\leqq T.$$

Then $v(t,x,y)$ satisfies a backward Kolmogorov equation which can written (in this case) as

$$\begin{aligned}0=\frac{\partial v}{\partial t}&+(\nabla_x v\,|\,a(t,x))+(\nabla_y v\,|\,g(t,x))+\tfrac{1}{2}\operatorname{sp}(\nabla_x^2 v\,b(t,x)\,b^*(t,x))\\&+\tfrac{1}{2}\operatorname{sp}(\nabla_y^2 v\,h(t,x)\,h^*(t,x))+\tfrac{1}{2}\operatorname{sp}(\nabla_y\nabla_x v\,[b h^*+h b^*])\\&+\int\big[v(t,x+c(t,x,u),y+q(t,x,u))-v(t,x,y)-(\nabla_x v\,|\,c(t,x,u))\\&\qquad-(\nabla_y v\,|\,q(t,x,u))\big]\,\Pi(du),\end{aligned}\tag{8}$$

where ∇_x is the gradient vector w.r.t. x, and ∇_y that w.r.t. y.

Let

$$F(x,y)=f(x)\,e^{i(\lambda|y)}.$$

Then, obviously,

$$v(t,x,y)=e^{i(\lambda|y)}\,v(t,x),$$

where

$$v(t,x)=v(t,x,0)=\mathbb{M}f(\xi_{xt}(T))\,e^{i(\lambda|\eta_{0t}(T))}.$$

Consequently,

$$\nabla_y v(t,x,y)=i\lambda e^{i(\lambda|y)}v(t,x),$$

$$\nabla_y^2 v(t,x,y)=-\lambda\lambda^* e^{i(\lambda|y)}v(t,x),$$

$$\begin{aligned}v(t,x+c,y+q)&-v(t,x,y)-(\nabla_x v(t,x,y)|c)-(\nabla_y v(t,x,y)|q)\\&=e^{+i(\lambda|y)}\{[v(t,x+c)-v(t,x)-(\nabla_x v(t,x)|c)]\\&\quad+[e^{i(\lambda|q)}-1-i(\lambda|q)]\,v(t,x)+(e^{i(\lambda|q)}-1)(v(t,x+c)-v(t,x))\}.\end{aligned}$$

Eq. (8) thus takes the form

$$\frac{\partial v(t,x)}{\partial t}+(L+L_1)\,v(t,x)=0, \tag{9}$$

where

$$\begin{aligned}L_1 v(t,x)=\big[&i(\lambda|g(t,x))-\tfrac12(\lambda h(t,x)|\lambda h(t,x))\\&+\textstyle\int[e^{i(\lambda|q(t,x,u))}-1-i(\lambda|q(t,x,u))]\,\Pi(du)\big]\,v(t,x)\\&+i(\lambda h(t,x)|\{\nabla v(t,x)\}\,b(t,x))\\&+\textstyle\int(e^{i(\lambda|q)}-1)(v(t,x+c)-v(t,x))\,\Pi(du),\end{aligned}\tag{10}$$

or in coordinate form,

$$\begin{aligned}L_1 v(t,x)=\Bigg[&i\sum_{k=1}^n\lambda_k g_k(t,x)-\tfrac12\sum_{j=1}^n\Big|\sum_{k=1}^n\lambda_k h_{kj}(t,x)\Big|^2\\&+\int\Big[e^{i\sum_{k=1}^n\lambda_k q_k(t,x,u)}-1-i\sum_{k=1}^n\lambda_k q_k(t,x,u)\Big]\,\Pi(du)\Bigg]\,v(t,x)\\&+i\sum_{k,j,r}\lambda_k\Big[h_{kj}(t,x)\,b_{rj}(t,x)+b_{kj}(t,x)\,h_{rj}(t,x)\Big]\frac{\partial v}{\partial x_r}\\&+\int(e^{i(\lambda|q(t,x,u))}-1)(v(t,x+c(t,x,u))-v(t,x))\,\Pi(du).\end{aligned}$$

For stochastic equations of continuous type $c(t,x,u)=q(t,x,u)\equiv 0$ and we have

$$\begin{aligned}L_1 v(t,x)=&\Big[i\sum_{k=1}^n\lambda_k g_k(t,x)-\frac12\sum_{k,r=1}^n H_{kr}(t,x)\,\lambda_k\lambda_r\Big]\,v(t,x)\\&+i\sum_{k,r=1}^n\lambda_k H^*_{kr}(t,x)\frac{\partial v}{\partial x_r},\end{aligned}\tag{11}$$

where

$$H_{kr}(t,x)=\sum_{j=1}^n h_{kj}(t,x)\,h_{rj}(t,x);$$

$$H^*_{kr}(t,x)=\sum_{j=1}^n[b_{kj}(t,x)\,h_{rj}(t,x)+h_{kj}(t,x)\,b_{rj}(t,x)].$$

The function $v(t,x)$ satisfies the boundary condition $v(T,x)=f(x)$. In particular, the boundary condition

$$v(T,x)=1 \tag{12}$$

corresponds to the characteristic function of the distribution of the random vector $\eta_{x0}(T)$.

If, on the other hand, we are interested in the joint characteristic function of the distribution of the integrals

$$\int_0^T g_k(s,\xi_{x0}(s))\,ds, \qquad \int_0^T \sum_{r=1}^n h_{kr}(s,\xi_{x0}(s))\,dw_r(s),$$

and

$$\int_0^T \int q(s,\xi_{x0}(s),u)\,\tilde{\nu}(ds,du), \qquad k=1,2,\dots,n,$$

then in (10) we can carry out the substitutions

$$\lambda_k\to 1, \qquad g_k(t,x)\to\lambda_k g_k(t,x), \qquad h_{kr}(t,x)\to\mu_k h_{kr}(t,x),$$
$$q_k(t,x,u)\to\nu_k q_k(t,x,u).$$

The characteristic function then equals $v_{\lambda\mu\nu}(0,x)$, where $v_{\lambda\mu\nu}(t,x)$ satisfies (9) and the boundary condition (12) and the operator L_1 is given by

$$\begin{aligned} L_1 v(t,x) = {} & \left[i(\lambda|g)-\frac{1}{2}\sum_{k,r=1}^n \mu_k\mu_r H_{kr}(t,x)+\int\left(e^{i(\nu|q)}-1-i(\nu|q)\right)\Pi(du)\right] \\ & \times v(t,x)+i\sum_{k,r=1}^n \mu_k H^*_{kr}(t,x)\frac{\partial v}{\partial x_r} \\ & +\int\left(e^{i(\nu|q)}-1\right)\left(v(t,x+c)-v(t,x)\right)\Pi(du). \end{aligned} \tag{13}$$

§ 11. Some Problems Connected with Homogeneous Stochastic Differential Equations

Consider

$$d\xi(t)=a(\xi(t))\,dt+b(\xi(t))\,dw(t)+\int c(\xi(t),u)\,\tilde{\nu}(dt,du), \tag{1}$$
$$\xi(0)=x,$$

whose coefficients $a(x)$, $b(x)$ and $c(x,u)$ are independent of time and satisfy the usual existence and uniqueness conditions. We will assume that the solution of this equation has been chosen in such a way that it has no discontinuities of the second kind and is right continuous. The solution of (1) with the nonrandom initial value x will be denoted by $\xi_x(t)$.

Let $g(x)$, $x \in X$ be bounded and twice continuously differentiable on some open set. We recall the generalized Itô formula

$$g(\xi_x(t)) - g(x) = \int_0^t (Lg)(\xi_x(s))\,ds + \int_0^t (\nabla g(\xi_x(s)) \,|\, b(\xi_x(s))\,dw(s)) \\ + \int_0^t \int [g(\xi_x(s) + c(\xi_x(s), u)) - g(\xi_x(s))]\,\tilde{\nu}(ds, du), \tag{2}$$

where

$$(Lg)(x) = (\nabla g(x) \,|\, a(x)) + \tfrac{1}{2}\,\mathrm{sp}\,[\nabla^2 g(x)\, b(x)\, b^*(x)] \\ + \int [g(x + c(x, u)) - g(x) - (\nabla g(x) \,|\, c(x, u))]\,\Pi(du),$$

valid for arbitrary $g(x) \in D_L$ and for all $t < \tau$ if $\xi_x(t) \in G$ for $0 \leqq t \leqq \tau$. D_L will be understood to be the class of functions $g(x)$ defined and continuous in X, twice differentiable in G and for which the function $g(x + c(x, u)) - g(x) - (\nabla g(x) \,|\, c(x, u))$ is Π-integrable for any x.

Let $\tau_x(F) = \inf\{t;\ \xi_x(t) \in F\}$, F a closed set. If $\xi_x(t)$ never hits F, then we will take $\tau_x(F) = \infty$.

Let

$$\tau_T = \min[T, \tau_x(F)].$$

Replacing t by τ_T in (2) we get

$$-\int_0^{\tau_T} L(g(\xi(\tau)))\,d\tau = g(x) - g(\xi_x(\tau_T)) + \alpha(\tau_T),$$

where

$$\alpha(t) = \alpha_g(t) = \int_0^t (\nabla g(\xi_x(s)) \,|\, b(\xi_x(s))\,dw(s)) \\ + \int_0^t \int \{g[\xi_x(s) + c(\xi_x(s), u)] - g(\xi_x(s))\}\,\tilde{\nu}(ds, du). \tag{3}$$

Let

$$Lg(x) \leqq -1, \qquad x \notin F, \tag{4}$$

then

$$\tau_T \leqq g(x) - g(\xi_x(\tau_T)) + \alpha(\tau_T).$$

Lemma 1. *If there exists a function $g(x)$ which is in D_L on $X \setminus F$, bounded on X and satisfies* (4), *then $\mathbf{M}\,\tau_x(F) < \infty$ and, consequently, $\tau_x(F)$ is finite w.p.1.*

Indeed,

$$\mathbf{M}\,\tau_T \leqq g(x) - \mathbf{M}\,g(\xi_x(\tau_T)) \tag{5}$$

and τ_T is a nondecreasing function of the parameter T with $\lim_{T \to \infty} \tau_T = \tau_x(F)$. From Lebesgue's theorem

$$\mathbf{M}\,\tau_x(F) = \lim_{T \to \infty} \mathbf{M}\,\tau_T < \infty.$$

Lemma 2. *If $g(x)$ satisfies the conditions of Lemma 1 and $g(x) \geqq 0$ in F, then*

$$\mathbb{M}\, \tau_x(F) \leqq g(x).$$

Actually, since τ_T becomes equal w.p. 1 to $\tau_x(F)$ for some T, we have $\xi_x(\tau_T) \to \xi_x(\tau_x(F))$ for $T \to \infty$. Since $\xi_x(t)$ is right continuous, we have $\xi_x(\tau_x(F)) \in F$ and $g(\xi_x(\tau_T)) \to g(\xi_x(\tau_x(F))) \geqq 0$. The proof now follows from (5).

Lemma 3. *If there exists a bounded function $v(x) \in D_L$ for which $Lv \geqq -1$ when $x \notin F$ and $v(x) \leqq 0$ when $x \in F$, then*

$$\mathbb{M}\, \tau_x(F) \geqq v(x).$$

The proof is analogous to those of Lemmas 1 and 2.

Corollary 1. *If there exists a $v(x) \in D_L$ bounded in X and such that*

$$Lv(x) = -1, \quad \text{for } x \in X \setminus F,$$
$$v(x) = 0, \quad \text{for } x \in F,$$

then

$$\mathbb{M}\, \tau_x(F) = v(x).$$

Lemma 4. *If $v_i(x) \in D_L$ $(i = 0, 1)$ exist with $v_0(x) = v_1(x) = 0$ for $x \in F$ and*

$$Lv_0(x) = -1, \quad Lv_1(x) = -2v_0(x), \quad x \in X \setminus F, \tag{6}$$

and $v_0(x)$ is bounded in X, then

$$\mathbb{M}\, \tau_x^2(F) = v_1(x). \tag{7}$$

Indeed, it follows from the above that $\tau_x(F)$ is finite. Setting $g(x) = v_0(x)$ and $t = \tau_x(F)$ in (2) we find

$$\tau_x(F) = \alpha(\tau_x(F)) + v_0(x),$$

whence

$$\begin{aligned} \mathbb{M}\, \tau_x^2(F) = {} & \mathbb{M} \left[\int_0^{\tau_x(F)} (\nabla v_0 | b\, dw) \right]^2 \\ & + \mathbb{M} \left[\int_0^{\tau_x(F)} \int \{ v_0(\xi_x(t) + c(\xi_x(t), u)) - v_0(\xi_x(t)) \} \tilde{v}(dt, du) \right]^2 \\ & + g^2(x) = \mathbb{M} \int_0^{\tau_x(F)} |\nabla v_0\, b|^2\, dt \\ & + \mathbb{M} \int_0^{\tau_x(F)} \int [v_0(\xi_x(t) + c(\xi_x(t), u)) - v_0(\xi_x(t))]^2\, \Pi(du) + v_0^2(x). \end{aligned} \tag{8}$$

Set

$$Z(x) = v_1(x) - v_0^2(x).$$

Easy calculations prove that for arbitrary $g(x)\in D_L$

$$Lg^2(x)=2g(x)\,Lg(x)+2\,|\nabla g(x)\,b(x)|^2+\int[g(x+c)-g(x)]^2\,\Pi(du),$$

so that

$$\begin{aligned}LZ(x)&=+2v_0(x)+2v_0(x)\,Lv_0(x)+2\,|\nabla v_0\,b|^2\\&\quad+\int[v_0(x+c)-v_0(x)]^2\,\Pi(du)\\&=2\,|\nabla v_0\,b|^2+\int[v_0(x+c)-v_0(x)]^2\,\Pi(du).\end{aligned}\tag{9}$$

From the generalized Itô formula

$$Z[\xi_x(\tau^x(F))]-Z(x)=\int_0^{\tau_x(F)}LZ[\xi_x(t)]\,dt+\alpha_{v_0}(\tau_x(F)),$$

whence

$$\begin{aligned}Z(x)&=\mathbb{M}\int_0^{\tau_x(F)}\left[|2\nabla v_0 b(\xi_x(t))|^2+\int\big(v_0(\xi_x(t)+c(\xi_x(t),u))-v_0(\xi_x(t))\big)^2\,\Pi(du)\right]dt\\&=\mathbb{M}\,\tau_x^2(F)-v_0^2(x),\end{aligned}$$

or

$$\mathbb{M}\,\tau_x^2(F)=v_0^2(x)+Z(x)=v_1(x).\quad\square$$

Let A and B be two closed disjoint sets. Let ${}_BP_x(A)$ denote the probability that $\xi_x(t)$ enters the set A before B.

We will call a function $v(x)$ *L-superharmonic* in the region G if $v(x)\in D_L$ and

$$Lv(x)\leqq 0\qquad\text{for } x\in G.$$

Similarly a function $v(x)\in D_L$ will be called *L-subharmonic* in G if

$$L(v(x))\geqq 0\qquad\text{for } x\in G.$$

If $v(x)$ is both L-super- and L-subharmonic in G, it satisfies

$$Lv(x)=0,\qquad x\in G,$$

and will be called *L-harmonic.*

Lemma 5. *If* $U(x)$ *is* L*-superharmonic in* $X\setminus(A\cup B)$ *and* $U(x)|_A\geqq 1$, $U(x)|_B\geqq 0$, *then*

$${}_B\mathbb{P}_x(A)\leqq U(x).$$

If, however, $V(x)$ *is L-subharmonic in* $X\setminus(A\cup B)$ *and* $V(x)|_A\leqq 1$, $V(x)|_B\leqq 0$, *then*

$${}_B\mathbb{P}_x(A)\geqq V(x).$$

The proof follows from the generalized Itô formula. Let $\tau=\tau_x(A\cup B)$. Then

$$U(\xi_x(\tau))-U(x)=\int_0^{\tau}LU(\xi(t))\,dt+\alpha_U(\tau),$$

where $\alpha_U(\tau)$ is defined at (3). Taking expectations on both sides of this we get

$$\mathbb{M}\, U(\xi_x(\tau)) \leqq U(x),$$

and

$$\mathbb{M}\, U(\xi_x(\tau)) \geqq {}_B P_x(A).$$

This proves the first part of the lemma. The second goes through similarly.

Corollary 2. *If there exists an L-harmonic function $U(x)$ in $X \smallsetminus (A \cup B)$ for which $U(x)|_A = 1$; $U(x)|_B = 0$, then*

$${}_B\mathbb{P}_x(A) = U(x).$$

In an analogous manner we prove

Lemma 6. *Let $U(x)$ be L-superharmonic in G and*

$$Lu(x) \leqq 0, \qquad x \in G,$$

$$U(x) \geqq f(x), \qquad x \in X \smallsetminus G.$$

Then

$$\mathbb{M}\, f(\xi_x(\tau_x)) \leqq U(x).$$

Let $U_*(x) = \inf U(x)$, where the infimum is taken over all L-superharmonic functions in G satisfying $U(x) \geqq f(x)$ in $X \smallsetminus G$. Similarly, $V^*(x)$ is the sup of all L-subharmonic functions $V(x)$ in G for which $V(x) \leqq f(x)$ in $X \smallsetminus G$. If

$$U_*(x) = V^*(x) = u(x),$$

then

$$\mathbb{M}\, f(\xi_x(\tau_x)) = u(x).$$

We make some remarks on the results obtained above in the special case of equations of continuous type $(c(x, u) \equiv 0)$. L has the form

$$Lv = \sum_{k=1}^{n} a_k(x) \frac{\partial v}{\partial x_k} + \frac{1}{2} \sum_{k,j=1}^{n} \sigma_{kj}(x) \frac{\partial^2 v}{\partial x_k \, \partial x_j},$$

where

$$\sigma_{kj}(x) = \sum_{r=1}^{n} b_{kr}(x)\, b_{jr}(x)$$

and

$$\sum_{k,j=1}^{n} \sigma_{kj}(x)\, \lambda_k \lambda_j = \sum_{r=1}^{n} \left(\sum_{k=1}^{n} \lambda_k b_{kr}(x) \right)^2 \geqq 0, \tag{10}$$

i.e., L is a second-order elliptic differential operator in the region in which equality in (10) is possible only when $\lambda_1 = \lambda_2 = \cdots = \lambda_n = 0$. Let G be an open, connected, bounded region with sufficiently smooth boundary. In the theory of partial differential equations of elliptic type, under certain assumptions on the smoothness of the coefficients of the operator L and of the boundary G_Γ of G one can prove the existence of a solution of the Dirichlet problem for the operator L in G. From this follows, in particular, the existence of functions $v_0(x)$ and $v_1(x)$ continuous in the closure of G and satisfying

$$\begin{aligned} Lv_0 &= -1, \qquad v_0 | G_\Gamma = 0, \\ Lv_1 &= -2v_0, \quad v_1 | G_\Gamma = 0. \end{aligned} \tag{11}$$

Since in the case under consideration the trajectories of the solutions are continuous w.p. 1, $\tau_x(G)$ is the time a trajectory leaving the point $x \in G$ at time $t=0$ reaches the boundary of G. The mean value of this time, i.e., $\mathbb{M}\,\tau_x(G) = v(x)$, considered as a function of x, coincides with the simplest solution of the Dirichlet problem (11) for the operator L and $v_1(x)$ yields the 2nd order moment.

For $x \in G$, the infimum of all functions L-superharmonic in G satisfying $u(x) \geqq f(x)$ on G_Γ coincides in G with the supremum of all functions L-subharmonic in G for which $v(x) \leqq f(x)$. The common value of these extremes is the generalized solution of the Dirichlet problem $Lu=0$, $x \in G$, $u=f(x)$ on the boundary of G.

Thus if the corresponding hypotheses are fulfilled, we have from Lemma 6

$$u(x) = \mathbb{M}\, f(\gamma_x),$$

where γ_x is the point of contact at the boundary of the region G of a solution trajectory starting at $x \in G$ at time $t=0$.

If as $f(x)$ we take a function equal to one on Γ' and zero for $x \in \Gamma''$ where $\Gamma' \cup \Gamma'' = G_\Gamma$, then $u(x) = \mathbb{P}\{\gamma_x \in \Gamma'\}$ is the probability that the corresponding trajectory hits the boundary of G for the first time at a point belonging to Γ'.

Chapter 3. Asymptotic Behavior of the Solutions of Stochastic Differential Equations

§ 12. Stability of Solutions

When we say that the solution of an ordinary differential equation is "stable" we mean that it varies little in the course of an unbounded time interval when the variation in the initial conditions is small, or when the coefficients of the equation vary by a small amount (stability under constantly acting disturbances).

In the theory of ordinary differential equations it is proved that in many cases the stability problem reduces to the study of the stability of a stationary (equilibrium) solution of the equation, i.e., to the investigation of the solution $x \equiv 0$ of

$$\frac{dx}{dt} = a(t, x), \qquad a(t, 0) \equiv 0. \tag{1}$$

The analogous problem for stochastic differential equations will now be treated. In the stochastic case one can also distinguish between several interesting stability notions.

We now consider several possible notions of stability.

Let the stochastic differential equation

$$d\xi = \alpha(t, \xi, dt) \tag{2}$$

be obtained as a result of disturbing (1), i.e.,

$$\alpha(t, x, h) = a(t, x)\, h + \beta(t, x, h), \tag{3}$$

where $\beta(t, x, h)$ is a random field for which the point $x \equiv 0$ is stationary, $\beta(t, 0, h) \equiv 0$.

We are interested in conditions under which for arbitrary $\varepsilon_i > 0$ $(i = 1, 2)$ there will exist a $\delta > 0$ such that from $|\xi(0)| \leqq \delta$ it follows that with probability greater than $1 - \varepsilon_1$, $\xi(t)$ does not leave a ε_2-neighborhood of the point 0 during the infinite time span $t \geqq 0$.

If a stochastic equation possesses this property, then the stationary solution $\xi \equiv 0$ will be called *uniformly stochastically stable*. If, in this connection

$$\lim_{t\to\infty} \xi(t)=0,$$

with probability greater $1-\varepsilon_1$, then the solution $\xi(t)\equiv 0$ will be called *asymptotically uniformly stable.*

Instead of requiring that

$$|\xi(t)|<\varepsilon_2 \tag{4}$$

for all t one could consider $\xi(t)$ as differing little from zero if for arbitrary $\varepsilon_1>0$, $\varepsilon_2>0$ there is a $\delta>0$ such that w.p. 1

$$\sup_{0\leqq t\leqq\infty} \mathbb{P}\{|\xi(t)|>\varepsilon_2/\xi(0)\}\leqq\varepsilon_1$$

for $|\xi(0)|\leqq\delta$ or w.p. 1 for $|\xi(0)|<\delta$

$$\mathbb{M}\{|\xi(t)|^2/\xi(0)\}\leqq\varepsilon \quad \text{for all } t\geqq 0.$$

In the first case we will say that the solution $\xi(0)\equiv 0$ is *stochastically stable* and in the second that its second order moments are stable.

To these definitions we will add the adverb "asymptotically" if

$$\mathbb{P}\{|(\xi(t))|>\varepsilon_1/\xi(0)\}\to 0 \quad \text{for } t\to\infty$$

and arbitrary $\varepsilon_1>0$ in the first case and

$$\mathbb{M}\{|\xi(t)|^2/\xi(0)\}\to 0 \quad \text{for } t\to\infty.$$

in the second.

From Chebychev's inequality it follows immediately that the stability of the second order moments (the asymptotic stability) implies stochastic stability (stochastic asymptotic stability).

We turn now to some other aspects. Assume that $\beta(t,0,h)\not\equiv 0$, i.e., that the disturbance is not identically zero at the stationary point of the unperturbed equation (1) but that $\beta(t,x,h)$ is small (in some sense) in a certain neighborhood of $x=0$. Smallness of the field $\beta(t,x,h)$ can be taken, for example, as smallness of the second order moment:

$$\mathbb{M}\,|\beta(t,x,h)|^2\leqq\varepsilon \quad \text{for } |x|\leqq\delta,\ h\leqq h_0,\ 0\leqq t<\infty.$$

Since in the most interesting examples of random disturbances the sample functions of the field $\beta(t,x,h)$ have w.p. 1 no finite upper bound on $[0,\infty)$ even for small moments, and even have arbitrarily large jumps w.p. 1 (coinciding with the jumps of $\xi(t)$), one cannot expect that the solutions of a stochastic equation will be uniformly stable. However, questions of stochastic stability and stability of second order moments with modified definitions are still important.

The solution $x(t)\equiv 0$ of (1) will be called stochastically stable (or has stable moments of second order) with respect to a constantly operating random perturbation in a given class K of random fields if for arbitrary $\varepsilon_i > 0$ $(i=1,2)$ there is a $\delta > 0$ such that, w.p. 1

$$\sup_{0 \leqq t < \infty} \mathbb{P}\{|\xi(t)| > \varepsilon_2 / \xi(0)\} \leqq \varepsilon_1$$

for arbitrary random perturbation $\beta(t,x,h) \in K$ provided that

$$\mathbb{M}|\beta(t,x,h)|^2 < \delta\big(t \in [0,\infty),\ x \in X,\ 0 \leqq h \leqq h_0\big), \quad |\xi(0)| < \delta$$

(or $\sup_{0 \leqq t < \infty} \mathbb{M}\{|\xi(t)|^2 / \xi(0)\} < \varepsilon_1$ under the same conditions).

Everything said about the perturbed ordinary differential equation (1) naturally applies also when we start with a stochastic equation.

We begin the study of the stability of stationary solutions of ordinary differential equations with respect to random perturbations with systems of linear equations. We will show that for a sufficiently wide class of perturbing random fields, questions concerning the stability of second order moments of perturbed equations can be reduced to the investigation of the stability of a certain system of ordinary differential equations (the order of this system equals $\frac{n(n+1)}{2}$, where n is the dimension of the space X). Such a result can be viewed as the answer to the stability question in this case.

Assume given a linear stochastic equation

$$d\xi(t) = [a_0(t) + a(t)\,\xi(t)]\,dt + \beta(t, \xi(t), dt), \tag{5}$$

where $a(t)$ is a nonrandom operator function of $t \geqq 0$ whose values map X into itself, $a_0(t)$ is a nonrandom vector function with values in X and $\beta(t,x,h)$ is a vector martingale depending on x. The functions $a_0(t)$ and $a(t)$ will be taken as continuous in t.

Eq. (5) can be considered as the outcome of a perturbation of the equation

$$\frac{dx}{dt} = a(t)\,x \tag{6}$$

by means of the random field $a_0(t)\,h + \beta(t,x,h)$, whereby $a_0(t)\,h$ is the mean value of this perturbation. More precisely,

$$\mathbb{M}\{a_0(t)\,h + \beta(t,x,h)/\mathfrak{F}_t\} = a_0(t)\,h.$$

We choose some orthonormal basis in X and assume that in this basis

$$a_0(t) = \{a_0^j(t);\ j=1,2,\dots,n\},$$

$$a(t) = \{a_j^k(t);\ k,j=1,2,\dots,n\},$$

$$\beta(t,x,h) = \left\{\sum_{s=1}^{n}\left[\sum_{j=1}^{n} b_{sj}^k(t)\,x^j + b_{s0}^k(t)\right](\beta^{(s)}(t+h) - \beta^{(s)}(t)),\ k=1,2,\dots,n\right\},$$

where b_{sj}^k $(k, s=1, 2, \ldots, n,\ j=0, 1, \ldots, n)$ is a nonrandom function, x^j are the coordinates of the vector x, $\beta(t)=\{\beta^{(1)}(t), \ldots, \beta^{(n)}(t)\}$ is a vector martingale with

$$\mathbb{M}\{\beta(t)/\mathfrak{F}_t\}=0, \tag{7}$$

$$\mathbb{M}\{[\beta^{(k)}(t+h)-\beta^{(k)}(t)][\beta^{(s)}(t+h)-\beta^{(s)}(t)]/\mathfrak{F}_t\}=l_s^k(t)\,h, \tag{8}$$

and the $l_s^k(t)$ are nonrandom. In the sequel we will use the usual terminology and notation for matrices. Thus, the vectors x, ξ, $a_0(t)$ and $\beta(t)$ will be viewed as matrices consisting of a single column and the symbol $b_s(t)$ will denote a square matrix with elements $b_{sj}^k(t)$ $(k, j=1, 2, \ldots, n)$ where k is the row index and j the column index. If C is a matrix, then C^* will denote its transpose, $C_j^{*k}=C_k^j$. In particular, if x is a column vector, then x^* is a row vector with elements $x_j^*=x^j$. The product $x\,y^*$, where x and y are column vectors with elements x^i, y^i is a square matrix with elements $z_j^k=x^k\,y_j^*$, and $y^*\,x=\sum_{j=1}^{n} y_j^*\,x^j$.

Relation (8) can be written as

$$\mathbb{M}\{[\beta(t+h)-\beta(t)][\beta(t+h)-\beta(t)]^*/\mathfrak{F}_t\}=L(t)\,h, \tag{8'}$$

where $L(t)$ is a matrix with entries $l_s^k(t)$.

Set

$$m(t)=\mathbb{M}\{\xi(t)/\mathfrak{F}_0\}.$$

From (5) and the properties of stochastic line integrals in which the integrable random field is a martingale it follows that

$$m(t)=\xi(0)+\int_0^t [a_0(\tau)+a(\tau)\,m(\tau)]\,d\tau, \tag{9}$$

since $m(t)$ is bounded, the differentiability of $m(t)$ follows from (8) and

$$\frac{dm}{dt}=a_0(t)+a(t)\,m(t), \qquad m(0)=\xi(0).$$

Hence for stability of the first order moment of the solution of (5) w.r.t. the class of perturbations under consideration it is necessary and sufficient that the linear system (6) be stable w.r.t. the constantly acting (nonrandom) perturbation $a_0(t)$. Necessary and sufficient conditions for the stability of the linear system (6) w.r.t. constantly operating perturbations are known. Using these results, we can formulate the following proposition: if $x\equiv 0$ is an asymptotically stable solution of (6) uniformly in x_0, t_0 (i.e., if the solution of (6) satisfying the initial condition $x(t_0)=x_0$ tends to zero for $t\to\infty$ uniformly in some region $|x_0|<\delta$, $t_0>0$), then for any $\varepsilon>0$ there exists a $\delta>0$ such that for an arbitrary perturbation field

$a_0(t)\,h+\beta(t, h, x)$, where $\beta(t, h, x)$ is an arbitrary martingale, we will have $|m(t)|<\varepsilon$, $0<t<\infty$ provided that $|x_0|+|a_0(t)|<\delta$ for all $t>0$.

We remark that in this rather trivial result it is not assumed that the martingale $\beta(t, x, h)$ is a linear function of x.

We proceed now to the second order moment. Introduce the matrix Q of second order moments

$$Q(t)=\mathbb{M}\{\xi(t)\,\xi^*(t)/\mathfrak{F}_0\}.$$

We will show that $Q(t)$ satisfies a certain differential equation. Fixing t, we set $\Delta\xi=\xi(t+h)-\xi(t)$, $h>0$. Then

$$Q(t+h)-Q(t)=\mathbb{M}\{\Delta\xi\,\xi^*(t)+\xi(t)\Delta\xi^*+\Delta\xi\,\Delta\xi^*/\mathfrak{F}_0\}.$$

For the sake of notational convenience we set

$$z_1=\int_t^{t+h}\big(a_0(\tau)+a(\tau)\,\xi(\tau)\big)\,d\tau,$$

$$\tilde{z}_1=\big(a_0(t)+a(t)\,\xi(t)\big)\,h,$$

$$z_2=\int_t^{t+h}\beta\big(\tau,\xi(\tau),d\tau\big).$$

From the continuity of $a_0(t)$ and $a(t)$ and of the second moments of $\xi(t)$ it follows that

$$\mathbb{M}\{|z_1-\tilde{z}_1|^2/\mathfrak{F}_0\}=o(h).$$

Moreover,

$$\mathbb{M}\{z_2\,\xi^*(t)/\mathfrak{F}_0\}=0,$$

so that

$$\begin{aligned}\mathbb{M}\{\Delta\xi\,\xi^*(t)/\mathfrak{F}_0\}&=\int_t^{t+h}[a_0(\tau)\,m^*(t)+a(\tau)\,Q(t)]\,d\tau\\&\quad+\int_t^{t+h}a(\tau)\,\mathbb{M}\{[\xi(\tau)-\xi(t)]\,\xi^*(t)/\mathfrak{F}_0\}\,d\tau\\&=[a_0(t)\,m^*(t)+a(t)\,Q(t)]\,h+o(h).\end{aligned}$$

We also have

$$\Delta\xi\,\Delta\xi^*=z_1\,z_1^*+z_1\,z_2^*+z_2\,z_1^*+z_2\,z_2^*,$$

$$|\mathbb{M}\{z_1\,z_1^*/\mathfrak{F}_0\}|\leqq\mathbb{M}\{|z_1|^2/\mathfrak{F}_0\}=o(h),$$

$$|\mathbb{M}\{z_1\,z_2^*/\mathfrak{F}_0\}|\leqq|\mathbb{M}\{\tilde{z}_1\,z_2^*/\mathfrak{F}_0\}+[\mathbb{M}\{|z_1-\tilde{z}_1|^2/\mathfrak{F}_0\}\,\mathbb{M}\{|z_2|^2/\mathfrak{F}_0\}]^{\frac{1}{2}}=o(h).$$

From the properties of integrals w.r.t. martingales

$$\begin{aligned}\mathbb{M}\{z_2\,z_2^*/\mathfrak{F}_0\}=\int_t^{t+h}\sum_{s,r}\{&b_{s\cdot}^{\bullet}(\tau)\,Q(\tau)\,b_{r\cdot}^{*}(\tau)+b_{s0}^{\bullet}(b_{r\cdot}^{\bullet}(\tau)\,m(\tau))^*\\&+(b_{s\cdot}^{\bullet}(\tau)\,m(\tau))\,b_{r0}^{\bullet *}+b_{s0}(\tau)\,b_{r0}^{\bullet *}(\tau)\}\,l_r^s(\tau)\,d\tau.\end{aligned}$$

These relationships prove the differentiability of $Q(t)$ and show that

$$\frac{dQ}{dt}=\sum_{s,r} b_{s\cdot}^{\bullet} Q b_{r\cdot}^{\bullet *} l_r^s + a(t)\,Q(t)+Q(t)\,a^*(t)+r(t), \tag{10}$$

where

$$\begin{aligned} r(t)=a_0(t)\,m^*(t)+m(t)\,a_0^*(t)+\sum_{s,r}\big[b_{s0}^{\bullet}(t)\,(b_{r\cdot}^{\bullet}(t)\,m(t))^* \\ +(b_{s\cdot}^{\bullet}(t)\,m(t))\,b_{r0}^{\bullet *}(t)+b_{s0}^{\bullet}(t)\,b_{r0}^{\bullet *}(t)\big]\,l_r^s(t). \end{aligned} \tag{11}$$

The matrix $Q(t)$ is symmetric. Hence (10) can be viewed as a system of $\frac{n(n+1)}{2}$ linear differential equations for the elements of $Q(t)$. The stability condition for the second moments follows in particular from all of the previously obtained equations if the random perturbation is independent of the spatial coordinates. Indeed, if the point 0 is uniformly asymptotically stable for the system (6), then the second order moments of the solution of (5) are stable w.r.t. constantly operating perturbations of the form

$$a_0(t)\,h+\beta(t,h), \qquad \beta(t,h)=\sum_{s=1}^{n} b_{s0}^{\bullet}(t)\,[\beta^{(s)}(t+h)-\beta^{(s)}(t)].$$

For the proof we make the following remarks. Let $z(t,t_0)$ be a matrix whose columns are solutions of (6) and let

$$z(t_0,t_0)=I,$$

where I is the identity matrix. From the uniform asymptotic stability of the solution $x\equiv 0$ of (6) it follows that $|z(t,t_0)|\leqq c$ for all $t\geqq t_0$ and that $z(t,t_0)\to 0$ for $t\to\infty$.

Set

$$S(t,t_0)=z(t,t_0)\,S_0\,z^*(t,t_0),$$

where S_0 is a constant matrix. $S(t,t_0)$ satisfies

$$\frac{dS}{dt}=a(t)\,S+S\,a^*(t) \tag{12}$$

with initial condition

$$S(t_0,t_0)=S_0.$$

It thus follows that the zero-solution of (12) is uniformly asymptotically stable, i.e.,

$$S(t,t_0)\to 0 \quad \text{for } t\to\infty \quad \text{and} \quad |S(t,t_0)|\leqq\varepsilon \quad \text{if} \quad |S_0|<\frac{\varepsilon}{c^2}.$$

Consequently, the solution $s\equiv 0$ of (12) is stable w.r.t. constantly operating random perturbations. Since the first order moments are also stable, for any $\varepsilon>0$ and $c>0$ there exists a $\delta>0$ such that when

$$|a_0(t)|+|l(t)|\leqq c, \quad |b^{\bullet}_{s0}(t)|<\delta,$$

the second order moments of the solution of

$$d\xi=[a_0(t)+a(t)\,\xi(t)]\,dt+\beta(t,dt)$$

satisfy

$$|Q(t)|<\varepsilon, \quad 0\leqq t\leqq\infty,$$

provided that $|Q(t)|<\delta$, independently of the remaining properties of the continuous functions $a_0(t)$, $l(t)$ and $b_{s0}(t)$.

It is not difficult to show that under the same assumptions, the solution $x\equiv 0$ of (5) is stable w.r.t. constantly acting disturbances of the more general from

$$a_0(t)\,h+\beta(t,x,h).$$

From (10) we can also deduce the following: if for the matrix differential equation

$$\frac{dQ}{dt}=a(t)\,Q(t)+Q(t)\,a^*(t)+\sum_{s,r} b^{\bullet}_{s\bullet}(t)\,Q(t)\,b^{\bullet}_{r}{}^*(t)\,l^s_r(t) \tag{13}$$

the solution $Q\equiv 0$ is asymptotically uniformly stable, then the second order moments of the stochastic system

$$d\xi=a(t)\,\xi(t)\,dt+\sum_s (b^{\bullet}_{s\bullet}\,d\beta^s)\,\xi(t) \tag{14}$$

are stable w.r.t. constantly acting perturbations of the form

$$a_0(t)\,dt+\sum_{s=1}^{n} b^{\bullet}_{s0}(t)\,d\beta^s.$$

We dwell on some special cases. If the external disturbances w.r.t. the system are missing, i.e., $a_0(t)=b_{s0}(t)=0$, then the system has the form (14) and the system for the second order moments the form (13). The latter can be viewed as a system of $\frac{n(n+1)}{2}$ ordinary differential equations for the various elements of Q. Renumbering these in a certain order and considering them as vectors in an $\frac{n(n+1)}{2}$-dimensional space, we can write (13) as

$$\frac{dQ}{dt}=\mathfrak{A}Q.$$

If the matrices a, $b^{\bullet}_s$ and l are constant, then $\mathfrak{A}$ is independent of t and a necessary and sufficient condition for asymptotic stability of the second order moments is that the eigen-values of the matrix $\mathfrak{A}$ have negative real parts. A necessary and sufficient condition for stability is that the eigen-values have nonpositive real parts and that the elementary devisors corresponding to the purely imaginary eigen-values be simple.

Example 1. An equation of second order with constant coefficients.

Consider

$$d\frac{d\xi}{dt}+(b_0\,dt+b\,d\beta_1)\frac{d\xi}{dt}+(a_0\,dt+a\,d\beta)\,\xi=0, \tag{15}$$

where b_0, b, a_0 and a are constants and $\beta_1=\beta_1(t)$ and $\beta=\beta(t)$ are non-correlated martingales. (15) is to be understood as the system of linear stochastic equations

$$d\xi=\xi_1\,dt,$$
$$d\xi_1=-(a_0\,dt+a\,d\beta)\,\xi-(b_0\,dt+b\,d\beta_1)\,\xi_1.$$

Let

$$\mathbb{M}\{[\beta(t+h)-\beta(t)]^2/\mathfrak{F}_t\}=\mathbb{M}\{[\beta_1(t+h)-\beta_1(t)]^2/\mathfrak{F}_t\}=h.$$

The system for the second order moments can be written as

$$Q'_{00}(t)=2Q_{01},$$
$$Q'_{01}(t)=-a_0\,Q_{00}-b_0\,Q_{01}+Q_{22},$$
$$Q'_{11}(t)=a^2\,Q_{00}-2a_0\,Q_{01}-(2b_0-b^2)\,Q_{22},$$

where

$$Q_{00}(t)=\mathbb{M}\,\xi^2(t),\quad Q_{01}(t)=\mathbb{M}\,\xi(t)\,\xi'(t),\quad Q_{11}(t)=\mathbb{M}\,\xi'(t)^2 \qquad (\xi'(t)=\xi_1(t)).$$

The characteristic equation for this system is

$$\lambda^3+\lambda^2(3b_0-b^2)+\lambda[b_0(2b_0-b^2)+4a_0]+2[a_0(2b_0-b^0)-a^2]=0. \tag{16}$$

It is not difficult (with the help of the Routh-Hurwitz theorem) to find conditions which guarantee that all roots of (16) lie in the left half-plane. Necessary and sufficient for this is that

$$b^2<2b_0 \quad\text{and}\quad a^2<(2b_0-b^2)\,a_0. \tag{17}$$

Example 2. Consider

$$\frac{d^2\xi}{dt^2}+[a(t)+b(t)\,\beta'(t)]\,\xi=0, \tag{18}$$

which is to be understood as the system

$$\begin{aligned} d\xi&=\xi_1\,dt,\\ d\xi_1&=-\xi\,[a(t)\,dt+b(t)\,d\beta(t)], \end{aligned} \tag{19}$$

where $\beta(t)$ is a martingale for which

$$\mathbb{M}\{[\beta(t+h)-\beta(t)]^2/\mathfrak{F}_t\}=\int_t^{t+h} l(\tau)\,d\tau.$$

For this equation one can establish various results analogous to those known for the ordinary differential equation [3]

$$u''+a(t)u=0. \tag{20}$$

Let $u_1(t)$ and $u_2(t)$ be a linearly independent system of solutions of (20). It is known that the Wronskian of this system

$$w(t)=\begin{vmatrix} u_1(t) & u_2(t)\\ u_1'(t) & u_2'(t)\end{vmatrix}$$

is constant. Without loss of generality we can assume that $w(t)=1$. Set

$$W(t,\tau)=\begin{vmatrix} u_1(\tau) & u_2(\tau)\\ u_1(t) & u_2(t)\end{vmatrix}.$$

We will show that $\xi(t)$ coincides with the solution of

$$\eta(t)=z(t)+\int_0^t W(t,\tau)b(\tau)\eta(\tau)\,d\beta(\tau), \tag{21}$$

where $z(t)$ is a solution of (20) for which

$$z(0)=\xi(0), \qquad z'(0)=\xi_1(0).$$

To this end we introduce

$$\eta_1(t)=z'(t)+\int_0^t W_t'(t,\tau)b(\tau)\eta(\tau)\,d\beta(\tau)$$

and verify that

$$\begin{aligned}\eta(t)&=z(0)+\int_0^t \eta_1(\tau)\,d\tau,\\ \eta_1(t)&=z'(0)-\int_0^t \eta(\tau)\,[a(\tau)\,d\tau+b(\tau)\,d\beta(\tau)].\end{aligned} \tag{22}$$

The latter also means that $\eta(t)$ and $\eta_1(t)$ yield the solution of (19). Since the solution of (19) is unique, we have

$$\xi(t)=\eta(t), \qquad \xi_1(t)=\eta_1(t).$$

The verification of (22) is carried through by means of integration by parts. In this connection, we note that the correctness of the formula

$$\int_a^b u'(t)\zeta(t)\,dt+\int_a^b u(t)\,d\zeta(t)=u(t)\zeta(t)\big|_a^b. \tag{23}$$

where $\zeta(t)$ is a martingale of the type under consideration and $u(t)$ is a continuously differentiable function, is established exactly as in Lemma 1 § 6.

Using (23) we get

$$\int_0^t d\tau \int_0^\tau W_\tau'(\tau,\theta)\, b(\theta)\,\eta(\theta)\, d\beta(\theta)$$
$$= \int_0^t \left\{ u_2'(\tau) \int_0^\tau u_1(\theta)\, b(\theta)\,\eta(\theta)\, d\beta(\theta) - u_1'(\tau) \int_0^\tau u_2(\theta)\, b(\theta)\,\eta(\theta)\, d\beta(\theta) \right\} d\tau$$
$$= -\int_0^t [u_2(\tau)\,u_1(\tau) - u_1(\tau)\,u_2(\tau)]\, b(\tau)\,\eta(\tau)\, d\beta(\tau)$$
$$+ \int_0^t (u_2(t)\,u_1(\tau)\, b(\tau)\,\eta(\tau)\, d\beta(\tau) - u_1(t)\,u_2(\tau)\, b(\tau)\,\eta(\tau)\, d\beta(\tau))$$
$$= \eta(t) - z(t).$$

Hence

$$\int_0^t \eta_1(\tau)\, d\tau = z(t) - z(0) + \int_0^t d\tau \int_0^\tau W_t'(\tau,\theta)\, b(\theta)\,\eta(\theta)\, d\beta(\theta) = \eta(t) - z(0).$$

This proves the first of the equations at (22). The second is verified analogously:

$$\eta_1(t) = z'(t) + u_2'(t) \int_0^t u_1(\tau)\,\eta(\tau)\, b(\tau)\, d\beta(\tau) - u_1'(t) \int_0^t u_2(\tau)\,\eta(\tau)\, b(\tau)\, d\beta(\tau)$$
$$= z'(t) + \int_0^t \left\{ u_2''(\tau) \int_0^\tau u_1(\theta)\,\eta(\theta)\, b(\theta)\, d\beta(\theta) \right.$$
$$\left. - u_1''(\tau) \int_0^\tau u_2(\theta)\,\eta(\theta)\, b(\theta)\, d\beta(\theta) \right\} d\tau$$
$$+ \int_0^t [u_2'(\tau)\,u_1(\tau) - u_1'(\tau)\,u_2(\tau)]\,\eta(\tau)\, b(\tau)\, d\beta(\tau)$$
$$= z'(t) - \int_0^t a(\tau) \int_0^\tau W(\tau,\theta)\,\eta(\theta)\, b(\theta)\, d\beta(\theta)\, d\tau + \int_0^t \eta(\theta)\, b(\theta)\, d\beta(\theta)$$
$$= z'(t) - \int_0^t z''(\tau)\, d\tau - \int_0^t \eta(\tau)\,[a(\tau)\, d\tau + b(\tau)\, d\beta(\tau)]$$
$$= z'(0) - \int_0^t \eta(\tau)\,[a(\tau)\, d\tau + b(\tau)\, d\beta(\tau)].$$

The second equality is thus also established. We still have to prove that (21) has a unique solution. This is easy and is shown (using the properties of the stochastic integral) as in the uniqueness and existence theorem for stochastic differential equations.

We turn to the first and second order moments of the solution of (18). If

$$m(t) = \mathbb{M}\{\xi(t)/\mathfrak{F}_0\},$$

then from (21)

$$m(t) = z(t).$$

Setting $Q(t) = \mathbb{M}\{\xi^2(t)/\mathfrak{F}_0\}$ and using (21), we obtain

$$Q(t) = z^2(t) + \int_0^t W^2(t,\tau)\, b^2(\tau)\, Q(\tau)\, l(\tau)\, d\tau.$$

This has the unique solution

$$Q(t) = \sum_{n=0}^{\infty} Q_n(t),$$

where

$$Q_0(t) = z^2(t), \qquad Q_{n+1}(t) = \int_0^t W^2(t,\tau)\, b^2(\tau)\, l(\tau)\, Q_n(\tau)\, d\tau.$$

Since $W^2(t,\tau) \leqq v(t)\, v(\tau)$, where $v(t) = u_1^2(t) + u_2^2(t)$, we have

$$Q_{n+1}(t) \leqq v(t) \int_0^t Q_n(\tau)\, v(\tau)\, b^2(\tau)\, l(\tau)\, d\tau.$$

Assume that for $t > 0$

$$v(t)\, b^2(t)\, l(t) \leqq C_1.$$

Then

$$Q_{n+1}(t) \leqq C_1\, v(t) \int_0^t Q_n(\tau)\, d\tau.$$

Noting that $z^2(t) \leqq C_0\, v(t)$, where C_0 is some constant, we find

$$Q_1(t) \leqq C_0\, C_1\, v(t) \int_0^t v(\tau)\, d\tau = \frac{1}{2}\, C_0\, C_1 \frac{d}{dt} \left[\int_0^t v(\tau)\, d\tau\right]^2$$

and by induction

$$Q_n(t) \leqq \frac{C_0\, C_1^n}{(n+1)!} \cdot \frac{d}{dt} \left[\int_0^t v(\tau)\, d\tau\right]^{n+1} = \frac{C_0\, C_1^n}{n!}\, v(t) \left[\int_0^t v(\tau)\, d\tau\right]^n.$$

We hence obtain the estimate

$$Q(t) \leqq C_0\, v(t)\, e^{C_1 \int_0^t v(\tau)\, d\tau},$$

which proves

Theorem 1. *Assume that* $l(t)\, b^2(t)$ *is bounded,* $0 < t < \infty$, *and that for any arbitrary solution of* (20)

$$\int_0^{\infty} u^2(\tau)\, d\tau < \infty.$$

Then

$$\int_0^\infty Q(\tau)\,d\tau<\infty .$$

If in addition, a) any solution $u(t)$ *of* (20) *tends to zero for* $t\to 0$ *or b) any solution is bounded* $(0<t<\infty)$, *then, in case a)* $Q(t)\to 0$ *for* $t\to\infty$ *uniformly in any region* $|\xi(0)|+|\xi_1(0)|\leqq R$ *and, in case b)* $Q(t)$ *is bounded uniformly in any region* $|\xi(0)|+|\xi_1(0)|\leqq R$.

This theorem is interesting for the following reason. Random perturbations of a general linear system of differential equations tend, generally speaking, to work against the stability of second order moments. On the other hand, Theorem 1 says that under certain conditions on the moment $Q(t)=\mathbb{M}\{\xi^2(t)\,|\,\mathfrak{F}_0\}$, several stability properties hold which are analogous to those of the nonperturbed equation.

We address ourselves to the question of uniform stochastic stability. We start with some special equations.

Consider the simplest one-dimensional, homogeneous equation

$$d\xi=\left(a(t)\,dt+b(t)\,dw(t)+\int c(t,u)\,\tilde{v}(dt,du)\right)\xi, \tag{24}$$

where $w(t)$ is a Wiener process, $v(t,A)$ a Poisson measure with parameter $t\,\Pi(A)$, $\tilde{v}(t,A)=v(t,A)-t\,\Pi(A)$, and $w(t)$ and $v(t,A)$ are independent. The solution of (24) is (see § 6)

$$\begin{aligned}\xi(t)=\xi_0\exp\Bigg\{&\int_0^t\left(a(\tau)-\tfrac{1}{2}b^2(\tau)\right)d\tau+\int_0^t\int\left(\ln|1+c(t,u)|-c(t,u)\right)\Pi(du)\,dt\\&+\int_0^t b(\tau)\,dw(\tau)+\int_0^t\int_{-\infty}^{\infty}\ln|1+c(t,u)|\,\tilde{v}(dt,du)\Bigg\}.\end{aligned} \tag{25}$$

Assume

$$\frac{1}{T^2}\left\{\int_0^T b^2(t)\,dt+\int_0^T\int_{-\infty}^{\infty}\ln^2|1+c(t,u)|\,\Pi(du)\,dt\right\}\to 0\qquad\text{for } T\to\infty .$$

Then

$$\frac{1}{T}\left\{\int_0^T b(t)\,dw(t)+\int_0^T\int\ln|1+c(t,u)|\,\tilde{v}(dt,du)\right\}\to 0 \tag{26}$$

when $T\to\infty$ (in the sense of mean square convergence). On the other hand, suppose that

$$\lim_{T\to\infty}\frac{1}{T}\int_0^T a(t)\,dt=A,$$

$$\lim_{T\to\infty}\frac{1}{T}\int_0^T b^2(t)\,dt=B^2,$$

and

$$\lim_{T\to\infty} \frac{1}{T} \int_0^T \int [\ln|1+c(t,u)| - c(t,u)]\, \Pi(du)\, dt = C$$

exist.

Under these assumptions the asymptotic behavior of $\xi(t)$ for $t\to\infty$ is determined by

$$z(t) = \xi_0 \exp\{t(A + C - \tfrac{1}{2}B^2)\}. \tag{27}$$

This means that in the mean square convergence sense

$$\frac{\xi(t)}{z(t)} \to 1 \quad \text{for } t\to\infty. \tag{28}$$

If (26) is strengthened somewhat, i.e., by requiring that the function

$$b^2(t) + \int_{-\infty}^{\infty} \ln^2|1+c(t,u)|\, \Pi(du) \tag{29}$$

be bounded, or that the series

$$\sum_{n=1}^{\infty} \frac{1}{n^2} \int_{n-1}^{n} \left(b^2(t) + \int_{-\infty}^{\infty} \ln^2|1+c(t,u)|\, \Pi(du)\right) dt,$$

converges, then it follows from the theorem of Kolmogorov on the strong law of large numbers for independent random variables with finite variance that (28) holds w.p.1.

Hence, if the above assumptions hold and

$$2(A+C) < B^2, \tag{30}$$

then $\xi(t)\to 0$ w.p.1 for $t\to\infty$. If, however,

$$0 \leqq B^2 < 2(A+C),$$

then $\xi(t)\to 0$ for $t\to\infty$ w.p.1. Noteworthy for the equation under consideration is then the circumstance that an increase in the variance of the continuous part of the random perturbation improves the stability. In addition, for arbitrary $a(t)$ and $c(t,u)$ (for $a(t)=a>0$, $c(t,u)=0$ and $b(t)=0$, the solution of (24) tends exponentially to ∞) one can add a term $b\,dw(t)$ with b so large that the equation becomes stable.

Concerning the second order moments, from the equation

$$dQ = Q(2a + b^2 + \int c^2(t,u)\, \Pi(du)\, dt)$$

it follows that $Q(t) = \mathbb{M}\{\xi^2(t) | \mathfrak{F}_0\}$ is stable iff

$$\overline{\lim_{T\to\infty}} \int_0^T (2a(t) + b^2(t) + \int c^2(t,u)\, \Pi(du))\, dt \leqq 0,$$

In particular, uniform stochastic stability can also obtain when the second order moments are unstable.

As our next example we consider (18):

$$\frac{d^2\xi}{dt^2}+[a(t)+b(t)\,\beta'(t)]\,\xi=0,$$

where $\beta(t)$ is a martingale satisfying the conditions of Example 2.

As already established

$$\xi(t)=z(t)+u_2(t)\int_0^t u_1(\tau)\,b(\tau)\,\xi(\tau)\,d\beta(\tau)-u_1(t)\int_0^t u_2(\tau)\,b(\tau)\,\xi(\tau)\,d\beta(\tau).$$

Theorem 2. *If Hypothesis a) of Theorem 1 holds, then the solution* $\xi\equiv 0$ *of* (18) *is uniformly stochastically stable.*

Proof. Let $\sup\{|z(t)|,\ 0\leqq t<\infty\}<\frac{\varepsilon}{3}$ and set

$$v(t_0)=\sup\{|u_i(t)|;\ i=1,2,\ t_0\leqq t<\infty\}.$$

We will estimate the probability that $|\xi(t)|>\varepsilon$ for at least one t in $[t_0, T]$. We have

$$\mathbb{P}\{\sup_{t_0\leqq t\leqq T}|\xi(t)|>\varepsilon/\mathfrak{F}_0\}\leqq\sum_{j=1}^{2}\mathbb{P}\left\{\sup_{t_0\leqq t\leqq T}\left|\int_0^t u_j(\tau)\,b(\tau)\,\xi(\tau)\,d\beta(\tau)\right|>\frac{\varepsilon}{3}\Big/\mathfrak{F}_0\right\}.$$

We recall that we previously agreed to choose the values of stochastic integrals in such a way that they be separable processes as functions of the upper limit of integration. Taking this and Eq. (28) § 1, Part I into account we find

$$\mathbb{P}\{\sup_{t_0\leqq t\leqq T}|\xi(t)|>\varepsilon/\mathfrak{F}_0\}\leqq\frac{9v^2(t_0)\,c}{\varepsilon^2}\int_0^T \mathbb{M}\{|\xi(t)|^2/\mathfrak{F}_0\},$$

provided that

$$|(u_1^2(t)+u_2^2(t))\,b^2(t)\,l(t)|\leqq c.$$

Thus,

$$\mathbb{P}\{\sup_{t_0\leqq t<\infty}|\xi(t)|>\varepsilon/\mathfrak{F}_0\}\leqq\frac{9v_0^2\,c(t)}{\varepsilon^2}\int_0^\infty Q(\tau)\,d\tau,$$

so that

$$\mathbb{P}\{\sup_{0\leqq t<\infty}|\xi(t)|>\varepsilon/\mathfrak{F}_0\}\leqq\frac{9v_0^2\,c}{\varepsilon^2}\int_0^{t_0}Q(\tau)\,d\tau+\frac{9v^2(t_0)\,c}{\varepsilon^2}\int_{t_0}^{\infty}Q(\tau)\,d\tau. \quad (31)$$

Choosing t_0 large enough that the second term on the right side of (31) does not exceed $\frac{\delta}{2}$ where δ is an arbitrary given positive number, and $\xi(0)$ in (10) so small (i.e., $Q(t)$ arbitrarily small, $0\leqq t\leqq t_0$) that a similar estimate holds for the first term, we obtain $\mathbb{P}\{\sup_{0\leqq t<\infty}|\xi(t)|>\varepsilon/\mathfrak{F}_0\}<\delta$, provided that $|\xi(0)|<\delta'$. □

Condition b) of Theorem 1 corresponds to the following statement on the uniform stochastic boundedness of the solutions of (18).

Theorem 3. *If an arbitrary solution* $u(t)$ *of*

$$u''(t)+a(t)\,u(t)=0$$

is bounded and

$$\int_0^\infty u^2(t)\,dt<\infty,$$

then for arbitrary $\delta>0$ *there is an* $A>0$ *and a* $\delta'>0$ *such that*

$$\mathbb{P}\{\sup_{0\leqq t<\infty}|\xi(t)|>A/\mathfrak{F}_0\}<\delta, \quad \text{if } |\xi(0)|<\delta'.$$

This also follows from the preceding estimates.

We now turn to general stochastic equations without after-effect. A series of results on the stability of such equations (i.e., of their solutions) can be obtained by applying the second method of Liapunov.

Consider

$$d\xi(t)=a(t,\xi)\,dt+b(t,\xi(t))\,dw(t)+\int f(t,\xi(t),u)\,\tilde{v}(dt,du), \tag{32}$$

where a, b, f and $\tilde{v}$ are defined, and satisfy the same conditions, as in § 7. Let

$$a(t,0)=f(t,0,u)\equiv 0, \qquad b(t,0)\equiv 0.$$

For an arbitrary twice continuously differentiable function $v(t,x)$ ($t\geqq 0$, $x\in X$) we introduce the operator

$$L(v)=\frac{\partial v}{\partial t}+(\nabla v|a)+\frac{1}{2}\operatorname{sp}[\nabla^2 v\,b\,b^*]$$
$$+\int\left[v(t,x+f(t,x,u))-v(t,x)-(\nabla v|f(t,x,u))\right]\Pi(du).$$

If $b\equiv c\equiv 0$, i.e., the stochastic equation goes over into a system of ordinary differential equations, then the operator

$$L(v)=L_0(v),$$

with

$$L_0(v)=\frac{\partial v}{\partial t}+(\nabla v|a),$$

represents the "total derivative of $v(t,x)$ w.r.t. the equation of motion" the use of which forms the idea of Liapunov's second method. If the random disturbance is continuous (i.e., $f \equiv 0$), then

$$L(v) = L_0(v) + \frac{1}{2} \sum_{j,k=1}^{n} \frac{\partial^2 v}{\partial x^j \partial x^k} \sigma^{jk}, \qquad \sigma^{jk} = \sum_{s=1}^{n} b_s^k(t,x)\, b_s^j(t,x),$$

is a local operator, i.e., the value of $L(v)$ at an arbitrary point (t_0, x_0) depends only on the value of $v(t,x)$ in an arbitrarily small neighborhood of this point.

In the general case when the random disturbance has jumps and the solution of the stochastic equation is discontinuous, the operator L is certainly not local and the value of $L(v)$ at (t_0, x_0) depends on the behavior of $v(t,x)$ in the entire space $x \in X$.

We introduce a definition. We will say that $v(t,x)$ is *positive definite in the sense of Liapunov* if $v(t,0)=0$ and

$$\inf_{t \geqq 0} v(t,x) \geqq g(x)$$

where $g(x) \geqq 0$ is continuous and $g(x)=0$ only if $x=0$. If $v(t,x)$ is twice continuously differentiable, positive definite in Liapunov's sense and

$$L(v) \leqq 0, \tag{33}$$

then we will call $v(t,x)$ a *generalized Liapunov function.*

Theorem 4. *If there exists a generalized Liapunov function $v(t,x)$ and $|\nabla v(t,x)| \leqq c(1+|x|)$, then the solution $\xi \equiv 0$ of* (32) *is uniformly stochastically stable. If, however, there exists a $k>0$ such that*

$$L(v) \leqq -k\, v, \tag{34}$$

then for arbitrary $\delta > 0$ there is a constant δ_1, such that for all $|\xi(0)| < \delta_1$, $\xi(t) \to 0$ for $t \to \infty$ with probability greater than $1-\delta$.

Proof. Set $\eta(t) = v(t, \xi(t))$. By the generalized Itô formula (§ 6, (19))

$$\begin{aligned} d\eta &= L(v)\, dt + (\nabla v | b(t, \xi(t))\, dw(t)) \\ &\quad + \int \left[v(t, \xi(t) + f(t, \xi(t), u)) - v(t, \xi(t)) \right] \tilde{\nu}(dt, du). \end{aligned} \tag{35}$$

Since $\eta(t) \geqq 0$,

$$\zeta(t) \geqq -\eta_0,$$

where

$$\begin{aligned} \zeta(t) = \int_0^t \{ &(\nabla v | b(\tau, \xi(\tau))\, dw(\tau)) \\ &+ \int \left[v(\tau, \xi(\tau) + f(\tau, \xi(\tau), u)) - v(\tau, \xi(\tau)) \right] \tilde{\nu}(d\tau, du) \}. \end{aligned}$$

Because of the assumption, $\nabla v(t,x)$ grows no faster than $|x|$. Hence, the coefficients in front of dw and $\tilde{\nu}(d\tau, du)$ possess finite moments.

Therefore,

$$\mathbb{M}\{\zeta(t)-\zeta(s)/\mathfrak{F}_s\}=0, \qquad t>s.$$

Taking into account Lemma 5 § 16, Part I, we find

$$\mathbb{P}\left\{\sup_{0\leqq t<\infty}\zeta(t)>\frac{\varepsilon}{2}\right\}\leqq\frac{2\eta_0}{\varepsilon}$$

and, consequently, $\sup\{\eta(t),\ 0\leqq t<\infty\}\leqq\eta_0+\frac{\varepsilon}{2}$ with probability greater than $1-\frac{2\eta_0}{\varepsilon}$. Since $v(t,x)$ is positive definite in Liapunov's sense, $\eta(t)=v(t,\xi(t))\geqq g(\xi(t))\geqq\varepsilon$ if $|\xi(t)|\geqq\varepsilon_1(\varepsilon)$.

Hence, $\sup\{\xi(t),\ 0\leqq t<\infty\}<\varepsilon_1$ with probability larger than $1-\delta$ provided that $|\xi(0)|<\delta_1$, whereby ε_1 and δ can be made arbitrarily small by suitable choice of ε and δ_1. The first part of the proof is finished.

Now assume that (34) holds. Then for the previously chosen constants it follows from (35) that, with probability at least $1-\delta$, we have

$$\eta(t)\leqq\eta_0+\frac{\varepsilon}{2}-k\int_0^t\eta(\tau)\,d\tau, \qquad \text{if } |\xi(0)|<\delta_1$$

for all $t>0$. Hence,

$$\eta(t)\leqq\left(\eta_0+\frac{\varepsilon}{2}\right)e^{-kt}.$$

Hence, it follows that $\xi(t)\to 0$ for $t\to\infty$ with probability greater than $1-\delta$ for all $\xi(0)$ for which $|(\xi(0))|<\delta_1$, i.e., in this case the solution $\xi\equiv 0$ is stochastically asymptotically stable. □

Remark 1. In the assumptions of the theorem it was assumed that $\nabla v(t,x)$ grows no faster than $|x|$ for $|x|\to\infty$. This was necessary to guarantee the existence of the first order moment of the stochastic integrals appearing in Itô's formula. If the solution of the stochastic differential equation (32) possesses moments of $2m$-th order, then it is sufficient to assume that $\nabla v(t,x)$ grows no faster than $|x|^{2m-1}$ for $x\to\infty$ (uniformly w.r.t. t). If the discontinuous part of the random field is missing $(f(t,x,u)\equiv 0)$, then it is sufficient that $\nabla v(t,x)$ grow no faster than $|x|$ at infinity.

Remark 2. If the hypotheses of the last theorem hold, then each trajectory which is a solution of a stochastic equation with initial conditions near zero lies (with probability arbitrarily close to unity) in an arbitrarily small neighborhood of zero. Hence, we need only require that (33) hold in some neighborhood of zero. If the discontinuous part of the equation is not present, then it is sufficient that $v(t,x)$ be defined for all $t\geqq 0$ in some neighborhood of zero.

In the theory of ordinary differential equations, the theorem corresponding to Theorem 4 is employed in two ways. On the one hand, for a given differential equation (or a class of differential equations), a Liapunov function is to be found or its existence demonstrated; on the other hand, one can give a function $v(t, x)$, positive definite in Liapunov's sense, and try to find a differential equation (or class of them) for which it is the Liapunov function. In this way one can establish certain stability characteristics.

We will use these two approaches to find conditions for uniform stochastic stability taking $|x|^2$ as the function $v(t, x)$. We then find

Corollary 1. *If for all* $t \geqq 0$ *and* $x \in X$

$$2x^* a(t, x) + \operatorname{sp}(b^*(t, x)\, b(t, x)) + \int |f(t, x, u)|^2\, \Pi(du) \leqq 0,$$

then the solution $\xi \equiv 0$ *of* (32) *is stable.*

We now show that if the second order moments of the solution of a stochastic equation are asymptotically stable and tend to zero sufficiently fast for $t \to \infty$ (exponential stability), then the solution $\xi \equiv 0$ is uniformly stochastically stable.

Consider the equation

$$\begin{gathered} d\xi = a(t, \xi)\, dt + \beta(t, \xi, dt), \\ a(t, 0) \equiv 0, \qquad \beta(t, 0, h) \equiv 0, \end{gathered} \tag{36}$$

where the field $\beta(t, x, h)$ is martingale. Assume that (36) fulfills the conditions for uniqueness and existence. Then it follows that for some constant C

$$|a(t, x)| \leqq C|x|. \tag{37}$$

Let

$$\mathbb{M}\{|\beta(t, x, h)|^2 / \mathfrak{F}_t\} \leqq C|x|^2 h. \tag{38}$$

Note that if

$$\begin{gathered} \beta(t, x, h) = b(t, x)[w(t+h) - w(t)] + \int f(t, x, u)\, \tilde{\nu}(dt, du), \\ b(t, 0) = f(t, 0, u) \equiv 0, \end{gathered}$$

Eq. (36) satisfies the existence and uniqueness conditions and $\beta(t, x, h)$ is a field with independent increments, then (38) is automatically fulfilled.

Theorem 5. *If* (36) *satisfies* (37) *and* (38) *and there exist positive numbers* t_0, N, q *and* δ_0 *such that for all* x, $|x| < \delta_0$ *and* t_1, t_2, $t_2 > t_1 > t_0$

$$\mathbb{M}|\xi_{xt_1}(t_2)|^2 \leqq N e^{-q(t_2 - t_1)}(1 + |x|^2), \tag{39}$$

is satisfied, then the solution $\xi(t) \equiv 0$ *of* (36) *is uniformly stochastically stable.*

Proof. Let $A_\varepsilon(T)$ be the event

$$\sup_{0\leqq t\leqq T}|\xi(t)|>\varepsilon.$$

In order to estimate the probability of $A_\varepsilon(T)$ we set $\eta(t)=\sup_{0\leqq\tau\leqq t}|\xi(\tau)|^2$ and use the inequality

$$\eta(t)\leqq 3\,|\xi(0)|^2+3\,C\int_0^t\eta(\tau)\,d\tau+3\sup_{0<t'\leqq t}\left|\int_0^{t'}\beta(\tau,\xi(\tau),d\tau)\right|^2. \tag{40}$$

Taking expectations on both sides of (40) and using (5) § 2 (in which we set $F(t)\equiv 0$), we have

$$\mathbb{M}\{\eta(t)/\mathfrak{F}_0\}\leqq 3\,|\xi(0)|^2+K\int_0^t\mathbb{M}\{\eta(\tau)/\mathfrak{F}_0\}\,d\tau.$$

This implies (see Lemma 1 § 1, Part I) that for some constants C_1 and C_2

$$\mathbb{M}\{\eta(t)/\mathfrak{F}_0\}\leqq C_1\,|\xi(0)|^2\,e^{C_2 t}.$$

Hence,

$$\mathbb{P}\{A_\varepsilon(2t_0)/\mathfrak{F}_0\}\leqq\frac{\mathbb{M}\{\eta(t)/\mathfrak{F}_0\}}{\varepsilon^2}<\delta,$$

provided that

$$|\xi(0)|^2<\frac{\delta\varepsilon^2}{C_1\,e^{2C_2t_0}}, \tag{41}$$

where δ is an arbitrary positive number.

Now consider the event $A_\varepsilon(2t_0,t)$: $\sup_{2t_0<\tau<T}|\xi(\tau)|>\varepsilon$. Since

$$|\xi(t)|\leqq|\xi(2t_0)|+\left|\int_{2t_0}^t a(\tau,\xi(\tau))\,d\tau\right|+\left|\int_{2t_0}^t\beta(\tau,\xi(\tau),d\tau)\right|,\qquad t>2t_0,$$

we have

$$\begin{aligned}&\mathbb{P}\{A_{3\varepsilon}(2t_0,T)/\mathfrak{F}_0\}\\ &\quad\leqq\mathbb{P}\{|\xi(2t_0)|>\varepsilon/\mathfrak{F}_0\}+\mathbb{P}\left\{\sup_{2t_0\leqq t\leqq T}\left|\int_{2t_0}^t a(\tau,\xi(\tau))\,d\tau\right|>\varepsilon/\mathfrak{F}_0\right\}\\ &\quad+\mathbb{P}\left\{\sup_{2t_0\leqq t\leqq T}\left|\int_{2t_0}^t\beta(\tau,\xi(\tau))\right|>\varepsilon/\mathfrak{F}_0\right\}.\end{aligned} \tag{42}$$

If the complementary event $\overline{A_\varepsilon(2t_0)}$ occurs, then $|\xi(2t_0)|<\varepsilon$ so that

$$\mathbb{P}\{|\xi(2t_0)|>\varepsilon/\mathfrak{F}_0\}<\delta,$$

if (41) holds. Moreover,

$$\mathbb{P}\left\{\sup_{2t_0\leqq t\leqq T}\left|\int_{2t_0}^{t} a(\tau,\xi(\tau))\,d\tau\right|>\varepsilon/\mathfrak{F}_0\right\}$$
$$\leqq \mathbb{P}\left\{\int_{2t_0}^{T}|a(\tau,\xi(\tau))|\,d\tau>\varepsilon/\mathfrak{F}_0\right\}\leqq\frac{1}{\varepsilon}\int_{2t_0}^{T} C\,\mathbb{M}\{\xi(\tau)|/\mathfrak{F}_0\}\,d\tau.$$

From

$$\int_{2t_0}^{T}\mathbb{M}\{|\xi(\tau)|/\mathfrak{F}_0\}\,d\tau\leqq\mathbb{M}\left\{\int_{2t_0}^{T}[\mathbb{M}\{|\xi(\tau)|^2/\mathfrak{F}_{t_0}\}]^{\frac{1}{2}}\,d\tau/\mathfrak{F}_0\right\}$$
$$\leqq\mathbb{M}\left\{\int_{2t_0}^{T}[N\,e^{-q(\tau-t_0)}(1+|\xi(t_0)|^2)]^{\frac{1}{2}}\,d\tau/\mathfrak{F}_0\right\}$$
$$\leqq N^{\frac{1}{2}}\,\mathbb{M}\{(1+|\xi(t_0)|^2)^{\frac{1}{2}}/\mathfrak{F}_0\}\int_{2t_0}^{T}e^{-\frac{q}{2}(\tau-t_0)}\,d\tau$$
$$\leqq N^{\frac{1}{2}}(1+\varepsilon^2\delta)^{\frac{1}{2}}\cdot\frac{2}{q}\left(e^{-\frac{q}{2}t_0}-e^{-\frac{q}{2}(T-t_0)}\right)$$

it follows that for arbitrary positive δ ($\delta<1$) and $\varepsilon>0$ we can choose t_0 so large that

$$\mathbb{P}\left\{\sup_{2t_0<t\leqq T}\left|\int_{2t_0}^{t}a(\tau,\xi(\tau))\,d\tau\right|>\varepsilon/\mathfrak{F}_0\right\}<\delta,$$

provided (41) holds. A similar conclusion can be reached concerning the third term on the right side of (42). Indeed, using (28) § 1, we get

$$\mathbb{P}\left\{\sup_{2t_0\leqq t\leqq T}\left|\int_{2t_0}^{t}\beta(\tau,\xi(\tau),d\tau)\right|>\varepsilon/\mathfrak{F}_0\right\}$$
$$\leqq\mathbb{P}\left\{\sup_{2t_0\leqq t\leqq T}\left|\int_{2t_0}^{T}\beta(\tau,\xi(\tau),d\tau)\right|^2>\varepsilon^2/\mathfrak{F}_0\right\}$$
$$\leqq\frac{C}{\varepsilon^2}\int_{2t_0}^{T}\mathbb{M}\{|\xi(\tau)|^2/\mathfrak{F}_0\}\,d\tau\leqq\frac{C}{\varepsilon^2}\,\mathbb{M}\left\{\int_{2t_0}^{T}\mathbb{M}[|\xi(\tau)|^2/\mathfrak{F}_{t_0}]\,d\tau\right\}$$
$$\leqq\frac{CN}{\varepsilon^2}\int_{2t_0}^{T}e^{-q(\tau-t_0)}\,\mathbb{M}\{1+|\xi(t_0)|^2\}\,d\tau$$
$$\leqq\frac{NC(1+\varepsilon^2\delta)}{\varepsilon^2 q}\left(e^{-qt_0}-e^{-q(T-t_0)}\right)<\delta$$

for sufficiently large t_0.

Hence, given arbitrarily small positive ε and δ, one can determine sufficiently large t_0 and $\delta_0>0$ such that for any $T>2t_0$

$$\mathbb{P}\{A_{3\varepsilon}(T)/\mathfrak{F}_0\}\leqq\mathbb{P}\{A_\varepsilon(2t_0)/\mathfrak{F}_0\}+\mathbb{P}\{A_{3\varepsilon}(2t_0,T)/\mathfrak{F}_0\}<4\delta,$$

provided that

$$|\xi(0)|<\delta_0.$$

The theorem is proved.

Corollary 2. *If the second order moments of the solution of a linear stochastic equation with constant coefficients of the form* (36) *are asymptotically stable, then the solution* $\xi\equiv 0$ *of this equation is uniformly stochastically stable.*

§ 13. Boundedness of the Solutions of Stochastic Differential Equations

We treat solutions of stochastic equations which remain bounded in some sense during an unbounded time interval. Some results in this direction have already been mentioned because they were closely connected with the stability properties of the solutions and can be investigated by the same methods.

Theorem 1. *Assume there exists a nonnegative function* $g(t, x)$ *which is twice continuously differentiable in* x *and continuously differentiable in* t *for which*

$$\begin{aligned} Lg\equiv \frac{\partial g}{\partial t}+(\nabla g\,|\,a)+\frac{1}{2}\,\mathrm{sp}(\nabla^2 g\,b\,b^*) \\ +\int\left[g(t, x+f(t,\xi,u))-g(t,x)-(\nabla g\,|\,f(t,\xi,u))\right]\Pi(du)\leqq 0 \end{aligned} \tag{1}$$

and

$$\lim_{T,\,|x|\to\infty}\ \inf_{t\geqq T} g(t,x)=+\infty. \tag{2}$$

Then the solution of the stochastic equation without after-effect

$$d\xi=a(t,\xi)\,dt+b(t,\xi)\,dw+\int f(t,\xi,u)\,\tilde{\nu}(dt,du) \tag{3}$$

has, w.p.1 a bounded solution in time:

$$\mathbb{P}\{\sup_{0\leqq t<\infty}|\xi(t)|<\infty/\mathfrak{F}_0\}=1.$$

A boundedness property, which will be called *stochastic boundedness*, can be obtained if the requirement that (1) be fulfilled in the entire space is replaced by additional assumptions on the behavior of $g(t,x)$ at infinity.

Theorem 2. *Let* $g(t,x)$ *be a nonnegative function satisfying the smoothness conditions of Theorem 1 and let there exist constants* $c_i>0$ $(i=1,2)$ *for which*

$$Lg\leqq -c_1\,g+c_2. \tag{4}$$

Then

$$\sup_{0 \leqq t < \infty} \mathbb{P}\{g(t, \xi(t)) \geqq C/\mathfrak{F}_0\} < \frac{g(0, \xi(0)) + \frac{c_2}{c_1}}{C}. \tag{5}$$

Proof. Setting

$$\eta(t) = g(t, \xi(t))$$

and using the generalized Itô formula, we arrive at

$$\eta(t) \leqq \eta(s) + \int_s^t (-c_1 \eta(\tau) + c_2)\, d\tau + \alpha(t) - \alpha(s),$$

where

$$\alpha(t) = \alpha_g(t) = \int_0^t (\nabla g(\tau, \xi(\tau)) \,|\, b(\tau, \xi(\tau))\, dw(\tau))$$
$$+ \int_0^t \{g(\tau, \xi(\tau) + f(\tau, \xi(\tau), u)) - g(\tau, \xi(\tau))\}\, \tilde{v}(d\tau, du).$$

Hence,

$$\mathbb{M}\{\eta(t)/\mathfrak{F}_0\} \leqq g(0, \xi(0)) + \int_0^t [-c_1 \mathbb{M}\{\eta(\tau)/\mathfrak{F}_0\} + c_2]\, d\tau. \tag{6}$$

Set

$$m(t) = \mathbb{M}\{\eta(t)/\mathfrak{F}_0\}.$$

Then, (6) has the form

$$m(t) \leqq g(0, \xi(0)) + c_2 t - \int_0^t c_1 m(\tau)\, d\tau,$$

whence (see Lemma 1 § 6, Part I)

$$m(t) \leqq g(0, \xi(0))\, e^{-c_1 t} + \frac{c_2}{c_1}(1 - e^{-c_1 t}).$$

Using Chebychev's inequality

$$\mathbb{P}\{g(t, \xi(t)) \geqq C/\mathfrak{F}_0\} \leqq \frac{m(t)}{C},$$

we obtain the proof.

Let us dwell on the question of boundedness of the moments of the solutions of stochastic equations. Assume that for (1) the conditions (2), (3) and (6) § 7 are satisfied on an arbitrary finite interval $[0, T]$. Then the solution is defined for all $t \geqq 0$ and possesses finite moments up to and including $2m$-th order. We will find conditions guaranteeing the boundedness of these moments on an infinite time interval.

From (5) § 6

$$\mathbb{M}|\xi(t)|^{2m} = \mathbb{M}|\xi(0)|^{2m} + \int_0^t \mathbb{M}\,\varphi(\tau)\, d\tau,$$

where

$$\begin{aligned}\varphi(t) = {} & 2m|\xi(t)|^{2m-2}\big(a(t,\xi(t))\,|\,\xi(t)\big)\\ & + 2m(m-1)|\xi(t)|^{2m-4}|b^*(t,\xi(t))\,\xi(t)|^2\\ & + m|\xi(t)|^{2m-2}\,\mathrm{sp}\big[b(t,\xi(t))\,b^*(t,\xi(t))\big]\\ & + \int\big\{|\xi(t)+f(t,\xi(t),u)|^{2m} - |\xi(t)|^{2m}\\ & \qquad - 2m|\xi(t)|^{2m-2}\big(\xi(t)\,|\,f(t,\xi,u)\big)\big\}\,\Pi(du).\end{aligned}$$

Assume there exist nonnegative constants A_j $(j=0,1)$, B_j $(j=0,1,2)$ and F_j $(j=0,1,\dots,2m)$ independent of time and such that

$$\big(a(t,x)\,|\,x\big) \leqq -A_0|x|^2 + A_1|x|, \qquad A_0>0, \tag{7}$$

$$\mathrm{sp}\big[b(t,x)\,b^*(t,x)\big] \leqq B_0|x|^2 + B_1|x| + B_2, \tag{8}$$

$$\begin{aligned}\int\big\{|x+f(t,x,u)|^{2m} - |x|^{2m} - 2m|x|^{2m-2}\big(x\,|\,f(t,x,u)\big)\big\}\,\Pi(du)\\ \leqq \sum_{p=0}^{2m} F_{2m-p}|x|^p.\end{aligned} \tag{9}$$

Then

$$|b^*(t,x)\,x|^2 \leqq (B_0|x|^2 + B_1|x| + B_2)|x|^2$$

and

$$\varphi(t) \leqq (-2m A_0 + 2m(m-1)B_0 + mB_0 + F_0)|\xi(t)|^{2m} + \sum_{k=0}^{2m-1} C_k|\xi(t)|^k,$$

where the C_k are some constants.

Let

$$\mathbf{M}|\xi(t)|^{2m} = M(t).$$

From Hölder's inequality

$$\mathbf{M}|\xi(t)|^k \leqq [M(t)]^{\frac{k}{2m}},$$

so that

$$\mathbf{M}\,\varphi(t) \leqq -a_0 M(t) + \sum_{k=0}^{2m-1} C_k [M(t)]^{\frac{k}{2m}},$$

where

$$a_0 = 2m A_0 - m(2m-1)B_0 - F_0.$$

On the other hand, for any $p\in[0,1]$ and $c>0$

$$a^p \leqq p\,c\,a + (1-p)\frac{1}{c^{\frac{p}{1-p}}},$$

so that

$$[M(t)]^{\frac{k}{2m}} \leqq \frac{ck}{2m}M(t) + \frac{2m-k}{2m}\cdot\frac{1}{c^{\frac{k}{2m-k}}}.$$

We thus obtain

$$M(t) \leqq M(0) - \int_0^t \left[\left(a_0 - \sum_{k=0}^{2m-1} C_k \frac{ck}{2m} \right) M(\tau) - \sum_{k=0}^{2m-1} C_k \frac{2m-k}{2m} \cdot \frac{1}{c^{\frac{k}{2m-k}}} \right] d\tau .$$

Assume $a_0 > 0$. Choose C so small that

$$C \sum_{k=0}^{2m-1} C_k \frac{k}{2m} = \delta < a_0 .$$

If we set

$$\frac{1}{a_0 - \delta} \sum_{k=0}^{2m-1} C_k \frac{2m-k}{2m} \cdot \frac{1}{c^{\frac{k}{2m-k}}} = d, \qquad M(t) - d = M_1(t),$$

then

$$M_1(t) \leqq M_1(0) - (a_0 - \delta) \int_0^t M_1(\tau)\, d\tau ,$$

whence

$$M_1(t) \leqq M_1(0)\, e^{-(a_0-\delta)t},$$

i.e.,

$$M(t) < d, \qquad 0 \leqq t < \infty .$$

This proves

Theorem 3. *If the conditions of Theorem 1 §7 and (7)–(9) hold, and*

$$A_0 > \frac{2m-1}{2} B_0 + \frac{F_0}{2m}, \tag{10}$$

then the 2-th *order moments of the solutions of* (1) *remain bounded in an unbounded time interval.*

In particular, it is sufficient for the boundedness ($0 \leqq t < \infty$) of 2-nd order moments that

$$A_0 > \frac{B_0 + F_0}{2} .$$

We note that it does not in general follow from the boundedness of moments that the solutions are bounded w.p.1.

§ 14. Limit Theorems for Stochastic Differential Equations

Let Λ denote a numerical set and let $0 \in \Lambda$ be a limit point of Λ.

Definition 1. A family of random functions $\{\beta_\lambda(\theta), \theta \in \Theta\}$, $\lambda \in \Lambda$ will be said to *converge weakly* for $\lambda \to 0$ to the random function $\{\beta(\theta), \theta \in \Theta\}$ if for arbitrary s, $s = 1, 2, \ldots, n, \ldots$, and $\theta_1, \theta_2, \ldots, \theta_s$ ($\theta_j \in \Theta$, $j = 1, \ldots, s$) the joint distribution of $\{\beta_\lambda(\theta_1), \beta_\lambda(\theta_2), \ldots, \beta_\lambda(\theta_s)\}$ converges weakly for $\lambda \to 0$ to the joint distribution of the sequence $\{\beta_0(\theta_1), \beta_0(\theta_2), \ldots, \beta_0(\theta_s)\}$.

Suppose the family of random fields $\{\alpha_\lambda(t, x, h), t \in [0, T], h \geqq 0, x \in X\}$, $\lambda \in \Lambda$ converges weakly for $\lambda \to 0$ to $\alpha_0(t, x, h)$ and let $\xi^{(\lambda)}(t)$ $(0 \leqq t \leqq T)$ be solutions of

$$d\xi^{(\lambda)}(t) = \alpha_\lambda(t, \xi^{(\lambda)}(t), dt), \qquad \xi^{(\lambda)}(0) = x^{(\lambda)}. \tag{1}$$

Under what conditions do the processes $\xi^{(\lambda)}(t)$ converge weakly for $\lambda \to 0$ to $\xi^{(0)}(t)$?

The following lemma shows that in a number of cases the Euler approximations $\xi_\delta^{(\lambda)}(t)$ are useful in finding an answer to this question (see Definition 2 § 3).

Lemma 1. *Let the random fields* $\{\alpha_\lambda(t, x, h), \lambda \in \Lambda\}$ *satisfy* (1)–(6) § 3 *with constants L and C and function* $g(h)$ $(h \geqq 0)$ *independent of* λ. *If the random processes* $\xi_\delta^{(\lambda)}(t)$ $(t \in [0, T])$ *converge weakly to* $\xi_\delta^{(0)}(t)$ *for* $\lambda \to 0$ *for any* δ, *then the processes* $\xi^{(\lambda)}(t)$ *converge weakly to* $\xi^{(0)}(t)$ $(t \in [0, T])$ *for* $\lambda \to 0$.

Indeed, let $f(x_1, x_2, \ldots, x_s)$ be an arbitrary uniformly continuous and bounded function of the x_j $(j = 1, 2, \ldots, s)$. We have

$$\begin{aligned} &|\mathbb{M} f(\xi^{(0)}(\tau_1), \ldots, \xi^{(0)}(\tau_s)) - \mathbb{M} f(\xi^{(\lambda)}(\tau_1), \ldots, \xi^{(\lambda)}(\tau_s))| \\ &\quad \leqq |\mathbb{M} f(\xi^{(0)}(\tau_1), \ldots, \xi^{(0)}(\tau_s)) - \mathbb{M} f(\xi_\delta^{(0)}(\tau_1), \ldots, \xi_\delta^{(0)}(\tau_s))| \\ &\qquad + |\mathbb{M} f(\xi_\delta^{(0)}(\tau_1), \ldots, \xi_\delta^{(0)}(\tau_s)) - \mathbb{M} f(\xi_\delta^{(\lambda)}(\tau_1), \ldots, \xi_\delta^{(\lambda)}(\tau_s))| \\ &\qquad + |\mathbb{M} f(\xi_\delta^{(\lambda)}(\tau_1), \ldots, \xi_\delta^{(\lambda)}(\tau_s)) - \mathbb{M} f(\xi^{(\lambda)}(\tau_1), \ldots, \xi^{(\lambda)}(\tau_s))| \end{aligned} \tag{2}$$

and

$$\begin{aligned} &|\mathbb{M} f(\xi_\delta^{(\lambda)}(\tau_1), \ldots, \xi_\delta^{(\lambda)}(\tau_s)) - \mathbb{M} f(\xi^{(\lambda)}(\tau_1), \ldots, \xi^{(\lambda)}(\tau_s))| \\ &\quad \leqq 2N \sum_{k=1}^{s} \mathbb{P}\left\{|\xi_\delta^{(\lambda)}(\tau_k) - \xi^{(\lambda)}(\tau_k)| > \frac{h}{\sqrt{s}}\right\} + \omega_f(h), \end{aligned} \tag{3}$$

where $\omega_f(h)$ is the modulus of continuity of f on the sphere X^s of radius h and N is an upper bound of f.

From Chebychev's inequality and Lemma 2 § 3 it follows that the quantity

$$\mathbb{P}\{|\xi_\delta^{(\lambda)}(t) - \xi^{(\lambda)}(t)| > h\}$$

for arbitrary $h > 0$, $t \in [0, T]$ can be made as small as we like uniformly in $\lambda \in \Lambda$ by suitable choice δ $(\delta = \delta_0)$. Hence, the first and third terms on the right side of (2) for $\delta < \delta_0$ become arbitrarily small for all $\lambda \in \Lambda$. According to the choice of δ, the second term tends for $\lambda \to 0$ to zero because of the assumptions of the lemma. □

Now consider the family of random fields $\alpha_\lambda(t, x, h)$ of the special type

$$\alpha_\lambda(t, x, h) = a_\lambda(t, x)\, h + b_\lambda(t, x)\, [\alpha_\lambda(t+h) - \alpha_\lambda(t)], \qquad \lambda \in \Lambda,$$

where $a_\lambda(t, x)$ and $b_\lambda(t, x)$ are non-stochastic, the $a_\lambda(t, x)$ take values in X, $b_\lambda(t, x)$ is an operator function mapping some space X' into X and $a_\lambda(t)$ is a process with values in X'. With the aid of the preceding lemma we can prove the following theorem which claims that the solution of

$$d\xi^{(\lambda)}(t) = a_\lambda(t, \xi^{(\lambda)}(t))\, dt + b_\lambda(t, \xi^{(\lambda)}(t))\, d\alpha_\lambda(t), \qquad \xi^{(\lambda)}(0) = x^{(\lambda)}, \tag{4}$$

is under certain conditions, weakly continuous in λ.

Theorem 1. *Assume that the following conditions are satisfied:* a) *there exists a monotone nondecreasing sequence of σ-algebras $\mathfrak{F}_t$ $(0 \leqq t \leqq T)$ such that*

$$|\mathbb{M}\{\alpha_\lambda(t+h) - \alpha_\lambda(t)/\mathfrak{F}_t\}| \leqq Ch,$$

$$\mathbb{M}\{|\alpha_\lambda(t+h) - \alpha_\lambda(t)|^2/\mathfrak{F}_t\} \leqq Ch;$$

b) $|a_\lambda(t, 0)| + |b_\lambda(t, 0)| \leqq C$;

c) $|a_\lambda(t, x) - a_\lambda(t, y)| + |b_\lambda(t, x) - b_\lambda(t, y)| \leqq C|x - y|$;

d) $|a_\lambda(t, x) - a_\lambda(t+h, x)| \leqq g_\lambda(h)(1 + |x|)$

and $g_\lambda(h) \to 0$ *for* $h \to 0$;

e) $|b_\lambda(t, x) - b_\lambda(t+h, x)| \leqq g(h)(1 + |x|)$

and $g(h) \to 0$ *for* $h \to 0$, $g(h)$ *is independent of* λ;

f) $\lim_{\lambda \to 0} \int_{t_1}^{t_2} a_\lambda(t, x)\, dt = \int_{t_1}^{t_2} a_0(t, x)\, dt \quad (0 \leqq t_1 < t_2 \leqq T)$; *and*

g) *the family*

$$\{x^{(\lambda)}, \alpha_\lambda(t), 0 \leqq t \leqq T\} \qquad \lambda \in \Lambda,$$

converges weakly for $\lambda \to 0$ *to* $\{x^0, \alpha_0(t), 0 \leqq t \leqq T\}$.

Then the processes $\xi^{(\lambda)}(t)$ *converge weakly for* $\lambda \to 0$ *to* $\xi^{(0)}(t)$.

Proof. Besides (4) we consider the equation

$$d\eta^{(\lambda)} = \beta_\lambda(t, \eta^{(\lambda)}, dt), \qquad \eta^{(\lambda)}(0) = x^{(\lambda)}, \tag{5}$$

where

$$\beta_\lambda(t, x, h) = \int_t^{t+h} a_\lambda(\tau, x)\, d\tau + b_\lambda(t, x)\, [\alpha_\lambda(t+h) - \alpha_\lambda(t)]. \tag{6}$$

It is easy to verify that the fields $\beta_\lambda(t, x, h)$ and $\alpha_\lambda(t, x, h)$ satisfy the existence and uniqueness conditions of Theorem 1 § 3. From assumptions d) and e), the solutions of (5) and (4) are stochastically equivalent (Remark 2 to Theorem 1 § 3). From a)–c) and e) it follows that the fields $\beta_\lambda(t, x, h)$ satisfy (1)–(6) § 3 with the constant C, the functions $g(h)$ and

$L=C$ independent of λ. In order to prove the theorem it remains to show that $\eta_\delta^{(\lambda)}(t)$ converges weakly to $\eta_\delta^{(0)}(t)$ for arbitrary δ, i.e., that

$$\lim_{\lambda\to 0} \mathbb{M} f(\eta_\delta^{(\lambda)}(\tau_1), \eta_\delta^{(\lambda)}(\tau_2), \ldots, \eta_\delta^{(\lambda)}(\tau_s)) = \mathbb{M} f(\eta_\delta^{(0)}(\tau_1), \eta_\delta^{(0)}(\tau_2), \ldots, \eta_\delta^{(0)}(\tau_s))$$

for an arbitrary uniformly continuous, bounded function $f(x_1, x_2, \ldots, x_s)$ in X^s, arbitrary integers s $(s=1,2,\ldots)$ and partitions $\delta=\{t_1, t_2, \ldots, t_n\}$ of $[0, T]$.

To this end consider $f(\eta_\delta^{(\lambda)}(\tau_1), \eta_\delta^{(\lambda)}(\tau_2), \ldots, \eta_\delta^{(\lambda)}(\tau_s))$ as a function of the initial value $x=x^{(\lambda)}$ and of the differences

$$y_k = \alpha_\lambda(t_{k+1}) - \alpha_\lambda(t_k) \qquad (k=0, 1, \ldots, n-1).$$

For the sake of convenience (this is not actually essential) we assume that the points $\tau_1, \tau_2, \ldots, \tau_s$ coincide with certain of the points δ. Let $\eta_\delta^{(\lambda)}(t_k)$ and $\tilde{\eta}_\delta^{(\lambda)}(t_k)$ denote Euler approximations determined by the quantities $x, y_1, y_2, \ldots$ and $\tilde{x}_1, \tilde{y}_1, \ldots$, resp. There exists a function $\omega(h)\geqq 0$ with $\lim_{h\to 0} \omega(h)=0$ such that

$$\begin{aligned}&|f(\eta_\delta^{(\lambda)}(\tau_1), \ldots, \eta_\delta^{(\lambda)}(\tau_s)) - f(\tilde{\eta}_\delta^{(\lambda)}(\tau_1), \ldots, \tilde{\eta}_\delta^{(\lambda)}(\tau_s))| \\ &\qquad \leqq \omega(|\eta_\delta^{(\lambda)}(\tau_1) - \tilde{\eta}_\delta^{(\lambda)}(\tau_2)| + \cdots + |\eta_\delta^{(\lambda)}(\tau_s) - \tilde{\eta}_\delta^{(\lambda)}(\tau_s)|).\end{aligned}$$

Assume that $x, y_1, \ldots, y_n$ and $\tilde{x}, \tilde{y}_1, \ldots, \tilde{y}_n$ take values in spheres of radius N. By means of induction and the assumptions made on the functions $a_\lambda(t, x)$ and $b_\lambda(t, x)$ it is easy to show that

$$|\eta_\delta^{(\lambda)}(t_k) - \tilde{\eta}_\delta^{(\lambda)}(t_k)| \leqq C'_N \left(|x-\tilde{x}| + \sum_{K=1}^{n} |y_k - \tilde{y}_k|\right),$$

where C'_N depends only on N, n and the constant C (in conditions 1–6), but not on λ. Hence, the family of functions

$$F_\lambda(x, y_1, \ldots, y_n) = f(\eta_\delta^{(\lambda)}(\tau_1), \ldots, \eta_\delta^{(\lambda)}(\tau_s)) \qquad (\lambda\in\Lambda)$$

is uniformly bounded on $|x|\leqq N$, $|y_k|<N$, is uniformly continuous and

$$F_\lambda(x, y_1, \ldots, y_n) \to F_0(x, y_1, \ldots, y_n) \qquad \text{for } \lambda\to 0.$$

The proof of the theorem is completed by refering to the following

Lemma 2. *If Q_λ $(\lambda\in\Lambda)$ is a family of probability measures in a separable metric space X converging weakly to Q_0 for $\lambda\to 0$, then*

$$\int f(x)\, Q_\lambda(dx) \to \int f(x)\, Q_0(dx)$$

uniformly w.r.t. $f\in S$, where S is an arbitrary set of continuous functions on X which satisfies: a) *functions in S are uniformly bounded;* b) *on each compact $K\subset X$ the set S is compact in the uniform topology.*

Proof. Since the Q_λ converge weakly to Q_0, for $\varepsilon>0$ there exists a compact K_ε (see [27]) for which $Q_\lambda(X\setminus K_\varepsilon)<\varepsilon$ for all λ. Let $f_1,\dots,f_N$ be an ε'-net for the restriction of functions of S to K_ε, $\varepsilon'>0$ and

$$L=\sup_{f\in S}\sup_{x\in X}|f(x)|.$$

Then

$$\begin{aligned}\sup_{f\in S}&\left|\int f(x)\,Q_\lambda(dx)-\int f(x)\,Q_0(dx)\right|\\ &\leqq 2L\varepsilon+\sup_{f\in S}\left|\int_{K_\varepsilon} f(x)\,Q_\lambda(dx)-\int_{K_\varepsilon} f(x)\,Q_0(dx)\right|\\ &\leqq 2L\varepsilon+2\varepsilon'+\max_j\left|\int_{K_\varepsilon} f_j(x)\,Q_\lambda(dx)-\int_{K_\varepsilon} f_j(x)\,Q_0(dx)\right|.\end{aligned}$$

This inequality proves the lemma.

We now turn to a more general stochastic equation. Let

$$\alpha_\lambda(t,x,h)=\int_t^{t+h} a_\lambda(\tau,x)\,d\tau+b_\lambda(t,x)\,\alpha_\lambda(t,h)+\beta_\lambda(t,x,h) \tag{7}$$

and

$$d\xi^{(\lambda)}(t)=\alpha_\lambda(t,\xi^{(\lambda)}(t),dt),\qquad \xi^{(\lambda)}(0)=x^{(\lambda)}. \tag{8}$$

With respect to $a_\lambda(t,x)$ and $b_\lambda(t,x)$ we retain all the assumptions of Theorem 1. Assume that $\alpha_\lambda(t,h)$ satisfies the existence and uniqueness theorem. The field $\beta(t,x,h)$ will satisfy conditions to be given below which will essentially imply that β converges to zero. Corresponding to what was said in the proof of Theorem 1 we can consider

$$f(\xi_\delta^{(\lambda)}(\tau_1),\xi_\delta^{(\lambda)}(\tau_2),\dots,\xi_\delta^{(\lambda)}(\tau_s))$$

as a function of the form

$$F_\lambda(x^{(\lambda)},\alpha_\lambda(0,h_1),\alpha_\lambda(t_1,h_2),\dots,\alpha_\lambda(t_{n-1},h_n),\gamma_1^{(\lambda)},\dots,\gamma_n^{(\lambda)}),$$

where $t_1,t_2,\dots,t_n=T$ are the points comprising the partition δ, $\xi_\delta^{(\lambda)}(t)$ is the Euler approximation of the solution of (7)–(8) corresponding to the partition δ, $h_k=t_k-t_{k-1}$ and $\gamma_k^{(\lambda)}=\beta_\lambda(t_{k-1},\xi_\delta^{(\lambda)}(t_{k-1}),h_k)$. It is possible to apply the same considerations as in Theorem 1 to these functions provided that the joint distribution of the random variables

$$x^{(\lambda)},\ \alpha_\lambda(0,h_1),\ \alpha_\lambda(t_1,h_2),\dots,\alpha_\lambda(t_{n-1},h_n),\quad \gamma_1^{(\lambda)},\dots,\gamma_n^{(\lambda)}$$

converges weakly to that of

$$x^{(0)},\ \alpha_0(0,h_1),\ \alpha_0(t_1,h_2),\dots,\alpha_0(t_{n-1},h_n),\quad 0,\dots,0.$$

Theorem 2. *Assume that $a_\lambda(t,x)$ and $b_\lambda(t,x)$ satisfy the assumptions of Theorem 1, and the random fields $\alpha_\lambda(t,h)$ and $\beta_\lambda(t,x,h)$ the conditions* (1–6) § 3 *with constants L and C and function $g(h)$ independent of λ. Furthermore,* *assume that for some* $q>0$

$$\mathbb{M}\{|\beta_\lambda(t,x,h)|^q/\mathfrak{F}_t\} \leqq L_\lambda(1+|x|^2),$$

where $L_\lambda \to 0$ *for* $\lambda \to 0$. *Then the solution* $\xi^{(\lambda)}(t)$ *of* (7)–(8) *converges weakly for* $\lambda \to 0$ *to the solution of the stochastic equation*

$$d\xi = a_0(t,\xi)\,dt + b_0(t,\xi)\,\alpha(t,dt). \tag{9}$$

It suffices to show that

$$\mathbb{M}|\gamma_k^{(\lambda)}|^q \to 0 \quad \text{for } \lambda \to 0, \qquad k=1,\dots,n.$$

This follows from the relation

$$\begin{aligned}\mathbb{M}|\gamma_k^{(\lambda)}|^q &= \mathbb{M}\,\mathbb{M}\{|\beta_\lambda(t_{k-1},\xi_\delta^{(\lambda)}(t_{k-1}),h_k)|^q/\mathfrak{F}_{t_{k-1}}\}\\ &\leqq \mathbb{M}L_\lambda(1+|\xi_\delta^{(\lambda)}(t_{k-1})|^2)\end{aligned}$$

and the fact that $\mathbb{M}|\xi_\delta^{(\lambda)}(t)|^2$ is bounded uniformly w.r.t. λ, δ and t (Lemmata 1 and 2 § 2).

Theorem 3. *Assume that the random fields* $\alpha_\lambda(t,x,h)$ *satisfy* (1–6) § 3 *with constants L, C and function* $g(h)$ *independent of* λ *and that*

$$\alpha_\lambda(t,x,h) = a_\lambda(t,x)\,h + \beta_\lambda(t,x,h), \tag{10}$$

whereby $a_\lambda(t,x)$ *fulfills the hypotheses of Theorem 1,* $\beta_\lambda(t,x,h)$ *is a martingale and* $\beta_\lambda(t,x,h)$ *for fixed* (t,x,h) *converges weakly for* $\lambda \to 0$ *to* $\beta_0(t,x,h)$. *Then the solution of*

$$d\xi^{(\lambda)} = \alpha_\lambda(t,\xi^{(\lambda)},dt), \qquad \xi^{(\lambda)}(0) = x^{(\lambda)},$$

converges weakly for $\lambda \to 0$ *to that of*

$$d\xi^{(0)} = \alpha_0(t,\xi^{(0)},dt), \qquad \xi^{(0)}(0) = x^{(0)},$$

where x^0 *is the weak limit of* $x^{(\lambda)}$ *for* $\lambda \to 0$ *and* $\mathbb{M}|x^{(\lambda)}|^2 \leqq C$.

Proof. From the proof of Theorem 2 § 3 we have

$$\mathbb{M} \sup_{0 \leqq t \leqq T} |\xi^{(\lambda)}|^2 \leqq C,$$

where C is independent of λ. It is thus sufficient to prove the theorem for fields satisfying the assumptions of the theorem and vanishing outside some sphere $|x| \leqq R$. Let ε be an arbitrary positive number. Construct in the sphere $|x| \leqq R$ an ε-net $x_1,\dots,x_n$ and a system of functions $g_j(x)$ satisfying

a) $g_j(x) \geqq 0$ *and* $g_j(x) \equiv 0$ *outside an* ε*-neighborhood of the point* x_j;

b) $\sum_{j=1}^{N} g_j(x) = 1$; c) $g_j(x)$ *is continuously differentiable.*

We set

$$\gamma_\lambda(t,x,h)=\alpha_\lambda(t,x,h)-\sum_{j=1}^{N} g_j(x)\,\alpha_\lambda(t,x_j,h).$$

Then

$$|\mathbb{M}\{\gamma_\lambda(t,x,h)/\mathfrak{F}_t\}|\leqq \max_j \sup_{|x-x_j|\leqq\varepsilon} |\mathbb{M}\{\alpha_\lambda(t,x,h)-\alpha_\lambda(t,x_j,h)/\mathfrak{F}_t\}|\leqq C\varepsilon h,$$

where C is independent of λ. Analogously,

$$\mathbb{M}\{|\gamma_\lambda(t,x,h)|^2/\mathfrak{F}_t\}\leqq C\varepsilon h.$$

Let

$$\alpha_\lambda^{(\varepsilon)}(t,x,h)=\sum_{j=1}^{N} g_j(x)\,\alpha_\lambda(t,x_j,h).$$

It's easy to verify that the fields $\alpha_\lambda^{(\varepsilon)}(t,x,h)$ satisfy the conditions of Theorem 2 with constants independent of λ and ε and thus, for these fields, Theorem 3 is valid.

On the other hand, if $\xi^{(\lambda,\varepsilon)}(t)$ satisfies

$$d\xi=\alpha_\lambda^{(\varepsilon)}(t,\xi,dt),\qquad \xi(0)=x^{(\lambda)},$$

then

$$\begin{aligned}\mathbb{M}|\xi^{(\lambda)}(t)-\xi^{(\lambda,\varepsilon)}(t)|^2 &\leqq 4\mathbb{M}\left|\int_0^t \alpha_\lambda(t,\xi^{(\lambda)},dt)-\int_0^t \alpha_\lambda(t,\xi^{(\lambda,\varepsilon)},dt)\right|^2\\ &\quad+4\mathbb{M}\left|\int_0^t \alpha_\lambda(t,\xi^{(\lambda,\varepsilon)},dt)-\int_0^t \alpha_\lambda^{(\varepsilon)}(t,\xi^{(\lambda,\varepsilon)},dt)\right|^2\\ &\leqq C_1\int_0^t \mathbb{M}|\xi^{(\lambda)}(t)-\xi^{(\lambda,\varepsilon)}|^2\,dt+\varepsilon C_1\big(1+\sup_{0\leqq t\leqq T}\mathbb{M}|\xi^{(\lambda,\varepsilon)}(t)|^2\big),\end{aligned}$$

where C_1 is some constant independent of λ and ε. To derive this inequality we use Lemma 8 § 1 and Corollary 1 to Lemma 7 § 1.

It follows from Lemma 9 § 1 that

$$\sup_{0\leqq t\leqq T}\mathbb{M}|\xi^{(\lambda,\varepsilon)}(t)|^2\leqq C_2,$$

where C_2 is also independent of λ and ε. Therefore,

$$\mathbb{M}|\xi^{(\lambda)}(t)-\xi^{(\lambda,\varepsilon)}(t)|^2\leqq\varepsilon C_3+C_1\int_0^t \mathbb{M}|\xi^{(\lambda)}(\tau)-\xi^{(\lambda,\varepsilon)}(\tau)|^2\,d\tau.$$

Lemma 1 § 6, Part I implies that

$$\sup_{0\leqq t\leqq T}\mathbb{M}|\xi^{(\lambda)}(t)-\xi^{(\lambda,\varepsilon)}(t)|^2\leqq\varepsilon C_4.$$

Since the theorem holds for the processes $\zeta^{(\lambda,\varepsilon)}(t)$, the previous unequality implies the correctness of the theorem for the processes $\zeta^{(\lambda)}(t)$ as well.

Remark 1. The assertion of Theorem 3 also holds for arbitrary fields $\alpha_\lambda(t,x,h)$ which satisfy (1–6) with constants L, C and $g(h)$ independent of λ and which converge weakly for $\lambda \to 0$ to the field $\alpha_0(t,x,h)$ provided that

$$\mathbb{P}\{\sup_{0 \leqq t \leqq T} |\xi_\lambda(t)| > N\} \to 0 \qquad \text{for } N \to \infty \tag{11}$$

uniformly w.r.t. λ.

Actually, the assumption that $\alpha_\lambda(t,x,h)$ has the form (10), where $\beta(t,x,h)$ is a martingale, was only used to assure the validity of (11).

As an example of the application of the theorems we have proved we consider the "averaging principle" for stochastic equations. For ordinary differential equations of, the averaging principle was stated and justified in its general form by N. N. Bogolubov [6]. We present it now. Let a system of ordinary differential equations with small parameter ε be given:

$$\frac{dx}{dt} = \varepsilon\, a(t,x), \qquad x(0) = x_0. \tag{12}$$

Assume that the time average $a_0(x)$ of $a(t,x)$ exists:

$$\lim_{T\to\infty} \frac{1}{T} \int_c^{c+T} a(t,x)\,dt = a_0(x). \tag{13}$$

Then we claim that on a time interval of order $1/\varepsilon$, the solution of (12) differs "negligibly" from the solution of

$$\frac{dy}{dt} = \varepsilon\, a_0(y), \qquad y(0) = x_0, \tag{14}$$

or, more precisely, for arbitrary $L > 0$ and $\delta > 0$ there exists an $\varepsilon_0 = \varepsilon_0(\delta, L)$ such that

$$\max_{0 \leqq t \leqq \frac{L}{\varepsilon}} |x(t) - y(t)| < \delta$$

for all $\varepsilon < \varepsilon_0$. A detailed exposition of the averaging principle for ordinary differential equations and its applications can be found in the monograph by N. N. Bogolubov and Ju. S. Mitropol'skii [8].

For stochastic differential equations the principle has been investigated by R. L. Stratonovitch [55], R. Z. Has'minskii [32, 35] and I. I. Gihman [22, 25].

Theorem 4. *Let $\xi_\varepsilon(t)$ satisfy*

$$d\xi_\varepsilon=\varepsilon\,\alpha(t,\xi_\varepsilon,dt),\qquad \xi_\varepsilon(0)=x_0, \tag{15}$$

where $\alpha(t,x,h)=a(t,x)h+\beta(t,x,h)$, $\beta(t,x,h)$ is a martingale, $a(t,x)$ and $\beta(t,x,h)$ satisfy the existence and uniqueness conditions and

$$\lim_{T\to\infty}\frac{1}{T}\int_c^{c+T} a(t,x)\,dt=a_0(x).$$

Then $\xi_\varepsilon\left(\frac{t}{\varepsilon}\right)$ converges weakly for $\varepsilon\to 0$ to $x(t)$, where $x(t)$ satisfies the ordinary differential equation

$$\frac{dx}{dt}=a_0(x),\qquad x(0)=x_0.$$

To prove this, we note that the process $\zeta_\varepsilon(t)=\xi_\varepsilon\left(\frac{t}{\varepsilon}\right)$ satisfies

$$d\zeta_\varepsilon=\alpha_\varepsilon(t,\zeta_\varepsilon,dt),\qquad \zeta_\varepsilon(0)=x_0,$$

where $\alpha_\varepsilon(t,x,h)=\alpha_\varepsilon(0,x,t+h)-\alpha_\varepsilon(x,t)$ and

$$\alpha_\varepsilon(0,x,t)=\varepsilon\int_0^{t/\varepsilon}\alpha(\tau,x,d\tau)=\varepsilon\int_0^{t/\varepsilon}a(\tau,x)\,d\tau+\int_0^{t/\varepsilon}\beta(\tau,x,d\tau).$$

In this regard

$$\varepsilon\int_0^{t/\varepsilon}a(\tau,x)\,d\tau=t\,\frac{\varepsilon}{t}\int_0^{t/\varepsilon}a(\tau,x)\,d\tau\to t\,a_0(x)$$

and

$$\mathbb{M}\left|\varepsilon\int_0^{t/\varepsilon}\beta(\tau,x,d\tau)\right|^2\leqq\varepsilon^2\int_0^{t/\varepsilon}C(1+|x|^2)\,d\tau=\varepsilon t\,C(1+|x|^2)\to 0$$

for $\varepsilon\to 0$. An application of Theorem 3 now completes the proof.

Let

$$\beta(t,x,h)=b(t,x)\left[w(t+h)-w(t)\right]+\int_t^{t+h}\int c(t,x,u)\,\tilde{\nu}(dt,du), \tag{16}$$

where $w(t)$ is a Wiener process, $\tilde{\nu}(t,A)=\nu(t,A)-\mathbb{M}\nu(t,A)$ and $\nu(t,A)$ is a Poisson measure. Under the hypotheses of Theorem 4 we treat the question of the fluctuation of the process $\xi_\varepsilon\left(\frac{t}{\varepsilon}\right)$ w.r.t. $x(t)$. Set

$$\eta_\varepsilon(t)=\frac{\xi_\varepsilon\left(\frac{t}{\varepsilon}\right)-x(t)}{\sqrt{\varepsilon}}.$$

The process $\eta_\varepsilon(t)$ satisfies

$$\eta_\varepsilon(t)=\frac{1}{\sqrt{\varepsilon}}\int_0^t\left[a\left(\frac{\tau}{\varepsilon}, x(\tau)+\sqrt{\varepsilon}\,\eta_\varepsilon(\tau)\right)-a_0(x(\tau))\right]d\tau$$
$$+\sqrt{\varepsilon}\int_0^{t/\varepsilon}\beta(\tau, x(\varepsilon\tau)+\sqrt{\varepsilon}\,\eta_\varepsilon(\varepsilon\tau), d\tau). \tag{17}$$

Theorem 5. *Let the conditions of Theorem 4 be fulfilled, the field $\beta(t,x,h)$ have the form* (16) *and*

a) $\displaystyle\frac{1}{\sqrt{\varepsilon}}\int_0^t\left[a\left(\frac{\tau}{\varepsilon}, x(\tau)\right)-a_0(x(\tau))\right]d\tau\to 0$ *for* $\varepsilon\to 0$;

b) $\displaystyle\int_0^t \nabla a\left(\frac{\tau}{\varepsilon}, x(\tau)\right)d\tau\to\int_0^t g(\tau)\,d\tau$;

c) $\nabla^2 a(t,x)$ *exists and* $|\nabla^2 a(t,x)-\nabla^2 a(t,y)|\leqq c\,|x-y|$;

d) $$\int_0^t\left[b\left(\frac{\tau}{\varepsilon}, x(\tau)\right)b^*\left(\frac{\tau}{\varepsilon}, x(\tau)\right)+\int c\left(\frac{\tau}{\varepsilon}, x(\tau), u\right)\right.$$
$$\left.\times c^*\left(\frac{\tau}{\varepsilon}, x(\tau), u\right)\Pi(du)\right]d\tau\to\int_0^t d(\tau)\,d^*(\tau)\,d\tau;$$

e) $\displaystyle\left|c\left(\frac{\tau}{\varepsilon}, x(\tau), u\right)\right|\leqq r(u)$ *and* $\int r^2(u)\,\Pi(du)<\infty$,

where $g(t)$ and $d(t)$ are continuous matrix functions. Then the process $\eta_\varepsilon(t)$ converges weakly for $\varepsilon\to 0$ to the solution of the linear stochastic equation

$$d\eta=g(t)\,\eta+d(t)\,dw,\qquad \eta(0)=0. \tag{18}$$

Proof. Set

$$A_\varepsilon(t,y)=\frac{1}{\sqrt{\varepsilon}}\int_0^t\left[a\left(\frac{\tau}{\varepsilon}, x(\tau)+\sqrt{\varepsilon}\,y\right)-a_0(x(\tau))\right]d\tau.$$

Then

$$A_\varepsilon(t,y)=\frac{1}{\sqrt{\varepsilon}}\int_0^t\left[a\left(\frac{\tau}{\varepsilon}, x(\tau)+\sqrt{\varepsilon}\,y\right)-a\left(\frac{\tau}{\varepsilon}, x(\tau)\right)-\sqrt{\varepsilon}\,\nabla a\left(\frac{\tau}{\varepsilon}, x(\tau)\right)y\right]d\tau$$
$$+\frac{1}{\sqrt{\varepsilon}}\int_0^t\left[a\left(\frac{\tau}{\varepsilon}, x(\tau)\right)-a_0(x(\tau))\right]d\tau+\int_0^t\nabla a_0\left(\frac{\tau}{\varepsilon}, x(\tau)\right)y\,d\tau.$$

From a)–c) follows

$$A_\varepsilon(t,g)\to\int_0^t g(\tau)\,y\,d\tau.$$

Moreover, the properties of a stochastic integral w.r.t. a martingale and those of the coefficients imply that

$$\mathbb{M}\left|\sqrt{\varepsilon}\int_0^{t/\varepsilon}\beta(\tau, x(\varepsilon\tau)+\sqrt{\varepsilon}\,y(\varepsilon\tau), d\tau)-\sqrt{\varepsilon}\int_0^{t/\varepsilon}\beta(\tau, x(\varepsilon\tau), d\tau)\right|^2$$

$$\begin{aligned}&=\varepsilon\int_0^{t/\varepsilon}\{[b(\tau, x(\varepsilon\tau)+\sqrt{\varepsilon}\,y(\varepsilon\tau))-b(\tau, x(\varepsilon\tau))]\\
&\quad\times[b(\tau, x(\varepsilon\tau)+\sqrt{\varepsilon}\,y(\varepsilon\tau))-b(\tau, x(\varepsilon\tau))]^*\\
&\quad+\int[c(\tau, x(\varepsilon\tau)+\sqrt{\varepsilon}\,y(\varepsilon\tau), u)-c(\tau, x(\varepsilon\tau), u)]\\
&\quad\times[c(\tau, x(\varepsilon\tau)+\sqrt{\varepsilon}\,y(\varepsilon\tau), u)-c(\tau, x(\varepsilon\tau), u)]^*\,\Pi(du)\}\,d\tau\\
&\leqq\varepsilon\,C_1\,\varepsilon\frac{t}{\varepsilon}=\varepsilon\,t\,C_1,\end{aligned}$$

where C_1 is some constant. Now consider the process

$$\gamma_\varepsilon(t)=\sqrt{\varepsilon}\int_0^{t/\varepsilon}\beta(\tau, x(\varepsilon\tau), d\tau).$$

Let $g_\varepsilon(z)$ be the characteristic function of the random vector $\gamma_\varepsilon(t)$. Then

$$\begin{aligned}\ln g_\varepsilon(z)&=-\frac{\varepsilon}{2}\int_0^{t/\varepsilon}|b(\tau, x(\varepsilon\tau))\,z|^2\,d\tau\\
&\quad+\int^{t/\varepsilon}\int[e^{i\sqrt{\varepsilon}(c(\tau,x(\varepsilon\tau),u)|z)}-1-i\sqrt{\varepsilon}(c(\tau, x(\varepsilon\tau), u)|z)]\,\Pi(du)\,d\tau\\
&=-\frac{\varepsilon}{2}\int_0^{t/\varepsilon}[|b(\tau, x(\varepsilon\tau))\,z|^2+\int(c(\tau, x(\varepsilon\tau), u)|z)^2\,\Pi(du)]\,d\tau\\
&\quad+\int_0^{t/\varepsilon}\int\Big[e^{i\sqrt{\varepsilon}(c(\tau,x(\varepsilon\tau),u)|z)}-1-i\sqrt{\varepsilon}(c(\tau, x(\varepsilon\tau), u)|z)\\
&\quad+\frac{\varepsilon}{2}(c(\tau, x(\varepsilon\tau), u)|z)^2\Big]\,\Pi(du)\,d\tau.\end{aligned}$$

From assumption d), the first term here tends to $-\frac{1}{2}\int_0^t z^*\,d(\tau)\,z\,d\tau$. Substituting $\tau=\dfrac{\theta}{\varepsilon}$ in the second term and noting that the integrand there tends to zero for $\varepsilon\to 0$ and that e) allows passage to the limit under the integral, we see that the second term goes to zero for $\varepsilon\to 0$.

Hence,

$$\ln g_\varepsilon(z)\to-\tfrac{1}{2}\int_0^t z^*\,d(\tau)\,d^*(\tau)\,z\,d\tau,$$

i.e., $\gamma_\varepsilon(t)$ converges weakly to a Gaussian process. This process automatically has independent increments. Applying Theorem 3, we find that $\eta_\varepsilon(t)$ converges weakly for $\varepsilon \to 0$ to a process satisfying

$$d\eta = g(t)\,\eta\,dt + d(t)\,dw,$$

where $w = w(t)$ is Wiener.

Remark 2. Eq. (18) is solved by the usual methods. Let $h(t)$ denote a square matrix which satisfies

$$\frac{dh(t)}{dt} = g(t)\,h(t), \qquad h(0) = I.$$

The matrix $h(t)$ is nonsingular and

$$\eta(t) = h(t) \int_0^t h^{-1}(\tau)\,d(\tau)\,dw. \tag{19}$$

Corollary 1. *The process $\eta(t)$ is Gaussian with mean zero and covariance matrix*

$$\mathbb{M}\,\eta(t)\,\eta^*(t) = \int_0^t h(t)\,h^{-1}(\tau)\,d(\tau)\,d^*(\tau)\big(h(t)\,h^{-1}(\tau)\big)^*\,d\tau. \tag{20}$$

The following lemma can be useful for verifying condition d) of Theorem 5.

Lemma 3. *If*

$$\lim_{T\to\infty} \frac{1}{T} \int_A^{T+A} b(t, x)\,dt = \bar{b}(x)$$

uniformly w.r.t. A for each x, $\bar{b}(x)$ is continuous, the family of functions $b(t, x) = b_t(x)$ is uniformly continuous on an arbitrary compact $|x| \leqq c$ and $x(t)$ is continuous $(0 \leqq t \leqq a)$, then

$$\lim_{\varepsilon\to 0} \int_0^t b\left(\frac{\tau}{\varepsilon}, x(\tau)\right) d\tau = \int_0^t \bar{b}(x(\tau))\,d\tau.$$

Proof. Let $|x(t)| \leqq N$ for $0 \leqq t \leqq a$ and let δ_1 be any positive number. Choose δ_2 so that

$$|\bar{b}(x) - \bar{b}(y)| < \delta_1,$$

and for all $t \in [0, a]$

$$|b(t, x) - b(t, y)| < \delta_1$$

if $|x - y| < \delta_2$ and $|x| \leqq N$.

Consider a partition of $[0, t]$ by means of $t_0 = 0, t_1, \ldots, t_n = t$ with $\max|t_k - t_{k-1}| \leqq \delta_3$, where δ_3 is so chosen that $|x(t') - x(t'') < \delta_2$ when

$|t'-t''| \leqq \delta_3$. Then

$$\left|\int_0^t b\left(\frac{\tau}{\varepsilon}, x(\tau)\right) d\tau - \int_0^t \bar{b}(x(\tau))\, d\tau\right| \leqq \left|\varepsilon \sum_{k=1}^{n} \int_{\frac{t_{k-1}}{\varepsilon}}^{\frac{t_k}{\varepsilon}} \left[b(\tau, x(\varepsilon\tau)) - \bar{b}(x(t_{k-1}))\right] d\tau\right|$$

$$+\left|\sum_{k=1}^{n} \int_{t_{k-1}}^{t_k} \left[\bar{b}(x(\tau)) - \bar{b}(x(t_{k-1}))\right] d\tau\right|$$

$$\leqq \sum_{k=1}^{n} \delta_1 \Delta t_k + \varepsilon \left|\sum_{k=1}^{n} \int_{\frac{t_{k-1}}{\varepsilon}}^{\frac{t_k}{\varepsilon}} \left[b(\tau, x(\varepsilon\tau)) - b(\tau, x(t_{k-1}))\right] d\tau\right|$$

$$+\varepsilon \left|\sum_{k=1}^{n} \int_{\frac{t_{k-1}}{\varepsilon}}^{\frac{t_k}{\varepsilon}} \left[b(\tau, x(t_{k-1})) - \bar{b}(x(t_{k-1}))\right] d\tau\right|$$

$$\leqq 2t\,\delta_1 + \left|\sum_{k=1}^{n} \Delta t_k \frac{\varepsilon}{\Delta t_k} \int_{\frac{t_{k-1}}{\varepsilon}}^{\frac{t_k}{\varepsilon}} \left[b(\tau, x(t_{k-1})) - \bar{b}(x(t_{k-1}))\right] d\tau\right|,$$

where $\Delta t_k = t_k - t_{k-1}$. For $\varepsilon \to 0$ and arbitrary n, the entire sum on the right side of this inequality tends to zero. Since δ_1 is arbitrary, the lemma is proved. Of the remaining assumptions of Theorem 5, a) can cause some difficulty. The following lemma gives a rough sufficient condition for the validity of a).

Lemma 4. *If* $a(t, x)$ *satisfies Theorem 5,* $x(t)$ *has a bounded derivative and*

$$\frac{1}{T} \int_A^{A+T} a(\tau, x)\, d\tau - a(x) = \frac{1}{T} \delta\left(\frac{1}{T}\right),$$

where $\delta(h) \to 0$ *for* $h \to 0$ *uniformly in* $A \geqq 0$ *and* x, $|x| \leqq N$ *(*N *is arbitrary), then* a) *of Theorem 5 is valid.*

Proof. We have $(0 = t_0 < t_1 < \cdots < t_n = t)$

$$\frac{1}{\sqrt{\varepsilon}} \int_0^t \left[a\left(\frac{\tau}{\varepsilon}, x(\tau)\right) - a_0(x(\tau))\right] d\tau$$

$$= \frac{1}{\sqrt{\varepsilon}} \sum_{k=1}^{n} \left\{ \int_{t_{k-1}}^{t_k} \left[a\left(\frac{\tau}{\varepsilon}, x(\tau)\right) - a\left(\frac{\tau}{\varepsilon}, x(t_{k-1})\right)\right] d\tau \right.$$

$$+ \varepsilon \int_{\frac{t_{k-1}}{\varepsilon}}^{\frac{t_k}{\varepsilon}} \left[a(\tau, x(t_{k-1})) - a_0(x(t_{k-1}))\right] d\tau$$

$$\left. + \int_{t_{k-1}}^{t_k} \left[a_0(x(t_{k-1})) - a_0(x(\tau))\right] d\tau \right\}.$$

Since $x(t)$ has a continuous derivative and $a(t, x)$ satisfies a Lipschitz condition in x uniformly in t, the sums corresponding to the first and third terms on the right side of the last equality do not exceed the quantity

$$\frac{1}{\sqrt{\varepsilon}} C \sum_{k=1}^{n} (t_k - t_{k-1})^2 \leqq C t \frac{\max \Delta t_k}{\sqrt{\varepsilon}}, \qquad \Delta t_k = t_k - t_{k-1}.$$

Furthermore,

$$\left| \sqrt{\varepsilon} \sum_{k=1}^{n} \int_{\frac{t_{k-1}}{\varepsilon}}^{\frac{t_k}{\varepsilon}} \left[a\left(\tau, x(t_{k-1})\right) - a\left(x(t_{k-1})\right)\right] d\tau \right|$$

$$\leqq \sqrt{\varepsilon} \sum_{k=1}^{n} \frac{\varepsilon}{\Delta t_k} \delta\left(\frac{\varepsilon}{\Delta t_k}\right) \frac{\Delta t_k}{\varepsilon} \leqq t \max_{1 \leqq k \leqq n} \frac{\sqrt{\varepsilon}}{\Delta t_k} \delta\left(\frac{\varepsilon}{\Delta t_k}\right).$$

Set $\frac{\sqrt{\varepsilon}}{\Delta t_k} = \rho$ and assume that ρ and ε are related in such a way that $\rho \to \infty$ and $\rho\,\delta(\sqrt{\varepsilon}\,\rho) \to 0$. Then condition a) of Theorem 5 will be valid.

We note the following special case of Theorem 5.

Assume that the conditions of Theorem 4 hold, that

$$\frac{1}{T} \int_0^T a(t, x_0)\, dt = o\left(\frac{1}{\sqrt{T}}\right),$$

$$\lim_{T \to \infty} \frac{1}{T} \int_0^T \nabla a(t, x_0)\, dt = a_1,$$

that $\nabla^2 a_0(x)$ exists with $|\nabla^2 a_0(x) - \nabla^2 a_0(y)| \leqq c\,|x - y|$, that $r(u)$ is such that $|c(t, x_0, u)| \leqq r(u)$ and that

$$\lim_{T \to \infty} \frac{1}{T} \int_0^T [b(t, x_0)\, b^*(t, x_0) + \int c(t, x_0, u)\, c^*(t, x_0, u)\, \Pi(du)]\, dt = D$$

exists.

Then, if $\xi_\varepsilon(t)$ satisfies

$$d\xi_\varepsilon = \varepsilon [a(t, \xi_\varepsilon)\, dt + b(t, \xi_\varepsilon)\, dw + \int c(t, \xi_\varepsilon, u)\, \tilde{v}(dt, du)],$$

the process

$$\eta_\varepsilon(t) = \frac{\xi_\varepsilon\left(\frac{t}{\varepsilon}\right) - x_0}{\sqrt{\varepsilon}}$$

converges weakly for $\varepsilon \to 0$ to a Gaussian Markov process $\zeta(t)$ which satisfies

$$d\zeta = a_1 \zeta\, dt + \sqrt{D}\, dw. \tag{21}$$

The covariance matrix of $\zeta(t)$ is

$$\int_0^t e^{a_1(t-\tau)} D e^{a_1^*(t-\tau)} d\tau.$$

This result can be written as

$$\xi_\varepsilon\left(\frac{t}{\varepsilon}\right) \cong x_0 + \sqrt{\varepsilon}\,\zeta(t), \tag{22}$$

where the symbol $\cong$ means that the right side differs from the left by a quantity which is of smaller order than $\sqrt{\varepsilon}$. The variable $\sqrt{\varepsilon}\,\zeta(t)$ characterizes the oscillation of the moving point around its rest state under the influence of a random perturbation during an interval of time of the order $\frac{1}{\varepsilon}$. These fluctuations turn out to be Gaussian and are described by simple Markov processes.

The assertion of Theorem 5 can be paraphrased somewhat differently. Consider the motion of a system in a rapidly varying field. Let the equation of motion have the form

$$d\dot{\xi} = -\operatorname{grad} U(\xi)\, dt + \beta\left(\frac{t}{\varepsilon}, \xi, dt\right), \tag{23}$$

where the random perturbation field $\beta(t, x, h)$ has the form (16). The presence of the parameter $\frac{1}{\varepsilon}$ in front of t refers to the fact that the field is "rapidly varying".

In connection with Theorem 5, if the field $\beta(t, x, h)$ satisfies conditions d) and e), then we have

$$\xi(t) \cong x(t) + \sqrt{\varepsilon}\,\zeta(t), \tag{24}$$

where $x(t)$ satisfies the unperturbed equation

$$\ddot{x} = -\operatorname{grad} U(x), \qquad x(0) = x_0,$$

and $\zeta(t)$ satisfies

$$d\dot{\zeta} = g(t)\,\zeta + d(t)\, dw, \qquad d\zeta = \dot{\zeta}\, dt, \tag{25}$$

with

$$g(t) = -\nabla \operatorname{grad} U(x(t)). \tag{26}$$

Bibliography

1. Baklan, V. V.: Dokl. Akad. Nauk Uk.SSR. **10**, 1299–1303 (1963) [in Russian].
2. Baklan, V.V.: Dokl. Akad. Nauk SSSR. **4**, 159 (1964) [in Russian].
3. Bellman, R.: Stability theory of differential equations. New York: Mc Graw-Hill 1953.
4. Bernštein, S.N.: Trudy Fiz-mat. Inst. Steklov. **5**, 95–124 (1934) [in Russian].
5. Bernštein, S.N.: Probability theory, 4th. ed. Moscow-Leningrad: Gostechizdat 1946 [in Russian].
6. Bogolubov, N.N.: Statistical methods in mathematical physics. Kiev 1945 [in Russian].
7. Bogolubóv, N.N., Krylov, N.M.: Zap. Kaf. Mat. Fiz. Akad. Nauk. Uk.SSR. **4**, 5–158 (1939) [in Russian].
8. Bogolubov, N.N., Mitropol'skii, Ju.A.: Asymptotic methods in the theory of nonlinear oscillations. Moscow 1963 [in Russian].
9. Blagóveščenskii, Ju.N.: Teor. Verojatnost. i Primenen. **7**, 135–152 (1962) [in Russian].
10. Blagoveščenskii, Ju.N., Freidlin, M.I.: Dokl. Akad. Nauk SSSR. **138**, 508–511 (1961) [in Russian].
11. Čantladze, T.L.: Comm. Akad. Nauk Gruz. SSR. **33**, 3, 529–534 (1964) [in Russian].
12. Daleckii, Ju.L.: Dokl. Akad. Nauk SSSR. **166**, 4 (1966) [in Russian].
13. De Dju Gen: Vestnik KNDR (N. Korea) **3** (1963) [in Korean].
14. De Dju Gen: Stochastic integral equations; 598 p. Phenjan 1963 [in Korean].
15. Dynkin, E.B.: Dokl. Akad. Nauk SSSR. **104**, 691–694 (1965) [in Russian].
16. Dynkin, E.B.: Markov processes. Berlin-Heidelberg-New York: Springer 1965.
17. Doob, J.L.: Stochastic processes. New York: Wiley 1953.
18. Feller, W.: Trans. Amer. Math. Soc. **77**, 1–31 (1954).
19. Gihman, I.I.: Naučnye Zap. Kiev. Univ. **5**, 119–132 (1941) [in Russian].
20. Gihman, I.I.: Naučnye Zap. Kiev. Univ. **5**, 141–149 (1941) [in Russian].
21. Gihman, I.I.: Dokl. Akad. Nauk SSSR. **58**, 961–964 (1947) [in Russian].
22. Gihman, I.I.: Uskr. Math. Ž. **2**, 3, 45–69 (1950) [in Russian].
23. Gihman, I.I.: Uskr. Math. Ž. **2**, 4, 37–63 (1950); **3**, 317–339 (1951) [in Russian].
24. Gihman, I.I.: In: Limit theorems and statistical inference, p. 14–45. Taškent 1966 [in Russian].
25. Gihman, I.I.: In: Winter school on probability theory and math. statistics, p. 48–65. Užgorod 1964 [in Russian].
26. Gihman, I.I., Dorogovcev, A.Ja.: Uspehi Math. Ž. **17**, 6, 3–21 (1965) [in Russian].
27. Gihman, I.I., Skorohod, A.V.: Introduction to the theory of random processes. Philadelphia: Saunders 1969.
28. Girsanov, I.V.: Teor. Verojatnost. i Primenen. **5**, 314–330 (1960) [in Russian].
29. Girsanov, I.V.: Dokl. Akad. Nauk SSSR. **138** (1961) [in Russian].
30. Has'minskii, R.Z.: Teor. Verojatnost. i Primenen. **5**, 196–214 (1960) [in Russian].
31. Has'minskii, R.Z.: Prikl. Math. Meh. **24**, 809–823 (1960) [in Russian].
32. Has'minskii, R.Z.: Teor. Verojatnost. i Primenen. **8**, 3–25 (1963) [in Russian].
33. Has'minskii, R.Z.: Problemy Peredači Informacii. **1**, 1 (1965) [in Russian].

34. Has'minskii, R.Z.: In: Pattern recognition. Teor. Peredači Informacii p. 72–85 (1965) [in Russian].
35. Has'minskii, R.Z.: Teor. Verojatnost. i Primenen. **11**, 2, 240–259 (1966) [in Russian].
36. Has'minskii, R.Z.: Teor. Verojatnost. i Primenen. **11**, 3, 444–462 (1966) [in Russian].
37. Khintchine, A.: Asymptotische Gesetze der Wahrscheinlichkeitsrechnung. Berlin: Springer 1933.
38. Itô, K.: Proc. Imp. Acad. Tokyo **20**, 519–524 (1944).
39. Itô, K.: Mem. Amer. Math. Soc. **4** (1951).
40. Itô, K.: Nagoya Math. J. **3**, 55–65 (1951).
41. Itô, K.: Proc. Japan Acad. **22**, 2, 32–35 (1946).
42. Itô, K.: Nagoya Math. J. **1**, 35–47 (1950).
43. Itô, K., McKean, H. P.: Diffusion Processes and their sample paths. Berlin-Heidelberg-New York: Springer 1965.
44. Itô, K., Nisio, M.: J. Math. Kyoto Univ. **4**, 1, 1–75 (1964).
45. Katz, I., Krasovskii, N.N.: Prikl. Math. Meh. **24**, 5, 809–823 (1962).
46. Krasovskii, N.N.: Stability of motion. Stanford: Stanford Univ. Press 1963.
47. Kolmogorov, A.N.: Uspehi Math. Nauk. 5–41 (1938) [in Russian].
48. Malkin, I.G.: The theory of the stability of motion. Moscow: Nauka 1966 [in Russian].
49. Maruyama, G.: Rend. Circ. Mat. Palermo. **4**, 1–43 (1955).
50. Maruyama, G., Tanaka, H.: Mem. Sci. Fac. Kyushu Univ., Ser. A. **11**, 2, 117–141 (1957).
51. Skorohod, A.V.: Sibirsk. Mat. Ž. **2**, 129–137 (1961) [in Russian].
52. Skorohod, A.V.: Teor. Verojatnost. i Primenen. **6**, 287–298 (1961); **7**, 5–25 (1962) [in Russian].
53. Skorohod, A.V.: Studies in the theory of random processes. Reading, Mass.: Addison-Wesley 1965.
54. Skorohod, A.V.: Random processes with independent increments. Moscow: Nauka 1964 [in Russian].
55. Stratanovič, R.L.: Some questions in the theory of fluctuations in radio-electronics. Sov. Radio 1961 [in Russian].
56. Stratanovič, R.L.: Conditional Markov Processes and their applications to the theory of optimal control. New York: Elsevier 1968.
57. Volkonskii, V.A.: Teor. Verojatnost. i Primenen. **5**, 3–30 (1958) [in Russian].

Index

absolute continuity of measures 81
absolutely continuous transformation of measures 90
absorption at a boundary 158, 165ff.
after-effect, random vector field with 216ff.
—, s.d.e.'s without 246ff., 273ff.
asymptotic expansions w.r.t. small parameter 58
— behavior of solutions of s.d.e.'s 124ff.
attracting boundary 165
averaging principle 340
(see also Bogolubov)

backward equation of Kolmogorov 73, 104, 293, 298, 302
Bogolubov's method of averaging 2, 340ff.
Borel-Cantelli lemma 19, 151
boundary, absorbing 165
—, attracting 165
— conditions at ends of an interval 157ff.
—, delayed reflection at 193, 200
—, elastic 158, 159
—, jump reflection at 205ff.
—, lower, of a region 166ff.
—, natural 158
—, reflecting 158, 178ff.
—, regular 165
—, upper, of a region 165ff.
— -value problem 301
boundedness of solutions 330
Brownian motion 1

Cauchy problem 137
— sequence 14, 15
Chapman-Kolmogorov equation for transition density 100
characteristic, local, of a motion 215
Chebychev's inequality 21
continuity, absolute with respect to a measure 81
—, in mean square 53, 61
convergence, weak, of distributions 189
cylinder set 81

D-martingale 247
delayed reflection at boundary 193
density of a measure 81
differentiability of functions of solutions 62
— of solution w.r.t. a parameter 55, 275ff.
— of solution w.r.t. initial data 275
differential operator for diffusion 75
—, stochastic 21ff.
differentiation, linearity of 21
— of products and compositions of functions 21
— of polynomials 23, 261
diffusion as solution of stochastic equation 67
— coefficient 65
—, Markov 64ff.
—, transition density fct. of 91ff.
Dirichlet problem for elliptic operator 309
discontinuities of the second kind 236, 250, 255
displacement coefficient 65
domain of definition of infinitesimal operator 289
Doob, J. L. 8
drift coefficient 65

eigen-values and stability 317
elastic boundary 158
elliptic differential operator 308
ergodic distribution for solutions 134, 186ff.
— theorem 134ff., 200
— theorem for temporal means 204
Euler approximate solution 240, 334
— polygonal segment 242

existence and uniqueness theorem 48, 51, 237 ff.
— of derivatives of solutions 285
— of moments of solutions 287
— of solutions 39 ff.

field, quasi-differential 218 ff.
— with limited after-effect 216 ff.
—, perturbing 310
—, rapidly varying 347
first exit time from a region 241
— passage time 28, 108
Fokker-Planck equation 102
function, bounded from both sides 117
—, generalized Liapunov 325
—, Green's 177
—, L-harmonic 114, 307
—, L-subharmonic 307
—, L-superharmonic 307
—, one-sided bounded, L-harmonic 115
—, positive-definite in Liapunov's sense 325
fundamental solution of Kolmogorov's equation 105

Gaussian distribution 92
— increments 35
— process 7, 344, 346
generalized solution of Kolmogorov's equation 79, 171 ff.
Green's function 177

homogeneous Markov process 106
— stochastic differential equation 37

infinitesimal operator (generator) 289
integral, stochastic 11 ff., 247
—, stochastic, free of discontinuities of the second kind 236, 255
—, stochastic, with random limits 27
integral sum 220
integration by parts, formula for 260
Itô's formula 24 ff., 33, 187, 199, 206, 263, 267, 270, 307

Jensen's inequality 83, 95
jump reflection at boundary 205 ff.

Kolmogorov's backward equation 164, 293, 298, 302
— equations 73, 168 ff., 184, 293, 298, 302
— equations for transition probability density 99 ff.
— equations, generalized solution of 79
Kolmogorov's first equation for transition density 102

L-harmonic function 114
Laplace transform 176 ff., 198
law of the iterated logarithm for Wiener process 121, 169
law of large numbers 134, 141, 322
left-stability of solutions 145, 147
Liapunov's second method 324 ff.
limit theorems for solutions 151 ff., 333 ff.
line integral along a random curve 3, 216 ff., 220, 226
— integral as fct. of the upper limit 232
linear s.d.e. 36, 57
Lipschitz condition 217
local characteristic of a motion 215
— dependence of solutions on coefficients 44
— dependence of solutions on field 242

Markov chain 1
— diffusion 64
— process 63, 182, 288 ff.
— process, homogeneous 106
— systems 63
— time 28
— time and strong Markov processes 107
— time as first exit time 241
martingale 247, 313, 314, 338
—, D- 247
measure, density of 81 ff.
—, induced by a diffusion 80 ff.
—, random Poisson 245, 251
—, standard Poisson 246
method of successive approximations 239
moments of stochastic integrals and processes, boundedness of 331
— of stochastic integrals and processes, estimation of 261, 274
— of stochastic integrals and processes, stability of 311, 320
nonlinear order of growth of solutions 126
normal distribution 151, 154

operator, adjoint 105
—, domain of definition of 289
—, elliptic differential 308
—, infinitesimal 289

order of growth of solutions 124
— of growth of solutions as solution of ordinary d.e. 129

parabolic partial differential equation for densities 293ff., 301
Poisson measure, random 245, 251
— measure, standard 246, 251
polygonal segment, Euler 242
polynomial, differentiation of 23, 261
positive-definite Liapunov function 325
power-law growth of solutions 127, 132
process, coordinated 16
—, diffusion 64ff., 67
—, Markov 63, 182, 288ff.
—, separable 16, 20
—, stochastically equivalent 19
—, strong Markov 107
— with finite jump intensity 298
— without discontinuities of the second kind 250

quasi-differential random field 218ff.

random Poisson measure 245
— vector field 216
— vector field, quasi-differential 218ff.
— vector field with limited after-effect 216ff.
reducibility of s.d.e. to linearity 38
reflection at a boundary 158
— at a boundary, delayed 158, 193
— at a boundary, jump 205
— at a boundary, instantaneous 158, 178ff.
— at a boundary, instantaneous, ergodic theorem for 192
regular boundary 165
"regularity" property of a function 155
restoration scheme 159
right-stability of solutions 145, 147
Routh-Hurwitz theorem 317

separability set 18, 234
solution of diffusion equation by Laplace transforms 176ff.
—, asymptotically uniformly stable 311
—, fundamental, of Kolmogorov's equation 105
—, generalized, of diffusion equation 171ff.
— of stochastic differential equation as diffusion 33, 67
— of stochastic differential equation, asymptotic behavior of 124ff.
— of stochastic differential equation, asymptotic expansion of w.r.t. parameter 58
— of stochastic differential equation, bounded 114ff.
— of stochastic differential equation, dependence on initial data 59, 275ff.
— of stochastic differential equation, depending on a parameter 50, 275ff.
— of stochastic differential equation, differentiability w.r.t. initial data 61, 275ff.
— of stochastic differential equation, differentiability w.r.t. a parameter 55, 275ff.
— of stochastic differential equation, distribution of functionals of 300ff.
— of stochastic differential equation, equilibrium 310
— of stochastic differential equation, ergodic theorems for 134ff.
— of stochastic differential equation, Euler approximate 240
— of stochastic differential equation, and L-harmonic functions 114
— of stochastic differential equation, limit theorems for 151ff., 333ff.
— of stochastic differential equation, local dependence on coefficients 44
— of stochastic differential equation, local dependence on field of 242
— of stochastic differential equation, moments of 48, 274, 287
— of stochastic differential equation, order of growth of 124ff.
— of stochastic differential equation, power-law growth of 127, 132
— of stochastic differential equation, stability of 145ff., 310ff.
— of stochastic differential equation, stochastically bounded 330
— of stochastic differential equation, time-homogeneous 105, 304ff.
— of stochastic differential equation, uniformly stochastically stable 310
sp (German *Spur*) = trace of a matrix 272
stability of solutions 145ff., 310ff.
stationary point 145
standard Poisson measure 246, 251

stochastic boundedness 330
— differential 21
— integral 11ff.
— integral as function of the upper limit 16ff., 232
— integral considered as values of a Wiener process at random times 31
— integral continuity of 19
— integral distribution of 77ff.
— integral independent of choice of defining sequence 14
— integral moments of 25ff., 261
— integral with random limits 27ff.
— line integral 216ff., 220, 226
— stability 310ff.
strong law of large numbers 141, 322
— Markov property 30, 107
substitution of an undetermined function in a s.d.e. 34
sum, integral 220
systems of s.d.e.'s 215ff.

temporal means, ergodic theorem for 191, 204
time substitution 111, 194ff.
transition density function 91ff., 298
— probability 63
— probability, infinitesimal operator of family of 298

uniform Lipschitz condition 217
uniqueness of solutions 39ff., 45, 242

vector field, random 216ff.
— stochastic differential equations 215

weak convergence of distributions 189
— convergence of random functions 333
white noise 1
Wiener process 7ff., 71, 88, 245, 250, 264
— process, law of the iterated logarithm for 121, 169
Wronskian 318

Ergebnisse der Mathematik und ihrer Grenzgebiete

1. Bachmann: Transfinite Zahlen
2. Miranda: Partial Differential Equations of Elliptic Type
4. Samuel: Méthodes d'Algèbre Abstraite en Géométrie Algébrique
5. Dieudonné: La Géométrie des Groupes Classiques
6. Roth: Algebraic Threefolds with Special Regard to Problems of Rationality
7. Ostmann: Additive Zahlentheorie. 1. Teil: Allgemeine Untersuchungen
8. Wittich: Neuere Untersuchungen über eindeutige analytische Funktionen
11. Ostmann: Additive Zahlentheorie. 2. Teil: Spezielle Zahlenmengen
13. Segre: Some Properties of Differentiable Varieties and Transformations
14. Coxeter/Moser: Generators and Relations for Discrete Groups
15. Zeller/Beckmann: Theorie der Limitierungsverfahren
16. Cesari: Asymptotic Behavior and Stability Problems in Ordinary Differential Equations
17. Severi: Il teorema di Riemann-Roch per curve-superficie e varietà questioni collegate
18. Jenkins: Univalent Functions and Conformal Mapping
19. Boas/Buck: Polynomial Expansions of Analytic Functions
20. Bruck: A Survey of Binary Systems
21. Day: Normed Linear Spaces
23. Bergmann: Integral Operators in the Theory of Linear Partial Differential Equations
25. Sikorski: Boolean Algebras
26. Künzi: Quasikonforme Abbildungen
27. Schatten: Norm Ideals of Completely Continuous Operators
28. Noshiro: Cluster Sets
30. Beckenbach/Bellman: Inequalities
31. Wolfowitz: Coding Theorems of Information Theory
32. Constantinescu/Cornea: Ideale Ränder Riemannscher Flächen
33. Conner/Floyd: Differentiable Periodic Maps
34. Mumford: Geometric Invariant Theory
35. Gabriel/Zisman: Calculus of Fractions and Homotopy Theory
36. Putnam: Commutation Properties of Hilbert Space Operators and Related Topics
37. Neumann: Varieties of Groups
38. Boas: Integrability Theorems for Trigonometric Transforms
39. Sz.-Nagy: Spektraldarstellung linearer Transformationen des Hilbertschen Raumes
40. Seligman: Modular Lie Algebras
41. Deuring: Algebren
42. Schütte: Vollständige Systeme modaler und intuitionistischer Logik
43. Smullyan: First-Order Logic
44. Dembowski: Finite Geometries
45. Linnik: Ergodic Properties of Algebraic Fields
46. Krull: Idealtheorie
47. Nachbin: Topology on Spaces of Holomorphic Mappings
48. A. Ionescu Tulcea/C. Ionescu Tulcea: Topics in the Theory of Lifting
49. Hayes/Pauc: Derivation and Martingales
50. Kahane: Séries de Fourier Absolument Convergentes
51. Behnke/Thullen: Theorie der Funktionen mehrerer komplexer Veränderlichen
52. Wilf: Finite Sections of Some Classical Inequalities
53. Ramis: Sous-ensembles analytiques d'une variété banachique complexe
54. Busemann: Recent Synthetic Differential Geometry
55. Walter: Differential and Integral Inequalities

56. Monna: Analyse non-archimédienne
57. Alfsen: Compact Convex Sets and Boundary Integrals
58. Greco/Salmon: Topics in m-Adic Topologies
59. López de Medrano: Involutions on Manifolds
60. Sakai: C*-Algebras and W*-Algebras
61. Zariski: Algebraic Surfaces
62. Robinson: Finiteness Conditions and Generalized Soluble Groups. Part 1
63. Robinson: Finiteness Conditions and Generalized Soluble Groups. Part 2
64. Hakim: Topos annelés et schémas relatifs
65. Browder: Surgery on Simply-Connected Manifolds
66. Pietsch: Nuclear Locally Convex Spaces
67. Dellacherie: Capacités et processus stochastiques
68. Raghunathan: Discrete Subgroups of Lie Groups
69. Rourke/Sanderson: Introduction to Piecewise-Linear Topology
70. Kobayashi: Transformation Groups in Differential Geometry
71. Tougeron: Idéaux de fonctions différentiables
72. Gihman/Skorohod: Stochastic Differential Equations

MIX
Papier aus verantwortungsvollen Quellen
Paper from responsible sources
FSC® C105338

If you have any concerns about our products,
you can contact us on
ProductSafety@springernature.com

In case Publisher is established outside the EU,
the EU authorized representative is:
Springer Nature Customer Service Center GmbH
Europaplatz 3, 69115 Heidelberg, Germany

Printed by Libri Plureos GmbH
in Hamburg, Germany